T0219864

Lecture Notes in Computer Science   10787

Commenced Publication in 1973
Founding and Former Series Editors:
Gerhard Goos, Juris Hartmanis, and Jan van Leeuwen

More information about this series at http://www.springer.com/series/7407

Roberto Solis-Oba · Rudolf Fleischer (Eds.)

Approximation and Online Algorithms

15th International Workshop, WAOA 2017
Vienna, Austria, September 7–8, 2017
Revised Selected Papers

 Springer

Editors
Roberto Solis-Oba 🆔
The University of Western Ontario
London, ON
Canada

Rudolf Fleischer
German University of Technology in Oman
Muscat
Oman

ISSN 0302-9743   ISSN 1611-3349 (electronic)
Lecture Notes in Computer Science
ISBN 978-3-319-89440-9   ISBN 978-3-319-89441-6 (eBook)
https://doi.org/10.1007/978-3-319-89441-6

Library of Congress Control Number: 2018939427

LNCS Sublibrary: SL1 – Theoretical Computer Science and General Issues

Printed on acid-free paper

This Springer imprint is published by the registered company Springer International Publishing AG
part of Springer Nature
The registered company address is: Gewerbestrasse 11, 6330 Cham, Switzerland

Preface

The 15th Workshop on Approximation and Online Algorithms (WAOA 2017) took place in Vienna, Austria, September 7–8, 2017. The workshop was part of the ALGO 2017 event that also hosted ESA, IPEC, ALGOSENSORS, ATMOS, and ALGOCLOUD. The previous WAOA workshops were held in Budapest (2003), Rome (2004), Palma de Mallorca (2005), Zurich (2006), Eilat (2007), Karlsruhe (2008), Copenhagen (2009), Liverpool (2010), Saarbrücken (2011), Ljubljana (2012), Sophia Antipolis (2013), Wraclas (2014), Patras (2015), and Aarhus (2016). The proceedings of previous WAOA workshops have appeared in the *Lecture Notes in Computer Science* series.

The Workshop on Approximation and Online Algorithms focuses on the design and analysis of algorithms for online and computationally hard problems. Both kinds of problems have a large number of applications in a wide variety of fields. Topics of interest for WAOA 2017 were: graph algorithms, inapproximability results, network design, packing and covering, paradigms for the design and analysis of approximation and online algorithms, parameterized complexity, scheduling problems, algorithmic game theory, coloring and partitioning, competitive analysis, computational advertising, computational finance, cuts and connectivity, geometric problems, mechanism design, resource augmentation, and real-world applications.

In response to the call for papers, we received 50 submissions. Each submission was reviewed by at least three referees. The submissions were judged on originality, technical quality, and relevance to the topics of the conference. Based on the reviews, the Program Committee selected 23 papers for presentation at the workshop. In addition to the presentations of the 23 accepted papers, Prof. Kirk Pruhs from the University of Pittsburgh gave an invited talk on "The Itinerant List Update Problem." We are grateful to Prof. Pruhs for accepting our invitation and delivering an outstanding lecture.

The EasyChair conference system was used to manage the electronic submissions and the review process. It made our task much easier. We would like to thank all the authors who submitted papers to WAOA 2017, the members of the Program Committee, the external reviewers, and the local Organizing Committee of ALGO 2017.

October 2017

Rudolf Fleischer
Roberto Solis-Oba

Organization

Program Committee

Yossi Azar	Tel Aviv University, Israel
Danny Chen	University of Notre Dame, USA
Jose Correa	Universidad de Chile, Chile
Khaled Elbassioni	Masdar Institute Abu Dhabi, United Arab Emirates
Leah Epstein	University of Haifa, Israel
Rolf Fagerberg	University of Southern Denmark, Denmark
Rudolf Fleischer	German University of Technology in Oman, Oman
Martin Fürer	The Pennsylvania State University, USA
Klaus Jansen	University of Kiel, Germany
Li Jian	Tsinghua University, China
George Karakostas	McMaster University, Canada
Monaldo Mastrolilli	IDSIA, Switzerland
Nicole Megow	University of Bremen, Germany
Jiří Sgall	Charles University, Czech Republic
Roberto Solis-Oba	The University of Western Ontario, Canada
Frits Spieksma	KU Leuven, Belgium
Clifford Stein	Columbia University, USA
Denis Trystram	Grenoble Institute of Technology, France
Gerhard Woeginger	RWTH Aachen University, Germany
Qin Zhang	Indiana University Bloomington, USA

Additional Reviewers

Bampis, Evripidis
Böhm, Martin
Boyar, Joan
Brubach, Brian
Canon, Louis-Claude
Chen, Cong
Chiplunkar, Ashish
Deng, Shichuan
Ding, Hu
Elmasry, Amr
Fotakis, Dimitris

Fu, Hao
Golovach, Petr
Hajiaghayi, Mohammadtaghi
Hoeksma, Ruben
Huang, Lingxiao
Huang, Zengfeng
Jin, Yifei
Kesselheim, Thomas
Kisfaludi-Bak, Sándor
Klein, Kim-Manuel

Klimm, Max
Knop, Dušan
Kurpisz, Adam
Laue, Soeren
Leppänen, Samuli
Li, Jian
Li, Yi
Li, Zhize
Lu, Yiping
Lucarelli, Giorgio
Maack, Marten

Mikkelsen, Jesper With
Nguyen, Trung Thanh
Page, Daniel R.
Rau, Malin
Rohwedder, Lars
Sadeqi Azer, Erfan
Schewior, Kevin

Srivastav, Abhinav
Tamir, Tami
van Stee, Rob
Veselý, Pavel
Wang, Haitao
Westphal, Stephan
Wiese, Andreas

Wulff-Nilsen, Christian
Ye, Deshi
Zhang, Haoyu
Zhu, Shenglong

Contents

Capacitated Domination Problems on Planar Graphs 1
Amariah Becker

On Approximability of Connected Path Vertex Cover 17
Toshihiro Fujito

Improved PTASs for Convex Barrier Coverage. 26
Paz Carmi, Matthew J. Katz, Rachel Saban, and Yael Stein

Complexity and Approximation of the Longest Vector Sum Problem. 41
Vladimir Shenmaier

Deadline TSP 52
Boaz Farbstein and Asaf Levin

A Bicriteria Approximation Algorithm for the k-Center
and k-Median Problems 66
Soroush Alamdari and David Shmoys

Approximating Domination on Intersection Graphs of Paths on a Grid. 76
Saeed Mehrabi

Submodular Function Minimization with Submodular Set Covering
Constraints and Precedence Constraints 90
Naoyuki Kamiyama

Lower Bounds for Several Online Variants of Bin Packing 102
János Balogh, József Békési, György Dósa, Leah Epstein,
and Asaf Levin

The Online Multicommodity Connected Facility Location Problem 118
Mário César San Felice, Cristina G. Fernandes,
and Carla Negri Lintzmayer

A Match in Time Saves Nine: Deterministic Online
Matching with Delays 132
Marcin Bienkowski, Artur Kraska, and Paweł Schmidt

Online Packing of Rectangular Items into Square Bins. 147
Janusz Januszewski and Łukasz Zielonka

A Tight Lower Bound for Online Convex Optimization
with Switching Costs.   164
    Antonios Antoniadis and Kevin Schewior

A *k*-Median Based Online Algorithm for the Stochastic *k*-Server Problem . . .   176
    Abhijin Adiga, Alexander D. Friedman, and Sharath Raghvendra

On Packet Scheduling with Adversarial Jamming and Speedup.   190
    Martin Böhm, Łukasz Jeż, Jiří Sgall, and Pavel Veselý

Non-clairvoyant Scheduling to Minimize Max Flow Time
on a Machine with Setup Times   207
    Alexander Mäcker, Manuel Malatyali, Friedhelm Meyer auf der Heide,
    and Sören Riechers

On-line Search in Two-Dimensional Environment.   223
    Dariusz Dereniowski and Dorota Urbańska

Online Unit Clustering in Higher Dimensions.   238
    Adrian Dumitrescu and Csaba D. Tóth

On Conceptually Simple Algorithms for Variants of Online
Bipartite Matching   253
    Allan Borodin, Denis Pankratov, and Amirali Salehi-Abari

Efficient Dynamic Approximate Distance Oracles for Vertex-Labeled
Planar Graphs.   269
    Itay Laish and Shay Mozes

A Communication-Efficient Distributed Data Structure
for Top-*k* and *k*-Select Queries   285
    Felix Biermeier, Björn Feldkord, Manuel Malatyali,
    and Friedhelm Meyer auf der Heide

Strategyproof Mechanisms for Additively Separable Hedonic Games
and Fractional Hedonic Games   301
    Michele Flammini, Gianpiero Monaco, and Qiang Zhang

The Asymptotic Price of Anarchy for *k*-uniform Congestion Games   317
    Jasper de Jong, Walter Kern, Berend Steenhuisen, and Marc Uetz

Author Index   329

Capacitated Domination Problems
on Planar Graphs

Amariah Becker$^{(\boxtimes)}$

Department of Computer Science, Brown University, Providence, RI, USA
amariah_becker@brown.edu

Abstract. CAPACITATED DOMINATION generalizes the classic DOMINAT-ING SET problem by specifying for each vertex a required demand and an available capacity for covering demand in its closed neighborhood. The objective is to find a minimum-size set of vertices that can cover all of the graph's demand without exceeding any of the capacities. CAPACITATED r-DOMINATION further generalizes the problem by allowing vertices to cover demand up to a distance r away. In this paper we look specifically at domination problems with hard capacities (i.e. each vertex can appear at most once in the solution). Previous complexity results suggest that this problem cannot be solved (or even closely approximated) efficiently in general graphs. In this paper we present polynomial-time approximation schemes for CAPACITATED DOMINATION and CAPACITATED r-DOMINATION in unweighted planar graphs when the maximum capacity is bounded. We also show how this result can be extended to the closely-related CAPACITATED VERTEX COVER problem.

Keywords: Capacitated domination · Approximation algorithms
Planar graphs · r-Domination

1 Introduction

DOMINATING SET is a classic NP-complete optimization problem defined as follows. For an undirected graph G with vertex set V, a dominating set S is a subset of V such that every vertex $v \in V$ is either in S or adjacent to a vertex in S. The problem seeks a dominating set such that $|S|$ is minimized. DOMINATING SET has many applications in resource allocation [10] and social network theory [24].

The r-DOMINATION problem seeks, for a positive integer r, a dominating subset S such that every vertex $v \in V$ is within a hop-distance r of some vertex in S and $|S|$ is minimized. The r-DOMINATION problem is one optimization version of the (k, r)-CENTER decision problem of determining for a given k and r whether there is a set of k vertices (*centers*) such that every vertex is at most a distance r from the nearest center. It is straightforward to see that DOMINATING SET is a special case of r-DOMINATION in which $r = 1$.

Research supported by NSF Grant CCF-1409520.

© Springer International Publishing AG, part of Springer Nature 2018
R. Solis-Oba and R. Fleischer (Eds.): WAOA 2017, LNCS 10787, pp. 1–16, 2018.
https://doi.org/10.1007/978-3-319-89441-6_1

An α-approximation algorithm for a minimization problem ($\alpha \geq 1$) returns a solution whose cost is within an α factor of the optimum value. A polynomial-time approximation scheme (PTAS) is a family of approximation algorithms such that for any $\epsilon > 0$ there is a $(1 + \epsilon)$-approximation algorithm with a polynomial runtime. An *efficient* polynomial-time approximation scheme (EPTAS) is a PTAS in which for any $\epsilon > 0$ the $(1 + \epsilon)$-approximation algorithm has an $O(f(\epsilon)n^c)$ runtime in which c is a constant that does not depend on ϵ. Assuming P $\neq$ NP, a PTAS is in some sense the best solution we can hope to achieve for NP-hard problems in polynomial time.

For DOMINATING SET in general graphs, an $O(\ln n)$-approximation is known [13]. It has also been shown that unless NP $\subset$ TIME($n^{\log \log n}$), DOMINATING SET cannot be approximated within a factor of $(1 - \epsilon) \ln n$ [9], thus a PTAS is unlikely to exist. For restricted graph classes such as unit disk graphs [22], bounded-degree graphs [5], and planar graphs [20], improvements on this bound have been found. In particular, the planar case is known to admit a PTAS using the framework established by Baker [1] and described below.

For r-DOMINATION in general graphs, a $\ln n$-approximation is known [2]. Planar graphs and map graphs are known to admit a PTAS, again using Baker's approach [6].

For many optimization problems that have applications in resource allocation, a natural generalization is to consider capacity restrictions on the resources. For covering-type problems, for example, we may impose capacity limits on the number of times an edge or vertex can be included in the solution or on how much of a commodity can be stored at or move through a location. Capacitated versions of many common optimization problems have been studied, including FACILITY LOCATION [21], VERTEX COVER [12], and VEHICLE ROUTING [23]. For DOMINATING SET, the capacitated version restricts the *capacity* of vertices (i.e. the number of vertices that can claim a given vertex as their representative in the dominating set).

The capacitated DOMINATING SET problem, often referred to as CAPACITATED DOMINATION takes as input *demand*, *capacity*, and *weight* values for every vertex, and seeks a dominating set S of minimum weight such that every vertex in V has its demand *covered* by vertices in S, the total amount of demand covered by any vertex in S does not exceed its capacity, and any vertex can only cover itself or its neighbors (i.e. the *closed neighborhood*)[1]. We similarly define CAPACITATED r-DOMINATION with the difference being that any vertex can only cover vertices within a distance r.

We use (G, d, c) to denote the instance of (unweighted) CAPACITATED DOMINATION or CAPACITATED r-DOMINATION on graph G with demand function d and capacity function c.

CAPACITATED DOMINATION was shown to be W[1]-hard for general graphs when parameterized by treewidth [7]. Moreover, this hardness result extends to planar graphs, and the bidimensionality framework does not apply to this problem [3].

---

[1] In one common variant of the problem, a vertex can cover its own demand 'for free' and the capacity only limits coverage of neighbors.

Kao et al. characterize CAPACITATED DOMINATION by several attributes [16]. The *weighted* version has real, non-negative vertex weights whereas the *unweighted* version has uniform vertex weights. The *unsplittable*-demand version requires that the entire demand of any vertex is covered by a single vertex in the dominating set, while the *splittable*-demand version allows demand to be covered by multiple vertices in the dominating set. The *soft*-capacity version allows multiple copies of vertices to be included in the dominating set (effectively scaling both the capacity and the weight of vertex by any integral amount) whereas the *hard*-capacity version limits the number of copies of each vertex (usually to one). Lastly, the capacities and demands can be required to be (non-negative) integral.

For the soft-capacity version, many results are known. For general graphs, Kao et al. present a $(\ln n)$-approximation for the weighted unsplittable-demand case, a $(4 \ln n + 2)$-approximation for the weighted splittable-demand case, a $(2 \ln n + 1)$-approximation for the unweighted splittable-demand case, and a (Δ)-approximation for the weighted splittable-demand case, where Δ is the maximum degree in the graph [14,16]. As these are generalizations of the classic DOMINATING SET problem, there is little room for improvement for these approximations on general graphs.

For the soft-capacity problem in planar graphs, Kao and Lee developed a constant-factor approximation algorithm [15]. Additionally, Kao and Chen present a pseudo-polynomial-time approximation scheme for the weighted splittable-demand case by first showing how to solve the problem on graphs of bounded treewidth and then applying Baker's framework [14].

Substantially fewer results have been published on CAPACITATED DOMINATION with hard capacities. In this paper we consider the **unweighted, splittable-demand, hard-capacity** version of CAPACITATED DOMINATION with integral capacities and demands. Our main result for CAPACITATED DOMINATION is stated in Theorem 1.

Theorem 1. *For any positive integer $\hat{c}$ there exists a PTAS for* CAPACITATED DOMINATION *on the instance (G, d, c) where G is a planar graph, c is a capacity function with maximum value $\hat{c}$, and d is a demand function.*

This result for hard-capacities mirrors the PTAS presented by Kao and Chen for soft-capacities [14]. The challenge in extending the Baker framework to the hard-capacitated version is that the technique is agnostic to vertices being duplicated in the solution and does not necessarily return a feasible solution when we disallow duplicates (or higher multiplicities). Our main contributions to CAPACITATED DOMINATION are twofold. First we show how to *smooth* a capacity-exceeding solution into a capacity-respecting solution. Second we modify the traditional Baker framework to bound the cost of this smoothing process.

Unlike CAPACITATED DOMINATION, very few results have been published for CAPACITATED r-DOMINATION. For points on a unit sphere, a heuristic using network flow was developed [19]. In the closely related CAPACITATED k-CENTER problem, the number of centers, k, is fixed and the objective is to minimize the maximum distance of a vertex to a center, subject to each center having a limited number of clients (vertices) assigned to it. For uniform capacities and unit demand, a 6-approximation is known [17] for edge-weighted graphs. Note that k-CENTER and r-DOMINATION are two different perspectives on optimization for (k, r)-CENTER.

Extending our smoothing technique to CAPACITATED r-DOMINATION is fairly straightforward. However, Baker's framework depends on being able to efficiently solve instances on graphs of bounded branchwidth, and no algorithm for this problem was known. Our main contribution to CAPACITATED r-DOMINATION, therefore, is a dynamic-programming algorithm for planar graphs of bounded branchwidth, which combining with the smoothing technique gives the result stated in Theorem 2.

Theorem 2. *For any positive integers r and $\hat{c}$ there exists a PTAS for* CAPACITATED r-DOMINATION *on the instance (G, d, c) where G is a planar graph, c is a capacity function with maximum value $\hat{c}$, and d is a demand function.*

Section 2 introduces the problems and Sect. 3 provides an overview of Baker's framework. Section 4 presents the PTAS for CAPACITATED (r)-DOMINATION and gives proofs of Theorems 1 and 2. In Sect. 5 we give the dynamic-programming algorithm for solving CAPACITATED r-DOMINATION exactly for graphs of bounded branchwidth. Finally, Sect. 6 discusses possible extensions of the techniques presented.

2   Preliminaries

Let $G = (V, E)$ be an undirected graph with vertex set V and edge set E. For a subset of edges, $F \subseteq E$, we define the *boundary* of F, denoted $\partial_G(F)$, to be the set of vertices in V incident both to edges in F and $E \setminus F$. For a subset of vertices, $S \subseteq V$, we define the *cut* of S, denoted $\delta_G(S)$, to be the set of edges in E incident both to vertices in S and $V \setminus S$, and we define the *frontier* $\partial_G(S)$ to be the set of vertices in S incident to edges in $\delta_G(S)$ (refer to Fig. 1). When G in unambiguous, we omit the subscript.

We consider a demand function $d : V \to \mathbb{Z}^*$ and a capacity function $c : V \to \mathbb{Z}^*$. The capacity (resp. demand) of a set of vertices is the sum of the constituent vertex capacities (resp. demands). Namely, for $R \subseteq V, c(R) = \sum_{v \in R} c(v)$ and $d(R) = \sum_{v \in R} d(v)$.

Let $h(u, v)$ denote the hop-distance (number of edges in shortest path) from u to v. If $h(u, v) \leq r$ we say that u and v are *r-neighbors* and $N_r(v) = \{u \mid h(u, v) \leq r\}$ is the (closed) *r-neighborhood* of v.

For a given demand function d, capacity function c, and hop-distance r, an *assignment* $\mathcal{A}$ is a multiset of *facility-client* pairs (u, v) with $u, v \in V$ such that

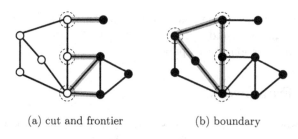

(a) cut and frontier   (b) boundary

Fig. 1. (a) Vertices in $S \subseteq V$ are shown in white; edges of cut $\delta_G(S)$ are highlighted gray; vertices of frontier $\partial_G(S)$ are circled. (b) Edges in $F \subseteq E$ are highlighted gray; vertices of boundary $\partial_G(F)$ are circled.

$h(u, v) \leq r$ and $\forall v \in V$, $|\{(x, v) \in \mathcal{A} : x \in V\}| \leq d(v)$ (i.e. v appears as a client at most $d(v)$ times).

An assignment is *proper* or *capacity-respecting* if $\forall u \in V, |\{(u, x) \in \mathcal{A} : x \in V\}| \leq c(u)$ (i.e. u appears as a facility at most $c(u)$ times). An assignment is *covering* or *demand-meeting* if $\forall v \in V$, $|\{(x, v) \in \mathcal{A} : x \in V\}| = d(v)$. The total unmet demand for an assignment is $t(\mathcal{A}) = \sum_{v \in V} (d(v) - |\{(x, v) \in \mathcal{A} : x \in V\}|)$. Observe that if $\mathcal{A}$ is covering then $t(\mathcal{A}) = 0$. A given assignment $\mathcal{A}$ has *dominating set* (or *facility set*) $S(\mathcal{A}) = \{u \in V : (u, v) \in \mathcal{A}\}$ and *size* $q(\mathcal{A}) = |S(\mathcal{A})|$. The objective of CAPACITATED DOMINATION and CAPACITATED r-DOMINATION is to find a proper, covering assignment $\mathcal{A}$ such that $q(\mathcal{A})$ is minimized.

Conceptually, in a solution to CAPACITATED r-DOMINATION, $\mathcal{A}$ describes an assignment of vertices to r-neighbors such that the demand of every vertex is met and the capacity of every vertex is not exceeded. This generalizes CAPACITATED DOMINATION in which $r = 1$ which itself generalizes the classic DOMINATING SET problem in which every vertex v has $d(v) = 1$ and $c(v) = \infty$.

We use superscripts to denote the multiplicity of a facility-client pair in an assignment. For example $\{(u, v)^4, (w, w)\}$ assigns four units of demand from the vertex v to the vertex u and one unit of demand from w to itself.

We use $\mathcal{A}_1 \uplus \mathcal{A}_2$ to denote the multiset union operation in which multiplicities are additive. For example $\{(u, v)^2, (u, w)\} \uplus \{(u, v)^3, (w, v)\} = \{(u, v)^5, (u, w), (w, v)\}$.

A *planar* graph is one that can be embedded on a sphere without any edge-crossings.

A *branch-decomposition* of G is a rooted binary tree T and a bijective mapping of edges of E to leaves of T. The nodes of T map to the subsets of E (called *clusters*) consisting of the corresponding edges of all descendant leaves. The root cluster therefore contains all edges in E, and leaf clusters are singleton edge sets. The *width* of a branch-decomposition is $\max_X |\partial(X)|$ over all clusters X. The *branchwidth* of G is the minimum width over all branch-decompositions of G. For planar graphs, an optimal branch decomposition can be computed in polynomial time [11]. The hierarchical structure of branch-decompositions lends itself well to dynamic-programming techniques. For many graph optimization problems, given a branch-decomposition with width w for the input graph, an optimal solution can be found in $2^{O(w)}n$ time [8]. We present such a dynamic program in the full version of this paper.

3   Baker's Framework

The PTAS presented in Sect. 4 is based on Baker's framework for designing PTASs in planar graphs [1]. The framework decomposes the input graph into subgraphs of bounded branchwidth, each of which can be solved efficiently, then forms an overall solution by combining the subgraph solutions. The error incurred can be charged to the interface between these subgraphs, and a shifting argument can be used to bound this error.

Specifically, Baker's approach has been applied to DOMINATING SET (and soft-CAPACITATED DOMINATION) on planar graphs as follows [14,20]. The input graph is partitioned into *levels* (vertex subsets) defined by minimum hop-distance (breadth-first search) from an arbitrary root v_0. The resulting subgraph induced by k consecutive levels has $O(k)$ branchwidth. The graph is decomposed into subgraphs of k consecutive levels such that every two adjacent subgraphs overlap in two *boundary* levels (see Fig. 2a). There are at most k different *shifts* that can divide the graph into k-level subgraphs in this way. For each shift, DOMINATING SET is solved for each (bounded-branchwidth) subgraph with the slight modification that the vertices of the topmost and bottommost levels of each subgraph require zero demand (but can still be part of the dominating set). This can be done in $2^{O(k)}n$ time. The overall solution is then the union of the solutions for the subgraphs.

Ignoring the demand of the outermost levels prevents a doubling-up of the demand requirements of the vertices in the boundary levels (adjacent-subgraph overlaps) while still allowing these vertices to satisfy demand from inner levels in the subgraph. Taking the union ensures that any ignored demand is met by some subgraph (no vertex's demand is ignored by all the subgraphs containing it). Each subgraph is solved optimally, and an optimum solution, $\mathcal{S}$, for the entire graph induces solutions on each subgraph, therefore the sum of the costs of the subgraph solutions is bounded by the cost of $\mathcal{S}$ plus some amount of error that can be bounded by the weight of $\mathcal{S}$ that intersects the boundary levels. For some *shift* this weight is a small fraction ($O(1/k)$) of the total weight of the optimum solution.

The problem with extending this approach to hard capacities emerges when a vertex in a boundary level is included in the solutions for two different subgraphs. For the classic DOMINATING SET problem, these vertices have unbounded capacities so taking the union of solutions is not an issue. Similarly, for soft-CAPACITATED (r)-DOMINATION, we are allowed to include multiple copies of these vertices, and since the weight is small we can afford to do so. However, for hard-CAPACITATED (r)-DOMINATION these vertices pose a problem since we may have *overloaded* them (exceeded their capacities) and are not allowed to simply pay the extra cost to use these vertices multiple times (see Fig. 2b).

In this paper, we address this issue by redirecting the assignment of overloaded vertices to underloaded vertices elsewhere in the graph. We may have to do this for every solution vertex that appears on the boundary levels. Unfortunately the above bound on the error arises from the sum of subgraph solution costs and does not bound the weight of the solution vertices that appear on boundary levels. We address this problem by modifying the way that the graph is decomposed into subgraphs: instead of every subgraph being constructed and used identically (as is typical in the Baker framework), we alternate between two different types of subgraphs. This facilitates accounting for the reassignment of overloaded vertices.

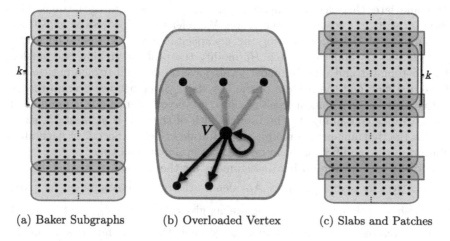

(a) Baker Subgraphs   (b) Overloaded Vertex   (c) Slabs and Patches

Fig. 2. (a) Subgraphs consist of k consecutive BFS levels. (b) An arc from u to v indicates that (u, v) is in the assignment. If $c(v) = 3$ the upper-subgraph assignment (gray arcs) and lower-subgraph assignment (black arcs) are each proper, but the union of solutions is not proper since vertex v is overloaded. (c) Patches are depicted by rectangles and slabs are depicted by rounded shapes.

4 PTAS

In this section we present a PTAS for the general problem of CAPACITATED r-DOMINATION. The CAPACITATED DOMINATION PTAS results from setting $r = 1$.

The PTAS for CAPACITATED r-DOMINATION uses the approach introduced by Baker [1] and described in Sect. 3. Breadth-first search from an arbitrary root vertex, v_0, partitions the graph into m levels (defined by hop-distance from v_0) such that the subgraph induced by any k consecutive levels has branchwidth at most $2k$.

For a given i, j, k, l, r such that $0 \leq i < k, 0 \leq j \leq \lfloor \frac{m}{k} \rfloor$, and $0 \leq l \leq \lfloor \frac{m}{k} \rfloor - 1$, let G_j^i denote the subgraph induced by levels $jk+i$ through $(j+1)k+i-1$ and H_l^i denote the subgraph induced by levels $(l+1)k+i-2r$ through $(l+1)k+i+2r-1$. For a fixed k and i we call G_j^i the jth *slab* and H_l^i the lth *patch* (see Fig. 2c). We call k the slab *height* and i the slab *shift*.

Consecutive slabs do not overlap each other and each slab has branchwidth at most $2k$. Each patch occurs at the interface between two consecutive slabs[2], overlapping with $2r$ levels from each, and for fixed r has constant branchwidth (at most $8r$).

The algorithm proceeds as follows (see the full version of this paper for pseudocode). Independently solve CAPACITATED r-DOMINATION restricted to each slab and each patch (with demands modified as described below). Then take the union of solutions over all slabs and patches. Since patches overlap the slabs, if a dominating vertex from a proper patch solution is also a dominating vertex from a proper slab solution, the union of solutions may no longer be proper. In fact the capacities for patch vertices have been effectively doubled. The algorithm alleviates this excess strain on patch vertices by reassigning the demand to *underfull* vertices elsewhere in the graph. We refer to this process as *smoothing*.

Similarly to the algorithm for the uncapacitated version described above, for each slab G_j^i and each patch H_l^i modify the problem so that the vertices of the r-topmost and r-bottommost levels have no demand (excluding the top of the first slab and the bottom of the last slab). Since patches overlap the $2r$-bottommost and $2r$-topmost levels of slabs, no vertex's demand is ignored by all the subgraphs containing it. The branchwidth of the slab is at most $2k$, so we can use the dynamic-programming algorithm described in Sect. 5 to generate a solution to the modified problem on the slab in polynomial time.

Let $\mathcal{A}_{G_j^i}$ denote the optimal assignment for the modified CAPACITATED r-DOMINATION for slab G_j^i and let $\mathcal{A}_{H_l^i}$ denote the optimal assignment for patch H_l^i. Combining slab and patch assignments gives $\tilde{\mathcal{A}}_i = \left(\biguplus_j \mathcal{A}_{G_j^i} \right) \uplus \left(\biguplus_l \mathcal{A}_{H_l^i} \right)$. Since $\tilde{\mathcal{A}}_i$ may no longer be a proper assignment, the algorithm uses a *smoothing* procedure, SMOOTH, to generate a final solution $\mathcal{A}_i'$. SMOOTH first removes facility-client pairs from an assignment $\mathcal{A}$ until it is proper by repeatedly finding an *overloaded* vertex v and removing a pair (v, u) from $\mathcal{A}$ until no overloaded vertex can be found. SMOOTH then reassigns any remaining unmet demand to *underfull* vertices in the graph with excess capacity by repeatedly finding alternating paths in the induced assignment graph (described in Lemma 2). This reassignment process is described in detail in the following lemmas. Finally, the algorithm returns the assignment $\mathcal{A}_i'$ with the minimum cost among all choices of shift i.

Lemma 1. *Given a graph G with demand and capacity functions d and c and a set of vertices $R \subseteq V$, if a proper, covering assignment exists for* CAPACITATED *1-DOMINATION on the instance (G, d, c), then $c(R) + d(\partial(R)) \geq d(R)$.*

Proof. Assume that a proper, covering assignment exists. For any subset of vertices $R \subseteq V$, the demand from non-frontier vertices of R must be met by vertices in R. Since a solution exists, R must have sufficient capacity. That is, $d(R \setminus \partial(R)) \leq c(R)$. Furthermore, $d(R \setminus \partial(R)) = d(R) - d(\partial(R))$. This gives, $d(R) - d(\partial(R)) \leq c(R)$, and the result follows.

---

[2] Note that in order to keep the patches themselves from overlapping, we must ensure that $k \geq 4r$.

Lemma 2. *Given a graph $G = (V, E)$ with demand and capacity functions d and c and a proper assignment $\mathcal{A}'$ with unmet demand $t(\mathcal{A}') > 0$, if a proper, covering assignment exists for CAPACITATED 1-DOMINATION on this instance then a new proper assignment $\mathcal{A}''$ can be found in polynomial time such that $q(\mathcal{A}'') \leq q(\mathcal{A}') + 1$ and $t(\mathcal{A}'') < t(\mathcal{A}')$.*

Proof. Let U be the set of vertices with unmet demand. We assume that these vertices are at full capacity, otherwise simply assign such an underfull vertex, v, to itself and satisfy the lemma by letting $\mathcal{A}'' = \mathcal{A}' \uplus \{(v, v)\}$. We also assume that a vertex $v \in U$ spends its entire capacity toward meeting its own demand, so (v, v) occurs $c(v)$ times in $\mathcal{A}'$. We can always reassign v's capacity in this way without increasing overall unmet demand or the size of $\mathcal{A}'$.

Use assignment $\mathcal{A}'$ to define an arc set on V such that $(u, v) \in \mathcal{A}'$ indicates an arc from facility u to client v, and for every vertex u, the outdegree $\Delta^-(u)$ is at most $c(u)$. The dominating set $S(\mathcal{A}')$ is $\{v : \Delta^-(v) > 0\}$, and a covering assignment is one in which, for every vertex u, the indegree $\Delta^+(u)$ equals $d(u)$.

Consider the directed *assignment graph* $\mathscr{A}$ defined by this arc set. We can assume that $\mathscr{A}$ has no directed cycles (except for self-loops), otherwise every arc in such a cycle can be changed to a self-loop without changing the total unmet demand or the size of $\mathcal{A}'$.

Since a proper, covering assignment exists, there must be a set W of vertices that are not at full capacity.

Let $G' = (V', E')$ denote $G = (V, E)$ augmented with self-loops at every vertex. Formally, $V' = V$ and $E' = E \cup \{vv : v \in V\}$. We define a *semi-alternating* directed path $P = p_0, p_1, \ldots, p_l$ to be a path in G' such that $\forall i$ $p_i p_{i+1} \in E'$ and $(p_i, p_{i+1}) \notin \mathcal{A} \Leftrightarrow (p_{i+1}, p_{i+2}) \in \mathcal{A}$. That is, P is a path through G' in which exactly every other edge is a *forward* arc in the assignment graph $\mathscr{A}$ (see Fig. 3). We show that there exists such a *semi-alternating* directed path $P = p_0, p_1, \ldots, p_l$ from a vertex $p_0 \in U$ to a vertex $p_l \in W$.

Assume to the contrary that no such path exists. Let U^+ be the set of all vertices $v \in V$ such that there exists a semi-alternating directed path from u to v for some $u \in U$. Clearly $U \subseteq U^+$, and by assumption $U^+ \cap W = \emptyset$, so all vertices of U^+ are at full capacity. Since each vertex in U spends all of its capacity on meeting its own demand, it has no outgoing arcs in $\mathscr{A}$, so the first edge of every semi-alternating directed path originating at some $u \in U$ is *not* a forward arc in $\mathscr{A}$. (If the first edge of the path is a self loop in $\mathscr{A}$, it has a subpath with the same endpoints that is semi-alternating whose first edge is not in $\mathscr{A}$). Therefore all of u's neighbors are in U^+ and u cannot appear on the frontier $\partial_{G'}(U^+)$, so $U \cap \partial_{G'}(U^+) = \emptyset$. Similarly, any arcs of $\mathscr{A}$ into vertices in $\partial_{G'}(U^+)$ must start in $V' \setminus U^+$ otherwise U^+ could be extended to include neighbors of such vertices.

Since all of the demand of every vertex in $\partial_{G'}(U^+)$ is met and all arcs of $\mathscr{A}$ into $\partial_{G'}(U^+)$ start in $V' \setminus U^+$, we infer that $V' \setminus U^+$ contributes a total of $d(\partial_{G'}(U^+))$ toward meeting the demand of U^+. Furthermore all vertices in U^+ are at full capacity and there are no outgoing arcs of $\mathscr{A}$ in $\delta_{G'}(U^+)$ so U^+ contributes a total of $c(U^+)$ toward meeting its own total demand. Yet $U \subseteq U^+$ so U^+ contains unmet demand. So $c(U^+) + d(\partial_{G'}(U^+)) < d(U^+)$ which by Lemma 1 contradicts our feasibility assumption.

Such a path can be found in linear time using breadth-first search from vertices with insufficient demand met.

Given a *semi-alternating* directed path $P = p_0, p_1, \ldots, p_l$ from a vertex $p_0 \in U$ to a vertex $p_l \in W$, we know that (p_0, p_1) is not in $\mathcal{A}'$ since p_0 is in U and does not contribute to meeting any neighbor's demand. We also assume that (p_{l-1}, p_l) is not in $\mathcal{A}'$ since if the path ends in an assignment arc we can append a self-loop. Reassign the arcs along this path as follows:

$$\mathcal{A}'' = (\mathcal{A}' \setminus \{(p_1, p_2), (p_3, p_4), \ldots, (p_{l-2}, p_{l-1})\}) \uplus \{(p_1, p_0), (p_3, p_2), \ldots, (p_{l-2}, p_{l-3}), (p_l, p_{l-1})\}$$

This gives $q(\mathcal{A}'') \leq q(\mathcal{A}') + 1$. All of the demand previously met along path P is still met although potentially by a different source, but $\mathcal{A}''$ also meets one additional unit of previously unmet demand at vertex p_0 (see Fig. 3). Therefore the resulting proper assignment $\mathcal{A}''$ satisfies $t(\mathcal{A}'') < t(\mathcal{A}')$.

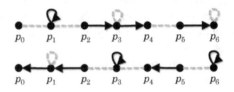

Fig. 3. Black arrows depict assignment arcs and dotted gray lines depict edges of G'. The upper path shows a semi-alternating directed path $P = p_0, p_1, p_1, p_2, p_3, p_3, p_4,$ p_5, p_6, p_6. The lower path shows the result of arc reassignment.

Lemma 3. *Given a graph $G = (V, E)$ with demand and capacity functions d and c and a proper assignment $\mathcal{A}'$ with unmet demand $t(\mathcal{A}') > 0$, if a proper, covering assignment exists for* CAPACITATED r-DOMINATION *then a new proper assignment $\mathcal{A}''$ can be found in polynomial time such that $q(\mathcal{A}'') \leq q(\mathcal{A}') + 1$ and $t(\mathcal{A}'') < t(\mathcal{A}')$.*

Proof. Construct $G^* = (V, E^*)$ from G by adding an edge between every pair of r-neighbors. That is, $E^* = E \cup \{uv \mid h_G(u, v) \leq r\}$ where h_G is hop-distance in G. Any assignment $\mathcal{A}$ for CAPACITATED r-DOMINATION in G is also an assignment for CAPACITATED 1-DOMINATION in G^*, and vice versa. Therefore applying Lemma 2 to G^* (using the same d and c) results in a new assignment with the desired properties.

We restate the main theorem here for convenience.

Theorem 1. *For any positive integers r and $\hat{c}$ there exists a PTAS for* CAPACITATED r-DOMINATION *on the instance (G, d, c) where G is a planar graph, c is a capacity function with maximum value $\hat{c}$, and d is a demand function.*

Proof. Let $\mathcal{A}^*$ denote an optimal assignment for CAPACITATED r-DOMINATION and let $OPT = q(\mathcal{A}^*)$. For a fixed shift i let $\mathcal{A}^*_{G^i_j} = \mathcal{A}^* \cap G^i_j$ be the intersection of the optimal assignment with the jth slab and $OPT_{G^i_j}$ denote the size of the optimal assignment for the modified problem on the jth slab. Similarly, let $\mathcal{A}^*_{H^i_l} = \mathcal{A}^* \cap H^i_l$ be the intersection of the optimal assignment with the lth patch and $OPT_{H^i_l}$ denote the size of the optimal assignment for the modified problem on the lth patch.

Because we modify the problem to ignore demand from the r outermost levels of a slab (resp. patch) $\mathcal{A}^*_{G^i_j}$ (resp. $\mathcal{A}^*_{H^i_l}$) is a covering assignment for the modified problem on G^i_j (resp. H^i_l) since all of the demand on the non-outermost levels will be covered by $\mathcal{A}^*_{G^i_j}$ (resp. $\mathcal{A}^*_{H^i_l}$). Therefore $OPT_{G^i_j} \leq q(\mathcal{A}^*_{G^i_j})$ and $OPT_{H^i_l} \leq q(\mathcal{A}^*_{H^i_l})$.

Our algorithm first determines the union, $\tilde{\mathcal{A}}$, of assignments on slabs and patches. The size of this union is at most the sum of the sizes of the assignments for each slab and each patch. The algorithm then removes and reassigns at most $\hat{c}$ assignment pairs for each duplicated vertex in the union (where $\hat{c}$ is the value of the maximum capacity). Duplicated vertices can only occur within the levels of the patches. By Lemma 3 each reassignment adds at most one vertex to the dominating set. Let $\mathcal{A}'$ be the final assignment output by the algorithm. Therefore

$$q(\mathcal{A}') \leq \sum_j OPT_{G^i_j} + \hat{c} \sum_l OPT_{H^i_l} \leq \sum_j q(\mathcal{A}^*_{G^i_j}) + \hat{c} \sum_l q(\mathcal{A}^*_{H^i_l}) \leq OPT + \hat{c} \sum_l q(\mathcal{A}^*_{H^i_l})$$

The final inequality comes from the slabs (and therefore their intersections with $\mathcal{A}^*$) being disjoint. Let i^* be the shift that minimizes the sum, $\sum_l q(\mathcal{A}^*_{H^i_l})$, of the intersection of patches with the optimal dominating set $S(\mathcal{A}^*)$.

If $\sum_l q(\mathcal{A}^*_{H^{i^*}_l})$ exceeds $(4r/k)OPT$ then the sum of patch intersections over all shifts, $\sum_i \sum_l q(\mathcal{A}^*_{H^i_l})$ exceeds $4rOPT$ since i^* was chosen from k possible shifts to give the minimum such sum. But each vertex in $\mathcal{A}^*$ appears in a patch for at most $4r$ different shifts so can contribute at most $4r$ to the sum, giving $\sum_i \sum_l q(\mathcal{A}^*_{H^i_l}) \leq 4rOPT$. Therefore, $\sum_l q(\mathcal{A}^*_{H^{i^*}_l}) \leq (4r/k)OPT$. This gives

$$q(\mathcal{A}') \leq OPT + (4r\hat{c}/k)OPT$$

Setting k to $4r\hat{c}/\epsilon$ gives a $(1+\epsilon)$-approximation.

For CAPACITATED DOMINATION the slabs can be solved in $O(2k\hat{c}^{6k}\hat{d}^{4k}n)$ time and patches can be solved in $O(8\hat{c}^{24}\hat{d}^{16}n)$ time, where $\hat{d}$ is the maximum demand value, using a dynamic program (see the full version of this paper for details). The algorithm then performs at most ϵn reassignments, each of which can be computed in linear time. This process is repeated k times, giving an overall $O(k^2(\hat{c}\hat{d})^{6k}n + \hat{c}n^2)$ runtime. The algorithm is therefore a PTAS since we take $\hat{c}$ and $1/\epsilon$ (and thus k) to be constant. Taking both $\hat{c}$ and $\hat{d}$ to be constant yields an EPTAS.

For CAPACITATED r-DOMINATION the slabs can be solved in $2k \cdot (\hat{c}n)^{O(kr)}$ time and patches can be solved in $8r \cdot (\hat{c}n)^{O(r^2)}$ time using the dynamic program described in Sect. 5.

Theorem 1, the analogous theorem for CAPACITATED DOMINATION, follows as a corollary.

As noted in the proof above, if the demand function d is also bounded by some maximum value $\hat{d}$, taken to be constant, the algorithm is in fact an EPTAS. Additionally, the capacity of a vertex never need exceed the sum of the demands in its closed neighborhood, so bounding both the maximum demand and vertex degree instead of capacity also gives an EPTAS.

CAPACITATED VERTEX COVER is a closely related problem in which each *edge* has demand that must be met by capacity from its endpoints. We can reduce CAPACITATED VERTEX COVER to CAPACITATED DOMINATION by bisecting each edge e by inserting a vertex v and assigning e's demand to v. A PTAS for CAPACITATED VERTEX COVER therefore follows as a corollary to Theorem 1 (refer to the full version of this paper for details and proof).

5 Solving Capacitated r-Domination for Bounded Branchwidth

In this section we describe a dynamic-programming algorithm for solving CAPACITATED r-DOMINATION for graphs with bounded branchwidth. Even though an algorithm for CAPACITATED DOMINATION follows from this algorithm, we present a much simpler and more efficient dynamic program for the former in the full version of this paper.

In any proper, covering assignment $\mathcal{A}$, each facility-client pair $(u, v) \in \mathcal{A}$ indicates one unit of demand at vertex v is met by one unit of capacity at vertex u within a distance r. We call the shortest v-to-u path in G the (directed) *resolution*[3] path for this unit of demand. We design an algorithm that finds the resolution paths corresponding to an optimal assignment.

Although designing a dynamic program for CAPACITATED DOMINATION is fairly straightforward, there are two main challenges with designing one for CAPACITATED r-DOMINATION. First, as described in Demaine et al. for *uncapacitated* r-DOMINATION, the resolution paths can wind up and down the branch decomposition tree, so the algorithm needs a way to convey information across levels [6]. The second challenge is that, even though the capacity of a vertex v on the boundary of a cluster may be bounded, the number of resolution paths that cross the boundary at v may be quite large, and unlike the uncapacitated version, we cannot merge these paths. We address these challenges by extending the dynamic program used for the uncapacitated version [6].

Consider an assignment $\mathcal{A}$ and the corresponding resolution paths. For each cluster in the branch decomposition, some number of these resolution paths cross

---

[3] We adopt this word from Demaine et al. [6].

the cluster boundary at each boundary vertex. Each such path has a *direction* (crossing *into* or *out of* the cluster) and a *magnitude* (path length remaining until facility is reached). Our dynamic-programming table, DP, is indexed by *configurations* (X, h_X) where X is a cluster and h_X is a function that specifies how many paths of each direction and magnitude cross at each vertex of a cluster boundary.

Specifically, the algorithm enumerates all functions of the form $h_X : \partial(X) \times [-r, r] \to [0, \hat{c}n]$, where $h_X(v, i)$ indicates the number of resolution paths that cross X at v with magnitude $|i|$ and direction $sign(i)$. If $i < 0$ the path crosses *into* X and if $i > 0$ the path crosses *out of* X. Note that the sign only indicates the direction of the next vertex in the resolution path and not necessarily the direction of the endpoint of the path (a path may enter and/or leave the cluster again elsewhere). If $i = 0$ the resolution path ends at v, so $h_X(v, 0) > 0$ indicates that v itself is in the dominating set. Therefore $h_X(v, 0)$ equals $|\{(v, u) \in \mathcal{A} : u \in V\}|$ and must not exceed $c(v)$.

We say that assignment $\mathcal{A}$ is *consistent* with h_X, if h_X exactly describes how the resolution paths corresponding to $\mathcal{A}$ cross the boundary of cluster X. The DP entry at index (X, h_X) is the $X \cap S(\mathcal{A})$ with minimum cardinality over all proper, covering assignments $\mathcal{A}$ that are consistent with h_X. The algorithm proceeds level-by-level through the decomposition, starting at the leaves and continuing rootward.

The leaves of the decomposition correspond to single edges in the graph. For the leaf cluster X of edge uv, a configuration is *valid* if it meets the following criteria.

- Any resolution path crossing *into* X at u (resp v) with i hops remaining must cross *out of* X at v (resp u) with $i - 1$ hops remaining. That is, for $i < 0, h_X(u, i) \le h_X(v, -i - 1)$ and $h_X(v, i) \le h_X(u, -i - 1)$.
- The resolution paths crossing X at u (resp v) must include a path for each unit of demand at u (resp v). That is, $\Sigma_i h_X(u, i) \ge d(u)$ and $\Sigma_i h_X(v, i) \ge d(v)$.
- The number of paths ending at u (resp v) cannot exceed the capacity of u (resp v). That is, $h_X(u, 0) \le c(u)$ and $h_X(v, 0) \le c(v)$.

Moving rootward, we show how to determine the table entry for a parent cluster given the table entries for the child clusters. Consider a cluster X_0 with child clusters X_1 and X_2. We divide the cluster boundaries into four regions and handle each region separately.

Let $\partial_{1,2} = \partial(X_1) \cap \partial(X_2)) \setminus \partial(X_0)$; $\partial_{0,1} = \partial(X_0) \cap \partial(X_1)) \setminus \partial(X_2)$; $\partial_{0,2} = \partial(X_0) \cap \partial(X_2)) \setminus \partial(X_1)$; and $\partial_{0,1,2} = \partial(X_0) \cap \partial(X_1) \cap \partial(X_2)$ (refer to Fig. 4).

Indices (X_1, h_{X_1}) and (X_2, h_{X_2}) are *compatible* with (X_0, h_{X_0}) if the following hold:

- Resolution paths crossing $\partial_{1,2}$ cross *into* one cluster and *out of* the other, so h_{X_1} and h_{X_2} agree on magnitude and disagree on sign. That is, $h_{X_1}(v, i) = h_{X_2}(v, -i), \forall i, v \in \partial_{1,2}$.
- Resolution paths crossing $\partial_{0,1}$ cross the X_1 boundary the same way they cross the X_0 boundary, so h_{X_0} must agree with h_{X_1}. That is $h_{X_0}(v, i) = h_{X_1}(v, i), \forall i, v \in \partial_{0,1}$. The same holds for $\partial_{0,2}$.

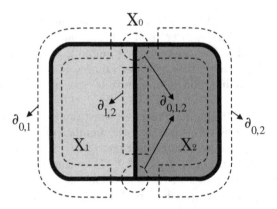

Fig. 4. Depiction of parent (X_0) and children (X_1 and X_2) boundary partitioning

- Resolution paths crossing vertices on $\partial_{0,1,2}$ are trickier to handle.
  - All three clusters must agree on the number of resolution paths ending on $\partial_{0,1,2}$, so $h_{X_1}(v,0) = h_{X_2}(v,0) = h_{X_0}(v,0), \forall i, v \in \partial_{0,1,2}$.
  - Each resolution path that crosses out of (resp into) X_0 must correspond to a path crossing out of (resp into) either X_1 or X_2, so $h_{X_0}(v,i) \leq h_{X_1}(v,i) + h_{X_2}(v,i)$.
  - Similarly, each path that crosses X_1 (resp X_2) must correspond to a path crossing either X_0 in the same direction or X_2 (resp X_1) in the opposite direction, so $h_{X_1}(v,i) \leq h_{X_0}(v,i) + h_{X_2}(v,-i)$ and $h_{X_2}(v,i) \leq h_{X_0}(v,i) + h_{X_1}(v,-i)$.
  - Finally, the net total of crossings in and out must agree. That is $h_{X_0}(v,i) - h_{X_0}(v,-i) = h_{X_1}(v,i) + h_{X_2}(v,i) - h_{X_1}(v,-i) - h_{X_2}(v,-i)$.

The entry for each valid configuration is the set of vertices in the cluster that are resolution path endpoints. For leaf cluster X corresponding to edge uv, this is $\{w \in \{u,v\} \mid h_X(w,0) > 0\}$. To determine the entry for configuration (X_0, h_{X_0}) the algorithm searches over all compatible pairs (X_1, h_{X_1}) and (X_2, h_{X_2}) and find $DP(X_1, h_{X_1}) \cup DP(X_2, h_{X_2})$ such that the cardinality is minimized.

Letting X_{G_0} denote the cluster containing all edges of input graph G_0 and $h_\emptyset$ be the trivial empty function, an optimal proper, covering dominating set is stored in $DP(X_{G_0}, h_\emptyset)$.

The assignment itself can then be found by setting the capacity of all vertices not in the dominating set to zero and then using the smoothing technique described in Sect. 4 to assign demand to the vertices in the dominating set.

The dynamic-programming algorithm described above gives us the following Theorem. The proof is deferred to the full version of this paper.

Theorem 3. *For any positive integers r, w and $\hat{c}$ there exists an exact polynomial time algorithm for* CAPACITATED r-DOMINATION *on the instance (G, d, c) where G is a planar graph with branch decomposition of width w, c is a capacity function with maximum value $\hat{c}$, and d is a demand function.*

6   Conclusion

Ultimately, we would like to find a PTAS that can accommodate arbitrary capacities. The main obstacle in extending our techniques toward this goal is that the slab height, and thus the branchwidth, depends on the capacity. Lower bound results for both DOMINATING SET [18] and r-DOMINATION [4] show that under the Strong Exponential-Time Hypothesis (SETH), the runtime for these problems has an exponential dependence on branchwidth. A Baker-like PTAS for arbitrary capacities would therefore require a substantially different approach to the slab decomposition.

Alternatively, we consider extending to the weighted case. The main challenge here is that our technique barters capacity for weight in unit quantities. While we can guarantee that there is some underfull vertex in the graph, we cannot guarantee a bound on the weight of this vertex. One idea to address this is to design a more nuanced smoothing process that considers the weights of the vertices chosen in the search. Alternatively a careful rounding strategy may lead to a bicriteria result in which either the capacities can be exceeded or the demands can be under-covered by a small (bounded) amount.

We would additionally like to extend the capacitated r-domination problem to arbitrary values of r, but here we are additionally hindered by our dynamic-programming approach. No efficient DP that accommodates capacities is known.

References

1. Baker, B.S.: Approximation algorithms for NP-complete problems on planar graphs. J. ACM **41**(1), 153–180 (1994)
2. Barilan, J., Kortsarz, G., Peleg, D.: How to allocate network centers. J. Algorithms **15**(3), 385–415 (1993)
3. Bodlaender, H.L., Lokshtanov, D., Penninkx, E.: Planar capacitated dominating set is $W[1]$-hard. In: Chen, J., Fomin, F.V. (eds.) IWPEC 2009. LNCS, vol. 5917, pp. 50–60. Springer, Heidelberg (2009). https://doi.org/10.1007/978-3-642-11269-0_4
4. Borradaile, G., Le, H.: Optimal dynamic program for r-domination problems over tree decompositions. In: LIPIcs-Leibniz International Proceedings in Informatics, vol. 63. Schloss Dagstuhl-Leibniz-Zentrum fuer Informatik (2017)
5. Chlebík, M., Chlebíková, J.: Approximation hardness of dominating set problems in bounded degree graphs. Inf. Comput. **206**(11), 1264–1275 (2008)
6. Demaine, E.D., Fomin, F.V., Hajiaghayi, M., Thilikos, D.M.: Fixed-parameter algorithms for (k, r)-center in planar graphs and map graphs. ACM Trans. Algorithms (TALG) **1**(1), 33–47 (2005)
7. Dom, M., Lokshtanov, D., Saurabh, S., Villanger, Y.: Capacitated domination and covering: a parameterized perspective. In: Grohe, M., Niedermeier, R. (eds.) IWPEC 2008. LNCS, vol. 5018, pp. 78–90. Springer, Heidelberg (2008). https://doi.org/10.1007/978-3-540-79723-4_9
8. Dorn, F.: Dynamic programming and planarity: Improved tree-decomposition based algorithms. Discret. Appl. Math. **158**(7), 800–808 (2010)
9. Feige, U.: A threshold of $\ln n$ for approximating set cover. J. ACM **45**(4), 634–652 (1998)

10. Fujita, S.: Vertex domination in dynamic networks. In: Nakano, S., Rahman, M.S. (eds.) WALCOM 2008. LNCS, vol. 4921, pp. 1–12. Springer, Heidelberg (2008). https://doi.org/10.1007/978-3-540-77891-2_1
11. Qian-Ping, G., Tamaki, H.: Optimal branch-decomposition of planar graphs in $O(n^3)$ time. ACM Trans. Algorithms **4**(3), 30 (2008)
12. Guha, S., Hassin, R., Khuller, S., Or, E.: Capacitated vertex covering. J. Algorithms **48**(1), 257–270 (2003)
13. Johnson, D.S.: Approximation algorithms for combinatorial problems. J. Comput. Syst. Sci. **9**(3), 256–278 (1974)
14. Kao, M.-J., Chen, H.-L.: Approximation algorithms for the capacitated domination problem. In: Lee, D.-T., Chen, D.Z., Ying, S. (eds.) FAW 2010. LNCS, vol. 6213, pp. 185–196. Springer, Heidelberg (2010). https://doi.org/10.1007/978-3-642-14553-7_19
15. Kao, M.-J., Lee, D.T.: Capacitated domination: constant factor approximations for planar graphs. In: Asano, T., Nakano, S., Okamoto, Y., Watanabe, O. (eds.) ISAAC 2011. LNCS, vol. 7074, pp. 494–503. Springer, Heidelberg (2011). https://doi.org/10.1007/978-3-642-25591-5_51
16. Kao, M.-J., Liao, C.-S., Lee, D.T.: Capacitated domination problem. Algorithmica **60**(2), 274–300 (2011)
17. Khuller, S., Sussmann, Y.J.: The capacitated k-center problem. SIAM J. Discret. Math. **13**(3), 403–418 (2000)
18. Lokshtanov, D., Marx, D., Saurabh, S.: Known algorithms on graphs of bounded treewidth are probably optimal. In: Proceedings of the Twenty-Second Annual ACM-SIAM Symposium on Discrete Algorithms, pp. 777–789. Society for Industrial and Applied Mathematics (2011)
19. Lupton, R., Maley, F.M., Young, N.: Data collection for the sloan digital sky survey–a network-flow heuristic. J. Algorithms **27**(2), 339–356 (1998)
20. Marzban, M., Qian-Ping, G.: Computational study on a PTAS for planar dominating set problem. Algorithms **6**(1), 43–59 (2013)
21. Melkote, S., Daskin, M.S.: Capacitated facility location/network design problems. Eur. J. Oper. Res. **129**(3), 481–495 (2001)
22. Nieberg, T., Hurink, J.: A PTAS for the minimum dominating set problem in unit disk graphs. In: Erlebach, T., Persinao, G. (eds.) WAOA 2005. LNCS, vol. 3879, pp. 296–306. Springer, Heidelberg (2006). https://doi.org/10.1007/11671411_23
23. Ralphs, T.K., Kopman, L., Pulleyblank, W.R., Trotter, L.E.: On the capacitated vehicle routing problem. Math. Program. **94**(2–3), 343–359 (2003)
24. Wang, F., Camacho, E., Xu, K.: Positive influence dominating set in online social networks. In: Du, D.-Z., Hu, X., Pardalos, P.M. (eds.) COCOA 2009. LNCS, vol. 5573, pp. 313–321. Springer, Heidelberg (2009). https://doi.org/10.1007/978-3-642-02026-1_29

On Approximability of Connected Path Vertex Cover

Toshihiro Fujito$^{(\boxtimes)}$

Toyohashi University of Technology, Toyohashi 441-8580, Japan
fujito@cs.tut.ac.jp

Abstract. This paper is concerned with the approximation complexity of the CONNECTED PATH VERTEX COVER problem. The problem is a connected variant of the more basic problem of path vertex cover; in k-PATH VERTEX COVER it is required to compute a minimum vertex set $C \subseteq V$ in a given undirected graph $G = (V, E)$ such that no path on k vertices remains when all the vertices in C are removed from G. CONNECTED k-PATH VERTEX COVER (k-CPVC) additionally requires C to induce a connected subgraph in G.

Previously, k-CPVC in the unweighted case was known approximable within k^2, or within k assuming that the girth of G is at least k, and no approximation results have been reported on the weighted case of general graphs. It will be shown that (1) unweighted k-CPVC is approximable within k without any assumption on graphs, and (2) weighted k-CPVC is as hard to approximate as the weighted set cover is, but approximable within $1.35 \ln n + 3$ for $k \leq 3$.

Keywords: Connected Path Vertex Cover
Approximation algorithms · Connected vertex cover

1 Introduction

The main subject of this paper is the approximation complexity of the CONNECTED PATH VERTEX COVER problem, which is a connected variant of the PATH VERTEX COVER problem.

A path on k vertices is called a k-*path*, and it is required in k-PATH VERTEX COVER (k-PVC) to compute a minimum vertex set $C \subseteq V$, called a k-*path vertex cover*, in a given undirected graph $G = (V, E)$ such that no k-path remains when all the vertices in C are removed from G. The CONNECTED k-PATH VERTEX COVER (k-CPVC) additionally requires C to induce a connected subgraph in G. The k-PATH VERTEX COVER problem is so called [3–5,14,16] as well as under different names such as VERTEX COVER P_k (VCP$_k$) [27–30], P_k-HITTING SET [6], and k-PATH TRANSVERSAL [18] (and all of them refer to the same problem). The k-path vertex cover number, $\psi_k(G)$, of G is the cardinality of a minimum k-path vertex cover for G. The k-PVC problem was introduced in [5] along with $\psi_k(G)$, motivated by its relation to secure connection

© Springer International Publishing AG, part of Springer Nature 2018
R. Solis-Oba and R. Fleischer (Eds.): WAOA 2017, LNCS 10787, pp. 17–25, 2018.
https://doi.org/10.1007/978-3-319-89441-6_2

in wireless networks [23]. The vertex weighted version of k-PVC was considered in [30] motivated by its applications in traffic control. Clearly, $\psi_2(G)$ is the vertex cover number of G. In addition $\psi_3(G)$ corresponds to another previously studied concept of the *dissociation number* of a graph, defined as follows. A subset of vertices in a graph G is called a *dissociation set* if it induces a subgraph with maximum degree at most 1. The maximum cardinality of a dissociation set in G is called the dissociation number of G and is denoted by diss(G). Clearly $\psi_3(G) = |V(G)| - \text{diss}(G)$. The problem of computing diss(G) was introduced by Yannakakis [31], who also proved it to be NP-hard in the class of bipartite graphs. See [24] for a survey on the dissociation number problem.

The k-CPVC problem was first studied by Liu et al. [20] as an extension of k-PVC with the connectivity requirement. They considered such an extension with applications in the area of wireless sensor networks in minds. The approximability of k-CPVC was further investigated by Li et al. [19]. We summarize below some of main results known for the approximation complexity of these problems.

k-PVC
- 2-PVC coincides with the well-known vertex cover problem. It is possible to approximate within a factor slightly better than 2 [2,13,15,21], while doing so better than $10\sqrt{5} - 21 \approx 1.36067$ is NP-hard [7].
- A linear time algorithm is presented for trees and upper bounds on $\psi_k(G)$ were investigated in [5], while lower bounds on $\psi_k(G)$ in regular graphs were given in [4].
- 3-PVC is approximable within 2 [9,29,30] (or within an expected ratio of 23/11 by a randomized algorithm [16]), and within 1.57 on cubic graphs [28].
- 4-PVC is approximable within 3 [6], and within 2 on regular graphs [25].
- k-PVC is approximable within $O(\log k)$ in time $2^{O(k^3 \log k)}n^{O(1)}$ in the unweighted case [18].

k-CPVC
- 2-CPVC coincides with the connected vertex cover (CVC) problem, which is as hard (to approximate) as the vertex cover problem [11]. Whereas unweighted CVC is approximable within 2 [1,26], weighted CVC is set cover hard and known approximable only within $\ln n + 3$ [10].
- PTAS was designed for unweighted k-CPVC when G is a unit disk graph, based on a k^2-approximation algorithm for general graphs [20].
- It was shown by Li et al. [19] that
  - weighted k-CPVC can be solved in time $O(n)$ when G is a tree, or in time $O(rn)$ when G has a unique cycle of length r, and
  - unweighted k-CPVC can be approximated within k when G has girth at least k.

  It is pointed out here not only that the performance ratio of k is tight for their algorithm but also that their algorithm may not return a feasible solution without the girth assumption.

Thus, k-CPVC in the unweighted case was previously known approximable within k^2, or within k assuming that the girth of G is at least k, and no approximation results have been reported on the weighted case of general graphs. This paper presents that

- unweighted k-CPVC, where k is a fixed parameter, is approximable within k on general graphs *without* any assumption on girth, and
- weighted k-CPVC is as hard to approximate as the weighted set cover problem, but approximable within $1.35 \ln n + 3$ for $k \leq 3$.

2   Unweighted Connected Path Vertex Cover

Let T be a dfs (depth-first-search) spanning tree of G rooted at $r \in V(G)$, and let T_v denote the subtree of T rooted at $v \in V(G)$. It is easy to observe that, even if T_v doesn't contain a k-path, the subgraph $G[V(T_v)]$ induced by the vertex set of T_v, may due to the existence of non-tree edges. So let us call a node v *safe* if $V(T_v)$ does not induce a k-path in G, and *unsafe* otherwise (i.e, if $G[V(T_v)]$ contains a k-path). Let S and U denote the sets of safe and unsafe nodes, respectively.

Our algorithm simply returns U as a solution for k-CPVC, and the next lemma confirms that U is a valid solution.

Lemma 1. *The set U of all the unsafe nodes in G is a connected k-pvc for G.*

Proof. Clearly, U is connected and in fact, the subtree $T[U]$ of T induced by U is connected because every ancestor of an unsafe node is unsafe. So, it suffices to show U is a k-pvc for G, and to do so, consider the set $\bar{S}$ of such safe nodes that are not descendants of any other safe nodes. Then, the nodes remaining after U is removed from G is exactly those nodes that are descendant of some node in $\bar{S}$, i.e., the union of $V(T_s)$'s for $s \in \bar{S}$. By definition of safe nodes each of $G[V(T_s)]$ is a k-path free. Besides, no edge can exist between T_s and $T_{s'}$ for $s, s' \in \bar{S}$ if $s \neq s'$ since T is a dfs tree. Therefore, the subgraph induced by all the remaining nodes, $\bigcup_{s \in \bar{S}} V(T_s)$, does not induce a k-path, and U is a feasible solution for k-CPVC.   □

So simple as it is, the algorithm can be seen as a natural generalization of the 2-approximation algorithm for connected vertex cover. The best known algorithm for CVC, due to Savage [26], works as follows. It first computes a dfs spanning tree T of a given graph, and then returns the set of all the non-leaves of T as a solution. In CVC ($\equiv$2-CPVC) a node v of T is safe iff $V(T_v)$ does not induce an edge iff v is a leaf of T. Thus, what the Savage's algorithm does is exactly to return the set of unsafe nodes.

A path on rooted T is called *straight* if one end is an ancestor of the other. Let $\mathcal{P}$ denote a packing of straight k-paths on T (i.e., a set of vertex disjoint straight k-paths on T). A node v in T is *covered* (by $\mathcal{P}$) if $v \in V(P)$ for some $P \in \mathcal{P}$, and it is *uncovered* otherwise. We say that $\mathcal{P}$ is *upper maximal*[1] if

---

[1] A related notion of *proper matchings* was used in [26].

1. it is a "maximal" packing (i.e., there remains no straight k-path on T disjoint from any in $\mathcal{P}$), and
2. all the ancestors of any covered node are covered.

Notice that the 2nd condition above implies that all the descendants of any uncovered node must be uncovered. An upper maximal packing of straight k-paths can be inductively constructed as follows: Suppose $\mathcal{P}$ satisfies the 2nd condition. If it is not maximal there must be a straight path P disjoint from any in $\mathcal{P}$. Let $v(P)$ be the "highest" ancestor of (any node of) P that is uncovered by $\mathcal{P}$ so that any "proper" ancestor of $v(P)$ is covered while all the descendants uncovered. Then, one can find a straight k-path starting at $v(P)$ and consisting of uncovered vertices only; add it to $\mathcal{P}$ and repeat as long as $\mathcal{P}$ is not maximal.

We say an unsafe node is *boundary* if all of its children are safe, that is, a "lowest unsafe node", and let $B = \{u \in U \mid u \text{ is boundary}\}$. Observe that T_{b_1} and T_{b_2} are vertex disjoint for all $b_1, b_2 \in B$ if $b_1 \neq b_2$. Let $B_{\mathcal{P}} = B \setminus \cup \mathcal{P}$, where $\cup \mathcal{P} = \bigcup_{P \in \mathcal{P}} V(P)$. We claim the following upper bound on the size of U:

Lemma 2. *Let U denote the set of unsafe nodes in T and $\mathcal{P}$ be an upper maximal packing of straight k-paths on T. Then,*

$$|U| \leq \left| \bigcup_{P \in \mathcal{P}} V(P) \cup \bigcup_{b \in B_{\mathcal{P}}} V(P_b) \right| \leq k|\mathcal{P}| + (k-1)|B_{\mathcal{P}}|,$$

where P_b for $b \in B_{\mathcal{P}}$ is the path on T running between b and the highest ancestor of b uncovered by $\cup \mathcal{P}$.

Proof. Suppose u is an unsafe node not covered by any $P \in \mathcal{P}$. Then, all the descendants of u must be uncovered by $\cup \mathcal{P}$, and at least one of them must be in $B_{\mathcal{P}}$. To cover uncovered unsafe nodes, consider P_b, the path on T between b and the highest ancestor of b uncovered by $\cup \mathcal{P}$. The length of P_b must be at most $k - 2$ since otherwise, it contradicts the maximality of $\mathcal{P}$. It is now easy to see that $\bigcup_{b \in B_{\mathcal{P}}} V(P_b)$ covers all the unsafe nodes not covered by $\cup \mathcal{P}$, and each of P_b's is an l-path for some $l \leq k - 1$. $\qquad\square$

It is also easy to obtain the following lower bound on the size of an optimal k-CPVC solution for G:

Lemma 3. *Any k-CPVC solution for G must be of size no less than $|\mathcal{P}| + |B_{\mathcal{P}}|$.*

Proof. Observe that all of P's in $\mathcal{P}$ and T_b's for $b \in B_{\mathcal{P}}$ are vertex disjoint, each of them containing a k-path. Thus, any k-PVC solution, connected or not, must contain at least one vertex from each of them. $\qquad\square$

Our algorithm for unweighted k-CPVC is now described as follows:

1. Pick any vertex v in G.
2. Compute a dfs spanning tree T of G rooted at v.
3. Compute the set U of unsafe nodes w.r.t. T, and output U.

By Lemmas 2 and 3, the size of U is no larger than that of any solution by a factor of k, and hence,

Theorem 1. *The algorithm above approximates unweighted k-CPVC within k.*

So the performance ratio of this algorithm is k and it is asymptotically tight as shown by the following example. Let $Q = \langle q_1, q_2, \cdots, q_{(l+1)k} \rangle$ be a simple cycle on $(l+1)k$ vertices, and construct a graph $\tilde{Q}$ by inserting l edges to Q, the ith edge between q_1 and q_{ik+1} for $i = 1, \cdots, l$. Observe here that, for any choice of the root, the dfs spanning tree to be computed could be a single path consisting only of edges of Q, which leads to $lk+1$ many nodes being categorized unsafe. So, the output could be of size $lk + 1$ whereas $\{q_1, q_{k+1}, \cdots, q_{lk+1}\}$ is a solution for k-CPVC ($(l + 1)$ many nodes). Hence, the ratio of a solution size computed by the algorithm to the optimal size could be as large as

$$\frac{lk+1}{l+1} = \frac{k(l+1) - (k-1)}{l+1} = k - \frac{k-1}{l+1},$$

which could become arbitrarily close to k.

3   Weighted Connected Path Vertex Cover

We consider here weighted k-CPVC, where vertices are arbitrarily weighted in a given graph and the goal is to minimize the weight of a solution for k-CPVC. To be observed first is that the weighted set cover problem can be reduced to this problem in an approximation preserving manner:

Theorem 2. *It is as hard to approximate weighted k-CPVC with $k \geq 2$ as the weighted set cover problem.*

Proof. Let $(U, \mathcal{S}, w)$ denote an instance of the weighted set cover problem, where U is the base set and $\mathcal{S} \subseteq 2^U$ is a family of subsets of U. We'll show how to reduce it to a weighted k-CPVC instance. Let H_u denote a copy of k-path for each $u \in U$. Let G be a graph with vertex set $\{c\} \cup U \cup \mathcal{S} \cup \bigcup_{u \in U} V(H_u)$, where c is a new vertex. Connect c with each of $S \in \mathcal{S}$ by an edge, and have an edge between $u \in U$ and $S \in \mathcal{S}$ iff $u \in S$. For each $u \in U$, let x_u be any vertex in $V(H_u)$, and attach H_u to $u \in U$ by superimposing x_u on u for each $u \in U$. Now the construction of a graph G is complete and define $w_G : V(G) \to \mathbb{Q}_+$ s.t. $w_G(S) = w(S)$ for each $S \in \mathcal{S}$ and $w_G(v) = 0$ for all the other vertices v.

For a vertex subset V' of G let $\Gamma(V')$ denote the set of vertices adjacent to a vertex in V'. Clearly, $\mathcal{S}' \subseteq \mathcal{S}$ is a set cover for $(U, \mathcal{S}, w)$ iff $U \subseteq \Gamma(\mathcal{S}')$ in G, and moreover, for any set cover $\mathcal{S}'$, $C = \mathcal{S}' \cup U \cup \{c\}$ is a connected k-pvc for G of the same weight since no k-path can remain in $G - C$. On the other hand, any connected k-pvc C for G must contain each $u \in U$ to hit H_u, and, in order for C to be connected on top of that, $C \cap \mathcal{S}$ must be a set cover for $(U, \mathcal{S}, w)$. Whereas $U \cup (C \cap \mathcal{S})$ could be already connected in G, having c in it costs nothing assuring its connectivity. Thus, it can be assumed that any connected k-pvc for G is of the form $\mathcal{S}' \cup U \cup \{c\}$ s.t. $\mathcal{S}'$ is a set cover for $(U, \mathcal{S}, w)$, with its weight equaling to that of $\mathcal{S}'$. Therefore, any algorithm approximating k-CPVC within a factor r can be used to compute a set cover of weight at most r times the optimal weight.   □

Due to the non-approximability of set cover [8], it follows that

Corollary 1. *The weighted k-CPVC cannot be approximated within a factor better than $(1 - \epsilon) \ln n$ for any $\epsilon > 0$, unless $NP \subset DTIME(n^{O(\log \log n)})$.*

Definition 1 (Node Weighted Group Steiner Tree). *Suppose that we are given a graph G with node-weight function $w_N : V(G) \to \mathbb{R}_+$ and a family of subsets of vertices $\mathcal{G} = \{g_1, g_2, \cdots, g_k\}, g_i \subseteq V$ which will be called groups. We have to find subtree T that minimizes cost function $\sum_{v \in V(T)} w_N(v)$ such that $V(T) \cap g_i \neq \emptyset$ for all $i \in \{1, \cdots, k\}$.*

The weighted k-CPVC can be seen reducible to the *node-weighted group Steiner tree* problem by taking the vertex set of each k-path existing in G as a group g. No polylogarithmic approximation is known for node-weighted group Steiner trees, however, and the best approximation guarantee known so far is $O(\sqrt{n} \log n)$ in polynomial time [17] or polylogarithmic in quasi-polynomial running time [22].

Let us recall that weighted 2-CPVC, better known under the more familiar name of CONNECTED VERTEX COVER, is approximable within $\ln n + 3$ [10]. One naive approach for approximating weighted k-CPVC suggested by the weighted CVC approximation is to compute first a k-pvc $C \subseteq V$ for $G = (V, E)$, within a reasonable approximation factor, and then to augment C to become connected, using additional vertices. It can be seen that, without further consideration, this strategy is too naive to be of any use as exemplified by the following simple instance of k-CPVC for $k \geq 3$. Let $P = \langle p_1, p_2, \cdots, p_{k+1} \rangle$ be a $(k + 1)$-path with weights $w(p_1) = w(p_{k+1}) = 1$ and $w(p_i) = w_i > 1$, $\forall i = 2, \cdots, k$. Since $w_2, \cdots, w_k$ can be arbitrarily large than 1, $C = \{p_1, p_4\}$ is the only choice for a k-pvc to be of bounded approximation. To make it connected, all of $p_2, \cdots, p_k$ need to be added, but then, it could be arbitrarily worse than the optimal solution consisting of a single vertex p_i of minimum weight within $\{p_2, \cdots, p_k\}$.

To circumvent this issue for 3-CPVC, we compute a 3-pvc C with more care. A 3-CPVC solution is then computed from C by approximating the node weighted Steiner tree problem with a terminal set C.

Definition 2 (Node Weighted Steiner Tree). *Suppose that we are given a graph G with node-weight function $w_N : V(G) \to \mathbb{R}_+$ and a set $Z \subseteq V$ of vertices called "terminals". We have to find a tree T in G that minimizes cost function $\sum_{v \in V(T)} w_N(v)$ such that $Z \subseteq V(T)$.*

The algorithm for 3-CPVC is now described as follows:

1. If G is a 3-path then return a singleton set $\{v\}$ of minimum weight vertex $v \in V(G)$ (and exist).
2. Let X be the set of vertices of degree 1 in G.

3. Compute a 3-pvc C s.t. $C \cap X = \emptyset$.
4. Compute a node weighted Steiner tree T, where terminal set is C, $w_N(v) = 0$ if $v \in C$ and $w_N(v) = w(v)$ if $v \in V - C$.
5. Output $V(T)$.

Let us elaborate more on the algorithm. Observe that it is redundant to have a degree 1 vertex, together with other vertices, in a k-CPVC solution, and hence, we may consider only such a solution with no degree 1 vertices in it or the one consisting only of a single degree 1 vertex. A singleton set of a degree 1 vertex can be a minimal solution iff a graph is a k-path, and that case is taken care of in step 1. After step 1 we may assume that any minimal k-CPVC solution does not include a degree 1 vertex. By excluding any degree 1 nodes, although bounded approximation can no longer be guaranteed for k-PVC, hitting only $V(P) - X$ rather than $V(P)$ for each k-path P is a valid requirement for k-CPVC and we have the following LP relaxation for 3-CPVC, where $\mathcal{P}_3$ is the set of 3-paths and X is the set of degree 1 vertices in G:

$$\min \sum_{v \in V} w(v)x_v$$
$$\text{(LP) subject to} : x(V(P) - X) \geq 1, \forall P \in \mathcal{P}_3$$
$$x_v \geq 0, \qquad \forall v \in V$$

By using the standard LP methods, rounding or primal-dual, based on (LP), one can compute a 3-PVC solution $C \subseteq V$ s.t. C contains no degree 1 vertex in it and $w(C)$ is within a factor of 3 from the minimum 3-CPVC solution (*not* from the minimum 3-PVC solution). In step 4 we compute a Steiner tree T s.t. $C \subseteq V(T)$. The node weighted Steiner tree problem is known approximable within a factor approaching $1.35 \ln k$, where k is the number of terminals [12].

Theorem 3. *The algorithm given above approximates weighted 3-CPVC within $1.35 \ln n + 3$.*

Proof. Let C^* denote the minimum weight 3-CPVC solution for G. As explained above, we have $w(C) \leq 3w(C^*)$ for C computed in step 3. We claim that $C \cup C^*$ induces a connected subgraph in G. Certainly $G[C^*]$ is connected and $G[V - C^*]$ consists of disjoint 2-paths and/or isolated vertices. Recall that C contains no vertex of degree 1, and hence, any vertex in $C - C^*$ must be adjacent to a vertex in C^*, which implies that $G[C \cup C^*]$ is connected. One can thus find a tree spanning $C \cup C^*$ only, which is a Steiner tree with terminal set C. Using the Guha-Khuller algorithm for computing T in our algorithm, we have

$$w(V(T) - C) \leq (1.35 \ln n)w(C^* - C) \leq (1.35 \ln n)w(C^*).$$

It follows that $w(V(T)) \leq (3 + 1.35 \ln n)w(C^*)$. $\square$

Acknowledgments. The author is grateful to the anonymous referees for a number of valuable comments and suggestions. This work is supported in part by JSPS KAKENHI under Grant Numbers 26330010 and 17K00013.

References

1. Arkin, E.M., Halldórsson, M.M., Hassin, R.: Approximating the tree and tour covers of a graph. Inf. Process. Lett. **47**(6), 275–282 (1993). https://doi.org/10.1016/0020-0190(93)90072-H
2. Bar-Yehuda, R., Even, S.: A local-ratio theorem for approximating the weighted vertex cover problem. In: Analysis and Design of Algorithms for Combinatorial Problems (Udine, 1982). North-Holland Mathematics Studies, vol. 109, pp. 27–45, North-Holland, Amsterdam (1985). https://doi.org/10.1016/S0304-0208(08)73101-3
3. Brešar, B., Krivoš-Belluš, R., Semanišin, G., Šparl, P.: On the weighted k-path vertex cover problem. Discrete Appl. Math. **177**, 14–18 (2014). https://doi.org/10.1016/j.dam.2014.05.042
4. Brešar, B., Jakovac, M., Katrenič, J., Semanišin, G., Taranenko, A.: On the vertex k-path cover. Discrete Appl. Math. **161**(13–14), 1943–1949 (2013). https://doi.org/10.1016/j.dam.2013.02.024
5. Brešar, B., Kardoš, F., Katrenič, J., Semanišin, G.: Minimum k-path vertex cover. Discrete Appl. Math. **159**(12), 1189–1195 (2011). https://doi.org/10.1016/j.dam.2011.04.008
6. Camby, E., Cardinal, J., Chapelle, M., Fiorini, S., Joret, G.: A primal-dual 3-approximation algorithm for hitting 4-vertex paths. In: 9th International Colloquium on Graph Theory and Combinatorics (ICGT 2014), p. 61 (2014)
7. Dinur, I., Safra, S.: The importance of being biased. In: Proceedings of the Thirty-Fourth Annual ACM Symposium on Theory of Computing, pp. 33–42. ACM, New York (2002). https://doi.org/10.1145/509907.509915
8. Feige, U.: A threshold of $\ln n$ for approximating set cover. J. ACM **45**(4), 634–652 (1998). https://doi.org/10.1145/285055.285059
9. Fujito, T.: A unified approximation algorithm for node-deletion problems. Discrete Appl. Math. **86**(2–3), 213–231 (1998). https://doi.org/10.1016/S0166-218X(98)00035-3
10. Fujito, T.: On approximability of the independent/connected edge dominating set problems. Inf. Process. Lett. **79**(6), 261–266 (2001). https://doi.org/10.1016/S0020-0190(01)00138-7
11. Garey, M.R., Johnson, D.S.: The rectilinear Steiner tree problem is NP-complete. SIAM J. Appl. Math. **32**(4), 826–834 (1977). https://doi.org/10.1137/0132071
12. Guha, S., Khuller, S.: Improved methods for approximating node weighted Steiner trees and connected dominating sets. Inf. Comput. **150**(1), 57–74 (1999). https://doi.org/10.1006/inco.1998.2754
13. Halperin, E.: Improved approximation algorithms for the vertex cover problem in graphs and hypergraphs. SIAM J. Comput. **31**(5), 1608–1623 (2002). https://doi.org/10.1137/S0097539700381097
14. Jakovac, M., Taranenko, A.: On the k-path vertex cover of some graph products. Discrete Math. **313**(1), 94–100 (2013). https://doi.org/10.1016/j.disc.2012.09.010
15. Karakostas, G.: A better approximation ratio for the vertex cover problem. ACM Trans. Algorithms **5**(4), 8 (2009). https://doi.org/10.1145/1597036.1597045. Article 41
16. Kardoš, F., Katrenič, J., Schiermeyer, I.: On computing the minimum 3-path vertex cover and dissociation number of graphs. Theor. Comput. Sci. **412**(50), 7009–7017 (2011). https://doi.org/10.1016/j.tcs.2011.09.009

17. Khandekar, R., Kortsarz, G., Nutov, Z.: Approximating fault-tolerant group-Steiner problems. Theor. Comput. Sci. **416**, 55–64 (2012). https://doi.org/10.1016/j.tcs.2011.08.021
18. Lee, E.: Partitioning a graph into small pieces with applications to path transversal. In: Proceedings of the Twenty-Eighth Annual ACM-SIAM Symposium on Discrete Algorithms (SODA 2017), pp. 1546–1558. Society for Industrial and Applied Mathematics, Philadelphia (2017)
19. Li, X., Zhang, Z., Huang, X.: Approximation algorithms for minimum (weight) connected k-path vertex cover. Discrete Appl. Math. **205**, 101–108 (2016). https://doi.org/10.1016/j.dam.2015.12.004
20. Liu, X., Lu, H., Wang, W., Wu, W.: PTAS for the minimum k-path connected vertex cover problem in unit disk graphs. J. Glob. Optim. **56**(2), 449–458 (2013). https://doi.org/10.1007/s10898-011-9831-x
21. Monien, B., Speckenmeyer, E.: Ramsey numbers and an approximation algorithm for the vertex cover problem. Acta Inf. **22**(1), 115–123 (1985). https://doi.org/10.1007/BF00290149
22. Naor, J., Panigrahi, D., Singh, M.: Online node-weighted Steiner tree and related problems. In: 2011 IEEE 52nd Annual Symposium on Foundations of Computer Science–FOCS 2011, pp. 210–219. IEEE Computer Society, Los Alamitos (2011). https://doi.org/10.1109/FOCS.2011.65
23. Novotný, M.: Design and analysis of a generalized canvas protocol. In: Samarati, P., Tunstall, M., Posegga, J., Markantonakis, K., Sauveron, D. (eds.) WISTP 2010. LNCS, vol. 6033, pp. 106–121. Springer, Heidelberg (2010). https://doi.org/10.1007/978-3-642-12368-9_8
24. Orlovich, Y., Dolgui, A., Finke, G., Gordon, V., Werner, F.: The complexity of dissociation set problems in graphs. Discrete Appl. Math. **159**(13), 1352–1366 (2011). https://doi.org/10.1016/j.dam.2011.04.023
25. Safina Devi, N., Mane, A.C., Mishra, S.: Computational complexity of minimum P_4 vertex cover problem for regular and $K_{1,4}$-free graphs. Discrete Appl. Math. **184**, 114–121 (2015). https://doi.org/10.1016/j.dam.2014.10.033
26. Savage, C.: Depth-first search and the vertex cover problem. Inf. Process. Lett. **14**(5), 233–235 (1982). https://doi.org/10.1016/0020-0190(82)90022-9
27. Tu, J.: A fixed-parameter algorithm for the vertex cover P_3 problem. Inf. Process. Lett. **115**(2), 96–99 (2015). https://doi.org/10.1016/j.ipl.2014.06.018
28. Tu, J., Yang, F.: The vertex cover P_3 problem in cubic graphs. Inf. Process. Lett. **113**(13), 481–485 (2013). https://doi.org/10.1016/j.ipl.2013.04.002
29. Tu, J., Zhou, W.: A factor 2 approximation algorithm for the vertex cover P_3 problem. Inf. Process. Lett. **111**(14), 683–686 (2011). https://doi.org/10.1016/j.ipl.2011.04.009
30. Tu, J., Zhou, W.: A primal-dual approximation algorithm for the vertex cover P_3 problem. Theor. Comput. Sci. **412**(50), 7044–7048 (2011). https://doi.org/10.1016/j.tcs.2011.09.013
31. Yannakakis, M.: Node-deletion problems on bipartite graphs. SIAM J. Comput. **10**(2), 310–327 (1981). https://doi.org/10.1137/0210022

Improved PTASs for Convex Barrier Coverage

Paz Carmi, Matthew J. Katz$^{(\boxtimes)}$, Rachel Saban, and Yael Stein

Department of Computer Science, Ben-Gurion University of the Negev,
8410501 Beer-Sheva, Israel
carmip@gmail.com, {matya,rachelfr,shvagery}@cs.bgu.ac.il

Abstract. Let R be a connected closed region in the plane and let S be a set of n points (representing mobile sensors) in the interior of R. We think of R's boundary as a barrier which needs to be monitored. This gives rise to the barrier coverage problem, where one needs to move the sensors to the boundary of R, so that every point on the boundary is covered by one of the sensors. We focus on the variant of the problem where the goal is to place the sensors on R's boundary, such that the distance (along R's boundary) between any two adjacent sensors is equal to R's perimeter divided by n and the sum of the distances traveled by the sensors is minimum. In this paper, we consider the cases where R is either a circle or a convex polygon. We present a PTAS for the circle case and explain how to overcome the main difficulties that arise when trying to adapt it to the convex polygon case. Our PTASs are significantly faster than the previous ones due to Bhattacharya et al. [4]. Moreover, our PTASs require efficient solutions to problems, which, as we observe, are equivalent to the circle-restricted and line-restricted Weber problems. Thus, we also devise efficient PTASs for these Weber problems.

1 Introduction

Given n sensors in the interior of a planar connected and closed region R, we consider the problem of moving the sensors to R's boundary to form a *barrier* in the sense that every point on R's boundary is covered by one of the sensors, where the objective is to minimize the total travel distance of the sensors. This problem arises when, for example, one uses mobile sensors to protect a region by monitoring its boundary and detecting intruders. In this paper we focus on the case where the underlying region is convex and in particular a disk. We present a PTAS for the disk version, which is significantly more efficient than the previous one. Moreover, we observe an interesting connection between a subproblem that must be addressed as part of our PTAS and a new Weber-type problem, for which we also give an efficient PTAS. Towards the end, we briefly discuss the convex polygon version.

R. Saban and Y. Stein were partially supported by the Lynn and William Frankel Center for Computer Sciences. M. Katz was partially supported by grant 1884/16 from the Israel Science Foundation.

We use the concept of *barrier coverage*, as defined in [6,10], rather than the more common concept of *full coverage*. While full coverage guarantees that the entire region is covered by sensors, barrier coverage only guarantees that its boundary is covered by sensors. Barrier coverage can be achieved with fewer sensors, and it is sufficient when the goal is, e.g., to detect attempts to penetrate into the region, a spread of a fire, or a leakage of toxic materials.

Even when barrier coverage is sufficient, the total cost of positioning the sensors, which is proportional to the total distance traveled, is an important issue. Thus, following Bhattacharya et al. [4] and Chen et al. [8], we consider the following two *barrier coverage* problems. Let C be a circle centered at the origin o, and let $S = \{s_1, \ldots, s_n\}$ be a set of n sensors (represented as points) in the interior of C.

Problem CIRCB: Move each sensor in S to a point on C, such that (i) the resulting set of points on C defines a regular n-gon, and (ii) the total distance traveled by the sensors is minimum, i.e., the sum $\sum_{i=1}^n |s_i s_i'|$ is minimum, where s_i' is the point to which s_i was moved and $|s_i s_i'|$ is the Euclidean distance between s_i and s_i'.

Similarly, let P be (the boundary of) a simple polygon, and let $S = \{s_1, \ldots, s_n\}$ be a set of n sensors in the interior of P.

Problem POLYB: Move each sensor in S to a point on P, such that (i) the distance along P between any two consecutive points is P's length divided by n, and (ii) the total distance traveled by the sensors is minimum, i.e., the sum $\sum_{i=1}^n |s_i s_i'|$ is minimum, where s_i' is the point to which s_i was moved.

It is unknown whether these problems are NP-hard. However, by establishing a connection between them and the circle-restricted Weber problem (for CIRCB) and the line-restricted Weber problem (for POLYB), see below, we tend to think that they are hard in the sense that the Weber problem and the line-restricted Weber problem are hard; see Bajaj [3].

1.1   Previous Work and Our Results

Bhattacharya et al. [4] studied both problems. For CIRCB they give an $O(n^2)$-time approximation algorithm with approximation ratio $1 + \pi$. Later, Chen et al. [8] improved this result by presenting an $O(n^2)$-time approximation algorithm with approximation ratio 3. Bhattacharya et al. also present a PTAS for CIRCB with running time $O(\frac{n^4}{\varepsilon})$, which can be reduced to $O(\frac{n^{3+\varepsilon'}}{\varepsilon})$, for any constant $\varepsilon' > 0$, by using the minimum-weight bipartite Euclidean matching algorithm of Agarwal et al. [1] instead of the Hungarian method. For POLYB, Bhattacharya et al. give a PTAS with running time $O(\frac{mn^5}{\varepsilon})$, which can be reduced (as above) to $O(\frac{mn^{4+\varepsilon'}}{\varepsilon})$. Here m is the number of vertices of the underlying polygon (which is not necessarily convex).

Bhattacharya et al. and Chen et al. also consider the corresponding min-max problems, where the objective is to minimize the maximum distance traveled by a sensor. For a circle, they present exact algorithms, where the faster one by Chen et al. runs in $O(n \log^3 n)$ time. For a simple polygon, Bhattacharya et al. give an exact $O(mn^{3.5} \log n)$-time algorithm.

In a related interesting problem, which received considerable attention, the barrier is the interval $B = [0, 1]$ and the sensors are located on the x-axis. The goal is to move the sensors so that every point in B is covered and the sum (alternatively, the maximum) of the moving distances is minimized. For the minsum problem (assuming n sensors, each of sensing range r), Andrews and Wang [2] presented an $O(n \log n)$-time algorithm. For the minmax problem (assuming n sensors, each with its own sensing range), Chen et al. [7] gave a clever $O(n^2 \log n)$-time algorithm. Other related problems have also been studied, see e.g. [9].

Our results. Our main result is a new and sophisticated PTAS for CIRCB with running time $O(\frac{1}{\varepsilon^{O(1)}} n^{2+\varepsilon'})$; an order of magnitude faster than the previous PTAS. Moreover, we show that one can compute a $(1 + \varepsilon)$-approximation for CIRCB in $O(n \operatorname{poly}(\log n, 1/\varepsilon))$ time with high probability.

Our PTAS requires us to address the following problem which we refer to as OPTPOS: Let C and S be as above, and let Q be a regular n-gon whose vertices lie on C. Let M be a perfect matching between the points in S and the vertices of Q. Find a position of Q (by rotating Q about the origin o) for which M's value is minimum, where M's value is the sum of the lengths of its edges. We observe that this problem is equivalent to the following Weber-type problem, which we refer to as WEBERC: Let C and S be as above, find a point p^* on C for which the sum of the distances from the points in S to p^* is minimum. We devise an $O(\frac{1}{\varepsilon^{O(1)}} n \log n)$-time PTAS for the latter problem.

The main ideas behind the PTAS for CIRCB are also relevant for devising an efficient PTAS for POLYB, assuming the underlying polygon is convex. However, some additional difficulties occur in this case. We discuss these difficulties and how to cope with them, but leave the tedious and technical details to the full version of this paper.

2 A PTAS for CIRCB

Let C be a circle centered at the origin o, and let $S = \{s_1, \ldots, s_n\}$ be a set of n points in the interior of C. In this section we present an efficient PTAS for CIRCB. We begin with a couple of definitions.

For a point q on C, let $\theta_q, 0 \le \theta_q < 2\pi$, be the angle between the x-axis and oq (see Fig. 1(a)). We say that θ_q is the angle *corresponding* to q. Let Q be a regular n-gon on C and denote its vertices in counterclockwise order $q_1, \ldots, q_n$, where q_1 is the vertex whose corresponding angle θ_{q_1} is the smallest. Notice that $0 \le \theta_{q_1} < \frac{2\pi}{n}$ and that Q is uniquely determined by the angle θ_{q_1}. We say that Q is in *position* θ_{q_1} (see Fig. 1(b)).

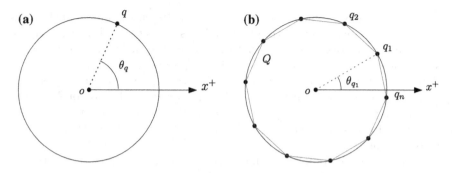

Fig. 1. (a) θ_q is the angle *corresponding* to q. (b) Q is in *position* θ_{q_1}.

2.1 Algorithm Description

We assume w.l.o.g. that the radius of C is such that the edge length of a regular n-gon on C is 1. Let $1 < (1 + \varepsilon) < 2$ be the desired approximation ratio and set $\delta = \frac{\varepsilon}{3(6+\varepsilon)}$. Let x_θ be the point on C whose corresponding angle is θ. Place $O(\frac{1}{\delta^2})$ points $p_1, p_2, \ldots, p_k$ on the arc of C between x_0 and $x_{\frac{2\pi}{n}}$, such that the distance between p_i and p_{i+1}, for $i = 1, \ldots, k-1$, as well as the distance between x_0 and p_1 and the distance between p_k and $x_{\frac{2\pi}{n}}$ is δ^2. Denote the angle corresponding to p_i by θ_i (instead of θ_{p_i}).

Let Q_i denote the regular n-gon Q in position θ_i. That is, q_1 coincides with p_i and the angle corresponding to q_j, the j'th vertex of Q_i, is $\theta_i + (j-1)\frac{2\pi}{n}$.

For each $1 \leq i \leq k$, we define a matching M_i between S and the vertices of Q_i as follows. First, for each vertex q_j of Q_i, let $s(q_j)$ be the closest point to q_j in S. If $|q_j s(q_j)| \leq \delta + \delta^2$ then add the edge $\{q_j, s(q_j)\}$ to M_i (see Fig. 2). (Notice that if $\{q_j, s(q_j)\}$ is added as an edge to M_i, then there is no other vertex $q_{j'}$ for which $|q_{j'} s(q_j)| \leq \delta + \delta^2$.)

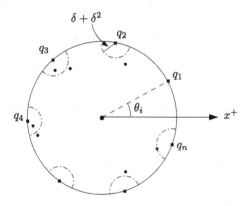

Fig. 2. Here, for each vertex q_j of $Q_i, |q_j s(q_j)| > \delta + \delta^2$, except for q_3, q_4 and q_n.

Next, compute a minimum-weight matching between the unmatched points in S and the unmatched vertices of Q_i and add its edges to M_i. Finally, compute the optimal position of Q w.r.t. M_i, that is, compute the angle ϕ_i, such that when Q is in position ϕ_i the value of M_i is minimum (see Sect. 3.4).

After computing the matchings $M_1, \ldots, M_k$ and the corresponding positions $\phi_1, \ldots, \phi_k$, we compute for each $1 \le i \le k$ the optimal value m_i^* of M_i. That is, for each $1 \le i \le k$, we compute the value of M_i when Q is in position ϕ_i. Return the value $m = \min\{m_1^*, \ldots, m_k^*\}$ and the corresponding matching M and optimal position ϕ.

Actually, since we do not know how to compute the positions ϕ_i exactly, we will approximate the optimal values m_i^* and return the smallest approximation and the matching and position corresponding to it. In Sect. 3.4 we describe how to approximate ϕ_i, and in Sect. 2.3 we show how to still obtain a PTAS.

2.2 The Algorithm's Output Approximates the Optimal Solution

Consider the regular n-gon Q^* defined by the optimal solution to our problem, and let ϕ^* be the angle for which Q^* is Q in position ϕ^*. Let M^* denote the matching between S and the vertices of Q^* induced by the optimal solution. (Obviously, M^* is the minimum weight matching between S and the vertices of Q^*.) Finally, let m^* be the optimal value of M^* (i.e., the value of M^* when Q is in position ϕ^*).

Let θ_j be the angle among $\theta_1, \ldots, \theta_k$ that is closest to ϕ^*. Recall that M_j is the matching defined between S and the vertices of Q_j and that ϕ_j is the optimal position of Q w.r.t. M_j.

Observation 1. For any $s \in S$ and vertex q of Q, set $d_j = |sq|$ w.r.t. Q_j and $d^* = |sq|$ w.r.t. Q^*. Then, (i) $d_j \le d^* + \delta^2$ and (ii) $d^* \le d_j + \delta^2$.

The observation follows from the fact that the distance between q w.r.t. Q_j and q w.r.t. Q^* is at most δ^2 and the triangle inequality.

For each point $s \in S$, we measure its distance to the vertex of Q_j to which it is matched in M_j. We then divide the points in S into two subsets, S_c and S_f, according to these distances.
$S_c = \{s \in S \mid |sq_i| \le \delta + \delta^2 \text{ w.r.t. } Q_j, \text{ where } \{s, q_i\} \in M_j\}$, and
$S_f = \{s \in S \mid |sq_i| > \delta + \delta^2 \text{ w.r.t. } Q_j, \text{ where } \{s, q_i\} \in M_j\}$.
Notice that if $s \in S_c$ and q_i is the vertex of Q_j to which it is matched in M_j, then $s(q_i) = s$.

Claim 1. Let $s \in S_c$ and let q_i be the vertex of Q_j to which s is matched in M_j, then

1. $|sq_i| \le \delta + 2\delta^2$ w.r.t. Q^*.
2. For each $k \ne i, |sq_k| \ge 1 - (\delta + 2\delta^2)$ w.r.t. Q^*.

Proof. The first part follows immediately from Observation 1(ii). Consider the second part, and let $k \ne i, 1 \le k \le n$. Then, w.r.t. $Q_j, |sq_k| \ge \min\{|sq_{i-1}|, |sq_{i+1}|\} \ge 1 - (\delta + \delta^2)$. Therefore, by Observation 1(i), $|sq_k| \ge 1 - (\delta + 2\delta^2)$ w.r.t. Q^*.

We further divide S_c into two subsets S_{c_1} and S_{c_2}.
$S_{c_1} = \{s \in S_c \mid M_j$ and M^* match s to the same vertex of $Q\}$, and
$S_{c_2} = \{s \in S_c \mid M_j$ and M^* match s to different vertices of $Q\}$.
Clearly, $\{S_{c_1}, S_{c_2}, S_f\}$ is a partition of S. Set $n_1 = |S_{c_1}|, n_2 = |S_{c_2}|, n_f = |S_f|$,
and set $x = \delta + 2\delta^2$.

We are now ready to state and prove our main theorem.

Theorem 1. $m_j^* \leq m^*(1 + 2\delta)(1 + \frac{2x}{1-x})$.

The proof consists of three stages. In the first stage, we define a matching M'
between S and the vertices of Q^*, such that, when restricted to S_c, M' and M_j
are identical, that is, for each $s \in S_c, M'$ and M_j match s to the same vertex
of Q. In the second stage, we show that $m' \leq m^*(1 + \frac{2x}{1-x})$, where m' is the
value of M' (when Q is in position ϕ^*). Finally, in the third stage, we show that
$m_j^* \leq m'(1 + 2\delta)$. By combining the last two inequalities we obtain the desired
result.

Stage I: Building M'. Informally, the matching M' is an 'intermediate'
matching, which we define for the purpose of the proof. It is obtained from
(the unknown) M^* by modifying it according to M_j. We build M' using
Algorithm 2.1.

Algorithm 2.1. Building M'

Input: The matchings M^* and M_j
Output: M', a matching between S and the vertices of Q^*, such that for each $s \in S_c$,
$\quad$ M_j and M' match s to the same vertex of Q
1: $M_0' \leftarrow M^*$
2: Let $s_1, \ldots, s_{n_2}$ be the points in S_{c_2}
3: **for** $i \leftarrow 1$ to n_2 **do**
4: $\quad$ Let q_a, q_b be the vertices of Q s.t. s_i is matched to q_a in M_j and to q_b in M_{i-1}'
5: $\quad$ **if** $q_a = q_b$ **then**
6: $\quad\quad$ $M_i' \leftarrow M_{i-1}'$
7: $\quad$ **else**
8: $\quad\quad$ Let $s \in S$ s.t s is matched to q_a in M_{i-1}'
9: $\quad\quad$ $M_i' \leftarrow (M_{i-1}' \setminus \{\{s_i, q_b\}, \{s, q_a\}\}) \cup \{\{s_i, q_a\}, \{s, q_b\}\}$
10: $M' \leftarrow M_{n_2}'$

Lemma 1. *For each $s \in S_c, M_j$ and M' match s to the same vertex of Q.*

Proof. Notice that whenever $q_a \neq q_b$, the two edges of M_{i-1}' that are not passed
on to M_i' are not in M_j. In other words, for any $s \in S_c$, if M_j and M_i' match
s to the same vertex of Q, for some $0 \leq i \leq n_2$, then the matching M' also
matches s to this vertex. Now, let $s \in S_c$. If $s \in S_{c_1}$, then by definition M_j and

$M_0 = M^*$ match s to the same vertex of Q, and if $s \in S_{c_2}$, then the iteration of the algorithm for s ensures that M_j and M'_i match s to the same vertex of Q.

Stage II: Bounding m' w.r.t. m^*. Let m'_i be the value of M'_i, for $i = 0, 1, \ldots, n_2$. Then, $m'_0 = m^*$ and $m'_{n_2} = m'$.

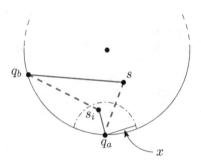

Fig. 3. Proof of Claim 2.

Claim 2. $m'_i \leq m'_{i-1} + 2x$, for $i = 1, \ldots, n_2$.

Proof. If $M'_i = M'_{i-1}$, then $m'_i = m'_{i-1}$ and we are done. Otherwise, $m'_i = m'_{i-1} - (|s_i q_b| + |s q_a|) + (|s_i q_a| + |s q_b|)$, and it is therefore enough to prove that $|s_i q_a| + |s q_b| \leq |s_i q_b| + |s q_a| + 2x$ (see Fig. 3). But, by the triangle inequality, $|s_i q_a| + |s q_b| \leq |s_i q_a| + |s q_a| + |q_a s_i| + |s_i q_b| = |s_i q_b| + |s q_a| + 2|s_i q_a|$, and, by Claim 1(1), $|s_i q_a| \leq x$. Combining these two observations, we obtain the desired inequality, i.e., $|s_i q_a| + |s q_b| \leq |s_i q_b| + |s q_a| + 2x$.

Lemma 2. $m' \leq m^*(1 + \frac{2x}{1-x})$.

Proof. From Claim 2, it follows that $m' \leq m'_0 + 2n_2 x = m^* + 2n_2 x = m^*(1 + \frac{2n_2 x}{m^*})$. Moreover, by Claim 1(2), the edge of M^* corresponding to any $s \in S_{c_2}$ is of length at least $1 - x$, and therefore $m^* \geq n_2(1-x)$. Using the latter inequality, we get that

$$m' \leq m^*(1 + \frac{2n_2 x}{m^*}) \leq m^*(1 + \frac{2xn_2}{n_2(1-x)}) = m^*(1 + \frac{2x}{1-x}),$$

as required.

Stage III: Bounding m^*_j w.r.t. m'.

Claim 3. Let s be any point in S_f and let q be the vertex of Q^* to which s is matched in M'. Then, $|sq| > \delta$.

Proof. From Lemma 1 it follows that M_j and M' match S_f to the same subset of vertices of Q. Therefore, the point $s' \in S$ that is matched to q in M_j also belongs to S_f. Assume now that $|sq| \leq \delta$ w.r.t. Q^*. By Observation 1, $|sq| \leq \delta + \delta^2$ w.r.t. Q_j, and in particular $|s(q)q| \leq \delta + \delta^2$ w.r.t. Q_j. (Recall that for a vertex q of Q_j, $s(q)$ is the closest point to q in S.) From M_j's construction it follows that $s' = s(q)$ and $|s'q| \leq \delta + \delta^2$. Hence $s' \in S_c$—contradiction.

Lemma 3. $m_j^* \leq m'(1 + 2\delta)$.

Proof. Let m_j be the value of M_j w.r.t. Q_j, and let m_{jc} and m_{jf} be the values of $\{\{s,q\} \in M_j | s \in S_c\}$ and $\{\{s,q\} \in M_j | s \in S_f\}$, respectively, w.r.t. Q_j (i.e., $m_j = m_{jc} + m_{jf}$). We define m_c' and m_f' in a similar manner (such that $m' = m_c' + m_f'$). Moreover, let $m_j(Q^*)$ (alternatively, $m'(Q_j)$) denote the value of M_j w.r.t. Q^* (resp., of M' w.r.t. Q_j).

From Lemma 1, we know that $m_{jc} = m_c'(Q_j)$ and $m_c' = m_{jc}(Q^*)$. Moreover, as mentioned above, this lemma also implies that M_j and M' match S_f to the same subset of vertices of Q. Since M_j matches S_f to this subset by computing a minimum weight matching, it holds that $m_{jf} \leq m_f'(Q_j)$. Also, by Observation 1,

$$m_{jf}(Q^*) \leq m_{jf} + n_f\delta^2 \leq m_f'(Q_j) + n_f\delta^2 \leq m_f' + 2n_f\delta^2.$$

Therefore,

$$m_j^* \leq m_j(Q^*) = m_{jc}(Q^*) + m_{jf}(Q^*) = m_c' + m_{jf}(Q^*) \leq m_c' + m_f' + 2n_f\delta^2 = m' + 2n_f\delta^2.$$

Finally, from Claim 3 it follows that $m_f' \geq n_f\delta$. Hence

$$m_j^* \leq m' + 2n_f\delta^2 = m'(1 + \frac{2n_f\delta^2}{m'}) \leq m'(1 + \frac{2n_f\delta^2}{n_f\delta}) = m'(1 + 2\delta).$$

Now, by Lemmas 2 and 3, we obtain that

$$m_j^* \leq m^*(1 + 2\delta)(1 + \frac{2x}{1 - x}),$$

thus completing the proof of Theorem 1. Since $m \leq m_j^*$, we conclude also that

$$m \leq m^*(1 + 2\delta)(1 + \frac{2x}{1 - x}).$$

2.3   Implementation and Analysis

We have assumed that we can compute the optimal positions ϕ_i and the corresponding values m_i^* exactly. However, we do not know how to do this. Moreover, it is unlikely that it is possible to do this, due to the connection between the problem of computing the optimal position of Q w.r.t. M_i and the Weber problem (see Sect. 3.4). Instead, we apply Theorem 4 below to approximate the values m_i^*. More precisely, for each $1 \leq i \leq k$, we compute a position of Q, such that the value of M_i for this position is at most $(1 + \frac{\delta}{2})$ times m_i^*; denote this value by $\widetilde{m_i^*}$.

We now show that $\widetilde{m_j^*} \le m^*(1+\varepsilon)$. By Theorem 1 we have

$$\widetilde{m_j^*} \le m_j^*(1+\frac{\delta}{2}) \le m^*(1+2\delta)(1+\frac{2x}{1-x})(1+\frac{\delta}{2}) = m^*(1+\frac{5}{2}\delta+\delta^2)(1+\frac{2x}{1-x}).$$

Notice that (i) $x \le 3\delta$ and (ii) $\frac{5}{2}\delta + \delta^2 \le 3\delta$ (since $\delta = \frac{\varepsilon}{3(6+\varepsilon)} \le \frac{1}{18}$). Thus,

$$(1+\frac{5}{2}\delta+\delta^2)(1+\frac{2x}{1-x}) \le \frac{(1+3\delta)^2}{1-3\delta} \le (\frac{1+3\delta}{1-3\delta})^2 = (1+\frac{6\delta}{1-3\delta})^2.$$

Since $(1+a)^2 \le 1+3a$ for any $0 \le a \le 1$, we get

$$(1+\frac{6\delta}{1-3\delta})^2 \le 1 + \frac{18\delta}{1-3\delta} \le (1+\varepsilon).$$

Hence $\widetilde{m_j^*} \le m^*(1+\varepsilon)$.

The running time of the above scheme is determined by the time needed to compute the $O(\frac{1}{\varepsilon^2})$ matchings $M_1, \ldots, M_k$ and the corresponding positions $\phi_1, \ldots, \phi_k$. Let $T(n)$ be the time needed to compute a minimum-weight bipartite Euclidean matching. Then, clearly, one can compute M_i in $O(n \log n + T(n))$ time, for $i = 1, \ldots, k$. By Theorem 4, one can compute ϕ_i (approximately) in $O(\frac{1}{\varepsilon^3} n \log n)$ time, for $i = 1, \ldots, k$. Currently, $T(n) = O(n^{2+\varepsilon'})$, see Agarwal et al. [1]. Sharathkumar and Agarwal [11] also presented a Monte Carlo algorithm that computes in $O(n \, \mathrm{poly}(\log n, 1/\varepsilon'))$-time a bipartite Euclidean matching of weight at most $(1+\varepsilon')$ times the minimum weight with high probability. We can easily adapt our scheme to use this algorithm. The following theorem summarizes the main result of this section.

Theorem 2. *For any $\varepsilon > 0$, (i) one can compute a $(1+\varepsilon)$-approximation for* CIRCB *in $O(\frac{1}{\varepsilon^2}n^{2+\varepsilon'} + \frac{1}{\varepsilon^5}n \log n) = O(\frac{1}{\varepsilon^{O(1)}}n^{2+\varepsilon'})$ time, for any $\varepsilon' > 0$, and (ii) one can compute a $(1+\varepsilon)$-approximation for* CIRCB *in $O(n \, \mathrm{poly}(\log n, 1/\varepsilon))$ time with high probability.*

3   Computing the Weber Point of S Restricted to C

In this section we present a PTAS for the following Weber-type problem.

Problem WEBERC: Let C be a circle centered at the origin o, and let S be a set of n points in the interior of C. Find a point p^* on C, such that the sum of the distances from p^* to the points in S is minimum, i.e., $\sum_{s \in S} |p^*s| \le \sum_{s \in S} |ps|$, for any point p on C.

3.1   Description

For each $s \in S$, let q_s be the endpoint (on C) of the radius of C passing through s; clearly, q_s is the closest point to s among the points of C. Let $q_1, \ldots, q_n$ be these endpoints in counterclockwise order. For $i = 1, \ldots, n$, let $r_i = \overline{oq_i}$ and let $a_i = \widehat{q_i q_{i+1}}$ be the arc between q_i and q_{i+1}, where $q_{n+1} = q_1$; see Fig. 4.

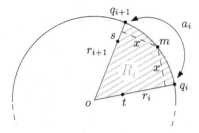

Fig. 4. The sector R_i defined by o, q_i, q_{i+1}. q_i is the endpoint of the radius r_i through $t \in S$ and q_{i+1} is the endpoint of the radius r_{i+1} through $s \in S$. m is the middle point of the arc a_i and x is the distance between m and r_i (and between m and r_{i+1}). The interior of R_i is empty of points of S.

Observation 2. For each i, let R_i be the semi-open sector of C bounded by r_i, r_{i+1} and a_i, where R_i contains a_i but does not contain r_i and r_{i+1}; see Fig. 4. Then R_i is empty of points of S.

Let $\varepsilon > 0$. For each $1 \leq i \leq n$, we find a point p_i on the arc a_i as follows. Denote by m the middle point of a_i, and by x the distance between m and r_i (which is also the distance between m and r_{i+1}). We place $O(\frac{1}{\varepsilon'})$ points $c_1, c_2, \ldots, c_k$ on a_i, for an appropriate parameter $\varepsilon' = O(\varepsilon)$, such that the distance between c_j and c_{j+1} is $\varepsilon'x$, for $j = 0, \ldots, k+1$, where $c_0 = q_i$ and $c_{k+1} = q_{i+1}$. (Notice that the length of a_i is θr, where $\theta = \angle q_i o q_{i+1}$, and $x = r\sin\frac{\theta}{2}$, assuming $\theta \leq \pi$, so the length of a_i is $O(x)$.) For a point p on C, let w_p be the sum $\sum_{s \in S} |ps|$. We compute the values $w_{c_0}, \ldots, w_{c_{k+1}}$. Now, let c_j be the point whose corresponding value is the smallest. We set $p_i = c_j$ and $w_i = w_{c_j}$.

After finding the points $p_1, \ldots, p_n$ on C and their values $w_1, \ldots, w_n$, we return the point $\bar{p} \in \{p_1, \ldots, p_n\}$ whose corresponding value $w_{\bar{p}}$ is the smallest.

3.2   Proof

Let p^* be the point on C for which the sum of distances to the points of S is minimum and set $w^* = w_{p^*}$. We prove that

Lemma 4. $w_{\bar{p}} \leq (1 + \varepsilon')w^*$.

Proof. Let a_j be the arc (among $a_1, \ldots, a_n$) to which p^* belongs. Notice that by $\bar{p}$'s definition, $w_{\bar{p}} \leq w_j$. Therefore, it is enough to prove that $w_j \leq (1 + \varepsilon')w^*$. Assume w.l.o.g. that p^* lies somewhere between m and q_{j+1}. Let c be the closest point to p^* among the points $c_0, \ldots, c_{k+1}$ that are between p^* and q_{j+1}; note that $|cp^*| \leq \varepsilon'x$. Let $h(p^*, m)$ be the halfplane containing p^* defined by the bisector between m and p^*, and let $h(p^*, c)$ be the halfplane containing p^* defined by the bisector between p^* and c. Finally, let $R(p^*)$ be the sector $h(p^*, m) \cap h(p^*, c)$; see Fig. 5.

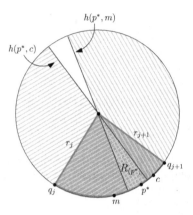

Fig. 5. $R(p^*) = h(p^*, m) \cap h(p^*, c)$. Notice that $R(p^*) \subseteq R_j$.

By Observation 2, for each $s \in S$ we have $|sm| \geq x$. Moreover, since $R(p^*) \subseteq R_j$, we know that $R(p^*)$ is empty of points of S. Therefore, for each $s \in S$ either (i) $|sm| \leq |sp^*|$ or (ii) $|sc| \leq |sp^*|$ (or both). In the former case (i.e., $|sm| \leq |sp^*|$), we get

$$|sc| \leq |sp^*| + \varepsilon'x = |sp^*|(1 + \frac{\varepsilon'x}{|sp^*|}) \leq |sp^*|(1 + \frac{\varepsilon'x}{|sm|}) \leq |sp^*|(1 + \frac{\varepsilon'x}{x}) = (1 + \varepsilon')|sp^*|,$$

and in the latter case (i.e., $|sc| \leq |sp^*|$), we get

$$|sc| \leq |sp^*| \leq (1 + \varepsilon')|sp^*|.$$

Hence,

$$w_j \leq w_c = \sum_{s \in S} |sc| \leq \sum_{s \in S} (1 + \varepsilon')|sp^*| = (1 + \varepsilon')w^*.$$

3.3 Running Time

After sorting the endpoints (on C) of the radii through the points in S, we need to compute the values w_c for $O(\frac{n}{\varepsilon'})$ points c on C. This can be done naively in $O(\frac{n^2}{\varepsilon'})$ time; however, since we are only approximating the Weber point of S (restricted to C), we can afford to approximate the values w_c. Specifically, Bose et al. [5] showed how to preprocess S in $O((1/\delta^2)n \log n)$ time and $O((1/\delta)n)$ space, where $\delta > 0$ is provided by the user, so that given any query point p (not necessarily on C), one can compute a value w'_p in $O((1/\delta^2) \log n)$ time, such that $(1 - \sqrt{2}\delta)w_p \leq w'_p \leq (1 + \sqrt{2}\delta)w_p$. (Their result holds for any constant dimension d.) Thus, instead of computing the values w_c, we compute the values w'_c in total time $O(\frac{1}{\varepsilon'\delta^2}n \log n)$ and choose the points $p_1, \ldots, p_n$ (and $\bar{p}$) in the algorithm above according to these values. By picking $\varepsilon' = O(\varepsilon)$ and $\delta = O(\varepsilon)$ appropriately, we obtain the following theorem.

Theorem 3. *Given* $\varepsilon > 0$, *one can find in* $O(\frac{1}{\varepsilon^3} n \log n)$ *time a point* $\bar{p}$ *on* C, *such that* $w_{\bar{p}} \leq (1 + \varepsilon)w^*$, *where* w^* *is the (exact) value corresponding to the Weber point of* S *restricted to* C.

In the line-restricted Weber problem (WEBERL), we are given a line L, instead of a circle C, and the goal is to find a point p^* on L, such that the sum of the distances from p^* to the points in S is minimum. It is easy to see that our ε-approximation algorithm for WEBERC also works for WEBERL, with obvious adjustments. We thus have.

Corollary 1. *Given* $\varepsilon > 0$, *one can find in* $O(\frac{1}{\varepsilon^3} n \log n)$ *time a point* $\bar{p}$ *on* L, *such that* $w_{\bar{p}} \leq (1 + \varepsilon)w^*$, *where* w^* *is the (exact) value corresponding to the Weber point of* S *restricted to* L.

3.4 Computing the Optimal Position of Q w.r.t M

Consider the following difficulty, which arises when trying to implement the algorithm described in Sect. 2.1. Let C be a circle centered at the origin o, let S be a set of n points in the interior of C, and let Q be a regular n-gon on the boundary of C. Let M be any matching between the points in S and the vertices of Q. Using the notation from Sect. 2, we wish to compute the optimal position of Q w.r.t. M, that is, we wish to compute the angle ϕ, such that when Q is in position ϕ the value of M is minimum. We refer to this problem as OPTPOS.

We show that given an instance of OPTPOS, one can transform it in linear time to an instance of WEBERC such that $m^* = w^*$. Moreover, if $\bar{p}$ is a point on C such that $w_{\bar{p}} \leq (1+\varepsilon)w^*$, then the value of M after rotating Q counterclockwise by $\theta_{\bar{p}}$ is $m = w_{\bar{p}}$ and $m \leq (1 + \varepsilon)m^*$, where $\theta_{\bar{p}}$ is the angle corresponding to $\bar{p}$.

We describe how to reduce OPTPOS to WEBERC[1]. Let a be the right intersection point between the x-axis and C. For each point $s_i \in S$, let s_i' be the point in the interior of C whose location w.r.t. a is the same as the location of s_i w.r.t. q_i; see Fig. 6. That is, s_i' is the point that satisfies (i) $|s_i'a| = |s_iq_i|$, (ii) $\angle oas_i' = \angle oq_is_i$, and (iii) when moving from o to a and then to s_i' one makes a right turn if and only if when moving from o to q_i and then to s_i one makes a right turn. The set $\{s_1', \ldots, s_n'\}$ (together with the circle C) constitute the input to WEBERC.

We thus conclude that

Theorem 4. *Given* $\varepsilon > 0$, *one can find in* $O(\frac{1}{\varepsilon^3} n \log n)$ *time a position* ϕ *of* Q, *such that* $m_\phi \leq (1 + \varepsilon)m^*$.

Proof. The position ϕ is the sum of the initial position of Q and the angle $\theta_{\bar{p}}$.

---

[1] Actually, one can show that the two problems are equivalent, but we are only interested in the one-sided reduction described.

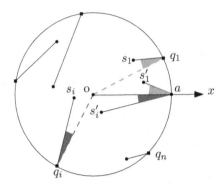

Fig. 6. The location of s_1' w.r.t. a is the same as the location of s_1 w.r.t. q_1.

4 A PTAS for POLYB

4.1 Description

Let P be a convex l-gon, and let $v_1, \ldots, v_l$ be its vertices in counterclockwise order. Given two points p_1 and p_2 on P, we denote the portion of P traversed when moving counterclockwise from p_1 to p_2 by $P_{[p_1,p_2]}$. Let S be a set of n sensors (i.e., points) in the interior of P. We assume w.l.o.g. that the origin o is also in the interior of P and that P's perimeter is n. Let a be the right intersection point between the x-axis and P, and let b be the point on P whose distance along P from a (moving counterclockwise) is 1. Denote by $Q_a = (q_1, \ldots, q_n)$ the following sequence of points on P: $q_1 = a$ and q_i is the point whose distance along P from q_1 (moving counterclockwise) is $i - 1$, for $i = 2, \ldots, n$. Similarly, for any point $p \in P_{[a,b)}$, let $Q_p = (q_1, \ldots, q_n)$ be the sequence of points on P satisfying $q_1 = p$ and q_i is at distance $i - 1$ from q_1 (moving counterclockwise along P), for $i = 2, \ldots, n$.

Given a point $p \in P_{[a,b)}$, let N_p be the minimum weight matching between S and Q_p. We are interested in the point $p^* \in P_{[a,b)}$, for which the value of the corresponding matching is minimum.

At first glance, one may think that the ε-approximation algorithm for CIRCB should also work (with minor modifications) for this problem: I.e., place $O(\frac{1}{\varepsilon^2})$ points on $P_{[a,b)}$; for each such point p_i, construct a matching M_i between S and the points of Q_{p_i}, and find the position of Q at which M_i's value is minimal; return the smallest of these $O(\frac{1}{\varepsilon^2})$ values and its corresponding position. Unfortunately, a few difficulties arise that require additional attention.

First, we need to add l special points to our sample of $O(\frac{1}{\varepsilon^2})$ points on $P_{[a,b)}$. These are the points $u_i, 1 \le i \le l$, for which Q_{u_i} coincides with a vertex of P. This guarantees that between any two consecutive sample points, the scene changes in a continuous manner.

A more serious obstacle is that Claim 1(2) is not true anymore. That is, a point $s \in S$ can be 'close' to more than one point in Q_p; see Fig. 7. In this case,

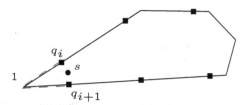

Fig. 7. The distance along P between q_i and q_{i+1} is 1, but still s is 'close' to both of them.

Lemma 2, whose proof is based on Claim 1(2), becomes false. However, since P is convex, the 'bad' situations are tractable. Specifically, we make the following observation.

Observation 3. For any three points x, y, z on P, such that the distance along P between any two of them is at least 1, it is impossible that all pairwise Euclidean distances ($|xy|, |xz|, |yz|$) are less than 1/2.

Corollary 2. 1. Let q, q' be two points in Q_p, for some $p \in P_{[a,b)}$, and set $\delta = |qq'|$. If $\delta < 1/2$, then, for any $q'' \neq q, q'$ in $Q_p, |qq''|, |q'q''| \geq 1/2 - \delta$. In particular if q and q' are 'close' to each other, then q'' is 'far' from both q and q'.

2. A point $s \in S$ can be 'close' to at most two points in Q_p. More precisely, if $|sq|, |sq'| \leq \delta < 1/4$, then, for any $q'' \neq q, q'$ in $Q_p, |sq''| \geq 1/2 - \delta$.

Hence, by dividing $P_{[a,b)}$ into even smaller parts (i.e., by placing $O(\frac{1}{\varepsilon^3}) + l$ points on $P_{[a,b)}$), and by detecting $O(n)$ additional intermediate points on $P_{[a,b)}$, which can be handled more efficiently, we obtain an efficient PTAS for POLYB (restricted to convex polygons). The following theorem summarizes our result, where the somewhat tedious and technical details are left for the full version of this paper.

Theorem 5. For any $\varepsilon > 0$, one can compute a $(1 + \varepsilon)$-approximation for POLYB, where P is a convex l-gon, in $O((\frac{1}{\varepsilon^3} + l)(\frac{n \log n}{\varepsilon^3} + n^{2+\delta})) = O(\frac{1}{\varepsilon^{O(1)}} ln^{2+\delta})$ time, for any $\delta > 0$.

References

1. Agarwal, P.K., Efrat, A., Sharir, M.: Vertical decomposition of shallow levels in 3-dimensional arrangements and its applications. SIAM J. Comput. **29**(3), 912–953 (1999)
2. Andrews, A.M., Wang, H.: Minimizing the aggregate movements for interval coverage. Algorithmica **78**(1), 47–85 (2017)
3. Bajaj, C.L.: The algebraic degree of geometric optimization problems. Discret. Comput. Geom. **3**, 177–191 (1988)
4. Bhattacharya, B.K., Burmester, M., Hu, Y., Kranakis, E., Shi, Q., Wiese, A.: Optimal movement of mobile sensors for barrier coverage of a planar region. Theor. Comput. Sci. **410**(52), 5515–5528 (2009)

5. Bose, P., Maheshwari, A., Morin, P.: Fast approximations for sums of distances, clustering and the Fermat-Weber problem. Comput. Geom. **24**(3), 135–146 (2003)
6. Chen, A., Kumar, S., Lai, T.: Designing localized algorithms for barrier coverage. In: Proceedings of the 13th Annual International Conference on Mobile Computing and Networking, MOBICOM 2007, pp. 63–74 (2007)
7. Chen, D.Z., Gu, Y., Li, J., Wang, H.: Algorithms on minimizing the maximum sensor movement for barrier coverage of a linear domain. Discret. Comput. Geom. **50**(2), 374–408 (2013)
8. Chen, D.Z., Tan, X., Wang, H., Wu, G.: Optimal point movement for covering circular regions. Algorithmica **72**(2), 379–399 (2015)
9. Dobrev, S., Durocher, S., Hesari, M.E., Georgiou, K., Kranakis, E., Krizanc, D., Narayanan, L., Opatrny, J., Shende, S.M., Urrutia, J.: Complexity of barrier coverage with relocatable sensors in the plane. Theor. Comput. Sci. **579**, 64–73 (2015)
10. Kumar, S., Lai, T., Arora, A.: Barrier coverage with wireless sensors. Wirel. Netw. **13**(6), 817–834 (2007)
11. Sharathkumar, R., Agarwal, P.K.: A near-linear time ϵ-approximation algorithm for geometric bipartite matching. In: Proceedings of the 44th Symposium on Theory of Computing, STOC 2012, pp. 385–394 (2012)

Complexity and Approximation of the Longest Vector Sum Problem

Vladimir Shenmaier[(✉)]

Sobolev Institute of Mathematics, Novosibirsk, Russia
shenmaier@mail.ru

Abstract. Given a set of n vectors in a d-dimensional normed space, consider the problem of finding a subset with the largest length of the sum vector. We prove that, for any ℓ_p norm, $p \in [1, \infty)$, the problem is hard to approximate within a factor better than $\min\{\alpha^{1/p}, \sqrt{\alpha}\}$, where $\alpha = 16/17$. In the general case, we show that the cardinality-constrained version of the problem is hard for approximation factors better than $1 - 1/e$ and is W[2]-hard with respect to the cardinality of the solution. For both original and cardinality-constrained problems, we propose a randomized $(1 - \varepsilon)$-approximation algorithm that runs in polynomial time when the dimension of space is $O(\log n)$. The algorithm has a linear time complexity for any fixed d and $\varepsilon \in (0, 1)$.

Keywords: Computational geometry · Vector sum · Normed space
Inapproximability bound · W[.]-hardness · Approximation algorithm

1 Introduction

In this work, we consider the following geometric optimization problem:

Longest vector sum (LVS). Given a set X of n vectors in a normed space $(\mathbb{R}^d, \|.\|)$, find a subset $S \subseteq X$ with the maximum value of

$$f(S) = \Big\| \sum_{x \in S} x \Big\|.$$

The problem has applications in diverse fields that include signal filtering, geophysics, and political science. We illustrate it on the following two examples.

Example 1. Suppose that we are given a set of measurements of the direction to some signal source (an acoustic or radio wave source, the magnetic field), each measurement is a unit vector in $\mathbb{R}^3$. The measurements contain an error and some of them may be received from outliers (noise, reflected waves, other random sources). How to determine the true direction? A reasonable answer is to find a subset of measurements with the maximum aggregate correlation, which equals to the Euclidean length of the sum vector.

© Springer International Publishing AG, part of Springer Nature 2018
R. Solis-Oba and R. Fleischer (Eds.): WAOA 2017, LNCS 10787, pp. 41–51, 2018.
https://doi.org/10.1007/978-3-319-89441-6_4

Example 2. shows that the LVS problem is also relevant for high-dimensional spaces. Suppose that the input vectors characterize the "political positions" of voters on the eve of democratic elections. The coordinates of each vector reflect the opinion of the corresponding voter on d major political issues and take values from -1 to 1. The voter's support for any political position equals to the correlation of this position with his own. An optimal solution of the Euclidean LVS problem on the set of normalized voters' political positions then determines a political position of a nominated candidate (or party) which has the maximum total support from an appropriate group of voters.

The variation of the LVS problem where the subset S is required to have a given cardinality $k \in [1, n]$ is called the *longest k-vector sum* problem (Lk-VS). It arises in the context of searching a quasiperiodically repeating fragment in a noisy numerical sequence [9] and 2-means clustering with the fixed center of one of the clusters [7].

Related work. The LVS and Lk-VS problems are special cases of the well-known *shaped partition* problem (see [13]) in which we are required to find a partition of a given set of n vectors into p subsets $S_1, \ldots, S_p$ with the maximum value of

$$c\Big(\sum_{x \in S_1} x, \ldots, \sum_{x \in S_p} x \Big),$$

where c is a given convex function of p vector variables while the cardinalities of $S_1, \ldots, S_p$ form a p-tuple that lies in some prescribed set. In the case of fixed d and p, this problem is solvable in time $O(n^{dp^2})$ [13]. If there are no restrictions on the cardinalities of S_i, an optimal partition can be computed in time $O(n^{d(p-1)})$ [16][1]. It follows that, if the dimension of space is fixed, the LVS and Lk-VS problems are solvable in polynomial time $O(n^d)$ and $O(n^{4d})$ respectively. Recently, the computational time for the first problem was reduced to $O(n^{d-1}(d + \log n))$ [22,23], for the second, to $O(dn^{d+1})$ [21,22].

In the case of a polyhedral norm, the LVS problem can be easily solved by using the simple algorithm from [4] with running time $O(Fdn)$, where F is the number of facets of the polyhedron determining the norm. A similar result holds for the cardinality-constrained problem. In particular, even if d is not fixed, both problems are polynomially solvable in the case of the ℓ_∞ norm and are fixed-parameter tractable with respect to d in the case of the ℓ_1 norm.

The Euclidean LVS and Lk-VS problems are strongly NP-hard if the dimension of space is not fixed [3,19]. In addition to the algorithms from [21–23], one can use the randomized algorithm from [10] which computes $(1-\varepsilon)$-approximate solutions of these problems in time $O(d^{3/2}(2\varepsilon - \varepsilon^2)^{-(d-1)/2} n \log \log n)$ with probability greater than $1 - 1/\log n$. A similar deterministic algorithm with running time $O\big(d^2\big(1 + \sqrt{\frac{d-1}{2\varepsilon}}\big)^{d-1} n\big)$ is proposed in [3].

---

[1] According to [17], this time bound is as pointed above and not $O(n^{d(p-1)-1})$ as is asserted in [16].

Our contributions. First, we obtain stronger hardness results. We prove that, for any ℓ_p norm, $p \in [1, \infty)$, the LVS problem is NP-hard to approximate within a factor better than $\alpha^{1/p}$, where α is the inapproximability bound for the maximum cut problem, $\alpha = 16/17$. For $p \in (2, \infty)$, the derived factor is improved to $\sqrt{\alpha}$ assuming NP $\not\subseteq$ RP. The obtained results can be easily extended for the cardinality-constrained version. Moreover, we show that, for some simple norm, this version contains the maximum coverage problem and, therefore, it is NP-hard to approximate within a factor better than $1 - 1/e$ and is W[2]-hard with respect to the parameter k. For any ℓ_p norm, $p \in [1, \infty)$, the Lk-VS problem, parameterized by k, is shown to be W[1]-hard.

On the other hand, we propose a randomized approximation algorithm for the case of an arbitrary norm. It is based on a randomized polyhedral approximation of the given norm and computes $(1 - \varepsilon)$-approximate solutions of both LVS and Lk-VS problems in time $O(d^{O(1)}(1 + \frac{2}{\varepsilon})^d n)$ with probability greater than $1 - 1/e$. In particular, we get a PRAS for instances with dimension $d = O(\log n)$ and a linear-time randomized approximation algorithm for any fixed d and $\varepsilon \in (0, 1)$. In the Euclidean case, the proposed algorithm coincides with the one from [10] and has the same running time.

2   Hardness Results

We start with proving that, unless P $=$ NP, the longest vector sum problem has an inapproximability bound. The proof is done by a reduction from the *maximum cut* problem:

Max-Cut. Given an undirected graph $G = (V, E)$, find a subset of vertices $S \subseteq V$ with the maximum value of $cut(S)$ that is the number of edges in E with one endpoint in S and the other in $V \backslash S$.

The complexity of this problem can be described as follows:

Fact 1. [11,14] *Max-Cut is NP-hard to approximate within a factor better than 16/17. Assuming the Unique Games Conjecture, there is no polynomial-time algorithm which approximates Max-Cut within a factor of 0.8786.*

Denote by LVS$_p$ the special case of the LVS problem for the ℓ_p norm, defined as $\|x\|_p = \left(\sum_{i=1}^{d} |x(i)|^p \right)^{1/p}$, where $p \in [1, \infty)$ and $x(i)$ is the ith coordinate of x.

Theorem 1. *For any $p \in [1, \infty)$ and $\alpha \in (0, 1]$, if there exists an α-approximation polynomial-time algorithm for the LVS$_p$, then there exists an α^p-approximation polynomial-time algorithm for Max-Cut.*

Proof. Given a graph $G = (V, E)$, we construct the set $X = X(G)$, an instance of the LVS$_p$ problem, in the following way. Let A be the incidence matrix of G. For each $e \in E$, the eth column of A contains exactly two 1s, at the rows indexed by the endpoints of e. Denote one of these endpoints by $a(e)$, the other by $b(e)$, and replace 1 at the row $b(e)$ with -1. Let x_v, $v \in V$, be the vth row of the matrix obtained in this way and $X = \{x_v \mid v \in V\}$.

Given a subset of vertices $S \subseteq V$, compute the vector $\xi = \sum_{v \in S} x_v$. For each $e \in E$, we have

$$\xi(e) = \begin{cases} 1 & \text{if } a(e) \in S \text{ and } b(e) \notin S, \\ -1 & \text{if } a(e) \notin S \text{ and } b(e) \in S, \\ 0 & \text{otherwise.} \end{cases}$$

So

$$\|\xi\|_p^p = \sum_{e \in E} |\xi(e)|^p = cut(S). \tag{1}$$

But $\|\xi\|_p = f(X_S)$, where $X_S = \{x_v \mid v \in S\}$. Therefore, X_S is an α-approximate solution of the LVS problem on the input set X if and only if S is an α^p-approximate solution of the Max-Cut problem on the graph G. The theorem is proved. $\qquad\square$

Corollary 1. *For any $p \in [1, \infty)$, the LVS_p problem is NP-hard to approximate within a factor better than $(16/17)^{1/p}$. Assuming the Unique Games Conjecture, there is no polynomial-time algorithm which approximates this problem within a factor of $0.8786^{1/p}$. Both statements hold even if $X \subseteq \{0, 1, -1\}^d$.*

Improvement for the Case $p > 2$

Next, we show that, for any $p > 2$, the derived inapproximability factors can be improved to the values of $\sqrt{16/17}$ and $\sqrt{0.8786}$ respectively. The idea of this improvement is based on the fact that almost any section of the ℓ_p normed space of appropriate polynomially large dimension is almost Euclidean by the randomised version of Dvoretzky's theorem (see [18]), so we can apply the inapproximability result for the Euclidean norm.

Definition 1. *A normed space $(V, \|.\|)$ is called δ-Euclidean, where $V \subseteq \mathbb{R}^d$ and $\delta \geq 1$, if there exists a constant $C > 0$ such that*

$$\|x\|_2 \leq C\|x\| \leq \delta\|x\|_2 \text{ for all } x \in V.$$

Fact 2. *[18] (Theorem 1.2, ii) For some constants $c_0, c, C > 0$, the following holds. For all large enough D, any $p \in (2, c_0 \log D)$, and any $\varepsilon \in (0, 1)$, there exists a value*

$$t(D, p, \varepsilon) \gtrsim \begin{cases} (Cp)^{-p}\varepsilon^2 D & \text{if } 0 < \varepsilon \leq (Cp)^{p/2}D^{-\frac{p-2}{2p-2}}, \\ p^{-1}(\varepsilon D)^{2/p} & \text{if } (Cp)^{p/2}D^{-\frac{p-2}{2p-2}} < \varepsilon \leq 1/p, \\ \varepsilon p D^{2/p}/\log\frac{1}{\varepsilon} & \text{if } 1/p < \varepsilon < 1 \end{cases}$$

such that any random subspace of $(\mathbb{R}^D, \ell_p)$ with dimension $d \leq t(D, p, \varepsilon)$ is $(1 + \varepsilon)$-Euclidean with probability greater than $1 - C\exp(-ct(D, p, \varepsilon))$.

This statement can be simplified to the following weaker form, which is more convenient for us:

Fact 3. *For any* $p \in (2, \infty)$, *there exist constants* $C_1, C_2, C_3 > 0$ *with the following property. If* $\varepsilon \in (0, 1)$, $D > C_1$, *and* $d \le d(D, p, \varepsilon) = C_2 \varepsilon^2 D^{2/p}$, *then a random* d-*dimensional subspace of* $(\mathbb{R}^D, \ell_p)$ *is* $(1 + \varepsilon)$-*Euclidean with probability greater than* $1 - \exp(-C_3 d(D, p, \varepsilon))$.

Theorem 2. *For any* $p \in (2, \infty)$, $\beta \in (0, 1]$, *and* $\varepsilon \in (0, 1)$, *if there exists a* β-*approximation polynomial-time algorithm for the* LVS_p, *then there exists a randomized* $\beta/(1 + \varepsilon)$-*approximation polynomial-time algorithm for the* LVS_2.

Proof. Let $\mathcal{A}_p$ be a polynomial-time β-approximation algorithm for the LVS_p. Consider an arbitrary finite set $X \subset \mathbb{R}^d$, an input of the LVS_2. Since this problem is polynomially solvable when d is fixed [16], we can assume that $d > C_1^{2/p} C_2 \varepsilon^2$, where the constants C_i are defined in Fact 3.

Put $D = \lceil C_2^{-p/2} \varepsilon^{-p} d^{p/2} \rceil$ and choose a random d-dimensional subspace V in $(\mathbb{R}^D, \ell_p)$. Note that $d \le d(D, p, \varepsilon)$ and $D > C_1$. Therefore, by Fact 3, the subspace V is $(1 + \varepsilon)$-Euclidean with probability $q = 1 - \exp(-C_3 d(D, p, \varepsilon)) \simeq 1 - \exp(-C_3 d) = 1 - o(1)$.

Define the map $\nu : \mathbb{R}^d \to V$ as $\nu(x) = \sum_{i=1}^d x(i) e_i$, where $(e_1, \ldots, e_d)$ is any orthonormal basis of V, and find a β-approximate solution S of the LVS_p problem on the input $\nu(X)$ by using the algorithm $\mathcal{A}_p$. Then, by Definition 1, the set S is a $\beta/(1+\varepsilon)$-approximate solution of the LVS_2 problem on that input. But the map ν is linear and isometric for the Euclidean norm, so $\nu^{-1}(S)$ is a $\beta/(1 + \varepsilon)$-approximate solution for the LVS_2 on the input X.

It remains to note that the dimension D is polynomially depends on d and $1/\varepsilon$ which follows that all the listed operations, including computing the set $\nu(X)$, the execution of the algorithm $\mathcal{A}_p$, and calculating the solution $\nu^{-1}(S)$, take polynomial time. The theorem is proved. $\qquad\square$

Corollary 2. *For any* $p \in (2, \infty)$, *assuming* $NP \not\subseteq RP$, *there is no polynomial-time algorithm which approximates the* LVS_p *problem within a factor better than* $\sqrt{16/17}$ *or within* $\sqrt{0.8786}$ *if the Unique Games Conjecture holds.*

Complexity of the Cardinality-Constrained Problem

All inapproximability factors for the LVS problem are also valid for its cardinality-constrained version since any T-time α-approximation algorithm for the Lk-VS gives a (Tn)-time α-approximation algorithm for the LVS. Moreover, by using the idea of the proof of Theorem 1, we can get a direct reduction to the Lk-VS from the *maximum bisection* problem which yields an inapproximability factor of $(15/16)^{1/p}$ under the assumption NP $\not\subseteq \cap_{\epsilon > 0}$DTIME$(2^{n^\epsilon})$ [12].

We now prove a stronger hardness result for the general Lk-VS problem.

Theorem 3. *For some norm* $\|.\|$ *computable in time* $O(d)$, *the* Lk-VS *problem is NP-hard to approximate within a factor better than* $1 - 1/e$ *and is* $W[2]$-*hard with respect to the parameter* k. *Both statements hold even if* $X \subseteq \{0, 1\}^d$.

Proof. We use a reduction from the *maximum coverage* problem:

Max-Coverage. Given a family of finite sets $F_1, \ldots, F_n$ and an integer $k \in [1, n]$, find a subset $S \subseteq \{1, \ldots, n\}$ such that $|S| \leq k$ and the value of $cover(S) = |\bigcup_{i \in S} F_i|$ is maximized.

The complexity of this problem can be described by the following facts:

Fact 4. [8] *Max-Coverage is NP-hard to approximate within a factor better than $1 - 1/e$.*

Fact 5. (See [5]) *Max-Coverage is $W[2]$-hard with respect to the parameter k.*

Define the norm $\|.\|$ as follows. Divide the set $\{1, \ldots, d\}$ into the $s = \lfloor \sqrt{d} \rfloor$ parts $\{a_1, \ldots, b_1\}, \ldots, \{a_s, \ldots, b_s\}$, where $a_j = \lfloor (j-1)\, d/s \rfloor + 1$ and $b_j = \lfloor jd/s \rfloor$. For any vector $x \in \mathbb{R}^d$, put

$$\|x\| = \sum_{j=1}^{s} \max \left\{ |x(a_j)|, \ldots, |x(b_j)| \right\}. \qquad (2)$$

Note that, for each j, the jth term in (2) equals to the value of the ℓ_∞ norm for the vector $(x(a_j), \ldots, x(b_j))$. So the function $\|.\|$ satisfies all the norm axioms.

Given a family $\mathcal{F}$ of finite sets $F_1, \ldots, F_n$, construct the set $X = X(\mathcal{F})$, an instance of the Lk-VS problem, in the following way. Without loss of generality, assume that each F_i is a subset of $\{1, \ldots, m\}$, where $m = cover(\{1, \ldots, n\})$. Put $s = \max\{m, n\}$, $d = s^2$, and, for each $i = 1, \ldots, n$, define the vector

$$x_i = (x_i(a_1), \ldots, x_i(b_1), \ldots, x_i(a_s), \ldots, x_i(b_s)),$$

where $a_j = \lfloor (j-1)\, d/s \rfloor + 1 = (j-1)s + 1$, $b_j = \lfloor jd/s \rfloor = js$, and

$$x_i(a_j + t - 1) = \begin{cases} 1 & \text{if } t = i \text{ and } j \in F_i, \\ 0 & \text{otherwise} \end{cases}$$

for all $i = 1, \ldots, n$, $j = 1, \ldots, s$, $t = 1, \ldots, s$. Let $X = \{x_1, \ldots, x_n\}$.

Consider any subset $S \subseteq \{1, \ldots, n\}$ and compute the vector $\xi = \sum_{i \in S} x_i$. It is easy to see that this vector is boolean and each sequence $(\xi(a_j), \ldots, \xi(b_j))$ contains 1s if and only if j belongs to one of the sets F_i, $i \in S$. So

$$\max \left\{ |\xi(a_j)|, \ldots, |\xi(b_j)| \right\} = \begin{cases} 1 & \text{if } j \in \bigcup_{i \in S} F_i, \\ 0 & \text{otherwise.} \end{cases}$$

Thus, we have $\|\xi\| = cover(S)$. It follows that any α-approximation polynomial-time algorithm for the Lk-VS gives an α-approximation polynomial-time algorithm for Max-Coverage. The same holds for FPT-algorithms. Then, by Facts 4 and 5, we obtain both statements of the theorem. □

In fact, the norm described in the proof of Theorem 3 is a simple combination of the ℓ_1 and ℓ_∞ norms. The following result shows that the Lk-VS problem, parameterized by k, is intractable for any ℓ_p norm, $p \in [1, \infty)$. Denote the special case of the Lk-VS problem for this norm by Lk-VS$_p$.

Theorem 4. *For any* $p \in [1, \infty)$, *the Lk-VS$_p$ problem is* $W[1]$-*hard with respect to the parameter* k *even if* $X \subseteq \{0, 1, -1\}^d$.

Proof. We use the $W[1]$-hardness of the *cardinality-constrained maximum cut* problem:

Max $(k, n - k)$-Cut. Given an n-vertex undirected graph $G = (V, E)$ and an integer $k \in [1, n]$, find a k-element subset of vertices $S \subseteq V$ with the maximum value of $cut(S)$.

Fact 6. [6] *Max $(k, n - k)$-Cut is $W[1]$-hard with respect to the parameter k.*

Given a graph G, construct the set $X(G)$ as described in the proof of Theorem 1. Then, by (1), the inequality $\max\{cut(S) \mid S \subseteq V, |S| = k\} \geq c$ is equivalent to the inequality $\max\{f(S) \mid S \subseteq X(G), |S| = k\} \geq c^{1/p}$, where c is any non-negative integer. In other words, (G, c, k) is a "yes" instance of the decision version of Max $(k, n-k)$-Cut if and only if $(X(G), c^{1/p}, k)$ is a "yes" instance of the decision version of the Lk-VS$_p$. Thus, we have an FPT-reduction from Max $(k, n-k)$-Cut to the Lk-VS$_p$ with respect to the parameter k. The theorem is proved. $\square$

An interesting open question is the parameterized complexity of the LVS$_2$ and Lk-VS$_2$ problems with respect to the dimension of space. Note that, in the case of the ℓ_1 norm, the LVS and Lk-VS problems can be solved in time $O(2^d dn)$ by using the approach from [4] for polyhedral norms. So the LVS$_1$ and Lk-VS$_1$ problems, parameterized by d, are in FPT.

3   Approximation Algorithm

In this section, we present a randomized approximation algorithm for the case of an arbitrary norm. The idea of this algorithm is to construct a polyhedral approximation of the given norm by using its gradients at some random points and to find a solution of the approximating problem which approximates a solution of the original problem.

Denote by $\mathbf{0}$ the origin, by xy the dot product of the vectors x and y, and by $B(x, r)$ and $S(x, r)$ the ball and the sphere of radius r centered at x respectively: $B(x, r) = \{y \in \mathbb{R}^d \mid \|y - x\| \leq r\}$, $S(x, r) = \{y \in \mathbb{R}^d \mid \|y - x\| = r\}$.

Algorithm $\mathcal{A}$

Step 1: Choose N independent random vectors $u_1, \ldots, u_N$ uniformly distributed in the ball $B(\mathbf{0}, 1)$, where N is a parameter of the algorithm.

Step 2: Put $I = \{1, \ldots, N\} \setminus \{i \mid u_i = \mathbf{0}\}$ and, for each $i \in I$, define an outward normal η_i to the sphere $S(\mathbf{0}, 1)$ at the point $v_i = u_i / \|u_i\|$ (Fig. 1).

Step 3: For each $i \in I$, construct the set $S_i = \{x \in X \mid x\eta_i > 0\}$ and a set $S_i(k)$ of k vectors from X with the maximum dot products with η_i.

Step 4: Select $S_{\mathcal{A}} = \arg\max_{i \in I} f(S_i)$ and $S_{\mathcal{A}}(k) = \arg\max_{i \in I} f(S_i(k))$, the outputs for the LVS and Lk-VS problems respectively.

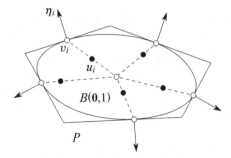

Fig. 1. The vectors u_i, v_i, and η_i

Remark 1. Generating random vectors uniformly distributed in the ball $B(\mathbf{0}, 1)$ depends on the specific norm. In particular, for the ℓ_p norm, these vectors can be computed by using the following fact:

Fact 7. [2] *Let $g(1), \ldots, g(d)$ be independent random variables with probability density $\phi(t) = \exp(-|t|^p)/(2\Gamma(1+1/p))$, where Γ is the gamma function, and let Z be an exponential random variable independent of $g(1), \ldots, g(d)$ with rate 1. Then the random vector $(g(1), \ldots, g(d))/(\sum_{i=1}^{d} |g(i)|^p + Z)^{1/p}$ is uniformly distributed in the unit ball of the ℓ_p norm.*

Remark 2. In the general case, to generate N independent random vectors uniformly distributed in the ball $B(\mathbf{0}, 1)$, we can execute N independent runs of the "ball-walk" algorithm from [15] for sampling from an arbitrary convex body K. The running time of this algorithm is $d^{O(1)}$ assuming that

(A1) more than 2/3 of the volume of $B_2(\mathbf{0}, 1)$ is contained in K, where $B_2(x, r)$ is the Euclidean ball of radius r centered at x;

(A2) more than 2/3 of the volume of K is contained in $B_2(\mathbf{0}, d^{3/2})$.

To meet these conditions, we need to make the following preliminary step. By using the randomized $d^{O(1)}$-time "rounding" algorithm from [15], find a linear transformation T such that the body $K = TB(\mathbf{0}, 1)$ satisfies (A1, A2). Replace each input vector $x \in X$ by the vector $x' = Tx$ and replace the norm $\|.\|$ by the norm $\|.\|'$ defined as $\|x\|' = \|T^{-1}x\|$, $x \in \mathbb{R}^d$. Then we obtain the equivalent instance of the LVS and Lk-VS problems where the unit ball of the given norm satisfies (A1, A2), therefore, sampling can be performed in time $d^{O(1)}$.

Remark 3. The outward normals η_i, $i \in I$, can be calculated as the gradients of the given norm at the points v_i. Indeed, any norm is a convex function and any convex function is differentiable almost everywhere (e.g., see [20], Theorem 25.5). Therefore, the given norm is almost surely is differentiable at all the points u_i and, by the homogeneity axiom, at all the points v_i. So the gradients of the norm are defined at these points and, by properties of gradients, give the outward normals to the corresponding level surface, i.e., the sphere $S(\mathbf{0}, 1)$.

Taking into account Remarks 2 and 3, we assume that generating a random vector which is uniformly distributed in the ball $B(\mathbf{0}, 1)$ and calculating an outward normal to the sphere $S(\mathbf{0}, 1)$ take time $d^{O(1)}$.

Theorem 5. *For any $\varepsilon \in (0, 1)$, there exists a value of the parameter N such that Algorithm $\mathcal{A}$ finds $(1 - \varepsilon)$-approximate solutions of the LVS and Lk-VS problems in time $O(d^{O(1)}(1 + \frac{2}{\varepsilon})^d n)$ with probability greater than $1 - 1/e$.*

Proof. Suppose that U^* is the sum of the vectors from an optimal solution of the LVS or Lk-VS problem. Without loss of generality, we will assume that $U^* \neq \mathbf{0}$ since otherwise the statement is obvious. Put $u^* = U^*/\|U^*\|$ and $\delta = \varepsilon/2$. Note that the vectors $y_i = (1 + \delta) u_i$ are independently and uniformly distributed in the ball $B(\mathbf{0}, 1 + \delta)$. Since $B(u^*, \delta) \subset B(\mathbf{0}, 1 + \delta)$, it follows that, for any $i = 1, \ldots, N$, the probability q of the event $y_i \in B(u^*, \delta)$ is $(\frac{\delta}{1+\delta})^d = (\frac{\varepsilon}{2+\varepsilon})^d$. Then, by the independency of the choice of y_i, at least one of these vectors belongs to $B(u^*, \delta)$ with probability

$$q_N = 1 - (1 - q)^N > 1 - \exp(-qN) = 1 - \exp\left(-(\tfrac{\varepsilon}{2+\varepsilon})^d N\right). \tag{3}$$

Let y_t be the nearest vector to u^* among $y_1, \ldots, y_N$. Then, by the above, $\|y_t - u^*\| \leq \delta$ with probability q_N. It follows that $y_t \neq \mathbf{0}$, i.e., $t \in I$. Assuming that $\|y_t - u^*\| \leq \delta$, we show that $\|y_t - v_t\| \leq \delta$. Indeed, $y_t = \|y_t\| v_t$ and $\|v_t\| = 1$, so $\|y_t - v_t\| = |\|y_t\| - 1|$. If $\|y_t\| > 1$, then $\|y_t - v_t\| = \|y_t\| - 1 \leq \delta$ since $\|y_t\| \in (1, 1 + \delta]$. If $\|y_t\| \leq 1$, then $\|y_t - v_t\| = 1 - \|y_t\| \leq \delta$ since, by the triangle inequality, $\|y_t\| \geq \|u^*\| - \|y_t - u^*\| \geq 1 - \delta$. In both cases, with probability at least q_N, we have $\|y_t - v_t\| \leq \delta$, therefore, by the norm axioms, $\|v_t - u^*\| \leq 2\delta = \varepsilon$.

Since the ball $B(\mathbf{0}, 1)$ is a convex body and the origin lies in its interior, we have $v_i \eta_i > 0$ for each $i \in I$. Consider the polyhedron P given by the inequality $\|x\|_+ \leq 1$, where

$$\|x\|_+ = \max_{i \in I} x \frac{\eta_i}{v_i \eta_i}.$$

By the construction of the vectors v_i and η_i, this polyhedron is the intersection of supporting half-spaces to the ball $B(\mathbf{0}, 1)$ at the points v_i (Fig. 1), so $B(\mathbf{0}, 1) \subseteq P$. Then, since the function $\|.\|_+$ is homogeneous for non-negative factors, we have $\|x\|_+ \leq \|x\|$ for all $x \in \mathbb{R}^d$. On the other hand, $\|.\|_+$ satisfies the triangle inequality and $\|v_i\|_+ = 1$ for all $i \in I$. Therefore,

$$\|u^*\|_+ \geq \|v_t\|_+ - \|v_t - u^*\|_+ \geq 1 - \|v_t - u^*\| \geq 1 - \varepsilon \tag{4}$$

with probability at least q_N as proved above.

At the same time, the maximum value of $\|\sum_{x \in S} x\|_+$ over all the subsets $S \subseteq X$ is attained on some set S_j, $j \in I$, which is constructed at Step 3 of Algorithm $\mathcal{A}$ since

$$\max_{S \subseteq X} \left\|\sum_{x \in S} x\right\|_+ = \max_{S \subseteq X} \max_{i \in I} \left(\sum_{x \in S} x\right) \frac{\eta_i}{v_i \eta_i} = \max_{i \in I} \max_{S \subseteq X} \left(\sum_{x \in S} x\right) \frac{\eta_i}{v_i \eta_i}$$

$$= \max_{i \in I} \left(\sum_{x \in S_i} x\right) \frac{\eta_i}{v_i \eta_i} \leq \max_{i \in I} \left\|\sum_{x \in S_i} x\right\|_+. \tag{5}$$

Put $U_A = \sum_{x \in S_A} x$ and $U_j = \sum_{x \in S_j} x$. Then, by the choice of the sets S_A, S_j and by (4), we have

$$\|U_A\| \geq \|U_j\| \geq \|U_j\|_+ \geq \|U^*\|_+ = \|U^*\| \, \|u^*\|_+ \geq \|U^*\| \, (1 - \varepsilon) \qquad (6)$$

with probability at least q_N. This means that Algorithm A returns a $(1 - \varepsilon)$-approximate solution of the LVS problem with such probability.

The same holds for the Lk-VS problem since, by the observations similar to inequalities (5), the maximum value of $\| \sum_{x \in S} x \|_+$ over all the k-element subsets $S \subseteq X$ is attained on some set $S_i(k)$, $i \in I$, which is constructed by Algorithm A. Therefore, we get the chain of inequalities of the type (6) for the vector $\sum_{x \in S_A(k)} x$.

We proceed to estimate the running time of the algorithm. Steps 1 and 2 can be performed in time $O(d^{O(1)} N)$ by the assumption above the statement of the theorem. For each $i \in I$, constructing the set S_i at Step 3 takes time $O(dn)$. The same holds for the set $S_i(k)$ since choosing k largest numbers from an n-element set can be performed in linear time (e.g., see [1]). Therefore, Algorithm A runs in time $O(d^{O(1)} Nn)$.

It remains to choose an appropriate value of the parameter N. We put $N = \lceil (1 + \frac{2}{\varepsilon})^d \rceil$. Then the running time of the algorithm is $O(d^{O(1)} (1 + \frac{2}{\varepsilon})^d n)$ and, by (3), the probability q_N is greater than $1 - 1/e$. The theorem is proved. $\square$

Remark 4. In the case of the Euclidean norm, the vectors v_i defined at Step 2 are independently and uniformly distributed on the sphere $S(\mathbf{0}, 1)$ and coincide with the normals η_i. Therefore, Steps 1 and 2 are reduced to generating random points on the unit sphere. In this case, Algorithm A works exactly as the algorithm from [10], so its running time is estimated as $O(d^{3/2}(2\varepsilon - \varepsilon^2)^{-(d-1)/2} n)$, which is less than in the general case.

By Theorem 5, Algorithm A implements a randomized fully polynomial-time approximation scheme (FPRAS) if the dimension of space is fixed. In the more general case when, for some constant $C > 0$, the dimension is bounded by the value of $C \log n$, we have $(1 + \frac{2}{\varepsilon})^d \leq n^{C \log(1 + \frac{2}{\varepsilon})} = n^{O(\log \frac{1}{\varepsilon})}$, so the algorithm gives a PRAS.

An interesting open question is the existence of a polynomial-time constant-factor approximation algorithm for the LVS and Lk-VS problems in the case of an arbitrary d.

Acknowledgments. This work is supported by the Russian Science Foundation under grant 16-11-10041.

References

1. Aho, A.V., Hopcroft, J.E., Ullman, J.D.: The Design and Analysis of Computer Algorithms. Addison-Wesley, Boston (1974)
2. Barthe, F., Guédon, O., Mendelson, S., Naor, A.: A probabilistic approach to the geometry of the l_p^n-ball. Ann. Probab. **33**(2), 480–513 (2005)

3. Baburin, A.E., Gimadi, E.K., Glebov, N.I., Pyatkin, A.V.: The problem of finding a subset of vectors with the maximum total weight. J. Appl. Ind. Math. **2**(1), 32–38 (2008)
4. Baburin, A.E., Pyatkin, A.V.: Polynomial algorithms for solving the vector sum problem. J. Appl. Ind. Math. **1**(3), 268–272 (2007)
5. Bonnet, É., Paschos, V.T., Sikora, F.: Parameterized exact and approximation algorithms for maximum k-set cover and related satisfiability problems. RAIRO-Theor. Inform. Appl. **50**(4), 227–240 (2016)
6. Cai, L.: Parameterized complexity of cardinality constrained optimization problems. Comput. J. **51**(1), 102–121 (2008)
7. Dolgushev, A.V., Kel'manov, A.V., Shenmaier, V.V.: Polynomial-time approximation scheme for a problem of partitioning a finite set into two clusters. Proc. Steklov Inst. Math. **295**(Suppl 1), 47–56 (2016)
8. Feige, U.: A threshold of $\ln n$ for approximating set cover. J. ACM **45**(4), 634–652 (1998)
9. Gimadi, E.K., Kel'manov, A.V., Kel'manova, M.A., Khamidullin, S.A.: A posteriori detecting a quasiperiodic fragment in a numerical sequence. Pattern Recogn. Image Anal. **18**(1), 30–42 (2008)
10. Gimadi, E., Rykov, I.: Efficient randomized algorithm for a vector subset problem. In: Kochetov, Y., Khachay, M., Beresnev, V., Nurminski, E., Pardalos, P. (eds.) DOOR 2016. LNCS, vol. 9869, pp. 148–158. Springer, Cham (2016). https://doi.org/10.1007/978-3-319-44914-2_12
11. Håstad, J.: Some optimal inapproximability results. J. ACM **48**(4), 798–859 (2001)
12. Holmerin, J., Khot, S.: A new PCP outer verifier with applications to homogeneous linear equations and max-bisection. In: 36th Annual ACM Symposium on Theory of Computing, pp. 11–20. ACM, New York (2004)
13. Hwang, F.K., Onn, S., Rothblum, U.G.: A polynomial time algorithm for shaped partition problems. SIAM J. Optim. **10**(1), 70–81 (1999)
14. Khot, S., Kindler, G., Mossel, E., O'Donnell, R.: Optimal inapproximability results for MAX-CUT and other 2-variable CSPs? SIAM J. Comput. **37**(1), 319–357 (2007)
15. Lovász, L., Simonovits, M.: Random walks in a convex body and an improved volume algorithm. Random Struct. Algorithms **4**(4), 359–412 (1993)
16. Onn, S., Schulman, L.J.: The vector partition problem for convex objective functions. Math. Oper. Res. **26**(3), 583–590 (2001)
17. Onn, S.: Personal communication, November 2016
18. Paouris, G., Valettas, P., Zinn, J.: Random version of Dvoretzky's theorem in ℓ_p^n. Stochast. Process. Appl. **127**(10), 3187–3227 (2017)
19. Pyatkin, A.V.: On the complexity of the maximum sum length vectors subset choice problem. J. Appl. Ind. Math. **4**(4), 549–552 (2010)
20. Rockafellar, R.T.: Convex Analysis. Princeton University Press, Princeton (1970)
21. Shenmaier, V.V.: Solving some vector subset problems by Voronoi diagrams. J. Appl. Ind. Math. **10**(4), 560–566 (2016). https://doi.org/10.1134/S199047891604013X
22. Shenmaier, V.: Complexity and algorithms for finding a subset of vectors with the longest sum. In: Cao, Y., Chen, J. (eds.) COCOON 2017. LNCS, vol. 10392, pp. 469–480. Springer, Cham (2017). https://doi.org/10.1007/978-3-319-62389-4_39
23. Shenmaier, V.V.: An exact algorithm for finding a vector subset with the longest sum. J. Appl. Ind. Math. **11**(4), 584–593 (2017). https://doi.org/10.1134/S1990478917040160

Deadline TSP

Boaz Farbstein and Asaf Levin$^{(\boxtimes)}$

Faculty of Industrial Engineering and Management, The Technion,
32000 Haifa, Israel
boazf@tx.technion.ac.il, levinas@ie.technion.ac.il

Abstract. We study the Deadline TSP problem. The input consists of a complete undirected graph $G = (V, E)$, a metric $c : E \to \mathbf{Z}_+$, a reward function $w : V \to \mathbf{Z}_+$, a non-negative deadline function $d : V \to \mathbf{Z}_+$, and a starting node $s \in V$. A feasible solution is a path starting at s. Given such a path and a node $v \in V$, we say that the path visits v by its deadline if the length of the prefix of the path starting at s until the first time it traverses v is at most $d(v)$ (in particular, it means that the path traverses v). If a path visits v by its deadline, it gains the reward $w(v)$. The objective is to find a path P starting at s that maximizes the total reward. In our work we present a bi-criteria $(1 + \varepsilon, \frac{\alpha}{1+\varepsilon})$-approximation algorithm for every $\varepsilon > 0$ for the Deadline TSP, where α is the approximation ratio for Deadline TSP with a constant number of deadlines (currently $\alpha = \frac{1}{3}$ by [5]) and thus significantly improving the previously best known bi-criteria approximation for that problem (a bi-criteria $(1 + \varepsilon, \frac{1}{O(\log(1/\varepsilon))})$-approximation algorithm for every $\varepsilon > 0$ by Bansal et al. [1]). We also present improved bi-criteria $(1+\varepsilon, \frac{1}{1+\varepsilon})$-approximation algorithms for the Deadline TSP on weighted trees.

1   Introduction

The Deadline TSP problem is defined as follows. The input consists of a complete undirected graph $G = (V, E)$ with n nodes, a metric $c : E \to \mathbf{Z}_+$, a reward function $w : V \to \mathbf{Z}_+$, a non-negative deadline function $d : V \to \mathbf{Z}_+$, and a starting node $s \in V$. A feasible solution is a path or a walk starting at s. Given such a path (or a walk) and a node $v \in V$, we say that the path visits v by its deadline if the length of the prefix of the path starting at s and ending at the first time it traverses v is at most $d(v)$ (in particular, it means that the path traverses v). If a path visits v by its deadline, it gains the reward $w(v)$ (and otherwise it gains no reward from v). The objective is to find a path P starting at s that maximizes the total reward. Notice that if $w(v) = 1$, $\forall v \in V$ the objective is to find a path P that visits the maximum number of nodes by their deadlines. Without loss of generality we assume $d(v) > 0$ for all $v \in V$, as otherwise we can add to the solution of the algorithm all nodes with zero deadline that are in the same position as s (i.e., all nodes in $\{v \in V : d(v) = 0$ and $c((s, v)) = 0\}$

This research was supported by a grant from the GIF, the German-Israeli Foundation for Scientific Research and Development (grant number I-1366-407.6/2016).

© Springer International Publishing AG, part of Springer Nature 2018
R. Solis-Oba and R. Fleischer (Eds.): WAOA 2017, LNCS 10787, pp. 52–65, 2018.
https://doi.org/10.1007/978-3-319-89441-6_5

immediately after s in an arbitrary order). Given a path P we denote the subpath of P that starts at $v_1 \in V$ and ends at the first time P traverses $v_2 \in V$ after v_1 by $P(v_1, v_2)$. We denote by $Z(P)$ the set of nodes that P visits by their deadline, that is, $Z(P) = \{v \in P : \sum_{e \in P(s,v)} c(e) \leq d(v)\}$. We denote by $w(P)$ the sum of the rewards collected by P, that is, $w(P) = \sum_{v \in Z(P)} w(v)$. Using this notation, the objective of the Deadline TSP is to find a path P starting at s that maximizes $w(P)$. Note that any solution can be transformed into a simple path without hurting its objective function value, however, it will be useful for the presentation of our algorithm to allow the repetition of nodes.

Before presenting the previous work on our problem, we define the notion of an approximation algorithm. A λ-approximation algorithm for a minimization problem is a polynomial time algorithm that always finds a feasible solution of cost at most λ times the cost of an optimal solution. A λ-approximation algorithm for a maximization problem is a polynomial time algorithm that always finds a feasible solution of value at least λ times the value of an optimal solution. Note that we use the convention of $\lambda < 1$ for maximization problems. In both cases the best value of λ is called the approximation ratio of the algorithm. In this paper we are interested in bi-criteria approximation algorithms. A bi-criteria (Δ, λ)-approximation algorithm for the Deadline TSP is a polynomial time algorithm collecting a total reward of at least λ times the value of an optimal solution from nodes that are visited by at most a factor of Δ of their deadlines. More formally, for $\Delta \geq 1$ and $0 < \lambda \leq 1$, let $Z_\Delta(P) = \{v \in P : \sum_{e \in P(s,v)} c(e) \leq \Delta \cdot d(v)\}$, $w_\Delta(P) = \sum_{v \in Z_\Delta(P)} w(v)$, and denote by O an optimal solution for a Deadline TSP instance. Then, a bi-criteria (Δ, λ)-approximation algorithm for Deadline TSP guarantees to return a solution P such that $w_\Delta(P) \geq \lambda \cdot w(O)$ in polynomial time.

Bansal et al. [1] were the first to show approximation results for this problem (in the general metric case). First they established a $\frac{1}{3 \log n}$-approximation algorithm, where n denotes the number of nodes in the graph. Then, motivated by the poor performance guarantee of this algorithm, the authors presented a bi-criteria $(1 + \varepsilon, \frac{1}{O(\log(1/\varepsilon))})$-approximation algorithm (for every $\varepsilon > 0$). This type of bi-criteria approximation guarantee has the bad property that if we let $\varepsilon \to 0$ so the violation of the deadline constraints will be minimized, the approximation guarantee grow unbounded. In this paper we significantly improve this bi-criteria approximation result. Chekuri and Kumar [5] presented a $\frac{1}{3}$-approximation algorithm for the special case of Deadline TSP with a constant number of (distinct) deadlines, the algorithm presented in this paper will use this algorithm as a subroutine. Bockenhauer et al. [2] showed an FPT $2\frac{1}{2}$-approximation algorithm for a different variation of the Deadline TSP where the objective is to find a minimal length Hamiltonian cycle visiting all nodes by their deadlines. The parameter used in (the exponent of) the running time of the approximation algorithm is the number of nodes that have finite deadlines in the input (all other nodes do not have deadlines and can be visited anytime).

A generalization of Deadline TSP is the following Vehicle Routing with Time Windows: Given an undirected graph $G = (V, E)$, a metric $c : E \to \mathbf{R}_+$, a non-negative time-window function $D : V \to \mathbf{R}_+^2$, and a starting node $s \in V$, the objective is to find a path P starting at s that visits a maximum number of nodes during their time-window. Note that Deadline TSP (when $w(v) = 1$ for every $v \in V$) is the special case of Vehicle Routing with Time Windows where all time windows have a common starting time of zero. Bansal et al. [1] showed by using the algorithm for Deadline-TSP that given an α_{DL}-approximation algorithm for Deadline-TSP, there is a α_{DL}^2-approximation algorithm for the Vehicle Routing with Time Windows. Chekuri et al. [3,4] showed an $O(\frac{\alpha_{OR}}{log(L_{max})})$-approximation algorithm for the case of integer time windows and an $O(\frac{\alpha_{OR}}{max\{log(\text{OPT}),log(\frac{L_{max}}{L_{min}})\}})$-approximation algorithm for the general case, where OPT is the number of nodes visited by an optimal solution during their time windows, α_{OR} is the approximation ratio for another related problem called Orienteering, L_{max} is the length of the longest time window and L_{min} is the length of the shortest time window. For a special case of Vehicle Routing with Time Windows, where every window is of unit length, Frederickson and Wittman [6] showed a $(\frac{1}{6+\varepsilon})$-approximation algorithm. They also showed a $(\frac{52}{219(2+\varepsilon)})$-approximation algorithm for the case where the length of the time windows is between 1 and 2.

Paper outline. In Sect. 2, we present a bi-criteria $(1 + \varepsilon, \frac{\alpha}{1+\varepsilon})$-approximation algorithm for every $\varepsilon > 0$ for the Deadline TSP, where α is the approximation ratio for Deadline TSP with a constant number of deadlines (currently $\alpha = \frac{1}{3}$ by [5]). In Sect. 3, we present an improved bi-criteria $(1+\varepsilon, \frac{1}{1+\varepsilon})$-approximation algorithm (for every $\varepsilon > 0$) for the Deadline TSP on weighted trees (i.e. a bi-criteria approximation scheme for this metric). We note that Deadline TSP is NP-hard for every metric in which TSP (for the version of a path with one endpoint in s) is NP-hard. The TSP on weighted tree metrics is polynomial time solvable and thus we conclude this work by showing in Sect. 4 that Deadline TSP is NP-hard on weighted trees even if there are only two distinct deadlines.

2 A Bi-criteria $(1 + \varepsilon, \frac{\alpha}{1+\varepsilon})$-Approximation Algorithm for Deadline TSP

The algorithm that we present (see Algorithm 1 below) uses the shifting technique of [7,8] and an α-approximation algorithm for the special case of Deadline TSP with a constant number of deadlines. We denote this special case of the problem by KDTSP, and we denote by A_k this α-approximation algorithm for KDTSP. Our algorithm uses a parameter $0 < \varepsilon \leq \frac{1}{2}$ such that $\frac{1}{\varepsilon}$ is an integer. Without loss of generality we assume $w(s) = 0$. First, the algorithm rounds up the different deadlines to the nearest integer power of $1+\varepsilon$ and then rounds down the resulting value to the nearest integer. As a result of this rounding, for every pair of positive integers L and U, the number of distinct rounded up deadlines in the interval $[L, U]$ is $O(\log_{1+\varepsilon} \frac{U}{L})$. We denote the rounding of $d(v)$ by $d'(v)$ for all $v \in V$. That is, for $v \neq s$, $d'(v) = \lfloor (1+\varepsilon)^{\lceil \log_{1+\varepsilon} d(v) \rceil} \rfloor$ is the rounded deadline of v (and $d'(s) = 0$). Note that for every v, $d'(v) \geq d(v)$. We set $t_1 = \frac{1}{\varepsilon^3}$ and $t_2 = \frac{1}{\varepsilon^2}$ to be the parameters for the shifting method. The shifting method will partition the

Algorithm 1. Main algorithm for a parameter ε, such that $0 < \varepsilon < \frac{1}{2}$ and $\frac{1}{\varepsilon}$ is an integer

---

Input: An undirected graph $G = (V, E)$, a metric $c : E \rightarrow \mathbf{Z}_+$, a deadline function $d : V \rightarrow \mathbf{Z}_+$, a reward function $w : V \rightarrow \mathbf{Z}_+$, and a starting node s such that $w(s) = 0$.
Output: A path P starting at s that maximizes $w_{1+2\varepsilon}(P)$.

1. For all $v \in V \setminus \{s\}$, let $d'(v) = \lfloor (1 + \varepsilon)^{\lceil \log_{1+\varepsilon} d(v) \rceil} \rfloor$ and $d'(s) = 0$.
2. Set $t_1 = \frac{1}{\varepsilon^3}$, $t_2 = \frac{1}{\varepsilon^2}$, and $k = \left\lceil \frac{\log_{1+\varepsilon} \max_v d'(v)}{t_1 + t_2} \right\rceil$.
3. For $i = 1, \ldots, t_1 + t_2$, $j = 0, \ldots, k$ do:
   (a) $V_i^j = \{s\} \cup \{v \in V : (1 + \varepsilon)^{i - t_1 + (j-1)(t_1 + t_2)} < d'(v) \le (1+\varepsilon)^{i + (j-1)(t_1 + t_2)}\}$,
   $E_i^j = \{(v_1, v_2) \in E : v_1, v_2 \in V_i^j\}$.
   (b) Apply A_k on graph $G_i^j = (V_i^j, E_i^j)$ and a starting node s to get a partial solution P_i^j.
   (c) Create cycle C_i^j by connecting the last node in P_i^j to s.
4. For $i = 1, \ldots, t_1 + t_2$ do:
   Concatenate all cycles ($\bigcup_j C_i^j$) in an increasing order of j and apply shortcuts to get a simple path P_i.
5. Return $P = \arg\max_{P_i : i = 1, 2, \ldots, t_1 + t_2} \{w_{1+2\varepsilon}(P_i)\}$.

---

instance into a collection of subproblems that we will consider separately. Each resulting subproblem will have up to $t_1 = \frac{1}{\varepsilon^3}$ distinct rounded deadlines (and in particular a constant number of deadlines), and there will be a (geometric) gap of at least $t_2 = \frac{1}{\varepsilon^2}$ between values of the rounded deadlines between different subproblems. Let $k = \lceil \frac{\log_{1+\varepsilon} \max_v d'(v)}{t_1 + t_2} \rceil$. For $i = 1, \ldots, t_1 + t_2$, $j = 0, 1, \ldots, k$, define $V_i^j = \{s\} \cup \{v \in V : (1 + \varepsilon)^{i - t_1 + (j-1)(t_1 + t_2)} < d'(v) \le (1 + \varepsilon)^{i + (j-1)(t_1 + t_2)}\}$, and let $G_i^j = (V_i^j, E_i^j)$ be the induced subgraph of G over V_i^j. We will use G_i^j for the (i, j) subproblem resulting from the shifting technique. For every $i = 1, \ldots, t_1 + t_2$ and every $j = 0, \ldots, k$, the algorithm uses A_k on graph $G_i^j = (V_i^j, E_i^j)$ with starting node s to get a solution P_i^j where we use $d'(v)$ as the deadline of v for every node $v \in V_i^j$. Then the algorithm adds an edge to P_i^j connecting the ending node in P_i^j to s to obtain C_i^j which is a cycle starting at s, traversing all the edges of P_i^j at the same time as P_i^j and returning back to s. For every $i = 1, \ldots, t_1 + t_2$, we create a candidate path P_i by concatenating all C_i^j in increasing order of the index j (considering only values of j for which V_i^j is not $\{s\}$) and applying shortcuts to obtain a simple path. The algorithm then returns the path $P \in \{P_1, P_2, \ldots, P_{t_1 + t_2}\}$ that maximizes $w_{1+2\varepsilon}(P)$. Recall that O is an optimal path for the problem (with the original deadlines). Without loss of generality, O is a simple path whose nodes are visited by O by their deadlines.

Our first lemma analyzes the impact of the concatenation step. The idea is that since the total length of P_i^j is at most the maximum rounded deadline of a node in V_i^j, the total length of $\bigcup_{l=0}^{j-1} C_i^l$ is tiny with respect to the value of the deadlines of nodes in $V_i^j \setminus \{s\}$, and thus the increase of the length of the prefix of the path from s to a given node in V_i^j while changing the solution from P_i^j to P_i is not significant. For a path or a cycle F we let $c(F)$ be the total length of the edges in F (i.e., if E_F is the set of edges in F, then $c(F) = \sum_{e \in E_F} c(e)$).

Lemma 1. *For every i, j and $v \in V_i^j$ such that v was visited by P_i^j by its rounded deadlines. Then, the path P_i visits v by time $(1 + 2\varepsilon)d(v)$.*

Proof. The claim trivially holds for $v = s$. Assume that $v \in V_i^j \setminus \{s\}$, and therefore $d(v) > (1 + \varepsilon)^{i - t_1 + (j-1)(t_1 + t_2)}$. Note that without loss of generality, for every $l \leq j$ the last node of P_i^l belongs to V_i^l and it is visited by its rounded deadline, and thus $c(P_i^l) \leq (1 + \varepsilon)^{i + (l-1)(t_1 + t_2)}$ and $c(C_i^l) \leq 2(1 + \varepsilon)^{i + (l-1)(t_1 + t_2)}$. We show that $\sum_{l=0}^{j-1} c(C_i^l) < \varepsilon \cdot d(v)$ and therefore any node that was visited by its rounded deadline by P_i^j will be visited by at most a factor of $(1 + 2\varepsilon)$ of its original deadline.

$$\sum_{l=0}^{j-1} c(C_i^l) \leq \sum_{l=0}^{j-1} 2(1 + \varepsilon)^{i + (l-1)(t_1 + t_2)} \leq 2 \cdot \frac{(1 + \varepsilon)^{i + (j-1)(t_1 + t_2)}}{(1 + \varepsilon)^{t_1 + t_2} - 1}$$
$$\leq \frac{2(1 + \varepsilon)^{i + (j-1)(t_1 + t_2)}}{(1 + \varepsilon)^{t_1 + \frac{7}{8}t_2}},$$

where the last inequality holds since $(1 + \varepsilon)^{t_1 + t_2} - 1 \geq (1 + \varepsilon)^{t_1 + \frac{7}{8}t_2}$ is equivalent to $(1 + \varepsilon)^{\frac{1}{8}t_2} - (1 + \varepsilon)^{-t_1 - 7/8t_2} \geq 1$ that we now establish. The inequality above holds for $0 < \varepsilon \leq \frac{1}{2}$ since it holds for $\varepsilon = \frac{1}{2}$, and $(1 + \varepsilon)^{\frac{1}{8}t_2}$ increases as ε decreases while $(1 + \varepsilon)^{-t_1 - 7/8t_2}$ decreases as ε decreases. Therefore,

$$\frac{\sum_{l=0}^{j-1} c(C_i^l)}{d(v)} < \frac{2(1 + \varepsilon)^{i + (j-1)(t_1 + t_2)}}{(1 + \varepsilon)^{i - t_1 + (j-1)(t_1 + t_2) + t_1 + \frac{7}{8}t_2}} = \frac{2}{(1 + \varepsilon)^{\frac{7}{8}t_2}} = \frac{2}{(1 + \varepsilon)^{\frac{7}{8}\frac{1}{\varepsilon^2}}} < \varepsilon,$$

where the last inequality holds for $\varepsilon \leq \frac{1}{2}$ since

$$\frac{1}{2}(1 + \varepsilon)^{\frac{7}{8}\frac{1}{\varepsilon^2}} = \frac{1}{2}\exp(\frac{7}{8\varepsilon^2}ln(1 + \varepsilon)) > \frac{1}{2}\exp(\frac{7}{8\varepsilon^2}(\varepsilon - \frac{1}{2}\varepsilon^2)) = \frac{1}{2}\exp(\frac{7}{8\varepsilon} - \frac{7}{16}) \geq \varepsilon,$$

for $\varepsilon \leq \frac{1}{2}$.   □

Next, we show that when our algorithm is allowed to visit the nodes by at most a factor of $1 + 2\varepsilon$ of their deadline, it collects a total reward of at least $\frac{\alpha}{1 + \varepsilon}$ of the total reward collected by an optimal solution (which has to visit the nodes by their original deadlines). In order to prove this we use the standard analysis for applications of the shifting technique of [7, 8]. For $i = 1 \ldots, t_1 + t_2$ and $j = 0, \ldots, k$, let $w'(P_i^j)$ be the total reward of nodes that P_i^j visits by their rounded deadline for the first time (that is, there does not exists an index $l < j$ where P_i^l visits any of those nodes).

Theorem 1. *The output of the algorithm, P, satisfies $w_{1+2\varepsilon}(P) \geq \frac{\alpha}{1+\varepsilon} w(O)$, and thus the algorithm is a bi-criteria $(1 + 2\varepsilon, \frac{\alpha}{1+\varepsilon})$-approximation algorithm.*

Proof. We write O as a node sequence, and let O_i^j be the subsequence of O consisting only of nodes in $V_i^j \setminus \{s\}$ (in the order that they appear along O). Using A_k, the algorithm guarantees that for every $i = 1, \ldots, t_1 + t_2$ and every $j = 0, \ldots, k$, $w'(P_i^j) \geq \alpha w(O_i^j)$. Observe that every node in O that is not s, appears in exactly t_1 such subsequences O_i^j. To see this let $v \neq s$ be a node in O. Without loss of generality let $d'(v) = (1 + \varepsilon)^{m+(\ell-1)(t_1+t_2)}$ such that ℓ is a non-negative integer and m is a positive integer satisfying $m \leq t_1 + t_2$. We look at two cases according to the value of m. If $m \leq t_2$ then the only possible value for j, for which there exists i such that O_i^j contains v, is $j = \ell$ and therefore $i - t_1 < m \leq i$ and as a result $m \leq i < m + t_1$. If $m > t_2$ ($m = t_2 + x$, $x > 0$) then for $\ell = j$, $m \leq i < m + t_1$ and as a result $t_2 + x \leq i \leq t_1 + t_2$ ($t_1 - x + 1$ different values for i), and for $j = \ell + 1$ we have $i + t_2 < m$ and therefore $i < x$. Overall, the number of different values of the pair (i, j) for the second case is $t_1 - x + 1 + x - 1 = t_1$. Thus, for both cases, there are exactly t_1 different values for the pair (i, j) such that $v \in O_i^j$. Therefore, using $w(s) = 0$, we have $\sum_{i=1}^{t_1+t_2} \sum_{j=0}^{k} w(O_i^j) = t_1 \cdot w(O)$. As a result, we have

$$w_{1+2\varepsilon}(P) \geq \frac{1}{t_1 + t_2} \sum_{i=1}^{t_1+t_2} w_{1+2\varepsilon}(P_i) \geq \frac{1}{t_1 + t_2} \sum_{i=1}^{t_1+t_2} \sum_{j=0}^{k} w'(P_i^j)$$

$$\geq \frac{\alpha}{t_1 + t_2} \cdot \sum_{i=1}^{t_1+t_2} \sum_{j=0}^{k} w(O_i^j) = \frac{t_1}{t_1 + t_2} \alpha \cdot w(O) = \frac{\alpha}{1+\varepsilon} w(O),$$

where the second inequality follows by Lemma 1. □

By scaling ε before applying Algorithm 1, we get a bi-criteria $(1 + \varepsilon, \frac{\alpha}{1+\varepsilon})$-approximation algorithm for Deadline TSP.

Remark 1. If there exists a bi-criteria $(1 + \varepsilon, \alpha)$-approximation algorithm for KDTSP, then we can use this bi-criteria approximation algorithm instead of A_k and by scaling ε we will get a bi-criteria $(1 + \varepsilon, \frac{\alpha}{1+\varepsilon})$-approximation algorithm for Deadline TSP.

3 A Bi-criteria $(1 + \varepsilon, \frac{1}{1+\varepsilon})$-Approximation Algorithm for Deadline TSP on Weighted Tree

In this section we show that we can derive a bi-criteria $(1+\varepsilon, \frac{1}{1+\varepsilon})$-approximation algorithm for deadline TSP on weighted tree using the methods of the previous section. Since the algorithm of the previous section uses a bi-criteria $(1 + \varepsilon, \alpha)$-approximation algorithm for KDTSP as a black box we will show how to obtain a bi-criteria $(1 + \varepsilon, 1)$ for the case of constant number of deadlines on weighted trees and by scaling ε, we get the claimed approximation guarantee. When the metric space is given by a weighted tree $T = (V, E_T)$ (rooted at s), the length of an edge $(v_1, v_2) \in E$ equals the length of the unique path connecting v_1 and v_2 in the given weighted tree. Without loss of generality we assume the metric is given by a binary tree. We show how an approximated solution for KDTSP can be found using dynamic programming in three main steps. First we show how an optimal solution for the problem can be found using dynamic programming (that runs in pseudo-polynomial time), then we present a polynomial time dynamic programming that traverses each of the nodes in the returned path only at rounded time-points. Last, we show that every element in the exponential dynamic programming table has a corresponding element in the polynomial dynamic programming table which does not extend the length of the returned path too much and every node that is visited by its deadline by the solution obtained by the exponential dynamic programming, is visited by the solution corresponding to it in the polynomial dynamic programming by a factor of at most $(1 + \varepsilon)$ of its deadline.

We root the tree at s. For every node $v \in V$ we denote by $L(v) \in V$ ($R(v) \in V$) the left (right) child of v in T and $L(v)$ or $R(v)$ are $\emptyset$ if no such child exists. For $v \in V$, we let T_v be the subtree of T induced by v and its descendants (i.e., the subtree rooted at v). We also denote by D the number of different deadlines of the nodes in V. In this section, we assume the optimal solution is a closed walk in the tree.

Lemma 2. *There exists an optimal solution that for every node $v \in V$ traverses each of the edges $(v, L(v))$ and $(v, R(v))$ at most $2D$ times.*

Proof. Let O be an optimal solution that traverses the minimum number of edges of T (among all optimal solutions that are closed walks in T). Assume by contradiction that O traverses an edge of T more than $2D$ times. Without loss of generality let that edge be $(v, L(v))$ and let $2m$ be the number of times O traverses that edge. For $i = 1, 2, \ldots m$, we define a cycle $C^i_{L(v)}$ by the subpath of O that starts with the $(2i - 1)$-th appearance of the edge $(v, L(v))$ along O, and ends with the $2i$-th appearance of this edge. We denote the minimum deadline of nodes in $C^i_{L(v)}$ by $d(C^i_{L(v)})$ when we consider only the nodes that O visits by their deadline during $C^i_{L(v)}$ for the first time. Since O traverses $(v, L(v))$ more than $2D$ times, there must exists $i < j$, $i, j \in \{1, 2, \ldots, m\}$ such that $d(C^i_{L(v)}) = d(C^j_{L(v)})$. Consider a solution P that traverses the same cycles as O but in a different order such that it skips $C^i_{L(v)}$ and traverses cycle $C^i_{L(v)}$ just before cycle $C^j_{L(v)}$. We also create a shortcut and unite the two cycles by deleting the last edge of $C^i_{L(v)}$ (that is $(L(v), v)$) and the first edge of $C^j_{L(v)}$ (that is $(v, L(v))$). Since $d(C^i_{L(v)}) = d(C^j_{L(v)})$, every node in $C^i_{L(v)}$ that was visited by its deadline in O is visited by its deadline in P. All the nodes in $C^j_{L(v)}$ or after it are visited by P not later than they were visited by O (due to the shortcut), all the nodes that O traverses before $C^j_{L(v)}$ and after $C^i_{L(v)}$ are visited by P before the time they were visited by O and the nodes that O traverses before $C^i_{L(v)}$ are visited by P at the same time they were visited by O. As a conclusion, P visits by their deadlines all the nodes visited by their deadline by O and traverses the edge $(v, L(v))$ two times less than O (and every other edge is traversed the same number of times by P and O). We received a contradiction to our choice of O.   □

We now describe the pseudo-polynomial time dynamic programming (DP_e) which returns an optimal path P'. Every node $v \in V$ has a table B'_v where every element $\{v, t'_{in}, t'_{out}\} \in B'_v$ encodes the following information (for $v \neq s$):

1. t'_{in} - a D dimensional vector of entrance times of a possible path P where an entrance time is the time v (for $v \neq s$) was traversed by P right after P traversed the parent of v. We denote the $i - th$ component in t'_{in} by $t'_{in}(i)$, $i = 1, \ldots, D$. If P traversed v right after its parent only $\ell < D$ times, then $t'_{in}(i) = \infty$ for all i such that $\ell < i \leq D$.
2. t'_{out} - a D dimensional vector of exit times of a possible path P where an exit time is the time v (for $v \neq s$) was traversed by P right before P traversed the parent of v. We denote the $i - th$ component in t'_{out} by $t'_{out}(i)$, $i = 1, \ldots, D$. If P traversed v right before its parent only $\ell < D$ times, then $t'_{out}(i) = \infty$ for all i such that $\ell < i \leq D$.

For every element $\{v, t'_{in}, t'_{out}\} \in B'_v$, DP_e calculates $r(v, t'_{in}, t'_{out})$ which is the maximum total reward collected from nodes in T_v that were visited before their deadline according to t'_{in} and t'_{out}. For the starting node s we let $t'_{in}(1)$ be 0 and $t'_{out}(D) = \infty$. For every $i = 1, \ldots, D - 1$, we let $t'_{out}(i) = t'_{in}(i+1)$ be the $(i+1)$-th time P traversed s. If s was traversed $\ell \leq D - 1$ times then $t'_{out}(i) = t'_{in}(i+1) = \infty$ for $\ell \leq i \leq D - 1$. For a leaf $v \in V$, $t'_{in} = t'_{out}$, and $r(v, t'_{in}, t'_{out}) = w(v)$ for every t'_{in} such that $t'_{in}(1) = 0, 1, \ldots, d(v)$ and otherwise $r(v, t'_{in}, t'_{out}) = 0$. DP_e calculates the different elements $\{v, t'_{in}, t'_{out}\}$ in B'_v after computing the tables of its children.

For every (non-leaf) node $v \in V$, the elements of B'_v are a result of united elements from the tables of the children of v in the following way. The union of two elements is done by first adjusting the entrance times and exit times of the time windows (a time window is the interval between the entrance time that is a component in t'_{in} and its corresponding exit time that is a component of the same index in t'_{out}) by subtracting the length of the edge from v to the child of v from all the entrance times and adjusting the exit times as well by adding the length of the edge from v to the child of v to all the exit times (this adjustment is crucial since the united element in B'_v should reflect the distances from v to the children of v). Then, DP_e sorts the time windows according to the adjusted entrance times and checks that no adjusted exit time is strictly greater than its successive adjusted entrance time. If two (or more) time windows have the same entrance times (this may happen if there are time windows of length zero), we sort them according to their exit times (breaking ties arbitrarily). Two (or more) consecutive time windows of the resulted list are united to one time window if the adjusted exit time of the earlier time window equals the adjusted entrance time of its successor time window and this is done until no such union of time windows can be applied. We require that the number of time windows in the resulting list is at most D. The potential reward of the united elements is the sum of the rewards of the two elements we unite together with possible reward of the node v. Formally, for every $v \in V$, we let $I'(v)$ be $w(v)$ if $t'_{in}(1) \leq d(v)$ and otherwise we let $I'(v) = 0$. Therefore, the total potential reward of a given output of the unite operation is the sum of $I'(v)$ and the rewards of the two elements we unite. If the union succeeds DP_e adds an element to B'_v with reward that is the maximum potential reward of this element (the maximum among all potential rewards corresponding to the element $\{v, t'_{in}, t'_{out}\}$, resulting from different pairs of united elements leading to the same t'_{in} and t'_{out}). The returned solution of DP_e is P' which corresponds to $\arg\max_{\{t'_{in}, t'_{out}\}}\{r(s, t'_{in}, t'_{out})\}$.

Corollary 1. DP_e *finds an optimal solution to KDTSP on a tree.*

Proof. First, we show that every feasible solution of DP_e encodes a feasible solution to KDTSP on a tree. Given a feasible solution of DP_e (an element from the table of each node) we can construct a path that is feasible to KDTSP by simply sorting all the entrance times of these elements and following the nodes according to the sorted entrance times. If two (or more) nodes have the same entrance times we follow them according to their exit times (breaking ties arbitrarily). By a trivial induction on the number of nodes in a prefix of this path, we conclude that the total length of this prefix equals the corresponding entrance time. Thus, the total reward of the nodes in this path equals the reward calculated by DP_e for this solution.

Next, we look at an optimal solution O that traverses each edge in T at most $2D$ times (by Lemma 2 we know that one exists) and show that O corresponds to a feasible solution of DP_e. For every node $v \in T$, O has at most D entrance times to v and D exit times from v. Since DP_e calculates the values of the total reward collected (from nodes visited before their deadlines) in all the paths that have at most D time windows for every v, there is a feasible solution of DP_e that corresponds to O (traverses the same edges as O in the same order as O). This corresponding solution is created by defining the set of time windows for each node according to the times O traverses the node. We denote the element in B'_s that corresponds to O by $\{s, t^o_{in}, t^o_{out}\}$. For every $v \in O$, we denote by $r(v)$ the total reward collected by DP_e in the subpaths corresponding to $O \cap T_v$ and therefore for every node $v \in O$, the collected reward from T_v is exactly $r(v) = r(L(v)) + r(R(v)) + I'(v) = \sum_{u \in O \cap T_v} I'(u)$. The reward calculated by DP_e for element $\{s, t^o_{in}, t^o_{out}\}$ is therefore $r(\{s, t^o_{in}, t^o_{out}\}) = \sum_{v \in O} I'(v)$ which is exactly the total reward of nodes in O that were visited by O before their deadline. The objective function value of the returned solution of DP_e is $\max\{r(s, t'_{in}, t'_{out})\} \geq r(\{s, t^o_{in}, t^o_{out}\})$ and therefore the value of the returned solution of DP_e is at least the value of O (and therefore is the value of O). □

In what follows, we refer to DP_e as the *exponential time dynamic programming*. We now present a polynomial time dynamic programming (DP_p) that is based on DP_e. The difference between DP_p and DP_e is that in DP_p the vectors t_{in} and t_{out} have a polynomial number of possible values and this also changes the unite operation. Let $DP_{times} = \{0, 1, \lceil 1 + \frac{\varepsilon}{(4nD)^2} \rceil, \lceil (1 + \frac{\varepsilon}{(4nD)^2})^2 \rceil, \ldots, \lceil (1 + \frac{\varepsilon}{(4nD)^2})^\nu \rceil, \infty\}$, where ν is the minimal integer value such that $\lceil (1 + \frac{\varepsilon}{(4nD)^2})^\nu \rceil \geq 2D \sum_{e \in E} c(e)$. In DP_p, for every $i = 1, \ldots, D$, $t_{in}(i), t_{out}(i) \in DP_{times}$. DP_p creates for every node $v \in V$ a table B_v. For a non-leaf node v, the elements of B_v are a result of united elements from the tables of the children of v in the following way. The union of two elements is done by first adjusting the entrance times and exit times of the elements. The adjusted entrance time is the result of rounding down to the nearest value in DP_{times} the subtraction of the length of the edge from v to the child of v from the entrance time. The adjusted exit time is the result of rounding up to the nearest value in DP_{times} the sum of the exit time and the length of the edge from v to the child of v. We use those rounded entrance times and rounded exit times to define the adjusted time windows of those elements. Then, DP_p sorts the adjusted time windows (at most $2D$ adjusted time windows) according to the adjusted entrance times (and in case of a tie, according to the adjusted exit times, breaking ties arbitrarily), and checks that no adjusted exit time exceeds its successive adjusted entrance time by a multiplicative factor greater than $(1 + \frac{\varepsilon}{(4nD)^2})^2$ (that is, for two adjacent adjusted time windows (a_{in}, a_{out}) and (b_{in}, b_{out}) where (a_{in}, a_{out}) is earlier than (b_{in}, b_{out}), we check if $a_{out} \leq (1 + \frac{\varepsilon}{(4nD)^2})^2 \cdot b_{in}$). Then DP_p unites adjusted time windows in the following way. Two (or more) consecutive adjusted time windows satisfying the above condition are candidates to be united to one time window if the adjusted exit time of the earlier adjusted time window is greater than or equals the adjusted entrance time of its successor adjusted time window. For every such candidate the algorithm creates two elements. In one element the adjusted time windows are left as is (not being united) and on the other, the algorithm unites the adjusted time windows such that the resulting time window has an adjusted entrance time that equals the adjusted entrance time of the first adjusted time window (among the adjusted time windows that are being united) and an adjusted exit time that equals the adjusted exit time of the last time window. If the time windows have the same entrance times and exit times they will always be united. We keep only the elements that have at most D time windows. As in DP_e, the potential reward of the united element is the sum of the rewards of the two elements we unite together with possible reward of the node v. For every $v \in V$ we let $I(v) = w(v)$ if $t_{in}(1) \leq d(v)$ and otherwise we let $I(v) = 0$. If the unite operation of $\{L(v), t_{in}^{L(v)}, t_{out}^{L(v)}\}$ and $\{R(v), t_{in}^{R(v)}, t_{out}^{R(v)}\}$ creates an element $\{v, t_{in}, t_{out}\} \in B_v$, then the potential reward of that element is the sum of $I(v)$ and the rewards of the two elements that were united. The reward that is associated with element $\{v, t_{in}, t_{out}\} \in B_v$ is the maximum among all potential rewards of that element.

Given a feasible solution of DP_p (for every $v \in V$, an element from B_v which is a result of a union of the elements of $L(v)$ and $R(v)$), we can construct a corresponding path in the tree, in the following way. We first sort all entrance times of all these elements (breaking ties arbitrarily) and then follow the nodes according to the sorted entrance time. Notice that elements $\{L(v), t_{in}^{L(v)}, t_{out}^{L(v)}\}$

and $\{R(v), t_{in}^{R(v)}, t_{out}^{R(v)}\}$ may have the same adjusted time windows (such that the adjusted entrance time and the adjusted exit time are the same). In such case either DP_p unites them into a single time window in $\{v, t_{in}, t_{out}\} \in B_v$ or they cannot be united at all. In both cases the resulting element $\{v, t_{in}, t_{out}\} \in B_v$ will not have more than one time window with the same adjusted entrance time and adjusted exit time. The returned solution of DP_p is P which corresponds to $\arg\max_{\{t_{in}, t_{out}\}} \{r(s, t_{in}, t_{out})\}$.

Lemma 3. *The returned solution of DP_p corresponds to a path P in T that satisfies $w_{(1+\varepsilon)^2}(P) \geq w(O)$.*

Proof. We say that an element $\{v, t_{in}^v, t_{out}^v\} \in B_v$ (in DP_p) dominates an element in $\{v, t_{in}'^v, t_{out}'^v\} \in B_v'$ (in DP_e) if the total reward collected by $\{v, t_{in}^v, t_{out}^v\}$ is at least the reward collected by $\{v, t_{in}'^v, t_{out}'^v\} \in B_v'$. Let $\{v, t_{in}'^v, t_{out}'^v\}_{v \in V}$ be a feasible solution for DP_e, that is, one element $\{v, t_{in}'^v, t_{out}'^v\}$ for every $v \in V$ such that $\{v, t_{in}'^v, t_{out}'^v\}$ is a result of the unite operation in DP_e of $\{L(v), t_{in}'^{L(v)}, t_{out}'^{L(v)}\}$ and $\{R(v), t_{in}'^{R(v)}, t_{out}'^{R(v)}\}$. Let $F(x)$ $(H(x))$ be a round up (round down) of x to the nearest value such that $F(x), H(x) \in DP_{times}$, if $x < 0$ then $H(x) = 0$ and if x is a vector then $F(x)$ and $H(x)$ round every component of x.

We claim that there is a feasible solution for DP_p whose corresponding path traverses the same nodes as the path corresponding to an optimal solution for DP_e and that for every $v \in V$, $\{v, t_{in}^v, t_{out}^v\}$ dominates $\{v, t_{in}'^v, t_{out}'^v\}$. First, we show that for every $v \in V$ the result of the union in DP_p of $\{L(v), t_{in}^{L(v)}, t_{out}^{L(v)}\}$ and $\{R(v), t_{in}^{R(v)}, t_{out}^{R(v)}\}$ is a feasible element $\{v, t_{in}^v, t_{out}^v\} \in B_v$. Then we will show that for every $v \in V$, DP_p collects the same reward from $\{v, t_{in}^v, t_{out}^v\}$ as DP_e collects from $\{v, t_{in}'^v, t_{out}'^v\}$.

A time window in an element in B_v (or in B_v') is received by either adjusting a time window of its child or by a union procedure of two or more time windows of its children. For the case where a time window in $\{v, t_{in}'^v, t_{out}'^v\}$ is the time window received by adjusting a time window of its child (without loss of generality its the $a \in \{1, \ldots D\}$ time window in $\{L(v), t_{in}'^{L(v)}, t_{out}'^{L(v)}\}$) then the time window created by DP_e for the element in B_v' is $\left(t_{in}'^{L(v)}(a) - c(v, L(v)), t_{out}'^{L(v)}(a) + c(v, L(v))\right)$ and the one created by DP_p for the element in B_v is $\left(H(t_{in}'^{L(v)}(a) - c(v, L(v))), F(t_{out}'^{L(v)}(a) + c(v, L(v)))\right)$. If DP_e unites two or more time windows of v's children (without loss of generality the time window $a, b \in \{1, \ldots D\}$ in $\{L(v), t_{in}'^{L(v)}, t_{out}'^{L(v)}\}$ and $\{R(v), t_{in}'^{R(v)}, t_{out}'^{R(v)}\}$ are united where $t_{in}'^{L(v)}(a) < t_{in}'^{R(v)}(b))$ then since

$$t_{out}'^{L(v)}(a) + c(v, L(v)) = t_{in}'^{R(v)}(b) - c(v, R(v)),$$

the resulting time window created by DP_e for the element in B_v' is

$$\left(t_{in}'^{L(v)}(a) - c(v, L(v)), t_{out}'^{R(v)}(b) + c(v, R(v))\right)$$

and the one created by DP_p is therefore

$$\left(H(t_{in}^{L(v)}(a) - c(v, L(v))), F(t_{out}^{L(v)}(b) + c(v, L(v))) \right).$$

For every $i \in \{1, \ldots, D\}$, the entrance time of $t_{in}^v(i)$ is equal to the entrance time of some time window of the child of v ($L(v)$ or $R(v)$) minus the edge connecting v to its child rounded down to the nearest value in DP_{times}. Denote this corresponding child of $t_{in}^v(i)$ by $m \in V$ and the index of the vector t_{in}^m corresponding to $t_{in}^v(i)$ by j. Then, $t_{in}^v(i) = H(t_{in}^m(j) - c(v, m)) \geq \left(\frac{1}{1 + \frac{\varepsilon}{(4nD)^2}} \right) (t_{in}^m(j) - c(v, m))$.
We let

$$last(v, i) = \arg \max_{u \in T_v} \{ c(v, u) | \exists f \in \{1, \ldots D\}, t_{in}^u(f) - c(v, u) = t_{in}^v(i) \}$$

and let num be the number of nodes on the simple path between v and $last(v, i)$ on T_v. Using induction we receive that

$$t_{in}^v(i) \geq \left(\frac{1}{1 + \frac{\varepsilon}{(4nD)^2}} \right)^{num} (t_{in}^{last(v,i)}(f) - c(v, last(v, i)))$$

$$\geq \left(\frac{1}{1 + \frac{\varepsilon}{(4nD)^2}} \right)^n t_{in}'^v(i) \geq \left(\frac{1}{1 + \frac{\varepsilon}{4nD}} \right) t_{in}'^v(i).$$

We use similar arguments to receive

$$t_{out}^v(i) \leq \left(1 + \frac{\varepsilon}{(4nD)^2} \right)^n \cdot t_{out}'^v(i)$$

$$\leq \left(1 + \frac{\varepsilon}{4nD} \right) \cdot t_{out}'^v(i).$$

For every two time windows, $t_{in}'^u(a) \leq t_{in}'^g(b)$, $a, b \in \{1, \ldots D\}$, $u, g \in V$, that are united in DP_e where $t_{in}'^u(a) < t_{in}'^g(b)$,

$$t_{out}^u(a) \leq \left(1 + \frac{\varepsilon}{(4nD)} \right) \cdot t_{out}'^u(a) = \left(1 + \frac{\varepsilon}{(4nD)} \right) \cdot t_{in}'^g(b) \leq \left(1 + \frac{\varepsilon}{(4nD)} \right)^2 \cdot t_{in}^g(b).$$

(1)

Therefore, $\{u, t_{in}^u, t_{out}^u\}$ and $\{g, t_{in}^g, t_{out}^g\}$ will be candidates of the union operation and thus will belong to an element where they are united in DP_p. In case $\{u, t_{in}^u, t_{out}^u\}$ and $\{g, t_{in}'^g, t_{out}'^g\}$ are not united in DP_e, then there exists an element in DP_p where $\{u, t_{in}^u, t_{out}^u\}$ and $\{g, t_{in}^g, t_{out}^g\}$ are not united. Notice that for every $v \in V$ the reward collected by DP_p is at least the reward collected by DP_e since by rounding down the entrance time of every time window, the algorithm ensures that $(t_{in}^v(1) \leq t_{in}'^v(1))$ for every $v \in V$. Therefore every element $\{v, t_{in}'^v, t_{out}'^v\} \in B_v'$ has an element $\{v, t_{in}^v, t_{out}^v\} \in B_v$ which dominates $\{v, t_{in}'^v, t_{out}'^v\}$ and gains reward from the exact same nodes.

To complete the proof we show that the calculated reward of DP_p is at most the sum of the rewards of nodes visited by P (that is, the path returned by DP_p) by time that is upper bounded by $(1 + \varepsilon)^2$ times their deadline. We let $\{v, t_{in}^v, t_{out}^v\}$ be the element in B_v corresponding to P. Since in the unite process of DP_p, time windows that are united may have an overlap (that is, the t_{out} of a time window may be greater than the t_{in} of its successor, up to a multiplicative factor of $(1 + \frac{\varepsilon}{4nD})^2)$, the length of the returned path P in graph G may be larger than the length as calculated by DP_p (that is, for every $v \in V$, the subpath of P starting at s and ending at the first time P traversed v may be larger than $t_{in}^v(1)$). For every $v \in V$, $c(P(s,v))$ is the length of the prefix of P, starting at s and ending at the first time P traversed v. We denote the time windows, corresponding to P, sorted by their entrance time, by $tw_1 = (t_{in(1)}, t_{out(1)}), tw_2 = (t_{in(2)}, t_{out(2)}), \cdots, tw_m = (t_{in(m)}, t_{out(m)})$ where $m \leq 2D \cdot n$. When we multiply a time window by a constant, we multiply both its entrance time and its exit time by this constant. For every $j \in \{1, \cdots, m\}$, we extend those time windows to receive extended time windows $\bar{tw}_j$ such that $\bar{tw}_j = (\bar{t}_{in(j)}, \bar{t}_{out(j)}) = (1 + \frac{\varepsilon}{4nD})^{2j} \cdot tw_j$. Thus, for every $j = 1, \ldots, m$,

$$\bar{t}_{in(j)} = (1 + \frac{\varepsilon}{4nD})^{2j} \cdot t_{in(j)} \geq (1 + \frac{\varepsilon}{4nD})^{2j} \cdot \frac{1}{\left(1 + \frac{\varepsilon}{(4nD)}\right)^2} \cdot t_{out(j-1)}$$

$$= (1 + \frac{\varepsilon}{4nD})^{2j} \cdot \frac{1}{\left(1 + \frac{\varepsilon}{(4nD)}\right)^2} \cdot \frac{1}{\left(1 + \frac{\varepsilon}{(4nD)}\right)^{2(j-1)}} \cdot \bar{t}_{out(j-1)} = \bar{t}_{out(j-1)},$$

where the first inequality holds by inequality (1). Therefore, the extended time windows do not overlap. For every $v \in V$, there is a time window such that its entrance time is $t_{in}^v(1)$. Denote that window by $tw_{j(v)}$. As a result, for every node $v \in P$ such that its reward was collected by DP_p,

$$c(P(s,v)) \leq \sum_{i=1}^{j(v)-1} \bar{tw}_i \leq (1 + \frac{\varepsilon}{4nD})^{2j(v)} \cdot t_{in}^v(1)$$

$$\leq (1 + \frac{\varepsilon}{4nD})^{4D \cdot n} \cdot t_{in}^v(1) \leq (1+\varepsilon)^2 \cdot t_{in}^v(1) \leq (1+\varepsilon)^2 \cdot d(v),$$

where the last inequality holds since DP_p collected a reward from v (and thus $I(v) = w(v)$). We conclude that every node $v \in P$ such that $I(v) = w(v)$ (in DP_p) was traversed by P for the first time by time at most $(1 + \varepsilon)^2 d(v)$. $\square$

By scaling ε we receive the following corollary,

Corollary 2. *There is a bi-criteria $(1 + \varepsilon, \frac{1}{1+\varepsilon})$ approximation algorithm for Deadline TSP on a tree metric and a bi-criteria $(1 + \varepsilon, 1)$ approximation algorithm for KDTSP on a tree metric.*

4   Deadline TSP is NP-Hard Even on Tree Metric

Theorem 2. *Deadline TSP problem and the KDTSP on a tree metric are NP-hard.*

Proof. We show a reduction from the decision variant of the Knapsack problem to Deadline TSP on a tree metric with only two distinct deadlines. An input to the decision variant of the Knapsack problem consists of a set of items $i = 1, \ldots, n$, each has a cost c_i and a value w_i. A total cost C and a total reward W is also given. The objective is to answer whether there is a set of items T such that the sum of the costs of items in T is less than or equal to C ($\sum_{i \in T} c_i \leq C$) and the sum of the values of items in T is at least W ($\sum_{i \in T} w_i \geq W$). The decision variant of the knapsack problem is NP-complete. We create the following input to Deadline TSP. The nodes of the tree are s, i (for every $i = 1, \ldots, n$), and a. That is, $V = \{s, a\} \cup \{i : 1 \leq i \leq n\}$. The rewards are $w(s) = 0$, $w(i) = w_i$, $\forall i = 1, \ldots n$ and $w(a) = n \cdot \max_i \{w_i\} + 1$. For every $i = 1, \ldots, n$ we create an edge (s, i) with length equal to $\frac{c_i}{2}$. We also create an edge (s, a) with length equal to C. The deadline of a is $2C$ and the deadlines of all the other nodes are equal to C. Let OPT be an optimal solution to the deadline TSP on the graph created and assume without loss of generality that OPT is a walk in the tree with node set that equals the nodes visited by OPT by their deadlines. It can easily be seen that the last node in OPT is a. For every node $i \in \{1, \ldots, n\}$ that is visited by OPT (except node a), OPT must traverse edge (s, i) twice adding to its length exactly c_i and therefore if the rewards of the nodes traversed by OPT are at least $W + w(a)$ then the answer to the decision variant of the knapsack problem is yes, otherwise the answer is no. □

References

1. Bansal, N., Blum, A., Chawla, S., Meyerson, A.: Approximation algorithms for deadline-TSP and vehicle routing with time-windows. In: Proceedings of STOC 2004, pp. 166–174 (2004)
2. Bockenhauer, H., Hromkovic, J., Kneis, J., Kupke, J.: The parameterized approximability of TSP with deadlines. Theory Comput. Syst. **41**(3), 431–444 (2007)
3. Chekuri, C., Korula, N.: Approximation algorithms for orienteering with time windows. CoRR, abs/0711.4825 (2007)
4. Chekuri, C., Korula, N., Pál, M.: Improved algorithms for orienteering and related problems. ACM Trans. Algorithms **8**(3), 23:1–23:27 (2012)
5. Chekuri, C., Kumar, A.: Maximum coverage problem with group budget constraints and applications. In: Proceedings of APPROX 2004, pp. 72–83 (2004)
6. Frederickson, G.N., Wittman, B.: Approximation algorithms for the traveling repairman and speeding deliveryman problems. Algorithmica **62**(3), 1198–1221 (2012)
7. Hochbaum, D.S. (ed.): Approximation Algorithms for NP-Hard Problems. PWS Publishing Co., Boston (1997)
8. Hochbaum, D.S., Maass, W.: Approximation schemes for covering and packing problems in image processing and VLSI. J. ACM **32**(1), 130–136 (1985)

A Bicriteria Approximation Algorithm for the k-Center and k-Median Problems

Soroush Alamdari$^{(\boxtimes)}$ and David Shmoys

Cornell University, New York, NY 14853, USA
{sh954,dbs10}@cornell.edu

Abstract. The k-center and k-median problems are two central clustering techniques that are well-studied and widely used. In this paper, we focus on possible simultaneous generalizations of these two problems and present a bicriteria approximation algorithm for them with constant approximation factor in both dimensions. We also extend our results to the so-called incremental setting, where cluster centers are chosen one by one and the resulting solution must have the property that the first k cluster centers selected must simultaneously be near-optimal for all values of k.

1 Introduction

Clustering problems have been studied from a range of algorithmic perspectives, due to the breadth of application domains for which they serve as appropriate optimization models. This is particularly true from the context of approximation algorithms, in which a wide variety of these optimization models have been shown to have polynomial-time algorithms with strong performance guarantees. In this paper, we will consider an important generalization of many of these problems for which no constant performance guarantee had been known previously, and give the first constant performance guarantee for a central special case that is a common generalization of two of the most well-studied models, the k-center and k-median problems.

In the metric k-supplier problem, we are given a set $\mathcal{D}$ of n demand points and a set $\mathcal{F}$ of m service points in a common metric space. The goal is to open k of these service points in order to serve the demand in $\mathcal{D}$, while minimizing a connection cost. Typically, when there are no capacities on how many demand points that each service point can serve, each demand point is served by its closest opened service point. Although in this paper we only focus on cases where $\mathcal{F} = \mathcal{D}$, for the sake of readability we try to distinguish between the two roles that any of the given points may play at any point.

We use $c_{jj'}$ denote the distance between points j and j'. Similarly, for sets S and D of points and a point $j \in D$, c_{jS} refers to the distance between j

S. Alamdari and D. Shmoys—Supported in part by NSF CCF-1526067, CMMI-1537394, and CCF-1522054.

and a point in S that is closest to it, and c_{DS} is the array of length $|D|$ of non-decreasing distances c_{jS} for each $j \in D$. For a solution S with $|S| \leq k$, the connection cost of the k-supplier problem can be expressed as function of c_{DS}.

In the *ordered median problem*, we are given a weight vector w, where $w(i)$ for $i \in [n]$ is the weight on the ith longest connection, and the cost of a solution $S \subset F$ with $|S| \leq k$ is simply $w^T c_{DS}$. Very recently, Aouad and Segev [1] presented an $O(\log n)$ approximation algorithm for the case when weights in w are non-decreasing. This result intensifies the question of finding conditions for w under which the ordered median problem can be approximated with a constant quality guarantee.

For example, again when $F = D$, the ordered median problem captures the k-center problem, by setting the cost vector w to be 0s for each index, except for the last index $w(n) > 0$. So the problem here is to select a set $S \subset F$ of k centers in order to minimize the maximum distance in c_{DS}. The first approximation algorithms for this problem were presented by Hochbaum and Shmoys [4] and Gonzalez [3], with an approximation factor of 2. The latter algorithm starts with $S = \{i\}$ for some arbitrary $i \in F$, and in each iteration adds the point $i' \in F$ that maximizes $c_{i'S}$ to S, provided $|S| < k$. Suppose $j \in D$ is the furthest demand point from S at the end of this algorithm. Then, if you consider the balls of radius $c_{jS}/2$ around each point in S, these balls must be disjoint, or else j should have been added to S at some point by the algorithm. If we assume that there is a set S^* with $\max(c_{DS^*}) < c_{jS}/2$, then the ball of radius $c_{jS}/2$ centered at each of $i \in S \cup \{j\}$ must intersect S^*, and since these $k+1$ balls are disjoint, it must be that $|S^*| > k$. Therefore, c_{jS} is within a factor 2 of any solution with size at most k. Hsu and Nemhauser [5] showed that, for any $\epsilon > 0$, there is no polynomial-time algorithm with a performance guarantee of $2 - \epsilon$ for this problem, unless $P = NP$.

Another example of a cost vector that is approximable within a constant factor would be that of the k-median problem. In this case, the cost vector w is 1 for each index, and therefore the cost is simply the sum of the distances between each demand point and its closest open service point. Although the k-median problem is arguably driven by the simplest cost vector, no constant approximation algorithms were known for it until Charikar *et al.* [2] presented an LP-based algorithm with a factor of $6\frac{2}{3}$. Indeed, our main result extends their approach. Subsequent to [2], there has been a series of improvements leading to the recent $1 + \sqrt{3} + \epsilon$ approximation algorithm by Li and Svensson [6].

In this paper, we push the envelope a little further by showing that even if the cost vector is a convex combination of the cost vectors of k-center and k-median, one can still efficiently find solutions within a constant factor of the optimal. More precisely, given an instance of k-supplier with an optimal solution S_{opt}, our algorithm presented in Sect. 2 is able to find a solution S such that $\mathrm{sum}(c_{DS}) \leq 8\mathrm{sum}(c_{DS_{\mathrm{opt}}})$ and $\max(c_{DS}) \leq 4\max(c_{DS_{\mathrm{opt}}})$. In the literature, this is referred to as a *bicriteria approximation algorithm*. Note that this result implies that for the special case of the ordered median problem in which w is obtained from a convex combination of the k-median and k-center problems, we obtain a constant approximation algorithm.

Fig. 1. An example showing that simultaneous approximation of k-center and k-median within a factor $o(\sqrt{n})$ is impossible. Here each of A and B represents a cluster of $\lfloor \frac{n}{2} \rfloor$ points that are located at the same position.

A stronger notion than a bicriteria approximation would be a simultaneous approximation; that is, an algorithm that produces one common solution S that is simultaneously compared to an optimal solution S_1^* for its k-center objective and to an optimal solution S_2^* for its k-median objective. However, we can show that producing such a simultaneous approximation of factor $o(\sqrt{n})$ is not possible. For this, suppose n is odd and $n - 1$ of the points are divided equally between two positions with distance 1 from each other. There is also a single outlier that is of distance $\sqrt{n}$ from everyone else. See Fig. 1. Let $k = 2$ and suppose S_1^* contains the outlier and one other point, whereas S_2^* contains one point at each of the densely populated locations. Therefore, $\max(c_{\mathcal{D}S_1^*}) = 1$ and $\operatorname{sum}(c_{\mathcal{D}S_2^*}) = \sqrt{n}$. In such a case, any solution S would have $\operatorname{sum}(c_{\mathcal{D}S}) \geq \lfloor \frac{n}{2} \rfloor$ if it opens the outlier service point and $\max(c_{\mathcal{D}S}) \geq \sqrt{n}$ otherwise. In either case, $\max(\frac{\max(c_{\mathcal{D}S})}{\max(c_{\mathcal{D}S_1^*})}, \frac{\operatorname{sum}(c_{\mathcal{D}S})}{\operatorname{sum}(c_{\mathcal{D}S_2^*})})$ would be in $\Omega(\sqrt{n})$.

We also extend our result to the incremental setting, when k is not known a priori and one has to add service points one by one to the solution such that when the size of the solution reaches k, the solution that is built is still within a constant factor of the optimal. For example, the algorithm that we mentioned for the k-center problem satisfies this constraint, since it opens service points greedily and oblivious of k. For the k-median problem, Mettu and Plaxton [8] presented the first constant-factor approximation in the incremental setting. Later, Lin et al. [7] presented a general framework for incremental approximations that also applies to the k-median problem with an improved approximation factor. In Sect. 3, we use this framework to develop an incremental approximation algorithm for any objective that is a convex combination of that of k-center and k-median problems.

2 A (4, 8)-Approximation Algorithm for k-Center and k-Median

Suppose we are given an instance of the k-supplier problem with $\mathcal{D} = \mathcal{F}$, and we wish to find a good solution with respect to both the k-median and k-center objectives. Let S_{opt} be the "ideal" solution in this context, and consider $\max(c_{\mathcal{D}S_{\text{opt}}})$ and $\operatorname{sum}(c_{\mathcal{D}S_{\text{opt}}})$, which are its k-center and k-median objectives, respectively. As is the convention in bi-criteria approximations, we assume that $\max(c_{\mathcal{D}S_{\text{opt}}})$ is given as an objective measure. Note that $\max(c_{\mathcal{D}S_{\text{opt}}})$ is equal to one of the $\binom{n}{2}$ possible distances between pairs of given points, and hence we can guess $\max(c_{\mathcal{D}S_{\mathrm{opt}}})$ with a multiplicative overhead of $O(n^2)$. This guess would allow us to achieve the desired guarantees, however, we would not necessarily be able to decide which guess corresponds to the true value of $\max(c_{\mathcal{D}S_{\mathrm{opt}}})$ and would need to choose among produced solutions based some other criteria.

Our algorithm works based on the (k-supplier LP). In this formulation, for a service point $i \in \mathcal{F}$ there is a variable y_i that represents whether service point i is opened or not, and a variable x_{ij} for each $j \in \mathcal{D}$ that represents the fraction of j's demand that is satisfied by i. Here we are assuming that each demand point has one unit of demand that must be satisfied. However, the algorithm will maintain a more general instance in which demand is concentrated at nodes in $\mathcal{D}$, and so we will let d_j denote the total demand at node j.

Suppose that variables x_{ij} and y_{ij} are restricted to be integers and $\mathcal{L}$ is set to infinity. Then the (k-supplier LP) formulation represents an linear relaxation of the k-median problem: the objective is to minimize the sum of distances between each demand point and the service point that serves it. Constraint (c1) ensures that a demand point $j \in \mathcal{D}$ relies on exactly one service point $i \in \mathcal{F}$, constraint (c2) verifies that no demand point uses an unopened service point to satisfy its demand, and constraint (c3) makes sure that k centers are opened.

$$\min \quad \sum_{i,j \in \mathcal{D}} d_j c_{ij} x_{ij} \qquad\qquad\qquad \text{(k-supplier LP)}$$

$$
\begin{aligned}
\text{s.t.} \quad & \sum_{i \in \mathcal{F}} x_{ij} = 1 && \forall j \in \mathcal{D} && \text{(c1)} \\
& x_{ij} \le y_i && \forall i \in \mathcal{F}, j \in \mathcal{D} && \text{(c2)} \\
& \sum_{i \in \mathcal{F}} y_i = k && && \text{(c3)} \\
& 0 \le y_i \le 1 && \forall i \in \mathcal{F} \\
& 0 \le x_{ij} \le 1 && \forall i \in \mathcal{F}, j \in \mathcal{D} \\
& x_{ij} = 0 && \forall i \in \mathcal{F}, j \in \mathcal{D} : c_{ij} > \mathcal{L} && \text{(cc)}
\end{aligned}
$$

Now suppose we set $\mathcal{L}$ to $\max(c_{\mathcal{D}S_{\mathrm{opt}}})$ and then solve for an optimal fractional solution (x^*, y^*). Then constraint (cc) guarantees that x_{ij} is 0 for any pair $i \in \mathcal{F}$ and $j \in \mathcal{D}$ that are further than $\max(c_{\mathcal{D}S_{\mathrm{opt}}})$ from each other. Also, by the optimality of (x^*, y^*) and since S_{opt} corresponds to a feasible solution for (k-supplier LP), we have that $\sum_{i,j \in \mathcal{D}} d_j c_{ij} x_{ij}^* \le \mathrm{sum}(c_{\mathcal{D}S_{\mathrm{opt}}})$. Thus, we can use (x^*, y^*) to bound both the k-center and k-median objectives in the optimal solution.

Algorithm 1 uses a solution (x^*, y^*) to construct an integral solution that is within a constant factor of both the k-center and k-median objectives of S_{opt}. First, a set S' of points is selected as candidates for being opened, and at the same time, the demand of each point $j \in \mathcal{D} \setminus S'$ is moved to a point $j' \in S'$ that is close enough to represent j (lines 5–10). Here, for each $j \in S'$, d_j' denotes

---

Algorithm 1. Finds an integral solution S with $\text{sum}(c_{DS})$ at most 8 times (k-supplier LP) objective and $\max(c_{DS})$ at most $4\mathcal{L}$.

INPUT: An instance of the k-supplier problem along with $\mathcal{L}$ and a feasible solution (x,y) of the (k-supplier LP).

OUTPUT: A set S with $|S| \leq k$ of service points that will be opened.

1: $(x',y') \leftarrow (x,y)$
2: $S' \leftarrow \emptyset$
3: $d'_j \leftarrow d_j \quad \forall j \in \mathcal{D}$ $\triangleright$ total demand that j represents
4: $\overline{C}_j \leftarrow \sum_{i \in \mathcal{F}} x_{ij} c_{ij} \quad \forall j \in \mathcal{D}$ $\triangleright$ cost of serving the demand of j
5: **for all** demand points $j \in \mathcal{D}$ in increasing order of $\overline{C}_j$ **do**
6:      **if** $c_{jS'} > \min(4\overline{C}_j, 2\mathcal{L})$ **then**
7:          $S' \leftarrow S' \cup \{j\}$
8:      **else**
9:          let j' be the closest point to j in S'
10:          $d'_{j'} \leftarrow d'_{j'} + d'_j, d'_j \leftarrow 0$ $\triangleright$ moving the demand of j to j'
11: **for all** $i \in \mathcal{F} \backslash S'$ **do**
12:      let i' be the closest point to i in S'
13:      $y'_{i'} \leftarrow y'_{i'} + y'_i, y'_i \leftarrow 0$ $\triangleright$ moving the y' value of j to j'
14: **while** there is $i, i' \in S'$ s.t. $y'_{i'} < 1, y'_i > 1$ **do**
15:      $\delta \leftarrow \min(y'_i - 1, 1 - y'_{i'})$
16:      $y'_{i'} \leftarrow y'_{i'} + \delta, y'_i \leftarrow y'_i - \delta$ $\triangleright$ redistributing the excess y' of i
17: let s_j be the closest point to j in $S' \backslash \{j\} \quad \forall j \in S'$
18: **while** there are $i, i' \in S'$ s.t. $y'_{i'} < 1, \frac{1}{2} < y'_i < 1$, and $d'_i c_{is_i} \leq d'_{i'} c_{i's_{i'}}$ **do**
19:      $\delta \leftarrow \min(y'_i - \frac{1}{2}, 1 - y'_{i'})$
20:      $y'_{i'} \leftarrow y'_{i'} + \delta, y'_i \leftarrow y'_i - \delta$
21: $S \leftarrow \{i \in S' : y'_i = 1\}$
22: let T be a forest of arbitrarily rooted trees with a node for each $j \in S'$ and an edge between each $j \in S'$ and s_j
23: let E and O be the set of points $j \in S'$ with $y'_j < 1$ that are of even and odd distance to their tree root in T, respectively
24: **if** $|O| < |E|$ **then**
25:      $S \leftarrow S \cup O$
26: **else**
27:      $S \leftarrow S \cup E$
28: **return** S

---

the total demand that point j represents. Then, another loop iterates over all points $i \in \mathcal{F} \backslash S'$ to create an updated solution y'; we move all of the weight y'_i to the closest point in S' so that $y'_i = 0$ for each $i \in \mathcal{F} \backslash S'$, and then a while loop redistributes among S' any excess of y' that might have been gathered in one $i' \in S'$ so that $y'_{i'} \leq 1$ (lines 11–16). By this point in the algorithm, all of the demand and (fractional) supply is concentrated in S'. The following lemma implies that the size of S' is at most $2k$.

Lemma 1. *After Line 16 of Algorithm 1 we have $y'_i > \frac{1}{2}$ for each $i \in S'$.*

Proof. By the order in which the points are considered in the loop at line 5 and the condition at line 6, we have that the balls centered at any $i \in S'$ with radius $\min(2\overline{C}_i, \mathcal{L})$ are disjoint. (Recall that $\overline{C}_j = \sum_{i \in \mathcal{F}} x_{ij} c_{ij}$ given our input fractional solution x.) By line 16 of the algorithm, all of the y value that is in any such a ball centered at an $i \in S'$ is moved to y'_i, and some of it is possibly redistributed to make sure $y'_i \leq 1$. If $\mathcal{L} < 2\overline{C}_i$ and the radius of the ball is $\mathcal{L}$, then by constraint (cc), the total y value in such a ball must be at least 1. Otherwise, by constraint (c2) and the definition of $\overline{C}_i$, at least half of the y value that i uses to satisfy its demand must be within distance $2\overline{C}_i$. □

Next, in a while loop, the algorithm moves any weight specified by y' from any point of S' to a more effective point in S' as long as all y' values remain in range $[\frac{1}{2}, 1]$ for all points in S' (lines 18–20).

Lemma 2. *By the time the while loop at line 18 of Algorithm 1 terminates, for any point $i \in S'$, we have $y'_i \in \{\frac{1}{2}, 1\}$.*

Proof. Since constraint (c3) holds with equality and (x^*, y^*) is a feasible solution to the (k-supplier LP), we have $\sum_{i \in \mathcal{F}} y^*_i = k$. Since y' is initialized to y^* and the total value of y' is conserved throughout (both when it is moved to S' and also in each iteration of the while loop,) therefore, any time the condition of the while loop at line 18 is being evaluated, we have

$$\sum_{i \in \mathcal{F}} y'_i = \sum_{i \in S'} y'_i = k.$$

So if the condition of the lemma is not satisfied, then there must be at least two points $i, i' \in S'$ such that $y'_i, y'_{i'} \in (\frac{1}{2}, 1)$. However, such a pair satisfies the condition of the while loop and therefore the loop would not yet terminate. □

After the while loop at line 18 terminates, all points $i \in S'$ with $y'_i = \frac{1}{2}$ are partitioned into two sets O and E, and the smaller of O and E along with all the points $i \in S'$ with $y'_i = 1$ are returned as the set S of points that are to be opened by the algorithm (lines 22–28). Since S is comprised of points $i \in S'$ with $y'_i = 1$ and at most half of the points $i' \in O \cup E$ with $y'_{i'} = \frac{1}{2}$, therefore the size of S is no larger than k. The following lemmas argue that this set S is indeed a bicriteria approximate solution for k-center and k-median.

Lemma 3. *Given an instance of k-supplier problem with $\mathcal{D} = \mathcal{F}$ and a corresponding optimal solution (x^*, y^*) of (k-supplier LP) with $\mathcal{L} = \max(c_{\mathcal{D} S_{\mathrm{opt}}})$, Algorithm 1 returns a set S of at most k centers such that $\max(c_{\mathcal{D} S}) \leq 4\max(c_{\mathcal{D} S_{\mathrm{opt}}})$.*

Proof. First we argue that for each $j \in \mathcal{D}$, the closest point to j that is added to S' is at most of distance $2\mathcal{L}$ from j. For this, note that when the for loop at line 5 of the algorithm iterates over j and the condition at line 6 is being evaluated, if the distance from j to all points in S' is greater than $2\mathcal{L}$, then j itself would have been added to S'.

Suppose that j' is the closest node in S' to j. Let j'' be the closest node in S' to j', other than j' itself. If the distance between j' and j'' is greater than $2\mathcal{L}$, then all of the y' value in the ball of radius $\mathcal{L}$ around j' must have been gathered at j' in lines 11–13, since j' is the closest point in S' to any point in this ball. Since the solution (x^*, y^*) is feasible and by constraint (cc), the total amount of y' value in such a ball should at least be 1 in order to satisfy the demand of j'. Therefore, in this case $y'_{j'}$ ends up being 1 and hence j' itself is opened by the algorithm.

Now suppose that the distance of j' and j'' is smaller than $2\mathcal{L}$. In this case, we show that at least one of j' and j'' is opened by the algorithm. By the triangle inequality, this would mean that there must be at least one opened service point within distance $4\mathcal{L}$ of j, concluding the proof of the lemma. By the construction of the forest T in line 22, there is an edge between nodes representing j' and j'' in T. If none of j' and j'' are added to S at line 21, then they are either in O or E. Since there is an edge between j' and j'' in T, the parity of their level in T must be different, and hence exactly one of them should belong to O and the other must be in E. One of O or E is added to S in lines 24–28, and hence one of j' or j'' must be opened by the algorithm.   □

Lemma 4. *Given an instance of k-supplier problem with $\mathcal{D} = \mathcal{F}$ and a corresponding optimal solution (x^*, y^*) of (k-supplier LP) with $\mathcal{L} = \max(c_{\mathcal{D}S_{\mathrm{opt}}})$, Algorithm 1 returns a set S of at most k centers such that $\mathrm{sum}(c_{\mathcal{D}S}) \leq 8\mathrm{sum}(c_{\mathcal{D}S_{\mathrm{opt}}})$.*

Proof. To show this we can break down the cost of rounding the LP solution into 3 parts. The first part (filtering phase at Lines 5–10) incurs an independent additive factor of 4, while the second (Lines 11–16) and third (Line 22 onward) parts each incur a multiplicative factors of 2, implying that $\mathrm{sum}(c_{\mathcal{D}S}) \leq (4 + 2 \times 2)\mathrm{sum}(c_{\mathcal{D}S_{\mathrm{opt}}})$.

We claim that moving the demands in Lines 5–10 induces an additive term of 4 to the approximation factor. To see this recall that $\mathrm{sum}(c_{\mathcal{D}S_{\mathrm{opt}}}) \geq \sum_{j \in \mathcal{D}} \sum_{i \in \mathcal{F}} x^*_{ij} c_{ij} = \sum_{j \in \mathcal{D}} \overline{C}_j$, and notice that the demand of each point j is moved a distance of at most $4\overline{C}_j$ and by the triangle inequality, returning this demand to its original position would incur at most a cost of $4\overline{C}_j$. Hence, the total overhead of moving the demands is within a factor 4 of the cost of the LP relaxation.

Moving the y' values to nodes at S' at Lines 11–16 incurs a factor of 2 to the approximation ratio. Suppose an ϵ amount of y is moved from i to $i' \in S'$ at Line 13. Note that at this point of the algorithm only points in S' have positive demand, and consider some $j \in S'$ that used to rely on i for an ϵ fraction of its demand, who now has to use i' to satisfy that ϵ fraction of its demand. By the choice of i' we have $c_{ii'} \leq c_{ji}$; therefore, by triangle inequality we have $c_{ji'} \leq c_{ji} + c_{ii'} \leq 2c_{ij}$.

The while loop at Line 18 chooses i and i' and manipulates the y' values such that the objective $\sum d'_j x'_{ij} c_{ij}$ does not increase; thus the bound on the approximation ratio is unchanged.

Suppose we update x' at this point in the algorithm (after termination of the while loop at Line 18) to get a feasible solution to the (k-supplier LP) while minimizing the objective $\sum d'_j x'_{ij} c_{ij}$ according to the new y' values produced by the algorithm. Then, by Lemma 2, at line 22 of the algorithm, each point j in S' with $y'_j < 1$ has $y'_j = \frac{1}{2}$, and therefore relies for exactly half of its demand on the closest other point in S'; that is, s_j as defined in Line 17. By the same arguments as in the proof of Lemma 3, at the end of the algorithm if j is not opened, then s_j must be opened, and since j already relies on s_j for half of its demand, this would at most double the cost of serving point j. With the additive factor of 4 that we reserved for moving back the demands to their original points, it all adds up to a factor of 8. $\qquad\square$

The following theorem immediately follows from Lemmas 3 and 4:

Theorem 5. *There is a polynomial-time bicriteria $(4, 8)$-approximation algorithm for k-center and k-median problems.*

3   Incremental Approximation for Convex Combination of k-Center and k-Median

In the incremental setting, the parameter k is not known *a priori*. In this case, we open service points one by one, and the true value of k is revealed only after the kth service point is opened. One can view this setting equivalently as producing an ordering of all nodes in $\mathcal{F}$, where each prefix of the first k nodes is the designated solution corresponding to the number k, and all such prefixes for all $1 \leq k \leq n$ must have the claimed performance guarantee. For this case, we use the framework developed by Lin et al. [7] to derive our result. The following is a corollary of their main theorem as it pertains to our problem.

Corollary 6 [7]. *If there is an efficient procedure that, given a set of centers S and a number $k' < |S|$, finds a set $S' \subset S$ with $|S'| \leq k'$ such that $\max(c_{DS'}) < \max(c_{DS}) + \alpha_1 \max(c_{DS_{\mathrm{opt}}})$ and $\mathrm{sum}(c_{DS'}) < \mathrm{sum}(c_{DS}) + \alpha_2 \mathrm{sum}(c_{DS_{\mathrm{opt}}})$ for any S_{opt} with $|S_{\mathrm{opt}}| \leq k'$, then there is a randomized $(e \max(\alpha_1, \alpha_2))$-approximation algorithm and a deterministic $(4 \max(\alpha_1, \alpha_2))$-approximation algorithm for the incremental version of the problem where the objective is a convex combination of the objectives for the k-center and k-median problems.*

To get a sense of how this algorithm works, suppose we wish to construct a chain $S_0 \subset S_1 \subset S_2 \subset S_3 \subset \ldots \subset S_l$ with $S_0 = \emptyset$ and $S_l = \mathcal{F}$, such that an incremental algorithm would only have to open the points in $S_i \backslash S_{i-1}$ before $S_{i+1} \backslash S_i$ for any $i \in \{1, 2, \ldots, l-1\}$ to be within the constant factor of the optimal solution for any intermediate value of k. To find this chain, one can start with S_l which is simply the set of all given points, and recursively use the procedure described in Corollary 6 to find a subset of the current set that is of cost roughly twice the current set. The following lemma describes such a procedure and proves its properties.

Lemma 7. *There is an efficient procedure that given a set of centers S and a number $k' < |S|$ finds a set $S' \subset S$ with $|S'| \leq k'$ such that $\max(c_{\mathcal{D}S'}) \leq \max(c_{\mathcal{D}S}) + 4\max(c_{\mathcal{D}S_{\mathrm{opt}}})$ and $\mathrm{sum}(c_{\mathcal{D}S'}) \leq \mathrm{sum}(c_{\mathcal{D}S}) + 16\mathrm{sum}(c_{\mathcal{D}S_{\mathrm{opt}}})$ for any S_{opt} with $|S_{\mathrm{opt}}| \leq k'$.*

Proof. To find S', we first use Theorem 5 with k set to k' and construct a solution $\hat{S}$ with $\max(c_{\mathcal{D}\hat{S}}) \leq 4\max(c_{\mathcal{D}S_{\mathrm{opt}}})$ and $\mathrm{sum}(c_{\mathcal{D}\hat{S}}) \leq 8\mathrm{sum}(c_{\mathcal{D}S_{\mathrm{opt}}})$. For each point j, let $n(j)$ and $\hat{n}(j)$ be the closest point to j in S and $\hat{S}$, respectively. Then we define S' simply as the union of all points in S that are closest to some point in $\hat{S}$, that is, $S' = \{n(j) : j \in \hat{S}\}$. Next we show that S' indeed satisfies the properties claimed in the lemma.

First note that since $n(\hat{n}(j)) \in S'$ for each point j, we have that $c_{jS'} \leq c_{jn(\hat{n}_j)}$, and, by the triangle inequality, we have that $c_{jn(\hat{n}_j)} \leq c_{j\hat{n}(j)} + c_{\hat{n}(j)n(\hat{n}_j)}$. Since $c_{j\hat{n}(j)} \leq \max(c_{\mathcal{D}\hat{S}})$ and $c_{\hat{n}(j)n(\hat{n}_j)} \leq \max(c_{\mathcal{D}S})$, therefore we have that

$$\max(c_{\mathcal{D}S'}) \leq \max(c_{\mathcal{D}S}) + \max(c_{\mathcal{D}\hat{S}}) \leq \max(c_{\mathcal{D}S}) + 4\max(c_{\mathcal{D}S_{\mathrm{opt}}}).$$

For the other side of the argument, note that since $n(\hat{n}_j)$ is by definition the closest point in S to $\hat{n}(j)$, we have that $c_{\hat{n}(j)n(\hat{n}_j)} \leq c_{\hat{n}(j)n(j)}$. Therefore, by the triangle inequality, we have that

$$c_{jn(\hat{n}(j))} \leq c_{j\hat{n}(j)} + c_{\hat{n}(j)n(\hat{n}(j))} \leq c_{j\hat{n}(j)} + c_{\hat{n}(j)n(j)} \leq c_{j\hat{n}(j)} + c_{j\hat{n}(j)} + c_{jn(j)}.$$

Putting these together, we get that $c_{jS'} \leq c_{jn(\hat{n}(j))} \leq c_{jn(j)} + 2c_{j\hat{n}(j)}$ for each $j \in \mathcal{D}$. Thus,

$$\mathrm{sum}(c_{\mathcal{D}S'}) \leq \mathrm{sum}(c_{\mathcal{D}S}) + 2\mathrm{sum}(c_{\mathcal{D}\hat{S}}) \leq \mathrm{sum}(c_{\mathcal{D}S}) + 16\mathrm{sum}(c_{\mathcal{D}S_{\mathrm{opt}}}).$$

□

The following theorem immediately follows from Corollary 6 and Lemma 7:

Theorem 8. *The incremental approximation of the problem of minimizing a convex combination of the objectives of the k-center and k-median problems admits a randomized $16e$-approximation algorithm and a deterministic 64-approximation algorithm.*

The k-center and k-median problems are, in many respects, antipodal extremes among possible weighting functions for the order median problem; since we have shown that any convex combination of those two weighting functions can be approximated within a constant factor of optimal, we believe that it is natural to conjecture that such a result can be obtained for the general ordered median problem as well.

References

1. Aouad, A., Segev, D.: The ordered k-median problem: surrogate models and approximation algorithms. In submission
2. Charikar, M., Guha, S., Tardos, É., Shmoys, D.B.: A constant-factor approximation algorithm for the k-median problem. J. Comput. Syst. Sci. **65**(1), 129–149 (2002)
3. Gonzalez, T.: Clustering to minimize the maximum intercluster distance. Theor. Comput. Sci. **38**, 293–306 (1985)
4. Hochbaum, D.S., Shmoys, D.B.: A best possible heuristic for the k-center problem. Math. Oper. Res. **10**(2), 180–184 (1985)
5. Hsu, W., Nemhauser, G.L.: Easy and hard bottleneck location problems. Discrete Appl. Math. **1**(3), 209–215 (1979)
6. Li, S., Svensson, O.: Approximating k-median via pseudo-approximation. SIAM J. Comput. **45**(2), 530–547 (2016)
7. Lin, G., Nagarajan, C., Rajaraman, R., Williamson, D.P.: A general approach for incremental approximation and hierarchical clustering. SIAM J. Comput. **39**(8), 3633–3669 (2010)
8. Mettu, R.R., Plaxton, C.G.: The online median problem. SIAM J. Comput. **32**(3), 816–832 (2003)

Approximating Domination
on Intersection Graphs of Paths on a Grid

Saeed Mehrabi[✉]

Cheriton School of Computer Science,
University of Waterloo, Waterloo, Canada
smehrabi@uwaterloo.ca

Abstract. A graph G is called B_k-*EPG* (resp., B_k-*VPG*), for some constant $k \geq 0$, if it has a string representation on an axis-parallel grid such that each vertex is a path with at most k bends and two vertices are adjacent in G if and only if the corresponding strings share at least one grid edge (resp., the corresponding strings intersect each other). If two adjacent strings of a B_k-VPG graph intersect each other exactly once, then the graph is called a *one-string* B_k-VPG graph.

In this paper, we study the MINIMUM DOMINATING SET problem on B_1-EPG and B_1-VPG graphs. We first give an $O(1)$-approximation algorithm on one-string B_1-VPG graphs, providing the first constant-factor approximation algorithm for this problem. Moreover, we show that the MINIMUM DOMINATING SET problem is APX-hard on B_1-EPG graphs, ruling out the possibility of a PTAS unless P = NP. Finally, to complement our APX-hardness result, we give constant-factor approximation algorithms for the MINIMUM DOMINATING SET problem on two nontrivial subclasses of B_1-EPG graphs.

1 Introduction

In this paper, we study the MINIMUM DOMINATING SET problem on B_1-VPG and B_1-EPG graphs. These are two special subclasses of *string graphs*, which are of interest in several applications such as circuit layout design and bioinformatics. A graph is called a B_k-*EPG graph* if it has an EPG representation (stands for Edge representation of Paths in a Grid) in which each path has at most k bends. In this paper, we are interested in B_k-EPG graphs for $k = 1$.

Definition 1 (B_1-EPG Graph). *A graph $G = (V, E)$ is called a B_1-EPG graph, if every vertex u of G can be represented as a path P_u on a grid $\mathcal{G}$ such that (i) P_u has at most one bend, and (ii) paths $P_u, P_v \in P$ share a grid edge of $\mathcal{G}$ if and only if $(u, v) \in E$.*

Similarly, a graph is said to have a VPG representation (stands for Vertex representation of Paths in a Grid), if its vertices can be represented as simple paths on an axis-parallel grid such that two vertices are adjacent if and only

© Springer International Publishing AG, part of Springer Nature 2018
R. Solis-Oba and R. Fleischer (Eds.): WAOA 2017, LNCS 10787, pp. 76–89, 2018.
https://doi.org/10.1007/978-3-319-89441-6_7

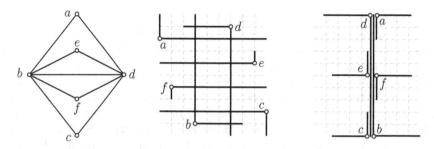

Fig. 1. A graph on six vertices (left) with its B_1-VPG (middle) and B_1-EPG (right) representations. Notice that the vertices e and f are not adjacent in the B_1-EPG representation as they only share a grid node (but not a grid edge).

if the corresponding paths share at least one grid node. Although these graphs were considered a while ago when studying string graphs [15], they were formally investigated by Asinowski et al. [2]. Similar to B_k-EPG graphs, a B_k-*VPG graph* is a VPG graph in which each path has at most k bends. In this paper, we are interested in B_k-VPG graphs for $k = 1$.

Definition 2 (B_1-VPG Graph). *A graph $G = (V, E)$ is called a B_1-VPG graph, if every vertex u of G can be represented as a path P_u on a grid $\mathcal{G}$ such that (i) P_u has at most one bend, and (ii) two paths P_u and P_v intersect each other at a grid node if and only if $(u, v) \in E$.*

We remark that by *intersecting* each other, we exclude the case where two paths only *touch* each other; that is, no endpoints of a path belongs to any other path. Figure 1 shows a graph with its B_1-VPG and B_1-EPG representations. A string graph is called *one-string* if it has a string representation in which curves intersect at most once [6]. By combining one-string and B_1-VPG representations, a *one-string* B_1-VPG graph is defined as a B_1-VPG graph in which two paths intersect each other exactly once whenever the corresponding vertices are adjacent.

In this paper, we study the MINIMUM DOMINATING SET problem on B_1-EPG and B_1-VPG graphs. Let $G = (V, E)$ be an unweighted, undirected graph. A set $S \subseteq V$ is a dominating set if every vertex in $V \backslash S$ is adjacent to some vertex in S. The objective of the MINIMUM DOMINATING SET problem is to compute a dominating set S of minimum size. MINIMUM DOMINATING SET is a fundamental optimization problem in graph theory, which arises in many applications such as wireless sensor networks, scheduling and resource allocation; the problem is well known to be NP-hard.

Related work and our results. It is known that every circle graph is a one-string B_1-VPG graph [2]. Since MINIMUM DOMINATING SET is APX-hard on circle graphs [8], the problem becomes APX-hard also on one-string B_1-VPG graphs. However, to the best of our knowledge, there is no approximation algorithm known for the problem. We note that there are $O(1)$-approximation algorithms for MINIMUM DOMINATING SET on circle graphs [9,10], but these algorithms do not work for B_1-VPG graphs as they heavily rely on the fact that the vertices of the input graph are modelled as chords of a circle. For B_1-EPG graphs, there exists a 4-approximation algorithm for the MINIMUM DOMINATING SET problem [5], as it is known that such graphs are a subclass of 2-interval graphs [12]. However, we were unable to find a reference on the complexity of the MINIMUM DOMINATING SET problem on B_1-EPG graphs. In fact, Epstein et al. [11] left open studying the MINIMUM DOMINATING SET problems on B_1-EPG graphs.

In this paper, we present the following results:

- We give a polynomial-time $O(1)$-approximation algorithm for the MINIMUM DOMINATING SET problem on *one-string* B_1-VPG graphs, providing the *first* constant-factor approximation algorithm for this problem on B_1-VPG graphs.
- We prove that the MINIMUM DOMINATING SET problem is APX-hard on B_1-EPG graphs, even if only two types of paths are allowed in the input graph. Thus, there exists no PTAS for this problem on B_1-EPG graphs unless $P = NP$.
- We give polynomial-time constant-factor approximation algorithms for the MINIMUM DOMINATING SET problem on two subclasses of B_1-EPG graphs.

Organization. We first give some notations and definitions in Sect. 2. We give our $O(1)$-approximation algorithm for the MINIMUM DOMINATING SET problem on one-string B_1-VPG graphs in Sect. 3. Then, we present our results for the MINIMUM DOMINATING SET problem on B_1-EPG graphs in Sect. 4. We conclude the paper with a discussion on open problems in Sect. 5.

2   Notation and Definitions

For a B_1-VPG (resp., B_1-EPG) graph G, we use $\langle \mathcal{P}_{vtx}, \mathcal{G} \rangle$ (resp., $\langle \mathcal{P}_{edg}, \mathcal{G} \rangle$) to denote a B_1-VPG (resp., B_1-EPG) representation of G, where $\mathcal{P}_{vtx}$ (resp., $\mathcal{P}_{edg}$) is the collection of paths corresponding to the vertices of G and $\mathcal{G}$ is the underlying grid. Since the recognition problem is NP-hard on such graphs [7, 12,17], we assume throughout this paper, that we are always given a string representation of a B_1-VPG or B_1-EPG graph as input (in addition to G). We sometimes violate the wording and say *path(s) in G* to actually refer to the vertices in G corresponding to the paths in $\mathcal{P}_{vtx}$ or $\mathcal{P}_{edg}$. We denote the x- and y-coordinates of a point p in the plane by $x(p)$ and $y(p)$, respectively.

Let P be a grid path with at most one bend. Since P has at most one bend, it is in one of the types $\{\ulcorner, \llcorner, \urcorner, \lrcorner, |, -\}$. We call a path P of type $x \in \{\ulcorner, \llcorner, \urcorner, \lrcorner\}$, a *$x$-type path*. Similar to [11], we complete the definition by referring to no-bend paths as $\llcorner$-type paths.

We denote the horizontal and vertical segments of P by $\mathtt{hPart}(P)$ and $\mathtt{vPart}(P)$, respectively. We call the common endpoint of $\mathtt{hPart}(P)$ and $\mathtt{vPart}(P)$

the *corner* of P and denote it by $\mathtt{corner}(P)$. Moreover, let $\mathtt{hTip}(P)$ (resp., $\mathtt{vTip}(P)$) denote the endpoint of $\mathtt{hPart}(P)$ (resp., $\mathtt{vPart}(P)$) that is not shared with $\mathtt{vPart}(P)$ (resp., not shared with $\mathtt{hPart}(P)$); see the figure on this page for an example. Let $N[P]$ denote the set of paths adjacent to P; we assume that $P \in N[P]$. Moreover, for a set S of paths, define $N[S] := \cup_{P \in S} N[P]$. We denote the set of neighbours of P that share at least one grid edge (or, a grid node) with $\mathtt{hPart}(P)$ (resp., with $\mathtt{vPart}(P)$) by $\mathtt{hNeighbor}(P)$ (resp., by $\mathtt{vNeighbor}(P)$).

3 Domination on B_1-VPG Graphs

Recall that the MINIMUM DOMINATING SET problem is known to be APX-hard on one-string B_1-VPG graphs. In the following, we give the first $O(1)$-approximation algorithm for MINIMUM DOMINATING SET on one-string B_1-VPG graphs. For the rest of this section, let G be a one-string B_1-VPG graph with n vertices, and let $\langle \mathcal{P}_{\mathtt{vtx}}, \mathcal{G} \rangle$ denote its string representation.

Our algorithm is based on first formulating the problem on G as a hitting set problem and then computing a small ε-net for the corresponding instance of the hitting set problem. Having such an ε-net along with the technique of Brönnimann and Goodrich [4] gives us an $O(1)$-approximation algorithm for MINIMUM DOMINATING SET on G.

A *set system* is a pair $\mathcal{R} = (\mathcal{U}, \mathcal{S})$, where $\mathcal{U}$ is a ground set of elements and $\mathcal{S}$ is a collection of subsets of $\mathcal{U}$. A *hitting set* for the set system $(\mathcal{U}, \mathcal{S})$ is a subset M of $\mathcal{U}$ such that $M \cap S \neq \emptyset$ for all $S \in \mathcal{S}$; we call each element of $\mathcal{U}$ a *hitting element*.

For the MINIMUM DOMINATING SET problem on G, we construct the set system $(\mathcal{U}, \mathcal{S})$ as follows. Let P be a ∟-type path. We associate a *cross* $c := (\ell_H, \ell_V)$ with P in which ℓ_H and ℓ_V denote its horizontal and vertical segments, respectively; we call ℓ_H and ℓ_V the *supporting segments* of c. The left endpoint of ℓ_H is

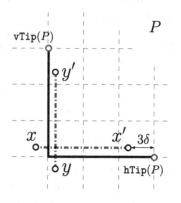

$$x := (x(\mathtt{corner}(P)) - \delta, y(\mathtt{corner}(P)))$$

and its right endpoint is

$$x' := ((x(\mathtt{hTip}(P)) - 3\delta, y(\mathtt{corner}(P)))$$

in which $\delta := w/4$, where w is the length of a grid edge. See the figure on the right. Analogously, the bottom endpoint of ℓ_V is $y := (x(\mathtt{corner}(P)), y(\mathtt{corner}(P)) - \delta)$ and its top endpoint is $y' := ((x(\mathtt{corner}(P)), y(\mathtt{vTip}(P)) - 3\delta)$. The cross of an $\mathbf{x}$-type path, where $\mathbf{x} \in \{ \ulcorner, \urcorner, \llcorner \}$, is defined analogously. We denote the cross of a path P by $\mathtt{cross}(P)$. The following observation is immediate by the construction of a cross.

Observation 1. *Let P_u, P_v be two paths in $\mathcal{P}_{vtx}$. Then, P_u and P_v intersect each other at a grid node if and only if $\mathrm{cross}(P_u)$ and $\mathrm{cross}(P_v)$ intersect each other.*

To see the hitting elements of the set system, for each path $P \in \mathcal{P}_{vtx}$, we add into $\mathcal{U}$ both of the supporting segments of $\mathrm{cross}(P)$. We say that a hitting element $e \in \mathcal{U}$ *hits* a cross c if e intersects one of the supporting segments of c; we assume that the hitting elements corresponding to supporting segments of c also hit c. To compute $\mathcal{S}$, consider the set of elements of $\mathcal{U}$ hitting a cross c and let $\mathcal{S}$ be the collection of these sets; notice that there is a set in $\mathcal{S}$ for each cross c. This forms the set system $(\mathcal{U}, \mathcal{S})$ corresponding to MINIMUM DOMINATING SET on G.

Lemma 1. *If there is a dominating set of size k on the one-string B_1-VPG graph G, then there is a hitting set of size at most $2k$ on the set system $(\mathcal{U}, \mathcal{S})$. Moreover, if there is a hitting set of size k on the set system $(\mathcal{U}, \mathcal{S})$, then there is a dominating set of size at most k on G.*

Proof. Let M be a dominating set of size k for G. For each path $P \in M$, add the supporting segments of $\mathrm{cross}(P)$ into M'. Clearly, $|M'| = 2k$. Now, consider a set $S \in \mathcal{S}$ and let P and c denote its corresponding path and cross, respectively. Since M is a dominating set, either $P \in M$ or $P' \in M$ for some path P' intersecting P. If $P \in M$, then clearly S is intersected by some segment in M'. If $P' \in M$, then c intersects $\mathrm{cross}(P')$ by Observation 1 and so at least one of the supporting segments of $\mathrm{cross}(P')$ (that is in M') intersects S.

Now, suppose that M' is a hitting set of size k for the set system $(\mathcal{U}, \mathcal{S})$. For each hitting element $e \in M'$: add P to M, where P is the path for which e is a supporting segment of $\mathrm{cross}(P)$. Clearly, $|M| \leq k$. To see why M is a dominating set for G, consider any path P and let $S \in \mathcal{S}$ denote its corresponding set. Since M' is a hitting set, there must be a hitting element $e' \in M'$ that intersects S. Let P' be the path for which e' is a supporting segment of $\mathrm{corner}(P')$; notice that $P' \in M$. If $P' = P$, then we are done. Otherwise, e' must hit the cross c corresponding to P since $c \in S$. This means that the cross corresponding to P' intersects c and so by Observation 1 P is intersected by P'. $\qquad\square$

An ε-*net* for a set system $\mathcal{R} = (\mathcal{U}, \mathcal{S})$ is a subset N of $\mathcal{U}$ such that every set S in $\mathcal{S}$ with size at least $\varepsilon \cdot |\mathcal{U}|$ has a non-empty intersection with N. Brönnimann and Goodrich [4] introduced an iterative-doubling approach to turn algorithms for finding ε-nets into approximation algorithms for hitting sets of minimum size (see the full version of the paper for a description of this approach).

Theorem 1. [4] *Let $\mathcal{R}$ be a set system that admits both a polynomial-time net finder and a polynomial-time verifier. Then, there exists a polynomial-time algorithm that computes a hitting set of size at most $s(4 \cdot \mathrm{OPT})$, where OPT is the size of a minimum hitting set, and $s(r)$ is the size of the $(1/r)$-net found by the net finder.*

Clearly, the hitting set problem corresponding to MINIMUM DOMINATING SET on G has a polynomial-time verifier. In the following, we compute an ε-net of size $O(1/\varepsilon)$ for the set system $(\mathcal{U}, \mathcal{S})$, which combined by Lemma 1 and Theorem 1 gives an $O(1)$-approximation algorithm for MINIMUM DOMINATING SET on G. To compute the ε-net, we first compute such an ε-net for a "one-dimensional" variant of our hitting set problem and then show that re-using such an ε-net twice would result in the desired ε-net. Recall the set system $(\mathcal{U}, \mathcal{S})$ corresponding to MINIMUM DOMINATING SET on G that we constructed before Lemma 1. Let $\mathcal{U}_H$ denote the set of only-horizontal hitting elements of $\mathcal{U}$, and consider the set system $(\mathcal{U}_H, \mathcal{S}_H)$ in which $\mathcal{S}_H$ is defined analogous to $\mathcal{S}$. Representing each cross by its vertical supporting segment only, the minimum hitting set problem on $(\mathcal{U}_H, \mathcal{S}_H)$ reduces to the following problem: given a set of horizontal line segments $\mathcal{H}$ and a set of vertical line segments $\mathcal{V}$, find a minimum-cardinality set $S \subseteq \mathcal{H}$ such that every line segment in $\mathcal{V}$ is intersected by S. This problem is known as the *orthogonal segment covering* problem and is shown to be NP-complete [13].

Lemma 2. *The minimum hitting set problem on $(\mathcal{U}_H, \mathcal{S})$ reduces to the orthogonal segment covering problem.*

Proof. ($\Rightarrow$) Let M be a feasible solution for the minimum hitting set problem on $(\mathcal{U}_H, \mathcal{S})$. If a hitting element e hits a cross c, then it intersects one of its supporting segments s. Let P denote the path in $\mathcal{P}_{\text{vtx}}$ corresponding to cross c. Since e is horizontal, s cannot be horizontal because if e intersects the horizontal supporting segment of P (i.e., s), then $\text{vPart}(P)$ and $\text{hPart}(P')$ must intersect each other in more than one point, where P' is the path in $\mathcal{P}_{\text{vtx}}$ for which e is a supporting segment. This contradicts the fact that G is a one-string B_1-EPG graph. Therefore, s is vertical and so for any feasible solution M, we have a feasible solution for the orthogonal segment covering problem with the same size.

($\Leftarrow$) Clearly, a feasible solution to the minimum segment covering problem is also a feasible solution to the minimum hitting set problem with the same size.   □

For the orthogonal segment covering problem, Biedl et al. [3] showed that there exists an ε-net of size $s(1/\varepsilon) \in O(1/\varepsilon)$. Therefore, by Lemma 2, we have the following result.

Lemma 3. *There exists an ε-net of size $O(1/\varepsilon)$ for the minimum hitting set problem on $(\mathcal{U}_H, \mathcal{S}_H)$.*

Lemma 4. *There exists a polynomial-time $O(1)$-approximation algorithm for the minimum hitting set problem on $(\mathcal{U}, \mathcal{S})$.*

Proof. To prove the lemma, it suffices by Theorem 1 to compute an ε-net of size $O(1/\varepsilon)$ for $(\mathcal{U}, \mathcal{S})$. Define the set system $(\mathcal{U}_V, \mathcal{S}_V)$ similar to $(\mathcal{U}_H, \mathcal{S}_H)$ (i.e., $\mathcal{U}_V$ is the set of only-vertical hitting elements of $\mathcal{U}$ and define $\mathcal{S}_V$ analogously). By Lemma 3, let N_H (resp., N_V) be an $(\varepsilon/2)$-net of size $O(1/\varepsilon)$ for the corresponding one-dimensional variant of the hitting set problem on $(\mathcal{U}_H, \mathcal{S}_H)$ (resp., $(\mathcal{U}_V, \mathcal{S}_V)$), and let $N := N_H \cup N_V$. Clearly, N has size $O(1/\varepsilon)$. We now prove that N is an ε-net.

Let $S \in \mathcal{S}$ such that $|S| \geq \varepsilon \cdot |\mathcal{U}|$. We need to show that $N \cap S \neq \emptyset$. Notice that having $|S| \geq \varepsilon \cdot |\mathcal{U}|$ means that there exists a cross c that is hit by $\varepsilon \cdot |\mathcal{U}|$ hitting elements. Assume w.l.o.g. that at least half of these hitting elements are horizontal. Then, the vertical supporting segment of c intersects at least $\varepsilon \cdot |\mathcal{U}/2|$ horizontal hitting elements of $\mathcal{U}$. By definition of an $(\varepsilon/2)$-net, there is a hitting element $e \in N_H$ that intersects the vertical supporting segment of c. Therefore, e hits cross c and so $e \in N \cap S$. □

Putting everything together, we can prove the main result of this section.

Theorem 2. *There exists a polynomial-time $O(1)$-approximation algorithm for* MINIMUM DOMINATING SET *on any one-string* B_1-VPG *graph.*

Proof. Let G be any one-string B_1-VPG graph with its string representation $\langle \mathcal{P}_{\mathtt{vtx}}, \mathcal{G} \rangle$. To prove the theorem, it is sufficient to show that there exists a polynomial-time algorithm that finds a constant-factor approximation of MINIMUM DOMINATING SET in G.

First, compute the set system $(\mathcal{U}, \mathcal{S})$ corresponding to $\mathcal{P}_{\mathtt{vtx}}$ as described above. Let OPT_{DS} (resp., OPT_{HS}) denote an optimal solution for MINIMUM DOMINATING SET on G (resp., for the minimum hitting set problem on $(\mathcal{U}, \mathcal{S})$). By Lemma 1, we know that $|\mathrm{OPT}_{HS}| \leq 2|\mathrm{OPT}_{DS}|$. By Lemma 4, let S_{HS} be a constant-factor approximation to the minimum hitting set problem on $(\mathcal{U}, \mathcal{S})$; that is, $|S_{HS}| \leq d \cdot |\mathrm{OPT}_{HS}|$ for some constant $d > 0$. Apply Lemma 1 to S_{HS} and let S_{DS} be a feasible solution for MINIMUM DOMINATING SET on G; notice that $|S_{DS}| \leq |S_{HS}|$. The set system $(\mathcal{U}, \mathcal{S})$, S_{HS} and S_{DS} can each be computed in polynomial time. Therefore,

$$\frac{|S_{DS}|}{|\mathrm{OPT}_{DS}|} \leq \frac{|S_{HS}|}{|\mathrm{OPT}_{DS}|} \leq \frac{2|S_{HS}|}{|\mathrm{OPT}_{HS}|} \leq 2d.$$

That is, S_{DS} is an $O(1)$-approximation of MINIMUM DOMINATING SET on G, which can be computed in polynomial time. □

4   Domination on B_1-EPG Graphs

In this section, we consider the MINIMUM DOMINATING SET problem on B_1-EPG graphs. We first show that the problem is APX-hard, even if only two types of paths are allowed in the graph; hence, ruling out the possibility of a PTAS for this problem unless $P = NP$. Recall that a 4-approximation algorithm is already known for MINIMUM DOMINATING SET on B_1-EPG graphs [5,12]. Here, in Sect. 4.2, we give c-approximation algorithms for this problem on two subclasses of B_1-EPG graphs, for small values of c.

4.1   APX-Hardness

To show the APX-hardness, we give an L-reduction [16] from the MINIMUM VERTEX COVER problem on graphs with maximum-degree three to the MINIMUM DOMINATING SET on B_1-EPG graphs; MINIMUM VERTEX COVER is known to be APX-hard on graphs with maximum-degree three [1].

Lemma 5. MINIMUM VERTEX COVER *on graphs with maximum-degree three is L-reducible to* MINIMUM DOMINATING SET *on* B_1-EPG *graphs.*

Proof. Consider an arbitrary instance I of MINIMUM VERTEX COVER on graphs of maximum-degree three; let $G = (V, E)$ be the graph corresponding to I and let k be the size of the smallest vertex cover in G. First, let $u_1, \ldots, u_n$ be an arbitrary ordering of the vertices of G, where $n = |V|$. In the following, we give a computable function f that takes I as input and outputs an instance $f(I)$ of MINIMUM DOMINATING SET in polynomial time, where $f(I)$ consists of a B_1-EPG graph such that its paths are of either $\ulcorner$-type or $\lrcorner$-type only.

We first describe the vertex gadgets. For each vertex u_i, $1 \leq i \leq n$, construct a horizontal $\ulcorner$-type Γ_i^h and a vertical $\ulcorner$-type Γ_i^v, and connect them as shown in Fig. 2. We call the big $\ulcorner$-type path used in the connection of Γ_i^h and Γ_i^v the *big connector* C_i of i, and the two small (blue, dashed) $\ulcorner$-type paths the *small connectors* of i. For each edge $(u_i, u_j) \in E$, where $i < j$, we add two small paths, one of $\ulcorner$-type and one of $\lrcorner$-type, at the intersection point of Γ_i^v and Γ_j^h such that each of them becomes adjacent to both Γ_i^v and Γ_j^h; see the two (red, dash-dotted) $\ulcorner$-type and $\lrcorner$-type paths at the intersection of Γ_1^v and Γ_2^h in Fig. 2. We denote this pair of paths by $E_{i,j}$; notice that the paths of $E_{i,j}$ are not adjacent to each other (they only share a grid node). This gives the instance $f(I)$ of

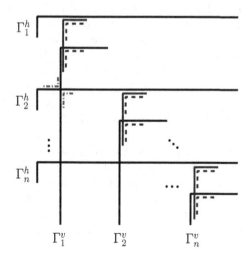

Fig. 2. An illustration in support of the construction in the proof of Lemma 5. (Color figure online)

MINIMUM DOMINATING SET on B_1-EPG graphs; let G' be the corresponding B_1-EPG graph. Notice that f is a polynomial-time computable function. In the following, we denote an optimal solution for the instance X of a problem by $s^*(X)$. We now prove that all the four conditions of L-reduction hold.

First, let M be a vertex cover of G of size k. Denote by $\Gamma^h[M] := \{\Gamma_i^h | u_i \in M\}$ the set of horizontal paths induced by M and define $\Gamma^v[M]$ analogously. Moreover, let $C[M] := \{C_i | u_i \notin M\}$ be the set of big connectors whose corresponding vertex is not in M. We show that $D := \Gamma^h[M] \cup \Gamma^v[M] \cup C[M]$ is a dominating set of G'. Let γ be a path. If γ is any of the paths in $E_{i,j}$ for some i, j, then $u_i \in M$ or $u_j \in M$ because M is a vertex cover; assume w.l.o.g. that $u_i \in M$. Then, $\Gamma_i^h, \Gamma_i^v \in D$ and so γ must be dominated. If γ is a connector of i for some i (either big or small), then there are two cases: if $u_i \in M$, then $\Gamma_i^h, \Gamma_i^v \in D$ and so the connector is dominated. If $u_i \notin M$, then $C_i \in C[M]$ and $C[M] \subseteq D$; hence, the connector is again dominated. Finally, suppose that γ is Γ_i^h (resp., Γ_i^v) for some i. If $u_i \in M$, then $\Gamma_i^h, \Gamma_i^v \in D$ and so Γ_i^h (resp., Γ_i^v) is dominated. If $u_i \notin M$, then $C_i \in C[M]$ and $C[M] \subseteq D$; hence, Γ_i^h (resp., Γ_i^v) is again dominated. This shows that D is a dominating set for G'.

Second, let D be an arbitrary dominating set on G'. First, notice that we can construct a dominating set D' for G' such that $|D'| \leq |D|$ and D' consists of only Γ_i^h and Γ_i^v for some i, or a big connector. This is because (i) any path dominated by a small connector is also dominated by some big connector, and (ii) any path dominated by a path from $E_{i,j}$ is also dominated by Γ_i^v or Γ_j^h. For (ii), in particular, if exactly one of the paths in $E_{i,j}$ is in D, then at least one of Γ_i^v or Γ_j^h must be in D in order to dominate the other path of $E_{i,j}$; hence, we can replace the path of $E_{i,j}$ in D with one of Γ_i^v or Γ_j^h arbitrarily. Otherwise, if both paths of $E_{i,j}$ are in D, then we can replace both of them with Γ_i^v and Γ_j^h. So, $|D'| \leq |D|$ and D' is a feasible dominating set for G'. Now, define $\Gamma_{\text{both}}[D'] = \{\Gamma_i^h, \Gamma_i^v | \Gamma_i^h, \Gamma_i^v \in D'\}$; i.e., the paths of a vertex u_i, where *both* its horizontal and its vertical copies appear in D'. Also, define $\Gamma_{\text{one}}[D']$ to be the remaining paths of type Γ_i^h and Γ_i^v; i.e., those of u_i, where *only* one of its copies appears in D'. Finally, let $C[D']$ be the set of big connectors in D'. We denote $\Gamma_{\text{both}}[D'] \cup \Gamma_{\text{one}}[D']$ by $\Gamma[D']$. Now, let $M := \{u_i | \Gamma_i^h \in D' \text{ or } \Gamma_i^v \in D'\}$. Since all $E_{i,j}$ are dominated by $\Gamma[D']$, M is a vertex cover.

Third, observe that $|\Gamma^h[M]| = |\Gamma^v[M]| = |M| = k$ and also $|C[M]| = n - k$. Given that G has degree three, $k \geq n/4$ and so $|s^*(f(I))| \leq n - k + k + k \leq 5k \leq 5|s^*(I)|$.

Finally, to dominate all connectors of i, we must have $C_i \in D'$ or $\Gamma_i^h, \Gamma_i^v \in D'$; this indeed holds for all i. Thus, $|C[D']| + |\Gamma_{\text{both}}[D']|/2 \geq n$. Moreover, $|\Gamma_{\text{one}}[D']| + |\Gamma_{\text{both}}[D']|/2 \geq k$ since M is a vertex cover of G. Therefore, $|D'| \geq |\Gamma_{\text{both}}[D']| + |\Gamma_{\text{one}}[D']| + |C[D']| \geq |\Gamma_{\text{one}}[D']| + |\Gamma_{\text{both}}[D']|/2 + n \geq k + n$. By this and our earlier inequality $|s^*(f(I))| \leq n - k + k + k$, we have $|s^*(f(I))| = n + k$. Now, suppose that $|D| = |s^*(f(I))| + c$ for some $c \geq 0$. Then,

$$|D| - |s^*(f(I))| = c$$
$$\Rightarrow |D| - (n + k) = c$$

$$\Rightarrow |D'| - (n + k) \le c$$
$$\Rightarrow |\Gamma_{\text{one}}[D']| + |\Gamma_{\text{both}}[D']|/2 + n - (n + k) \le c$$
$$\Rightarrow |\Gamma_{\text{one}}[D']| + |\Gamma_{\text{both}}[D']|/2 - k \le c$$
$$\Rightarrow |M| - |s^*(I)| \le c.$$

That is, $|M| - |s^*(I)| \le |D| - |s^*(f(I))|$. This concludes our L-reduction from MINIMUM VERTEX COVER on graphs of maximum-degree three to MINIMUM DOMINATING SET on B_1-EPG graphs with $\alpha = 5$ and $\beta = 1$.   □

Our reduction reveals that every path in the constructed B_1-EPG graph G' is a $\ulcorner$-type or a $\llcorner$-type path. Therefore, by Lemma 5, we have the following theorem.

Theorem 3. *The* MINIMUM DOMINATING SET *problem is* APX-*hard on* B_1-EPG *graphs, even if all the paths in the graph are of type* $\ulcorner$ *or* $\llcorner$. *Thus, there is no PTAS for this problem on* B_1-EPG *graphs unless* P = NP.

4.2   Approximation Algorithms

In this section, we give constant-factor approximation algorithms for the MINIMUM DOMINATING SET problem on two subclasses of B_1-EPG graphs. Let us first define these subclasses.

First, we consider a subclass of B_1-EPG graphs in which every path of each B_1-EPG graph intersects two axis-parallel lines that are normal to each other. We notice that this variant has already been considered, where Lahiri et al. [14] gave an exact solution for the MAXIMUM INDEPENDENT SET problem when the input graph is a B_1-VPG graph: they showed that the induced graph is a co-comparability graph and so solved the MAXIMUM INDEPENDENT SET problem exactly. However, the graph induced by this variant when considering B_1-EPG graphs is not necessarily a co-comparability graph; this is mainly because two paths intersecting in only one point in a B_1-EPG graph are not adjacent. Here, we give a 2-approximation algorithm for this problem on such B_1-EPG graphs. We call this subclass, the class of DOUBLE-CROSSING B_1-EPG graphs. Next, we consider a less-restricted subclass of B_1-EPG graphs in which every path of each B_1-EPG graph intersects only a vertical line. We show that the same algorithm is a 3-approximation algorithm for the problem on this subclass of B_1-EPG graphs, albeit considering a "non-containment" assumption. We call this subclass, the class of VERTICAL-CROSSING B_1-EPG graphs.

Before describing the algorithms, let us define an ordering $\prec$ on the paths in $\mathcal{P}_{\text{edg}}$ as follows. The paths appear in the ordering by the y-coordinate of their corners from bottom to top and then from left to right whenever they have the same y-coordinate; that is, $P \prec P'$ for two paths in $\mathcal{P}_{\text{edg}}$, if and only if $y(\text{corner}(P)) < y(\text{corner}(P'))$ or $y(\text{corner}(P)) = y(\text{corner}(P'))$ but $x(\text{corner}(P)) < x(\text{corner}(P'))$; we break ties arbitrarily to complete the ordering. For the rest of this section, we assume that every path in $\mathcal{P}_{\text{edg}}$ is a $\llcorner$-type path.

Algorithm 1. APPROXIMATELINEMDS(G, L_1, L_2)

1: $S \leftarrow \emptyset$;
2: **for** each path $P \in \mathcal{P}_{\text{edg}}$ in increasing order $\prec$ **do**
3:     $S \leftarrow S \cup P$;
4:     $\mathcal{P}_{\text{edg}} \leftarrow \mathcal{P}_{\text{edg}} \backslash N[P]$;
5: **return** S;

Double-crossing B_1-EPG *graphs.* For a DOUBLE-CROSSING B_1-EPG graph, we are given a B_1-EPG graph G, a horizontal line L_1 and a vertical line L_2 both on the grid $\mathcal{G}$ such that L_1 and L_2 intersect each other and P intersects both L_1 and L_2 for all $P \in \mathcal{P}_{\text{edg}}$ (hence, $\text{corner}(P)$ lies in the lower-left quadrant defined by L_1 and L_2). Our 2-approximation algorithm for the MINIMUM DOMINATING SET problem is as follows; let S be an initially-empty set. For each path P in the increasing order $\prec$: add P into S and set $\mathcal{P}_{\text{edg}}:=\mathcal{P}_{\text{edg}} \backslash N[P]$. See Algorithm 1. Clearly, the algorithm terminates in time polynomial in $|\mathcal{P}_{\text{edg}}|$, and S is a feasible solution for the problem. To see the approximation factor, let OPT be an optimal solution for the MINIMUM DOMINATING SET problem on G; notice that by deleting the paths in $S \cap OPT$ we can assume that $S \cap OPT = \emptyset$. This means that every path in S must be adjacent to at least one path in OPT.

Lemma 6. *Any path in OPT is adjacent to at most two distinct paths in S.*

Proof. Suppose for a contradiction that there exists a path $P \in OPT$ that is adjacent to three distinct paths P_1, P_2 and P_3 of S, where $P_1 \prec P_2 \prec P_3$. W.l.o.g., assume that $P_i, P_j \in \text{vNeighbor}(P)$ for some $i < j \in \{1, 2, 3\}$. This means that $x(\text{corner}(P)) = x(\text{corner}(P_i)) = x(\text{corner}(P_j))$. Since $i < j$, we have that $y(\text{corner}(P_i)) < y(\text{corner}(P_j))$. Moreover, since all the paths in $\mathcal{P}_{\text{edg}}$ intersect the horizontal line L_1, the three paths P, P_i and P_j must all intersect L_1 at the same point and so they share the top-most vertical grid edge below L_1 on which $\text{vPart}(P)$ lies; see the figure on the right. Thus, P_i and P_j are adjacent in G. Since $y(\text{corner}(P_i)) < y(\text{corner}(P_j))$, we have $P_i \prec P_j$ and so P_j is removed from $\mathcal{P}_{\text{edg}}$ when the algorithm adds P_i to S. So, $P_j \notin S$—a contradiction. $\square$

Since every path in S must be adjacent to at least one path in OPT and any path in OPT can be adjacent to at most two distinct paths in S by Lemma 6, we have $|S| \leq 2|OPT|$. Therefore, we have the following theorem.

Theorem 4. *There exists a polynomial-time 2-approximation algorithm for the MINIMUM DOMINATING SET problem on DOUBLE-CROSSING B_1-EPG graphs.*

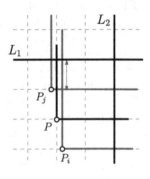

Vertical-crossing B_1-EPG *graphs.* Here, we are given a B_1-EPG graph G and a vertical line ℓ on the grid $\mathcal{G}$ such that $\texttt{hPart}(P)$ intersects ℓ for all $P \in \mathcal{P}_{\texttt{edg}}$. Moreover, we make a *non-containment assumption* in the sense that the vertical segment of no path is entirely contained in that of any other path in $\mathcal{P}_{\texttt{edg}}$; that is, for every two paths $P, P' \in \mathcal{P}_{\texttt{edg}}$ such that $P \in \texttt{vNeighbor}(P')$, neither $\texttt{vPart}(P) \subseteq \texttt{vPart}(P')$ nor $\texttt{vPart}(P') \subseteq \texttt{vPart}(P)$. We prove that Algorithm 1 is a 3-approximation algorithm for the MINIMUM DOMINATING SET problem on VERTICAL-CROSSING B_1-EPG graphs.

Theorem 5. *Algorithm 1 is a 3-approximation algorithm for the* MINIMUM DOMINATING SET *problem on* VERTICAL-CROSSING B_1-EPG *graphs under the non-containment assumption.*

Proof. Let G be any VERTICAL-CROSSING B_1-EPG graph. Moreover, let OPT be an optimal solution for the MINIMUM DOMINATING SET problem on G and let S be the solution returned by Algorithm 1. Again, we can assume that $S \cap OPT = \emptyset$. Thus, every path in S must be dominated by at least one path in OPT. In the following, we show that any path in OPT can dominate at most three distinct paths in S and so prove that $|S| \leq 3|OPT|$. Let P be any path in OPT. We show that $|S \cap \texttt{hNeighbor}(P)| \leq 1$ and $|S \cap \texttt{vNeighbor}(P)| \leq 2$.

First, suppose for a contradiction that there are two paths $P_1, P_2 \in S \cap \texttt{hNeighbor}(P)$; assume w.l.o.g. that $P_1 \prec P_2$. Notice that $y(\texttt{corner}(P)) = y(\texttt{corner}(P_1)) = y(\texttt{corner}(P_2))$. Thus, since all the three paths P, P_1 and P_2 intersect the vertical line ℓ, we conclude that they all intersect ℓ at the same point. Therefore, they share the rightmost horizontal grid edge to the left of ℓ on which $\texttt{hPart}(P)$ lies. This means that P_1 and P_2 are adjacent in G. Since $P_1 \prec P_2$, Algorithm 1 removes P_2 from $\mathcal{P}_{\texttt{edg}}$ when adding P_1 into S; that is, $P_2 \notin S$—a contradiction. So, $|S \cap \texttt{hNeighbor}(P)| \leq 1$.

Now, suppose for a contradiction that there are three paths $P_1, P_2, P_3 \in S \cap \texttt{vNeighbor}(P)$; assume w.l.o.g. that $P_1 \prec P_2 \prec P_3$. Notice that $x(\texttt{corner}(P)) = x(\texttt{corner}(P_i))$ for all $1 \leq i \leq 3$. Consider $y(\texttt{corner}(P))$. Then, for at least two paths P_i, P_j, where $i < j \in \{1, 2, 3\}$, we have

$$y(\texttt{corner}(P_i)), y(\texttt{corner}(P_j)) < y(\texttt{corner}(P))$$

or

$$y(\texttt{corner}(P_i)), y(\texttt{corner}(P_j)) \geq y(\texttt{corner}(P)).$$

- If $y(\texttt{corner}(P_i)), y(\texttt{corner}(P_j)) < y(\texttt{corner}(P))$, then we know $\texttt{vPart}(P_i)$ and $\texttt{vPart}(P_j)$ both share with P the bottom-most vertical grid edge on which $\texttt{vPart}(P)$ lies, implying that P_i and P_j are adjacent in G. Since $i < j$, we have $P_i \prec P_j$ and so Algorithm 1 removes P_j from G when adding P_i to S. So, $P_j \notin S$—a contradiction.
- If $y(\texttt{corner}(P_i)), y(\texttt{corner}(P_j)) \geq y(\texttt{corner}(P))$, then

$$y(\texttt{vTip}(P_i)) > y(\texttt{vTip}(P)) \text{ and } y(\texttt{vTip}(P_j)) > y(\texttt{vTip}(P)),$$

because otherwise $\texttt{vPart}(P_i) \subseteq \texttt{vPart}(P)$ or $\texttt{vPart}(P_j) \subseteq \texttt{vPart}(P)$, which is a contradiction to the non-containment assumption of paths in G. Since

$i < j$, we have $P_i \prec P_j$. Therefore, P_i and P_j share the top-most vertical grid edge on which $\text{vPart}(P)$ lies, meaning that P_i and P_j are adjacent in G. So, Algorithm 1 removes P_j when adding P_i into S—a contradiction to $P_j \in S$.

Therefore, $|S \cap \text{hNeighbor}(P)| \leq 1$ and $|S \cap \text{vNeighbor}(P)| \leq 2$. This completes the proof of the theorem. □

5   Conclusion

In this paper, we studied the MINIMUM DOMINATING SET problem on B_1-VPG and B_1-EPG graphs. For B_1-VPG graphs, we gave an $O(1)$-approximation algorithm for this problem when the input graph is one-string. For B_1-EPG graphs, we proved that MINIMUM DOMINATING SET is APX-hard (even if the graph has only two types of paths), ruling out the existence of a PTAS unless P = NP. We also gave c-approximation algorithms for this problem on two subclasses of B_1-EPG graphs, for $c \in \{2, 3\}$. We conclude the paper by the following open problems:

- Our $O(1)$-approximation algorithm for MINIMUM DOMINATING SET on one-string B_1-VPG graphs relies on the fact that the input graph is one-string (in the proof of Lemma 2, in particular); is there an $O(1)$-approximation algorithm for this problem on any B_1-VPG graph?
- Is the MINIMUM DOMINATING SET problem APX-hard on B_1-EPG graphs, if the graph consists of only one type of paths? We believe that a slight modification to our APX-hardness result would answer this question affirmatively.
- Our 2- and 3-approximation algorithms for MINIMUM DOMINATING SET work only on the described subclasses of B_1-EPG graphs; is there an α-approximation algorithm for the MINIMUM DOMINATING SET problem on any B_1-EPG graph, for some $\alpha < 4$? (Recall that a 4-approximation algorithm is already known [5,12].)
- Is the MINIMUM DOMINATING SET problem NP-hard on VERTICAL-CROSSING B_1-EPG graphs?

References

1. Alimonti, P., Kann, V.: Some APX-completeness results for cubic graphs. Theor. Comput. Sci. **237**(1–2), 123–134 (2000)
2. Asinowski, A., Cohen, E., Golumbic, M.C., Limouzy, V., Lipshteyn, M., Stern, M.: Vertex intersection graphs of paths on a grid. J. Graph Algorithms Appl. **16**(2), 129–150 (2012)
3. Biedl, T., Chan, T.M., Lee, S., Mehrabi, S., Montecchiani, F., Vosoughpour, H.: On guarding orthogonal polygons with sliding cameras. In: Poon, S.-H., Rahman, M.S., Yen, H.-C. (eds.) WALCOM 2017. LNCS, vol. 10167, pp. 54–65. Springer, Cham (2017). https://doi.org/10.1007/978-3-319-53925-6_5
4. Brönnimann, H., Goodrich, M.T.: Almost optimal set covers in finite VC-dimension. Discret. Comput. Geom. **14**(4), 463–479 (1995)

5. Butman, A., Hermelin, D., Lewenstein, M., Rawitz, D.: Optimization problems in multiple-interval graphs. ACM Trans. Algorithms **6**(2), 40:1–40:18 (2010)
6. Chalopin, J., Gonçalves, D., Ochem, P.: Planar graphs have 1-string representations. Discret. Comput. Geom. **43**(3), 626–647 (2010)
7. Chaplick, S., Jelínek, V., Kratochvíl, J., Vyskočil, T.: Bend-bounded path intersection graphs: sausages, noodles, and waffles on a grill. In: Golumbic, M.C., Stern, M., Levy, A., Morgenstern, G. (eds.) WG 2012. LNCS, vol. 7551, pp. 274–285. Springer, Heidelberg (2012). https://doi.org/10.1007/978-3-642-34611-8_28
8. Damian, M., Pemmaraju, S.V.: APX-hardness of domination problems in circle graphs. Inf. Process. Lett. **97**(6), 231–237 (2006)
9. Damian-Iordache, M., Pemmaraju, S.V.: Constant-factor approximation algorithms for domination problems on circle graphs. ISAAC 1999. LNCS, vol. 1741, pp. 70–82. Springer, Heidelberg (1999). https://doi.org/10.1007/3-540-46632-0_8
10. Damian-Iordache, M., Pemmaraju, S.V.: A $(2+\varepsilon)$-approximation scheme for minimum domination on circle graphs. J. Algorithms **42**(2), 255–276 (2002)
11. Epstein, D., Golumbic, M.C., Morgenstern, G.: Approximation algorithms for B_1-EPG graphs. In: Dehne, F., Solis-Oba, R., Sack, J.-R. (eds.) WADS 2013. LNCS, vol. 8037, pp. 328–340. Springer, Heidelberg (2013). https://doi.org/10.1007/978-3-642-40104-6_29
12. Heldt, D., Knauer, K.B., Ueckerdt, T.: Edge-intersection graphs of grid paths: the bend-number. Discret. Appl. Math. **167**, 144–162 (2014)
13. Katz, M.J., Mitchell, J.S.B., Nir, Y.: Orthogonal segment stabbing. Comput. Geom. **30**(2), 197–205 (2005)
14. Lahiri, A., Mukherjee, J., Subramanian, C.R.: Maximum independent set on B_1-VPG graphs. In: Lu, Z., Kim, D., Wu, W., Li, W., Du, D.-Z. (eds.) COCOA 2015. LNCS, vol. 9486, pp. 633–646. Springer, Cham (2015). https://doi.org/10.1007/978-3-319-26626-8_46
15. Middendorf, M., Pfeiffer, F.: The max clique problem in classes of string-graphs. Discret. Math. **108**(1–3), 365–372 (1992)
16. Papadimitriou, C.H., Yannakakis, M.: Optimization, approximation, and complexity classes. J. Comput. Syst. Sci. **43**(3), 425–440 (1991)
17. Pergel, M., Rzążewski, P.: On edge intersection graphs of paths with 2 bends. In: Heggernes, P. (ed.) WG 2016. LNCS, vol. 9941, pp. 207–219. Springer, Heidelberg (2016). https://doi.org/10.1007/978-3-662-53536-3_18

Submodular Function Minimization with Submodular Set Covering Constraints and Precedence Constraints

Naoyuki Kamiyama[1,2]([✉])

[1] Institute of Mathematics for Industry, Kyushu University, Fukuoka, Japan
kamiyama@imi.kyushu-u.ac.jp
[2] JST, PRESTO, Saitama, Japan

Abstract. In this paper, we consider the submodular function minimization problem with submodular set covering constraints and precedence constraints, and we prove that the algorithm of McCormick, Peis, Verschae, and Wierz for the precedence constrained covering problem can be generalized to our setting.

1 Introduction

Assume that we are given a finite set U. Then a real-valued function g with a domain 2^U is said to be *submodular*, if for every pair of subsets X, Y of U,

$$g(X) + g(Y) \geq g(X \cup Y) + g(X \cap Y).$$

Submodular functions play an important role in many fields, e.g., combinatorial optimization, machine learning, and game theory. One of the most fundamental problems related to submodular functions is the submodular function minimization problem. In this problem, we are given a submodular function g with a domain 2^U. Then the goal of the submodular function minimization problem is to find a subset X of U minimizing $g(X)$ among all subsets of U, i.e., to find a minimizer of the function g. It is known [1–4] that this problem can be solved in polynomial time if we are given a value oracle for the function g.

Furthermore, constrained variants of the submodular function minimization problem have been extensively studied in various fields [5–15]. For example, Iwata and Nagano [7] considered the submodular function minimization problem with vertex covering constraints, set covering constraints, and edge covering constraints, and gave approximability and inapproximability results. Goel et al. [5] considered the vertex cover problem, the shortest path problem, the perfect matching problem, and the minimum spanning tree problem with a monotone submodular cost function. Svitkina and Fleischer [14] also considered several optimization problems with a submodular cost function. Jegelka and Bilmes [11] considered the submodular function minimization problem with cut constraints. Zhang and Vorobeychik [15] considered the submodular function minimization problem with routing constraints. Hochbaum [6] considered

© Springer International Publishing AG, part of Springer Nature 2018
R. Solis-Oba and R. Fleischer (Eds.): WAOA 2017, LNCS 10787, pp. 90–101, 2018.
https://doi.org/10.1007/978-3-319-89441-6_8

the submodular minimization problem with linear constraints having at most two variables per inequality. Kamiyama [16] considered the submodular function minimization problem with covering type linear constraints. Koufogiannakis and Young [13] considered the monotone submodular function minimization problem with general covering constraints. Furthermore, Iyer and Bilmes [8] and Kamiyama [12] considered the submodular function minimization problem with submodular set covering constraints (see Sect. 2 for its formal definition).

In this paper, we consider the submodular function minimization problem with submodular set covering constraints and precedence constraints. This problem is inspired by the precedence constrained covering problem proposed by McCormick et al. [17]. This problem is an integer program with covering type inequalities and precedence constraints over variables. The submodular set cover problem with a linear cost function was introduced by Wolsey [18]. A greedy algorithm [18] and a primal-dual algorithm [19] were proposed for this problem. Applications of the submodular set cover problem include the capacitated supply-demand problem [20] and the bounded degree deletion problem [21]. See also [8] for its applications. The submodular function minimization problem with submodular set covering constraints was considered by Iyer and Bilmes [8] and Kamiyama [12]. In this paper, we prove that the algorithm of McCormick et al. [17] for the precedence constrained covering problem can be generalized to our setting by using the technique of Kamiyama [12] that is based on the result of Iwata and Nagano [7].

2   Preliminaries

Throughout this paper, we denote by $\mathbb{R}$ and $\mathbb{R}_+$ the sets of real numbers and non-negative real numbers, respectively. Assume that we are given a finite set U. Then for each vector x in $\mathbb{R}^U$ and each subset X of U, we define $x(X) := \sum_{i \in X} x(i)$. Furthermore, a function $g \colon 2^U \to \mathbb{R}$ is said to be *monotone*, if $g(X) \leq g(Y)$ for every pair of subsets X, Y of U such that $X \subseteq Y$.

In this paper, we consider the submodular function minimization problem with submodular set covering constraints and precedence constraints defined as follows. In this problem, we are given a non-empty finite set N and monotone submodular functions $\rho, \mu \colon 2^N \to \mathbb{R}_+$ such that $\rho(\emptyset) = 0$ and $\mu(\emptyset) = 0$. We assume that for every subset X of N, we can compute $\rho(X), \mu(X)$ in time bounded by a polynomial in $|N|$. Furthermore, we are given a partial order $\preceq$ over N (i.e., a reflexive, antisymmetric, and transitive order over N). Define $\mathcal{L}$ as the family of subsets X of N satisfying the condition that if $j \in X$, then $i \in X$ for every pair of elements i, j in N such that $i \preceq j$. Then the submodular function minimization problem with submodular set covering constraints and precedence constraints is defined as follows.

$$\begin{aligned}
\text{Minimize} \quad & \rho(X) \\
\text{subject to} \quad & \mu(X) = \mu(N) \\
& X \in \mathcal{L}.
\end{aligned} \tag{1}$$

We assume without loss of generality that $\mu(N) > 0$. Otherwise, $\mu(X) = 0$ for every subset X of N, and thus $\emptyset$ is an optimal solution of the problem (1).

For each subset X of N, let χ_X be the vector in $\{0,1\}^N$ satisfying the condition that $\chi_X(i) = 1$ for every element i in X, and $\chi_X(i) = 0$ for every element i in $N \setminus X$. For each pair of elements i, j in N such that $i \preceq j$ and $i \neq j$, we write $i \prec j$. For each subset S of N, we define the function $\mu_S : 2^{N \setminus S} \to \mathbb{R}$ by

$$\mu_S(X) := \mu(X \cup S) - \mu(S).$$

The *Lovász extension* $\widehat{\rho} : \mathbb{R}_+^N \to \mathbb{R}_+$ of ρ is defined as follows [22]. Assume that we are given a vector x in $\mathbb{R}_+^N$. Furthermore, we assume that for non-negative real numbers $\hat{x}_1, \hat{x}_2, \ldots, \hat{x}_s$ such that $\hat{x}_1 > \hat{x}_2 > \cdots > \hat{x}_s$, $\{\hat{x}_1, \hat{x}_2, \ldots, \hat{x}_s\} = \{x(i) \mid i \in N\}$ holds. For each integer p in $\{1, 2, \ldots, s\}$, we define N_p as the set of elements i in N such that $x(i) \geq \hat{x}_p$. Define $\widehat{\rho}(x)$ by

$$\widehat{\rho}(x) := \sum_{p=1}^{s} (\hat{x}_p - \hat{x}_{p+1}) \rho(N_p),$$

where we define $\hat{x}_{s+1} := 0$. It is not difficult to see that $\rho(X) = \widehat{\rho}(\chi_X)$ holds for every subset X of N. Define $\mathrm{P}(\rho)$ as the set of vectors x in $\mathbb{R}^N$ such that $x(X) \leq \rho(X)$ for every subset X of N.

Theorem 1 (Edmonds [23]). *For every vector x in $\mathbb{R}_+^N$,*

$$\widehat{\rho}(x) = \max_{z \in \mathrm{P}(\rho)} \sum_{i \in N} x(i) z(i). \tag{2}$$

By considering the dual problem of (2), Theorem 1 implies the following theorem.

Theorem 2 (See, e.g., [7]). *For every vector x in $\mathbb{R}_+^N$, $\widehat{\rho}(x)$ is equal to the optimal objective value of the following problem.*

$$\begin{aligned} \text{Minimize} \quad & \sum_{X \subseteq N} \rho(X) \xi(X) \\ \text{subject to} \quad & \sum_{X \subseteq N : i \in X} \xi(X) = x(i) \quad (i \in N) \\ & \xi \in \mathbb{R}_+^{2^N}. \end{aligned}$$

The following theorem plays an important role in our algorithm.

Theorem 3 (Wolsey [18]). *Assume that we are given a subset X of N. Then $\mu(X) = \mu(N)$ holds if and only if for every subset S of N,*

$$\sum_{i \in N \setminus S} \mu_S(\{i\}) \cdot \chi_X(i) \geq \mu_S(N \setminus S).$$

3   Algorithm

In this section, we propose a polynomial-time approximation algorithm for the submodular function minimization problem with submodular set covering constraints and precedence constraints. This algorithm is a natural generalization of the algorithm of McCormick et al. [17] for the precedence constrained covering problem.

Theorem 3 implies that the problem (1) is equivalent to the following problem.

$$
\begin{aligned}
&\text{Minimize} \quad \rho(X) \\
&\text{subject to} \quad \sum_{i \in N \setminus S} \mu_S(\{i\}) \cdot \chi_X(i) \geq \mu_S(N \setminus S) \quad (S \subseteq N) \\
&\qquad\qquad X \in \mathcal{L}.
\end{aligned}
\tag{3}
$$

Then the problem (3) is equivalent to the following problem.

$$
\begin{aligned}
&\text{Minimize} \quad \widehat{\rho}(x) \\
&\text{subject to} \quad \sum_{i \in N \setminus S} \mu_S(\{i\}) \cdot x(i) \geq \mu_S(N \setminus S) \quad (S \subseteq N) \\
&\qquad\qquad x(i) \geq x(j) \quad (i, j \in N \text{ such that } i \prec j) \\
&\qquad\qquad x \in \{0,1\}^N.
\end{aligned}
\tag{4}
$$

For each member S in $\mathcal{L}$, we define $\mathrm{Min}(S)$ as the set of elements i in $N \setminus S$ such that there does not exist an element j in $N \setminus S$ such that $j \prec i$. Furthermore, for each member S in $\mathcal{L}$, we define $\mathrm{Min}^+(S) := \mathrm{Min}(S) \cup S$. For each member S in $\mathcal{L}$ and each element i in $N \setminus \mathrm{Min}^+(S)$, we define $D_i(S)$ the set of elements j in $\mathrm{Min}(S)$ such that $j \prec i$. Notice that $D_i(S) \neq \emptyset$. For each member S in $\mathcal{L}$ and each element i in $\mathrm{Min}(S)$, we define

$$
\pi_S(i) := \min \left\{ \mu_S(N \setminus S), \; \mu_S(\{i\}) + \sum_{j \in N \setminus \mathrm{Min}^+(S): \, i \prec j} \frac{\mu_S(\{j\})}{|D_j(S)|} \right\}.
$$

Notice that $\pi_S(i) \geq 0$. Then we consider the following problem.

$$
\begin{aligned}
&\text{Minimize} \quad \widehat{\rho}(x) \\
&\text{subject to} \quad \sum_{i \in \mathrm{Min}(S)} \pi_S(i) \cdot x(i) \geq \mu_S(N \setminus S) \quad (S \in \mathcal{L}) \\
&\qquad\qquad x \in \{0,1\}^N.
\end{aligned}
\tag{5}
$$

The following lemma is almost the same as [17, Lemma 3].

Lemma 1. *Every feasible solution of the problem* (4) *is a feasible solution of the problem* (5).

Proof. Let x be a feasible solution of the problem (4). Assume that we are given a member S in $\mathcal{L}$. Define X as the set of elements i in N such that $x(i) = 1$. We first consider the case where there exists an element i^* in $\mathrm{Min}(S) \cap X$ such that $\pi_S(i^*) = \mu_S(N \setminus S)$. In this case,

$$\sum_{i \in \mathrm{Min}(S)} \pi_S(i) \cdot x(i) = \sum_{i \in \mathrm{Min}(S) \cap X} \pi_S(i) \geq \pi_S(i^*) = \mu_S(N \setminus S).$$

Next, we consider the case where $\pi_S(i) < \mu_S(N \setminus S)$ for every element i in $\mathrm{Min}(S) \cap X$. Then the second constraint of the problem (4) implies that for every element j in $(N \setminus \mathrm{Min}^+(S)) \cap X$, we have $D_j(S) \subseteq \mathrm{Min}(S) \cap X$. Thus, since $\mu_S(\{j\}) \geq 0$ follows from the monotonicity of μ for every element j in $N \setminus S$, we have

$$\sum_{i \in \mathrm{Min}(S) \cap X} \sum_{j \in N \setminus \mathrm{Min}^+(S) \,:\, i \prec j} \frac{\mu_S(\{j\})}{|D_j(S)|} \geq \sum_{i \in \mathrm{Min}(S) \cap X} \sum_{j \in (N \setminus \mathrm{Min}^+(S)) \cap X \,:\, i \prec j} \frac{\mu_S(\{j\})}{|D_j(S)|}$$

$$= \sum_{j \in (N \setminus \mathrm{Min}^+(S)) \cap X} \sum_{i \in D_j(S)} \frac{\mu_S(\{j\})}{|D_j(S)|} = \sum_{j \in (N \setminus \mathrm{Min}^+(S)) \cap X} \mu_S(\{j\})$$

$$= \sum_{j \in N \setminus \mathrm{Min}^+(S)} \mu_S(\{j\}) \cdot x(j).$$

This implies that

$$\sum_{i \in \mathrm{Min}(S)} \pi_S(i) \cdot x(i) = \sum_{i \in \mathrm{Min}(S) \cap X} \pi_S(i)$$

$$= \sum_{i \in \mathrm{Min}(S) \cap X} \mu_S(\{i\}) + \sum_{i \in \mathrm{Min}(S) \cap X} \sum_{j \in N \setminus \mathrm{Min}^+(S) \,:\, i \prec j} \frac{\mu_S(\{j\})}{|D_j(S)|}$$

$$\geq \sum_{i \in \mathrm{Min}(S)} \mu_S(\{i\}) \cdot x(i) + \sum_{j \in N \setminus \mathrm{Min}^+(S)} \mu_S(\{j\}) \cdot x(j)$$

$$= \sum_{i \in N \setminus S} \mu_S(\{i\}) \cdot x(i) \geq \mu_S(N \setminus S).$$

Notice that the last inequality follows from the first constraint of the problem (4). This completes the proof.   □

We are now ready to introduce a relaxation problem of our problem. The following technique is based on the technique proposed by Iwata and Nagano [7]. Lemma 1 implies that the following problem (6) is a relaxation problem of the problem (4).

$$
\begin{aligned}
\text{Minimize} \quad & \widehat{\rho}(x) \\
\text{subject to} \quad & \sum_{i \in \mathrm{Min}(S)} \pi_S(i) \cdot x(i) \geq \mu_S(N \setminus S) \quad (S \in \mathcal{L}) \\
& x \in \mathbb{R}_+^N.
\end{aligned}
\tag{6}
$$

Then Theorem 2 implies that the optimal objective value of the problem (6) is equivalent to that of the following problem.

$$\text{Minimize} \quad \sum_{X \subseteq N} \rho(X) \cdot \xi(X)$$

$$\text{subject to} \quad \sum_{i \in \text{Min}(S)} \pi_S(i) \cdot x(i) \geq \mu_S(N \setminus S) \quad (S \in \mathcal{L})$$

$$\sum_{X \subseteq N: \, i \in X} \xi(X) = x(i) \quad (i \in N)$$

$$(x, \xi) \in \mathbb{R}^N \times \mathbb{R}_+^{2^N}.$$

(7)

Here we neglect the redundant non-negativity constraint of x. The dual problem of the problem (7) is the following problem.

$$\text{Maximize} \quad \sum_{S \in \mathcal{L}} \mu_S(N \setminus S) \cdot y(S)$$

$$\text{subject to} \quad \sum_{S \in \mathcal{L}: \, i \in \text{Min}(S)} \pi_S(i) \cdot y(S) = z(i) \quad (i \in N)E$$

$$(y, z) \in \mathbb{R}_+^{\mathcal{L}} \times P(\rho).$$

(8)

Assume that we are given a vector z in $P(\rho)$. Define the function $\rho - z$ with a domain 2^N by $(\rho - z)(X) := \rho(X) - z(X)$. Then $\rho - z$ is submodular, and $\min_{X \subseteq N}(\rho - z)(X) = (\rho - z)(\emptyset) = 0$. Furthermore, it is not difficult to see that for every pair of minimizers X, Y of $\rho - z$, $X \cup Y$ is a minimizer of $\rho - z$. Thus, for every subset S of N, there exists the unique maximal subset X of S such that $\rho(X) = z(X)$.

For each member S in $\mathcal{L}$, we define the vector d_S in $\mathbb{R}_+^N$ by

$$d_S(i) := \begin{cases} \pi_S(i) & \text{if } i \in \text{Min}(S) \\ 0 & \text{if } i \in N \setminus \text{Min}(S). \end{cases}$$

We are now ready to propose our algorithm.

Algorithm 1
Step 1: Define y_1, z_1 as the zero vectors in $\mathbb{R}^{\mathcal{L}}$ and $\mathbb{R}^N$, respectively. Define $S_1 := \emptyset$. Set $t := 1$.
Step 2: If $\mu(S_t) = \mu(N)$, then output S_t and halt.
Step 3: Do the following steps **(3-a)** to **(3-e)**.
(**3-a**) Define the real number α_t by

$$\alpha_t := \min_{X \subseteq N: \, d_{S_t}(X) \neq 0} \frac{\rho(X) - z_t(X)}{d_{S_t}(X)}.$$

(**3-b**) Define the vector y_{t+1} in $\mathbb{R}^{\mathcal{L}}$ by

$$y_{t+1}(S) := \begin{cases} y_t(S) + \alpha_t & \text{if } S = S_t \\ y_t(S) & \text{otherwise.} \end{cases}$$

(3-c) Define $z_{t+1} := z_t + \alpha_t \cdot d_{S_t}$.

(3-d) Define S_{t+1} as the maximal subset of $\mathsf{Min}^+(S_t)$ such that $\rho(S_{t+1}) = z_{t+1}(S_{t+1})$.

(3-e) Set $t := t + 1$, and go back to **Step 2**.

4   Analysis

We first prove that Algorithm 1 is well-defined.

Lemma 2. *Assume that we are given a member S in $\mathcal{L}$ such that $\mu(S) < \mu(N)$. Then there exists an element i in $\mathsf{Min}(S)$ such that $\pi_S(i) \neq 0$.*

Proof. Assume that $\pi_S(i) = 0$ for every element i in $\mathsf{Min}(S)$. Since $\mu_S(N \setminus S) = \mu(N) - \mu(S) > 0$, we have

$$\mu_S(\{i\}) + \sum_{j \in N \setminus \mathsf{Min}^+(S): \, i \prec j} \frac{\mu_S(\{j\})}{|D_j(S)|} = 0$$

for every element i in $\mathsf{Min}(S)$. This implies that $\mu_S(\{i\}) = \mu_S(\{j\}) = 0$ for every element i in $\mathsf{Min}(S)$ and every element j in $N \setminus \mathsf{Min}^+(S)$ such that $i \prec j$. Thus, since for every element j in $N \setminus \mathsf{Min}^+(S)$, there exists an element i in $\mathsf{Min}(S)$ such that $i \prec j$, we have $\mu_S(\{i\}) = 0$ for every element i in $N \setminus S$. Thus, $\mu(S) = \mu(S \cup \{i\})$ holds for every element i in $N \setminus S$. Furthermore, since μ is a submodular function, we have

$$\sum_{i \in N \setminus S} \mu(S \cup \{i\}) \geq \mu(N) + (|N \setminus S| - 1)\mu(S).$$

Recall that $\mu(S) = \mu(S \cup \{i\})$ for every element i in $N \setminus S$. Thus, this implies that $\mu(S) \geq \mu(N)$. This contradicts the fact that $\mu(S) < \mu(N)$. This completes the proof.   □

Lemma 2 implies that for every member S in $\mathcal{L}$ such that $\mu(S) < \mu(N)$, there exists a subset X of N such that $d_S(X) \neq 0$. The following lemma implies that Algorithm 1 is well-defined.

Lemma 3. *Assume that we are given a vector z in $\mathrm{P}(\rho)$ and a member S of $\mathcal{L}$ such that $\mu(S) < \mu(N)$, $\rho(S) = z(S)$, and $z(i) = 0$ for every element i in $N \setminus \mathsf{Min}^+(S)$. Define*

$$\alpha := \min_{X \subseteq N: \, d_S(X) \neq 0} \frac{\rho(X) - z(X)}{d_S(X)}, \quad z' := z + \alpha \cdot d_S.$$

Then we have

(1) $z' \in \mathrm{P}(\rho)$.

Furthermore, we define S' as the maximal subset of $\mathsf{Min}^+(S)$ such that $\rho(S') = z'(S')$. Then the following statements hold.

(2) $S \subsetneq S'$ and $S' \in \mathcal{L}$.
(3) $z'(i) = 0$ for every element i in $N \setminus \mathrm{Min}^+(S')$.

Proof. (1) For every subset X of N such that $d_S(X) = 0$, we have $z'(X) = z(X) \leq \rho(X)$. For every subset X of N such that $d_S(X) \neq 0$,

$$z'(X) = z(X) + \alpha \cdot d_S(X) \leq z(X) + \frac{\rho(X) - z(X)}{d_S(X)} \cdot d_S(X) = \rho(X). \quad (9)$$

This completes the proof.

(2) Since $z(i) = z'(i)$ for every element i in S and $\rho(S) = z(S)$, we have $\rho(S) = z'(S)$. Thus, since $S \subseteq \mathrm{Min}^+(S)$, we have $S \subseteq S'$. Furthermore, since $S' \setminus S \subseteq \mathrm{Min}(S)$, we have $S' \in \mathcal{L}$. What remains is to prove that $S \neq S'$. For proving this, we first prove that there exists a subset Z of $\mathrm{Min}^+(S)$ such that $d_S(Z) \neq 0$ and

$$\alpha = \frac{\rho(Z) - z(Z)}{d_S(Z)}.$$

Let X be a subset of N such that $d_S(X) \neq 0$ and

$$\alpha = \frac{\rho(X) - z(X)}{d_S(X)}.$$

Furthermore, we assume that X minimizes $|X \setminus \mathrm{Min}^+(S)|$ among all subsets of N satisfying these conditions. If $|X \setminus \mathrm{Min}^+(S)| = 0$, then the proof is done. Assume that $|X \setminus \mathrm{Min}^+(S)| \neq 0$. Let j be an element in $X \setminus \mathrm{Min}^+(S)$. Then since $z(j) = 0$, $d_S(j) = 0$, and $\rho(X \setminus \{j\}) \leq \rho(X)$ follows from the monotonicity of ρ, we have $d_S(X \setminus \{j\}) = d_S(X) \neq 0$ and

$$\alpha = \frac{\rho(X) - z(X)}{d_S(X)} \geq \frac{\rho(X \setminus \{j\}) - z(X \setminus \{j\})}{d_S(X \setminus \{j\})} \geq \alpha.$$

This contradicts the definition of X. This completes the proof of the existence of Z. We are now ready to prove that $S \neq S'$. It is not difficult to see that (9) implies that $\rho(Z) = z'(Z)$. Since $Z \subseteq \mathrm{Min}^+(S)$ holds, the maximality of S' implies that $Z \subseteq S'$. Since $d_S(Z) \neq 0$, we have $Z \not\subseteq S$. Thus, we have $S \neq S'$. This completes the proof.

(3) Recall that for every element i in N, $d_S(i) > 0$ only if $i \in \mathrm{Min}(S)$. Since $S \subseteq S' \subseteq \mathrm{Min}^+(S)$, we have $\mathrm{Min}(S) \subseteq \mathrm{Min}^+(S')$. This completes the proof. $\square$

Lemma 3(2) implies that the number of iterations of Algorithm 1 is at most $|N| + 1$. It is known [24] that α_t can be computed in polynomial time. Furthermore, it is known (see, e.g., [25, Note 10.11] and [24, Lemma 1]) that we can find the maximal subset S_{t+1} of $\mathrm{Min}^+(S_t)$ such that $\rho(S_{t+1}) = z_{t+1}(S_{t+1})$ in polynomial time. These imply that Algorithm 1 is a polynomial-time algorithm.

Next, we evaluate the approximation ratio of Algorithm 1. Assume that Algorithm 1 halts when $t = k$.

Lemma 4. *For every integer t in $\{1, 2, \ldots, k\}$, (y_t, z_t) is a feasible solution of the problem* (8).

Proof. We prove this lemma by induction on t. Since $\rho(X) \geq 0$ for every subset X of N, (y_1, z_1) is a feasible solution of the problem (8). Assume that (y_t, z_t) is a feasible solution of the problem (8) for some integer t in $\{1, 2, \ldots, k-1\}$. Lemma 3(1) implies that $z_{t+1} \in P(\rho)$. Furthermore, since $\alpha_t \geq 0$ follows from $z_t \in P(\rho)$, $y_{t+1} \in \mathbb{R}_+^{\mathcal{L}}$. For every element i in $N \setminus \text{Min}(S_t)$, $z_{t+1}(i) = z_t(i)$ and

$$\sum_{S \in \mathcal{L}:\, i \in \text{Min}(S)} \pi_S(i) \cdot y_t(S) = \sum_{S \in \mathcal{L}:\, i \in \text{Min}(S)} \pi_S(i) \cdot y_{t+1}(S).$$

For every element i in $\text{Min}(S_t)$, $z_{t+1}(i) - z_t(i) = \alpha_t \cdot \pi_{S_t}(i)$ and

$$\sum_{S \in \mathcal{L}:\, i \in \text{Min}(S)} \pi_S(i) \cdot y_{t+1}(S) - \sum_{S \in \mathcal{L}:\, i \in \text{Min}(S)} \pi_S(i) \cdot y_t(S)$$
$$= \pi_{S_t}(i) \cdot (y_{t+1}(S_t) - y_t(S_t))$$
$$= \alpha_t \cdot \pi_{S_t}(i).$$

This completes the proof.   □

Define $\mathcal{A}$ as the family of subsets S of N such that $i \npreceq j$ for any pair of distinct elements i, j in S. Notice that for every member S in $\mathcal{L}$, we have $\text{Min}(S) \in \mathcal{A}$. Define $\Pi := \max_{S \in \mathcal{A}} |S|$.

Lemma 5. *For every member T in $\mathcal{L}$ such that $\mu(T) = \mu(N)$, we have $\rho(S_k) \leq \Pi \cdot \rho(T)$.*

Proof. Let T be a member in $\mathcal{L}$ such that $\mu(T) = \mu(N)$. Lemma 4 implies that

$$\sum_{S \in \mathcal{L}} \mu_S(N \setminus S) \cdot y_k(S) \leq \rho(T).$$

Thus, we have

$$\rho(S_k) = z_k(S_k)$$
$$= \sum_{i \in S_k} \sum_{S \in \mathcal{L}:\, i \in \text{Min}(S)} \pi_S(i) \cdot y_k(S) = \sum_{S \in \mathcal{L}} \sum_{i \in \text{Min}(S) \cap S_k} \pi_S(i) \cdot y_k(S)$$
$$\leq \sum_{S \in \mathcal{L}} \sum_{i \in \text{Min}(S) \cap S_k} \mu_S(N \setminus S) \cdot y_k(S) \leq \sum_{S \in \mathcal{L}} \Pi \cdot \mu_S(N \setminus S) \cdot y_k(S)$$
$$\leq \Pi \cdot \rho(T).$$

This completes the proof.   □

We are now ready to prove the main result of this paper.

Theorem 4. *Algorithm 1 is a Π-approximation algorithm for the submodular function minimization problem with submodular set covering constraints and precedence constraints.*

Proof. This theorem immediately follows from Lemmas 3(2) and 5.   □

5   Special Case

In this section, we consider the case where $i \not\preceq j$ for any pair of distinct elements i, j in N, i.e., $\mathcal{L} = 2^N$.

Assume that Algorithm 1 halts when $t = k$. Theorem 5 can be regarded as a generalization of [17, Corollary 2], and matches the result of [12] under the assumption that ρ is monotone.

Theorem 5. *Assume that* $i \not\preceq j$ *for any pair of distinct elements* i, j *in* N. *Then for every member* T *in* $\mathcal{L}$ *such that* $\mu(T) = \mu(N)$, *we have*

$$\rho(S_k) \leq \left[\max_{S \subseteq N: \, \mu(S) < \mu(N)} \frac{\sum_{i \in N \setminus S} \mu_S(\{i\})}{\mu_S(N \setminus S)} \right] \cdot \rho(T).$$

Proof. Since $\mu(N) > 0$, then $k > 1$. The assumption in this theorem implies that for every member S in $\mathcal{L}$ and every element i in $\text{Min}(S)$, $\pi_S(i) \leq \mu_S(\{i\})$ holds. Thus, since $y_k(S) = 0$ for every member S in $2^N \setminus \{S_1, S_2, \ldots, S_{k-1}\}$,

$$\rho(S_k) = z_k(S_k)$$
$$= \sum_{i \in S_k} \sum_{S \in \mathcal{L}: \, i \in \text{Min}(S)} \pi_S(i) \cdot y_k(S) = \sum_{S \in \mathcal{L}} \sum_{i \in \text{Min}(S) \cap S_k} \pi_S(i) \cdot y_k(S) \tag{10}$$
$$\leq \sum_{S \in \mathcal{L}} \sum_{i \in \text{Min}(S) \cap S_k} \mu_S(\{i\}) \cdot y_k(S) = \sum_{t=1}^{k-1} \sum_{i \in \text{Min}(S_t) \cap S_k} \mu_{S_t}(\{i\}) \cdot y_k(S_t).$$

Since $\mu_S(\{i\}) \geq 0$ for every member S in $\mathcal{L}$ and every element i in $N \setminus S$, (10) implies that

$$\rho(S_k) \leq \sum_{t=1}^{k-1} \sum_{i \in N \setminus S_t} \mu_{S_t}(\{i\}) \cdot y_k(S_t). \tag{11}$$

Let T be a member in $\mathcal{L}$ such that $\mu(T) = \mu(N)$. Lemma 4 implies that

$$\rho(T) \geq \sum_{S \in \mathcal{L}} \mu_S(N \setminus S) \cdot y_k(S) = \sum_{t=1}^{k-1} \mu_{S_t}(N \setminus S_t) \cdot y_k(S_t). \tag{12}$$

Define

$$\beta := \max_{t \in \{1, 2, \ldots, k-1\}} \frac{\sum_{i \in N \setminus S_t} \mu_{S_t}(\{i\})}{\mu_{S_t}(N \setminus S_t)}.$$

Then (11) and (12) imply that

$$\rho(S_k) \leq \sum_{t=1}^{k-1} \sum_{i \in N \setminus S_t} \mu_{S_t}(\{i\}) \cdot y_k(S_t) \leq \beta \cdot \sum_{t=1}^{k-1} \mu_{S_t}(N \setminus S_t) \cdot y_k(S_t) \leq \beta \cdot \rho(T).$$

Since $\mu(S_t) < \mu(N)$ for every integer t in $\{1, 2, \ldots, k-1\}$, we have

$$\beta \leq \max_{S \subseteq N: \, \mu(S) < \mu(N)} \frac{\sum_{i \in N \setminus S} \mu_S(\{i\})}{\mu_S(N \setminus S)}.$$

This completes the proof.   □

Acknowledgements. This research was supported by JST PRESTO Grant Number JPMJPR14E1, Japan.

References

1. Grötschel, M., Lovász, L., Schrijver, A.: The ellipsoid method and its consequences in combinatorial optimization. Combinatorica **1**(2), 169–197 (1981)
2. Grötschel, M., Lovász, L., Schrijver, A.: Geometric Algorithms and Combinatorial Optimization. Springer, Heidelberg (1988). https://doi.org/10.1007/978-3-642-78240-4
3. Iwata, S., Fleischer, L., Fujishige, S.: A combinatorial strongly polynomial algorithm for minimizing submodular functions. J. ACM **48**(4), 761–777 (2001)
4. Schrijver, A.: A combinatorial algorithm minimizing submodular functions in strongly polynomial time. J. Comb. Theory Ser. B **80**(2), 346–355 (2000)
5. Goel, G., Karande, C., Tripathi, P., Wang, L.: Approximability of combinatorial problems with multi-agent submodular cost functions. In: Proceedings of the 50th Annual Symposium on Foundations of Computer Science, pp. 755–764 (2009)
6. Hochbaum, D.S.: Submodular problems - approximations and algorithms. Technical report arXiv:1010.1945 (2010)
7. Iwata, S., Nagano, K.: Submodular function minimization under covering constraints. In: Proceedings of the 50th Annual Symposium on Foundations of Computer Science, pp. 671–680 (2009)
8. Iyer, R.K., Bilmes, J.A.: Submodular optimization with submodular cover and submodular knapsack constraints. In: Advances in Neural Information Processing Systems 26, pp. 2436–2444 (2013)
9. Iyer, R.K., Jegelka, S., Bilmes, J.A.: Curvature and optimal algorithms for learning and minimizing submodular functions. In: Advances in Neural Information Processing Systems 26, pp. 2742–2750 (2013)
10. Iyer, R.K., Jegelka, S., Bilmes, J.A.: Monotone closure of relaxed constraints in submodular optimization: connections between minimization and maximization. In: Proceedings of the 30th Conference on Uncertainty in Artificial Intelligence, pp. 360–369 (2014)
11. Jegelka, S., Bilmes, J.A.: Graph cuts with interacting edge weights: examples, approximations, and algorithms. Math. Program. **162**, 241–282 (2017)
12. Kamiyama, N.: Submodular function minimization under a submodular set covering constraint. In: Ogihara, M., Tarui, J. (eds.) TAMC 2011. LNCS, vol. 6648, pp. 133–141. Springer, Heidelberg (2011). https://doi.org/10.1007/978-3-642-20877-5_14
13. Koufogiannakis, C., Young, N.E.: Greedy Δ-approximation algorithm for covering with arbitrary constraints and submodular cost. Algorithmica **66**(1), 113–152 (2013)
14. Svitkina, Z., Fleischer, L.: Submodular approximation: sampling-based algorithms and lower bounds. SIAM J. Comput. **40**(6), 1715–1737 (2011)
15. Zhang, H., Vorobeychik, Y.: Submodular optimization with routing constraints. In: Proceedings of the 30th AAAI Conference on Artificial Intelligence, pp. 819–826 (2016)
16. Kamiyama, N.: A note on submodular function minimization with covering type linear constraints. Algorithmica (to appear)
17. McCormick, S.T., Peis, B., Verschae, J., Wierz, A.: Primal–dual algorithms for precedence constrained covering problems. Algorithmica **78**, 771–787 (2017)

18. Wolsey, L.A.: An analysis of the greedy algorithm for the submodular set covering problem. Combinatorica **2**(4), 385–393 (1982)
19. Fujito, T.: On approximation of the submodular set cover problem. Oper. Res. Lett. **25**(4), 169–174 (1999)
20. Fujito, T., Yabuta, T.: Submodular integer cover and its application to production planning. In: Persiano, G., Solis-Oba, R. (eds.) WAOA 2004. LNCS, vol. 3351, pp. 154–166. Springer, Heidelberg (2005). https://doi.org/10.1007/978-3-540-31833-0_14
21. Fujito, T.: Approximating bounded degree deletion via matroid matching. In: Fotakis, D., Pagourtzis, A., Paschos, V.T. (eds.) CIAC 2017. LNCS, vol. 10236, pp. 234–246. Springer, Cham (2017). https://doi.org/10.1007/978-3-319-57586-5_20
22. Lovász, L.: Submodular functions and convexity. In: Bachem, A., Korte, B., Grötschel, M. (eds.) Mathematical Programming–The State of the Art, pp. 235–257. Springer, Heidelberg (1983). https://doi.org/10.1007/978-3-642-68874-4_10
23. Edmonds, J.: Submodular functions, matroids, and certain polyhedra. In: Guy, R., Hanani, H., Sauer, N., Schönheim, J. (eds.) Combinatorial Structures and their Applications, pp. 69–87. Gordon and Breach (1970)
24. Nagano, K.: A faster parametric submodular function minimization algorithm and applications. Technical report METR 2007-43. The University of Tokyo (2007)
25. Murota, K.: Discrete convex analysis. SIAM Monographs on Discrete Mathematics and Applications, vol. 10. Society for Industrial and Applied Mathematics (2003)

Lower Bounds for Several Online Variants of Bin Packing

János Balogh[1], József Békési[1], György Dósa[2], Leah Epstein[3],
and Asaf Levin[4(✉)]

[1] Department of Applied Informatics, Gyula Juhász Faculty of Education,
University of Szeged, Szeged, Hungary
{balogh,bekesi}@jgypk.u-szeged.hu
[2] Department of Mathematics, University of Pannonia, Veszprem, Hungary
dosagy@almos.vein.hu
[3] Department of Mathematics, University of Haifa, Haifa, Israel
lea@math.haifa.ac.il
[4] Faculty of Industrial Engineering and Management, The Technion, Haifa, Israel
levinas@ie.technion.ac.il

Abstract. We consider several previously studied online variants of bin packing and prove new and improved lower bounds on the asymptotic competitive ratios for them. For that, we use a method of fully adaptive constructions. In particular, we improve the lower bound for the asymptotic competitive ratio of online square packing significantly, raising it from roughly 1.68 to above 1.75.

1 Introduction

In bin packing problems, there is an input consisting of a set of items, and the goal is to partition it into a minimum number of subsets called bins, under certain conditions and constraints. In the classic variant [20,21,25,29,32], items have one-dimensional rational numbers in $(0,1]$, called sizes, associated with them, and the total size of items of one bin cannot exceed 1. In online variants items are presented as a sequence and the partition is created throughout this process in the sense that any new item should be assigned to a bin before any information regarding the next item is provided. The conditions on the partition or packing remain as in the offline problem where the items are all given at once as a set. Using an algorithm A to partition the items into subsets, which is also seen as a process of packing items into bins, the number of partitions or bins used for the packing is defined to be the cost of A.

Gy. Dósa was supported by VKSZ_12-1-2013-0088 "Development of cloud based smart IT solutions by IBM Hungary in cooperation with the University of Pannonia" and by National Research, Development and Innovation Office – NKFIH under the grant SNN 116095. L. Epstein and A. Levin were partially supported by a grant from GIF - the German-Israeli Foundation for Scientific Research and Development (grant number I-1366-407.6/2016).

R. Solis-Oba and R. Fleischer (Eds.): WAOA 2017, LNCS 10787, pp. 102–117, 2018.
https://doi.org/10.1007/978-3-319-89441-6_9

Algorithms for bin packing problems are normally studied using the asymptotic approximation ratio, also called asymptotic competitive ratio for the case of online algorithms (and we will use this last term). For an algorithm A and an input I, let $A(I)$ denote the number of bins used by A for I, that is, the cost of A for I. Let $OPT(I)$ denote the number of bins that an optimal solution uses for I, that is, the cost of an optimal (offline) algorithm OPT for I. Consider the set of inputs J_Q of all inputs for which the number of bins used by OPT is Q. For the problems studied here (and non-empty inputs for them), Q will be a positive integer. Let $c(Q) = \max_{I \in J_Q} A(I)$ (where for reasonable algorithms this value is finite), and let $R_A = \limsup_{Q \to \infty} \frac{c(Q)}{Q}$. The absolute competitive ratio of A is defined by $\sup_I \frac{A(I)}{OPT(I)}$, that is, this is the supremum ratio between the cost of A and the optimal cost, over all inputs, and the asymptotic competitive ratio is the superior limit of the absolute competitive ratios for fixed values of $Q = OPT(I)$ when Q grows to infinity. Since the standard measures for online bin packing problems (and offline bin packing problems, respectively), are the asymptotic competitive ratio (and the asymptotic approximation ratio), we also use the terms *competitive ratio* (and *approximation ratio*) for them, and always use the word *absolute* when we discuss the absolute measures. To prove lower bounds on the (asymptotic) competitive ratio one can use inputs where the optimal cost is arbitrarily large, and we use this method. The study of lower bounds on the competitive ratio for a given problem characterizes the extent to which the performance of the system deteriorates due to lack of information regarding the future input items.

Here, we study three versions of the online bin packing problem, providing new lower bounds on the competitive ratio for them. Previous constructions used for proving such lower bounds were often inputs where items arrive in batches, such that the items of one batch all have the exact same size (and the input may stop after a certain batch or it can continue to another one). In the known lower bounds for classic bin packing, it is even known what the next batches will be, if they are indeed presented [7, 24, 30]. While it may be obvious that adaptive inputs where the properties of the next item are based on the packing of previous items are harder for an algorithm to deal with, it was not known until recently how to use this idea for designing lower bounds, except for special cases [2, 10, 18]. In cardinality constrained bin packing [2, 9, 13, 22, 23], items are one-dimensional, a fixed integer $t \geq 2$ is given, and the two requirements for a packed bin are that its total size of items is at most 1, and that it contains at most t items. The special case analyzed in the past [2, 10, 18] is $t = 2$, which can also be seen as a matching problem, as every bin can contain at most two items. In [4] we showed that the overall competitive ratio (supremum over all values of t) is 2 (an upper bound was known prior to that work [2, 9]), and provided improved lower bounds for relatively small values of t. For standard bin packing, the best known lower bound on the competitive ratio is 1.5403 [7, 30] and the best upper bound is 1.57829 [5].

Another lower bound presented in [4] is for the competitive ratio of vector packing in at least two dimensions. For an integer dimension $d \geq 2$, the items have d-dimensional vectors associated with them, whose components are rational numbers in $[0, 1]$ (none of which are all-zero vectors), and bins are all-one vectors of dimension d. A subset of items can be packed into a bin if taking no component exceeds 1 in their vector sum. This generalizes cardinality constrained bin packing, and we showed a lower bound of 2.03731129 on the competitive ratio of the online variant for any $d \geq 2$ (prior to that work, no lower bound strictly above 2 for a constant dimension was known).

Our main goal here is to exhibit how to exploit adaptive constructions with some connection to those used in [4] in order to obtain lower bounds for other variants. We focus on the following three variants. In all three variants of online bin packing which we study, the input consists of rational numbers in $(0, 1]$, however there is additional information received with the input in some of the cases and the input is interpreted in different ways. Two of the problems are one-dimensional and the input numbers are sizes of items. The third variant is two-dimensional, and the numbers are side lengths of squares. In our first variant called *bin packing with known optimal cost*, the cost of an optimal (offline) solution is given in advance, that is, it is known how many bins are required for packing the input. This problem is also called K-O (known-OPT). It is currently hard to find an appropriate way to use this additional piece of information for algorithm design, but in all lower bounds known for standard online bin packing [7, 30] the property that the optimal cost is different for different inputs is crucial for achieving the result. For K-O, a lower bound of 1.30556 on the competitive ratio was presented [15] and later improved to 1.32312 [3]. We show a new lower bound of $\frac{87}{62} \approx 1.4032258$ on the competitive ratio, improving the previous result significantly. This problem is related to the field of semi-online algorithms and to the so-called model of *online algorithms with advice* [1, 11], where the online algorithm is provided with some (preferably very small) pieces of information regarding the input.

In the square packing (SP) problem, the goal is to assign an input set of squares whose sides are rational numbers in $(0, 1]$ into bins that are unit squares in a non-overlapping and axis-parallel way, so as to minimize the number of non-empty bins. We use the standard definition of this packing problem, where two squares do not overlap if their interiors do not overlap (but they may have common points on the boundaries of the squares). The offline variant is well-studied [8, 16]. The history of lower bounds on the competitive ratio of online algorithms for this problem is as follows. Several such lower bounds were proved for the online version of SP, starting with a simple construction yielding a lower bound of $\frac{4}{3}$ on the competitive ratio by Coppersmith and Raghavan [12], and then there were several improvements [17, 19, 26], all showing bounds above 1.6. In 2016 a copy of the thesis of Blitz [10] from 1996 was found by the authors of [19]. This thesis contains a number of lower bounds for bin packing problems, including a lower bound of 1.680783 on the competitive ratio of online algorithms for SP. The result of Blitz [10] is now the previous best lower bound on the competitive ratio for the problem (prior to our work), and it is higher than the lower bounds of [17, 19, 26]. Here, we show a much higher lower bound, larger than 1.7515445, on the competitive ratio of this problem.

Finally, we consider class constrained bin packing (CLCBP) [14,27,28,31]. In this one-dimensional variant every item has a size and a color, and for a given parameter $t \geq 1$, any bin can receive items of at most t different colors (of total size at most 1), while the number of items of each color can be arbitrary. This problem generalizes standard bin packing, as for any input of standard bin packing, defining a common color to all items results in an instance of CLCBP for any t. It also generalizes bin packing with cardinality constraints, though here to obtain an instance of CLCBP one should assign distinct colors to all items. We provide improved lower bounds for $t = 2, 3$. For $t = 2$, the previous known lower bound was 1.5652 [14]. For $t = 3$, the previous lower bound was $\frac{5}{3} \approx 1.6667$ [27]. This last result was proved even for the special case with equal size items. Interestingly, it has elements of adaptivity, but with respect to colors (as all items have identical sizes), and the input moves to presenting items of a new color once the algorithm performs a certain action. We show that the competitive ratio of any online algorithm for CLCBP with $t = 2$ is at least 1.717668, and that the competitive ratio of any online algorithm for CLCBP with $t = 3$ is at least 1.808142.

The drawback of previous results for all those problems is that while the exact input was not known in advance, the set of sizes used for it was determined prior to the action of the algorithm. We show here that our methods for proving lower bounds can be combined with a number of other approaches to result in improved lower bounds for a variety of bin packing problems. We use the following theorem proved in [4] (see the construction in Sect. 3.1 and Corollary 3).

Theorem 1. *(i) Let $N \geq 1$ and $k \geq 2$ be large positive integers. Assume that we are given an arbitrary deterministic online algorithm for a variant of bin packing and a condition C_1 on the possible behavior of an online algorithm for one item (on the way that the item is packed). An adversary is able to construct a sequence of values a_i ($1 \leq i \leq N$) such that for any i, $a_i \in \left(k^{-2^{N+3}}, k^{-2^{N+2}} \right)$, and in particular $a_i \in \left(0, \frac{1}{k^4} \right)$. For any item i_1 satisfying C_1 and any item i_2 not satisfying C_1, it holds that $\frac{a_{i_2}}{a_{i_1}} > k$. Specifically, there are values β and γ such that for any item i_1 satisfying C_1, and any item i_2 not satisfying C_1, it holds that $a_{i_1} < \gamma < a_{i_2}$ and $\frac{a_{i_2}}{a_{i_1}} > \beta$.*

(ii) If another condition C' is given for stopping the input (it can be a condition on the packing or on the constructed input), it is possible to construct a sequence a_i consisting of N items such that C' never holds, or a sequence of $N' < N$ items, such that C' holds after N' items were introduced (but not earlier), and where the sequence satisfies the requirements above.

Examples for the condition C_1 can be the following: "the item is packed as a second item of its bin", "the item is packed into a non-empty bin", "the item is packed into a bin an item of size above $\frac{1}{2}$", etc. An example for the condition C' can be "the algorithm has at least a given number of non-empty bins".

The construction of such inputs is based on presenting items one by one, where there is an active (open) interval of sizes out of which future values a_i

are selected. When a new item is presented, and the algorithm packs it such that it does not satisfy C_1, all future items will be smaller. If the algorithm packs a new item such that it satisfies C_1, all future items will be larger. This reduces the length of the active interval. Thus, even though the active interval becomes shorter in every step where a new item arrives, it always has a positive length. One can see this as a kind of binary search on the value γ, which will always be contained in the remaining interval (as it remains non-empty). For example, Fujiwara and Kobayashi [18] used a similar approach and in their work the middle point of the active interval is the size of the next item, and the active interval has length that it smaller by a factor of 2 after every step. To obtain the stronger property that items whose sizes is at least the right endpoint of the active interval are larger by a factor of k than items no larger than the left endpoint of the active interval, the selection of the next size is performed by a process similar to geometrical binary search.

Note that an important feature is that the value a_i is defined *before* it is known whether C_1 holds for the ith item (the item corresponding to a_i, that is, the item whose size is a function of a_i). We will use this theorem throughout the paper. We study the problems in the order they were defined. Omitted proofs and additional details appear in the full version [6].

2   Online Bin Packing with Known Optimal Cost (K-O)

Here, we consider the problem K-O, and prove a new lower bound on the competitive ratio for it. We prove the following theorem.

Theorem 2. *The competitive ratio of any online algorithm for K-O is at least* $\frac{87}{62} \approx 1.4032258$.

Let M be a large integer that is divisible by 4 (M will be the value of the known optimal cost). We will create several alternative inputs, such that the optimal cost will be equal to M for each one of them.

We use the following construction. For $k = 10$ and $N = M$, define an input built using Theorem 1 as follows applied twice on different parts of the input as explained below. The outline of our lower bound construction is as follows. The first part of the input will consist of M items of sizes slightly above $\frac{1}{7}$ (such that some of them, those packed first into bins, are larger than the others). Then, there are M items of sizes slightly above $\frac{1}{3}$ (where items packed into new bins are larger than others, while those combined with items of sizes roughly $\frac{1}{7}$ or with another item of size roughly $\frac{1}{3}$, or both, are slightly smaller). Finally, the algorithm will be presented with a list of identical items of one of the three sizes 1 (exactly), or slightly above $\frac{1}{2}$, or slightly below $\frac{2}{3}$, such that every larger item of size slightly above $\frac{1}{3}$ cannot be packed together with such an item (of size slightly below $\frac{2}{3}$). Additionally, after the first M items arrive, it is possible that instead of the input explained here there are items of sizes slightly below $\frac{6}{7}$, either such that every such item can be packed with any item out of the first M items, or such that it can only be combined with the smaller items out of the first M items (due to the property that the size of an item will be just below $\frac{6}{7}$, in both cases it can be combined with at most one item of size just above $\frac{1}{7}$).

Next, we formally define our input sequences. Throughout this section, let the condition C_1 be that the item is not packed as a first item into a bin. The first M items are defined as follows. Using Theorem 1, we create M items such that the size of item i is $\frac{1}{7} + a_i$. These items are called S-items. The sizes of such items are in $(\frac{1}{7}, 0.143)$, and there is a value γ_1 such that any item whose packing satisfies condition C_1 has size below $\frac{1}{7} + \gamma_1$ and any item whose packing does not satisfy C_1 has size above $\frac{1}{7} + \gamma_1$. The first kind of items are called small S-items, and the second kind of items are called large S-items.

Let Y_7 denote the current number of bins used by the algorithm (after all S-items have arrived), and this is also the number of large S-items. Two possible continuations at this point are M items of sizes equal to $\frac{4}{5}$ (the first option), and $M - \lceil \frac{Y_7}{6} \rceil$ items of sizes equal to $\frac{6}{7} - \gamma_1$ (the second option).

Lemma 1. *In both options, an optimal solution has cost M.*

In the first case, the algorithm can use bins containing exactly one item to pack (also) an item of size $\frac{4}{5}$, but it cannot use any other bin again. In the second case, as every bin has exactly one item of size above $\frac{1}{7} + \gamma_1$, the algorithm uses an empty bin for every item of size $\frac{6}{7} - \gamma_1$.

We explain the continuation of the input in the case where none of the two continuations already defined is used. The next M items are defined using Theorem 1, and we create M items such that the size of the ith item of the current subsequence of M items is $\frac{1}{3} + a_i$ (the values a_i are constructed here again, and they are different from the values a_i constructed earlier). We call these items T-items. The sizes of T-items are in $(\frac{1}{3}, 0.33344)$, and there is a value γ_2 such that any item whose packing satisfies condition C_1 (defined in this section) has size below $\frac{1}{3} + \gamma_2$ and for any item whose packing does not satisfy C_1, it has size above $\frac{1}{3} + \gamma_2$. The first kind of items are called small T-items, and the second type items are called large T-items.

Here, there are three possible continuations. The first one is $\frac{M}{2}$ items, all of size 1. The second one is M items, each of size 0.52. Let Y_3 denote the number of new bins created for the T-items, which is also the number of large T-items (so after the T-items are packed the algorithm uses $Y_7 + Y_3$ bins). If $Y_3 \leq \frac{M}{2}$, the third continuation is with $\frac{3M}{4}$ items, each of size $\frac{2}{3} - \gamma_2$ (where $\frac{2}{3} - \gamma_2 > 0.66656$). Otherwise ($Y_3 > \frac{M}{2}$), the third continuation is with $M - \lceil \frac{Y_3}{2} \rceil$ items, each of size $\frac{2}{3} - \gamma_2$. Thus, in the third continuation, the sizes of items are the same (i.e., $\frac{2}{3} - \gamma_2$) in both cases, and the number of items is $M - \max\{\frac{M}{4}, \lceil \frac{Y_3}{2} \rceil\}$.

Lemma 2. *The optimal cost in all cases (i.e., after the packing of the items of each possible continuation has been completed) is exactly M.*

This completes the description of the input where we showed that in each case the optimal cost is exactly M. Next, we consider the behavior of the algorithm. Consider the kinds of bins the algorithm may have after all T-items have arrived.

The T-items do not necessarily arrive, but we will deduce the numbers of different kinds of bins the algorithm has after the S-items have arrived from the numbers of bins assuming that the T-items have arrived. This is an approach similar to that used in [30], where numbers of bins packed according to certain patterns (subsets of items that can be packed into one bin) at the end of the input are considered, and based on them, the number of bins already opened at each step of the input are counted. More precisely, if the input consists of batches of identical (or similar) items, given the contents of a bin it is clear when it is opened and at what times (after arrival of sub-inputs) it should be counted towards the cost of the algorithm.

A bin with no T-items can receive an item of size 0.52 if it has at most three S-items and it can receive an item of size $\frac{2}{3} - \gamma_2$ if it has at most two S-items. The only case where a bin with at least one S-item and at least one T-item can receive another item (out of a continuation of the input) is the case that a bin has one of each of these types of items, and it will receive an item of size 0.52.

Let X_{60} denote the number of bins with four or five or six S-items and no T-items. Such a bin cannot receive any further items in addition to its S-items. Let X_{30} denote the number of bins with three S-items and no T-items. Such a bin can receive an item of size 0.52 (but not a larger item). Let X_{20} and X_{10} denote the number of bins with two S-items and one S-item, respectively, and no T-items. Out of possible input items, such a bin can receive an item of size 0.52 or an item of size $\frac{2}{3} - \gamma_2$. We distinguish these two kinds of bins due to the possible other continuations after T-items have arrived. Let X_{41} denote the number of bins with two or three or four S-items and one T-item. Such bins cannot receive any further items out of our inputs. Let X_{11} denote the number of bins with one S-item and one T-item. Let X_{12} and X_{22} denote the numbers of bins with two T-items and one and two S-items, respectively. Obviously, there can be bins without S-items containing one or two T-items, and we denote their numbers by X_{01} (one T-item) and X_{02} (two T-items).

We have five scenarios based on the different options and continuations described above, and we use ALG_i to denote the cost of a given algorithm for each one of them, in the order they were presented. Let R be the (asymptotic) competitive ratio. Let $A_i = \limsup_{M \to \infty} \frac{ALG_i}{M}$, which is a lower bound on the competitive ratio R since the optimal cost is always M (by Lemmas 1 and 2), so for $i = 1, 2, 3, 4, 5$ we have the constraint $A_i \leq R$. The A_i (for $i = 1, 2, 3, 4, 5$) will not appear explicitly as variables in the forthcoming linear program. Instead, we will compute each A_i based on the other variables in the program and substitute the resulting expression in the constraint $A_i \leq R$. We use $y_i = \frac{Y_i}{M}$ and $x_{ij} = \frac{X_{ij}}{M}$ for those values of i and j such that Y_i and X_{ij} are defined. For all thirteen variables there is a non-negativity constraint. In addition, the number of items should satisfy $\sum_{i,j} j \cdot X_{ij} = M$ and $\sum_{i,j} i \cdot X_{ij} \geq M$ (the second constraint is not an equality as in some cases X_{ij} counts bins with at most (i) S-items). Using the definitions of Y_7 and Y_3 we have $Y_7 = X_{60} + X_{30} + X_{20} + X_{10} + X_{41} + X_{11} + X_{12} + X_{22}$ and $Y_3 = X_{01} + X_{02}$.

We get the following four constraints:

$$x_{41} + x_{11} + 2x_{12} + 2x_{22} + x_{01} + 2x_{02} = 1 \tag{1}$$

$$6x_{60} + 3x_{30} + 2x_{20} + x_{10} + 4x_{41} + x_{11} + x_{12} + 2x_{22} \geq 1 \tag{2}$$

$$y_7 - x_{60} - x_{30} - x_{20} - x_{10} - x_{41} - x_{11} - x_{12} - x_{22} = 0 \tag{3}$$

$$y_3 - x_{01} - x_{02} = 0 \tag{4}$$

The costs of the algorithm are as follows. We have $ALG_1 = M + X_{60} + X_{30} + X_{20} + X_{41} + X_{22}$, $ALG_2 = M - \lceil \frac{Y_7}{6} \rceil + Y_7$, $ALG_3 = Y_7 + Y_3 + \frac{M}{2}$, and $ALG_4 = X_{60} + X_{41} + X_{22} + X_{12} + X_{02} + M$.

If $Y_3 \leq \frac{M}{2}$, we have $ALG_5 = Y_7 + Y_3 - X_{20} - X_{10} + \frac{3M}{4}$, and if $Y_3 > \frac{M}{2}$, we have $ALG_5 = Y_7 + Y_3 - X_{20} - X_{10} + M - \lceil \frac{Y_3}{2} \rceil$.

The four first costs of the algorithm (for the first four scenarios) gives the constraints

$$R - x_{60} - x_{30} - x_{20} - x_{41} - x_{22} \geq 1 \tag{5}$$

$$6R - 5y_7 \geq 6 \tag{6}$$

$$2R - 2y_7 - 2y_3 \geq 1 \tag{7}$$

$$R - x_{60} - x_{41} - x_{22} - x_{12} - x_{02} \geq 1 \tag{8}$$

The two final constraints form two cases (according to the value of y_3), and therefore our list of constraints results in two linear programs (with all previous constraints and two additional ones). The inputs for the two cases are different, and therefore they are considered separately (due to the different inputs, there is one other different constraint except for the constraint on the value of y_3). For each one of the linear programs, the objective is to minimize the value of R.

One pair of constraints is $y_3 \leq \frac{1}{2}$ and $4R - 4y_7 - 4y_3 + 4x_{20} + 4x_{10} \geq 3$, and the alternative pair is $y_3 \geq \frac{1}{2}$ and $2R - 2y_7 - y_3 + 2x_{20} + 2x_{10} \geq 2$ (observe that the constraint $y_3 \geq \frac{1}{2}$ is a relaxation of the valid constraint $y_3 > \frac{1}{2}$, and thus the weaker constraint $y_3 \geq \frac{1}{2}$ is valid in this case).

Multiplying the first five constraints by the values 2, 1, 3, 2, 1, respectively, and taking the sum gives:

$$2x_{60} + 2x_{41} + 2x_{12} + 2x_{02} + 2x_{22} - 2x_{10} - x_{30} - 2x_{20} + 3y_7 + 2y_3 + R \geq 4. \tag{9}$$

For the first case, we take the sum of the sixth, eighth, and tenth constraints multiplied by the values 2, 20, 5, respectively, and get:

$$52R - 30y_7 - 20y_3 - 20x_{60} - 20x_{41} - 20x_{22} - 20x_{12} - 20x_{02} + 20x_{20} + 20x_{10} \geq 47.$$

Summing this with ten times (9) we get $62R - 10x_{30} \geq 87$, and by $x_{30} \geq 0$ we get $R \geq \frac{87}{62} \approx 1.4032258$.

For the second case, we take the sum of the seventh, eighth, and tenth constraints multiplied by the values 1, 4, 2, respectively, and get:

$$10R - 6y_7 - 4y_3 - 4x_{60} - 4x_{41} - 4x_{22} - 4x_{12} - 4x_{02} + 4x_{20} + 4x_{10} \geq 9.$$

Summing this with twice (9) we get $12R - 2x_{30} \geq 17$, and as $x_{30} \geq 0$, we have $R \geq \frac{17}{12} \approx 1.41666$. Thus, we have proved $R \geq 1.4032258$.

3   Online Square Packing (SP)

We continue with the online square packing (SP) problem. We prove the following theorem.

Theorem 3. *The competitive ratio of any online algorithm for* SP *is at least* 1.7515445.

Here, in the description of the input, when we refer to the size of an item, this means the length of the side of the square (and not its area). Consider the following input. For a large positive even integer M and $k = 10$, we define an input based on using Theorem 1 twice. The construction is similar to that of the previous section, though here we are not committed to a specific optimal cost, and we take into account the multidimensionality. Moreover, for one of the item types the number of such items is also determined by the action of the algorithm (which was difficult to implement in the previous section when the cost of an optimal packing is fixed in advance, and we did not use such an approach there as extensively as in the current section). Here, we only compute upper bounds on the optimal cost for each case.

The outline of the construction is as follows. The first part of the input will consist of M items of sizes slightly above $\frac{1}{4}$ (such that some of them, those packed first into bins, are larger than the others), then, there are items of sizes slightly above $\frac{1}{3}$ (where such items that are packed into bins containing relatively few items, where the exact condition is defined below, will be larger than other items of this last kind). Finally, there will be items of one of the sizes: $\frac{3}{5}$, and slightly below $\frac{2}{3}$ (all of them will have exactly the same size), such that every larger item of size slightly above $\frac{1}{3}$ cannot be packed together with such an item of size slightly smaller than $\frac{2}{3}$. Additionally, after the first M items arrive, it is possible that instead of the input explained here there are items of sizes slightly below $\frac{3}{4}$, such that it can be only be combined with the smaller items out of the first M items (any bin with an item of size slightly below $\frac{3}{4}$ may have at most five smaller items out of the first M items in a common bin).

Next, we formally define the construction. Let the condition C_{11} be that the item is not packed as a first item into a bin. This is the condition we will use for items of sizes slightly above $\frac{1}{4}$. For items of sizes slightly above $\frac{1}{3}$, let the condition C_{12} be that the item is either packed in a bin already containing an item of size above $\frac{1}{3}$, or that it contains at least five items whose sizes are in $(\frac{1}{4}, \frac{1}{3}]$.

The first M items are defined as follows. Using Theorem 1, we create M items such that the size of item i is $\frac{1}{4} + a_i$. These items are called F-items. The sizes of items are in $(0.25, 0.2501)$, and there is a value γ_1 such that any item whose packing satisfies condition C_{11} has size below $\frac{1}{4} + \gamma_1$ and for any item whose packing does not satisfy C_{11}, it has size above $\frac{1}{4} + \gamma_1$. The first kind of items are called small F-items, and the second type items are called large F-items. No matter how the input continues, as any packing of the first M items requires at least $\frac{M}{9}$ bins, the cost of an optimal solution is $\Omega(M)$.

Let Y_4 denote the current number of bins used by the algorithm, and this is also the number of large F-items. A possible continuation at this point is $\lceil \frac{M-Y_4}{5} \rceil$ items of (identical) sizes equal to $\frac{3}{4} - \gamma_1$. Note that such an item cannot be packed into a bin with an item of size above $\frac{1}{4} + \gamma_1$, as it cannot be packed next to it or below (or above) it, and the remaining space (not next to it or below it or above it) is too small (the sum of the diagonals of these two items is too large to be packed into a unit square bin).

Lemma 3. *There exists a packing of the items of the presented sequence (in this case) of cost at most $\frac{M}{5} - \frac{4Y_4}{45} + 2$.*

The algorithm has one large F-item in each of the first Y_4 bins and therefore it uses a new bin for every item of size $\frac{3}{4} - \gamma_1$. Thus, the total number of bins in the packing of the algorithm (in this case) is exactly $Y_4 + \lceil \frac{M-Y_4}{5} \rceil$.

We explain the continuation of the input in the case where the continuation defined above is not used. Here, for the construction, we state an upper bound on the number of items as the exact number of items is not known in advance and it will be determined during the presentation of the input. There will be at most $1.5M$ items of sizes slightly above $\frac{1}{3}$. We will use the variables S_3 and L_3 to denote the numbers of items for which condition C_{12} was satisfied and was not satisfied, respectively, in the current construction. Initialize $S_3 = L_3 = 0$, and increase the value of the suitable variable by 1 when a new item is presented. The ith item of the current construction has size $\frac{1}{3} + a_i$, and the sizes of items are in $(\frac{1}{3}, 0.33344)$. These items are called T-items. There is a value γ_2 such that any item whose packing satisfies condition C_{12} has size below $\frac{1}{3} + \gamma_2$ and any item whose packing does not satisfy C_{12} has size above $\frac{1}{3} + \gamma_2$. The first kind of items are called smaller T-items and the second type items are called larger T-items. Present items until $8S_3 + 15L_3 \geq 12M$ holds (this does not hold initially, so at least one item is presented, and this is defined to be condition C'). We show that indeed at most $1.5M$ items are presented. If $1.5M$ items were already presented, $8S_3 + 15L_3 \geq 8 \cdot (1.5M) = 12M$, and therefore the construction is stopped. In what follows, let S_3 and L_3 denote the final values of these variables. Before the last item of this part of the input was presented, it either was the case that $8(S_3 - 1) + 15L_3 < 12M$ or $8S_3 + 15(L_3 - 1) < 12M$ (as exactly one of S_3 and L_3 was increased by 1 when the last item was presented), so $8S_3 + 15L_3 - 15 < 12M$, or alternatively, $8S_3 + 15L_3 \leq 12M + 15$. Moreover, $S_3 + L_3 \geq \frac{4M}{5}$ as $12M \leq 8S_3 + 15L_3 \leq 15(S_3 + L_3)$. Let $M' = S_3 + L_3$ (and we have $M' = \Theta(M)$).

Here, there are two possible continuations. The first one is ($\lfloor \frac{M'}{3} \rfloor$) identical items, each of size exactly 0.6, and the second one is $\lfloor \frac{S_3}{3} \rfloor$ identical items, each of size $\frac{2}{3} - \gamma_2$.

Lemma 4. *The optimal cost in the first continuation is at most $\frac{M}{9} + \frac{7S_3}{27} + \frac{7L_3}{27} + 3$. The optimal cost in the second continuation is at most $\frac{S_3}{3} + \frac{L_3}{4} + 2$.*

Let Y_3 denote the number of new bins created for the T-items (where these bins were empty prior to the arrival of T-items). Here, there may be previously existing bins containing larger T-items (with at most four F-items), and $Y_3 \le L_3$. Consider the kinds of bins the algorithm may have after all T-items have arrived. Once again, T-items do not necessarily arrive, but we will deduce the numbers of different kinds of bins the algorithm has after all F-items have arrived based on number of bins existing after the arrival of T-items. After all T-items have arrived, a non-empty bin can receive an item of size 0.6 if it has at most five items, out of which at most three are T-items. The construction is such that any non-empty bin except for bins with at most five F-items has either at least six items in total (each of size above $\frac{1}{4}$) or it has an item of size above $\frac{1}{3} + \gamma_2$ (or both options may occur simultaneously), and therefore it cannot receive an item of size above $\frac{2}{3} - \gamma_2$.

Consider a given online algorithm for SP after the T-items were presented. Let X_{90} denote the number of bins with six, seven, eight, or nine F-items and no T-items. Such a bin cannot receive any further items in addition to its F-items in any of our continuations. Let X_{50} denote the number of bins with at least one and at most five F-items and no T-items. Such a bin can receive any item of size larger than $\frac{1}{2}$ that may arrive (but not an item of size $\frac{3}{4} - \gamma_1$). Let X_{81} denote the number of bins with five, six, seven, or eight F-items and one (small) T-item. Let X_{41} denote the number of bins with at least one and at most four F-items and one (large) T-item. Let X_{72} denote the number of bins with five, six, or seven F-items and two (small) T-items. Let X_{42} denote the number of bins with four F-items and two T-items (out of which one is small and one is large). Let X_{32} be the number of bins with at least one and at most three F-items and two T-items (out of which one is small and one is large). Let X_{63} denote the number of bins with five or six F-items and three T-items (all of which are small). Let X_{43} denote the number of bins with three or four F-items and three T-items (out of which two are small and one is large). Let X_{23} denote the number of bins with one or two F-items and three T-items (out of which two are small and one is large). Let X_{54} denote the number of bins with five F-items and four T-items (all of which are small). Let X_{44} denote the number of bins with two or three or four F-items and four T-items (out of which three are small and one is large). Let X_{14} denote the number of bins with one F-item and four T-items (out of which three are small and one is large).

Let X_{03} be the number of bins with no F-items and at least one and at most three T-items, one of which is a large T-item, while the others (at most two) are small. Let X_{04} be the number of bins with no F items and four T-items, one of which is large, while three are small.

We have three scenarios, and we use ALG_i to denote the cost of the algorithm for each one of them, in the order they were presented. Let $A_i = \limsup_{M \to \infty} \frac{ALG_i}{M}$. The optimal cost is always in $\Theta(M)$, and we let OPT_i denote our upper bounds on the optimal cost of the ith scenario, $O_i = \liminf_{M \to \infty} \frac{OPT_i}{M}$, and the ratio $\frac{A_i}{O_i}$ is lower bound on the competitive ratio R. We use the notation $y_i = \frac{Y_i}{M}$ and $x_{ij} = \frac{X_{ij}}{M}$ for those values of i and j such that Y_i and X_{ij} are defined. Let $\ell_3 = \frac{L_3}{M}$ and $s_3 = \frac{S_3}{M}$, so $12 \le 8s_3 + 15\ell_3 \le 12 + \frac{15}{M}$, and for M growing to infinity, $8s_3 + 15\ell_3 = 12$.

Let R be the (asymptotic) competitive ratio. For all twenty variables there is a non-negativity constraint. In addition, the number of items should satisfy $\sum_{i,j} j \cdot X_{ij} \geq S_3 + L_3$ and $\sum_{i,j} i \cdot X_{ij} \geq M$ (once again, the first constraint is inequality and not equality as X_{03} counts also bins with less than three T-items, and the second constraint is not an equality as in some cases X_{ij} counts bins with fewer than i F-items). Using the definitions of Y_4 and Y_3 we have $Y_4 = X_{90} + X_{50} + X_{81} + X_{41} + X_{72} + X_{42} + X_{32} + X_{63} + X_{43} + X_{23} + X_{54} + X_{44} + X_{14}$ and $Y_3 = X_{03} + X_{04}$.

We also have $ALG_1 = Y_4 + \lceil \frac{M-Y_4}{5} \rceil$ while $OPT_1 \leq \frac{M}{5} - \frac{4Y_4}{45} + 2$, so $R \geq \frac{A_1}{O_1} \geq \frac{1/5 + 4y_4/5}{1/5 - 4y_4/45} = \frac{9 + 36y_4}{9 - 4y_4}$. Additionally, $ALG_2 = Y_4 + Y_3 - X_{50} - X_{41} - X_{32} - X_{23} - X_{03} + \lfloor \frac{M'}{3} \rfloor \geq Y_4 + Y_3 - X_{50} - X_{41} - X_{32} - X_{23} - X_{03} + \frac{S_3 + L_3}{3} - 2$ while $OPT_2 \leq \frac{M}{9} + \frac{7S_3}{27} + \frac{7L_3}{27} + 3$, and $ALG_3 = Y_4 + Y_3 - X_{50} + \lfloor \frac{S_3}{3} \rfloor \geq Y_4 + Y_3 - X_{50} + \frac{S_3}{3} - 1$ while $OPT_3 \leq \frac{S_3}{3} + \frac{L_3}{4} + 2$, so $R \geq \frac{A_2}{O_2} \geq \frac{y_4 + y_3 - x_{50} - x_{41} - x_{32} - x_{23} - x_{03} + s_3/3 + \ell_3/3}{7s_3/27 + 7\ell_3/27 + 1/9}$ and $R \geq \frac{A_3}{O_3} \geq \frac{y_4 + y_3 - x_{50} + s_3/3}{s_3/3 + \ell_3/4}$.

Using these constraints we obtain a mathematical program of minimizing R whose optimal objective function value is approximately 1.751544578513 (and it is not smaller than this number). Thus, we have proved $R \geq 1.751544578513$.

4   Online Class Constrained Bin Packing (CLCBP)

In this section we exhibit our approach to proving lower bounds for the last variant of the bin packing problem which we study here, by improving the known lower bounds for the cases $t = 2$ and $t = 3$ of CLCBP. We will prove the following theorem.

Theorem 4. *The competitive ratios of online algorithms for CLCBP with $t = 2$ and $t = 3$ are at least 1.717668486 and at least 1.80814287, respectively.*

The constructions for $t = 2$ and $t = 3$ have clear differences, but the general idea is similar. The outline of the constructions is as follows. Start with a large number of tiny items, all of distinct colors, so every bin of any algorithm will contain at most t tiny items. Here, the construction is such that the items packed first into their bins are much larger than other items (large tiny items will be larger by at least a constant multiplicative factor than small tiny items, but they are still very small). One option at this point is to continue with huge items of sizes close to 1, all of distinct colors out of the colors of small tiny items, such that every item of size almost 1 can be packed into a bin with t small tiny items in an offline solution, one of which has the same color as the huge item packed with it. Note that no large tiny item can be combined with a huge item, so those items will be packed separately, t items per bin. The number of huge items is chosen in a way such that the optimal cost is not increased. Another option to continue the construction (instead of introducing the huge items) is with items of sizes slightly above $\frac{1}{3}$, where an item packed into a bin already containing an item of size above $\frac{1}{3}$ is smaller than an item packed into a bin with no such item (but it could possibly be packed with tiny items). It is ensured that bins of the algorithm already containing t (tiny) items will not be used again by the algorithm by never introducing items of their colors again. The sizes will be $\frac{1}{3}$ plus small values, where these small values are much larger than sizes of tiny items (including sizes of large tiny items). An interesting feature is that there will be exactly *two* items of sizes slightly above $\frac{1}{3}$ with each color which is used for such items, where the idea is to reuse (as much as possible) colors of tiny items packed by the algorithm into bins with at most $t-1$ tiny items (where those tiny items can be large or small), and never reuse colors of tiny items packed in bins of t items. In some cases (if there are too few such colors which can be reused), new colors are used as well for items of sizes slightly above $\frac{1}{3}$ (but there are still two items of sizes just above $\frac{1}{3}$ for each color). After these last items are presented, the final list of items will be items of sizes above $\frac{1}{2}$ whose colors will match exactly those of items of sizes in $(\frac{1}{3}, \frac{1}{2}]$ with the goal of packing such pairs of one color together into bins of offline solutions. There are two options for the final items. There are either such items not much larger than $\frac{1}{2}$, or there are items of sizes close to $\frac{2}{3}$, such that such an item having a color of an item of size slightly above $\frac{1}{3}$ can be combined into a bin with that item and with at most t tiny items coming from bins of the algorithm with at most $t-1$ items (no matter whether they are small or large, but one of them has to be of the same color). However, in the case of items of sizes almost $\frac{2}{3}$, only small items of sizes just above $\frac{1}{3}$ will be combined with them in good offline solutions while others are packed in pairs (of the same color whenever possible, and of different colors otherwise, combining tiny items where possible).

First, we present the parts of the constructions that are identical for $t = 2$ and $t = 3$. The condition C_1 will be that the current item is not the first item of its type packed into its bin, where a type consists of all items of similar size (the two relevant types are tiny items and items of sizes slightly above $\frac{1}{3}$). Let $M > 1$ be a large integer divisible by 6. The construction starts with the first type of items, where these items are called E-items or tiny items, consisting of M items constructed using Theorem 1. Let the value of k be 20, and the resulting values a_i are smaller than $20^{-2^{2M+2}}$. The number of tiny items presented is always exactly M (so the stopping condition is that there are M items), and the size of the ith item is simply a_i. Every E-item has its own color that may be reused in future parts of the construction but not for E-items. Let ε_1 and γ_1 be such that the size of any E-item satisfying C_1 (which we call a small E-item) is below $\frac{2\varepsilon_1}{20} < \frac{\varepsilon_1}{t}$ and the size of any E-item not satisfying C_1 (which we call a large E-item) is above $2\varepsilon_1$ (but smaller than $20^{-2^{2M+2}}$). Let X_j (for $1 \le j \le t$) be the number of bins of the algorithm with j E-items. Let X denote the total number of bins of E-items, i.e., $X = \sum_{j=1}^{t} X_j$.

If huge items arrive now, their number is $\lfloor \frac{M-X}{t} \rfloor$ and their colors are distinct colors out of colors of small E-items. The size of every huge item is $1-\varepsilon_1$. If $X_t \le \frac{M}{2t}$, there are no other continuations. In all other cases, there are two possible continuations except for the one with huge items, which was just discussed.

In all other continuations, items of a second type are presented such that their number is at most $2M$, and they will be called T-items. They are constructed using Theorem 1 with $k = 10$, so their values of a_i are in $(10^{-2^{2M+3}}, 10^{-2^{2M+2}})$. We have (by $M \geq 1$) $\frac{10^{-2^{2M+3}}}{20^{-2^{2M+2}}} = \frac{10^{2^{2M+2}} 2^{2^{2M+2}}}{10^{2^{2M+3}}} = \frac{2^{2^{2M+2}}}{10^4} > 6 > t$. The size of the ith T-item is $\frac{1}{3} + a_i$, and here condition C_1 means that the T-item is packed by the algorithm as the second T-item of its bin. Let ε_2 and γ_2 be such that a T-item satisfying C_1 (which we call a small T-item) has size smaller than $\frac{1}{3} + \frac{\varepsilon_2}{10}$ and a T-item not satisfying C_1 (which we call a large T-item) has size larger than $\frac{1}{3} + \varepsilon_2$. The number of T-items is even, and their colors are such that there are two T-items for each color. These colors are colors of E-items that are not packed in bins of t E-items by the algorithm. As the number of such E-items is $M - t \cdot X_t$, if the number of T-items is larger than $2(M - t \cdot X_t)$, new colors (which were not used for any earlier item) are used (and for the new colors there are also two T-items for each color). The variables Z_1 and Z_2 denote the numbers of bins with at least one T-item and with exactly two T-items, respectively, used by the algorithm (so $Z_2 \leq Z_1$). The algorithm may use bins with at most $(t - 1)$ E-items to pack T items (but not bins with (t) E-items, as no additional items have colors as those items).

For $t = 2$, the number of T-items is $\max\{2X_1, 2X_2\}$. Since $2X_2 \leq M$ and $2X_1 \leq 2M$, the number of T-items does not exceed $2M$. For $t = 3$, the stopping condition is defined as follows. First, present items until at least one of $Z_1 + Z_2 + 6X_3 \geq 2M - 1$, $3Z_1 + 4Z_2 \geq 2M - 7$ holds. Then, if the second condition holds, stop presenting items. If the first condition holds (and the second one does not hold), continue presenting items until $2Z_1 + 3Z_2 \geq 6X_3 - 5$ holds and stop. At this time, if the current number of T-items is odd, one additional item is presented. Thus, we guarantee that the value of $Z_1 + Z_2$ is an even number. Since the value X_3 is already fixed when T-items are presented, we analyze the increase in the value of each expression when a new T-item is presented. If a new item is packed into a bin with no T-item (and it is large), then the value of Z_1 increases by 1 while the value of Z_2 is unchanged. Otherwise (it is small), the value of Z_2 increases by 1 while the value of Z_1 is unchanged. Thus, the value of $Z_1 + Z_2$ can increase by at most 1, while that of $3Z_1 + 4Z_2$ can increase by at most 4, and that of $2Z_1 + 3Z_2$ can increase by at most 3. Thus, there are two cases. If the first condition that holds is $3Z_1 + 4Z_2 \geq 2M - 7$, when it started to hold, the value of the left hand side was increased by at most 4. If another item is presented to make the number of items even, it could increase by at most 4 again, so $3Z_1 + 4Z_2 \leq 2M$. If $Z_1 + Z_2 + 6X_3 \geq 2M - 1$ holds first (note that the two conditions could potentially start holding at the same time), then still $Z_1 + Z_2 + 6X_3 \leq 2M$. If in the current step it holds that $Z_1 + Z_2 + 6X_3 \geq 2M - 1$ and $3Z_1 + 4Z_2 \leq 2M - 8$, at that time, $2Z_1 + 3Z_2 \leq 6X_3 - 6$ holds (as otherwise, taking the sum of $Z_1 + Z_2 + 6X_3 \geq 2M - 1$ and $2Z_1 + 3Z_2 \geq 6X_3 - 5$ gives $3Z_1 + 4Z_2 \geq 2M - 6 > 2M - 7$). Therefore in the case the first condition holds first while the second one does not, additional items are presented and finally $2Z_1 + 3Z_2 \leq 6X_3$ (counting the last two items). Thus, after all T-items have arrived, it is either the case that $Z_1 + Z_2 \leq 3Z_1 + 4Z_2 \leq 2M$ or that

$Z_1 + Z_2 \leq 2Z_1 + 3Z_2 \leq 6X_3 \leq 2M$ (as $3X_3 \leq M$), so there are indeed at most $(2M)$ T-items.

A *matching item* for a T-item is an item of size above $\frac{1}{2}$ with the same color. There are two continuations as follows. In the first one, there are items of sizes 0.6, such that there is a matching item for every T-item (a different matching item for every item, i.e., $Z_1 + Z_2$ items of size 0.6 in total). In the second one, there are items of sizes $\frac{2}{3} - \frac{\varepsilon_2}{5}$, such that every small T-item has a matching item (once again, a different matching item for every item, i.e., Z_2 items in total). This concludes the description of our lower bounds constructions for the two cases of $t = 2$ and $t = 3$. Using these constructions we prove the theorem.

5   Summary

We showed that the method of designing fully adaptive instances, previously used for cardinality constrained bin packing and vector packing [4] (see also [2,10,18]) can be used to improve the known lower bounds for several additional bin packing problems. We analyzed its effect (together with many additional ideas) for several variants, and expect that it could be useful for a number of other variants as well.

References

1. Angelopoulos, S., Dürr, C., Kamali, S., Renault, M., Rosén, A.: Online bin packing with advice of small size. In: Dehne, F., Sack, J.-R., Stege, U. (eds.) WADS 2015. LNCS, vol. 9214, pp. 40–53. Springer, Cham (2015). https://doi.org/10.1007/978-3-319-21840-3_4
2. Babel, L., Chen, B., Kellerer, H., Kotov, V.: Algorithms for on-line bin-packing problems with cardinality constraints. Discret. Appl. Math. **143**(1–3), 238–251 (2004)
3. Balogh, J., Békési, J.: Semi-on-line bin packing: a short overview and a new lower bound. CEJOR **21**(4), 685–698 (2013)
4. Balogh, J., Békési, J., Dósa, G., Epstein, L., Levin, A.: Online bin packing with cardinality constraints resolved. The Computing Research Repository (CoRR) (2016). http://arxiv.org/abs/1608.06415. Also in ESA 2017 (to appear)
5. Balogh, J., Békési, J., Dósa, G., Epstein, L., Levin, A.: A new and improved algorithm for online bin packing. The Computing Research Repository (CoRR) (2017). http://arxiv.org/abs/1707.01728
6. Balogh, J., Békési, J., Dósa, G., Epstein, L., Levin, A.: Lower bounds for several online variants of bin packing. The Computing Research Repository (CoRR) (2017). http://arxiv.org/abs/1708.03228
7. Balogh, J., Békési, J., Galambos, G.: New lower bounds for certain classes of bin packing algorithms. Theoret. Comput. Sci. **440–441**, 1–13 (2012)
8. Bansal, N., Correa, J., Kenyon, C., Sviridenko, M.: Bin packing in multiple dimensions: inapproximability results and approximation schemes. Math. Oper. Res. **31**(1), 31–49 (2006)
9. Békési, J., Dósa, G., Epstein, L.: Bounds for online bin packing with cardinality constraints. Inf. Comput. **249**, 190–204 (2016)

10. Blitz, D.: Lower bounds on the asymptotic worst-case ratios of on-line bin packing algorithms. M.Sc. thesis, University of Rotterdam, Number 114682 (1996)
11. Boyar, J., Kamali, S., Larsen, K.S., López-Ortiz, A.: Online bin packing with advice. Algorithmica **74**(1), 507–527 (2016)
12. Coppersmith, D., Raghavan, P.: Multidimensional online bin packing: algorithms and worst case analysis. Oper. Res. Lett. **8**(1), 17–20 (1989)
13. Epstein, L.: Online bin packing with cardinality constraints. SIAM J. Discret. Math. **20**(4), 1015–1030 (2006)
14. Epstein, L., Imreh, C., Levin, A.: Class constrained bin packing revisited. Theoret. Comput. Sci. **411**(34–36), 3073–3089 (2010)
15. Epstein, L., Levin, A.: On bin packing with conflicts. SIAM J. Optim. **19**(3), 1270–1298 (2008)
16. Epstein, L., Levin, A.: Robust approximation schemes for cube packing. SIAM J. Optim. **23**(2), 1310–1343 (2013)
17. Epstein, L., van Stee, R.: Online square and cube packing. Acta Inform. **41**(9), 595–606 (2005)
18. Fujiwara, H., Kobayashi, K.: Improved lower bounds for the online bin packing problem with cardinality constraints. J. Comb. Optim. **29**(1), 67–87 (2015)
19. Heydrich, S., van Stee, R.: Improved lower bounds for online hypercube packing. The Computing Research Repository (CoRR) (2016). http://arxiv.org/abs/1607.01229
20. Johnson, D.S.: Fast algorithms for bin packing. J. Comput. Syst. Sci. **8**, 272–314 (1974)
21. Johnson, D.S., Demers, A., Ullman, J.D., Garey, M.R., Graham, R.L.: Worst-case performance bounds for simple one-dimensional packing algorithms. SIAM J. Comput. **3**, 256–278 (1974)
22. Kellerer, H., Pferschy, U.: Cardinality constrained bin-packing problems. Ann. Oper. Res. **92**, 335–348 (1999)
23. Krause, K.L., Shen, V.Y., Schwetman, H.D.: Analysis of several task-scheduling algorithms for a model of multiprogramming computer systems. J. ACM **22**(4), 522–550 (1975)
24. Liang, F.M.: A lower bound for on-line bin packing. Inf. Process. Lett. **10**(2), 76–79 (1980)
25. Seiden, S.S.: On the online bin packing problem. J. ACM **49**(5), 640–671 (2002)
26. Seiden, S.S., van Stee, R.: New bounds for multi-dimensional packing. Algorithmica **36**(3), 261–293 (2003)
27. Shachnai, H., Tamir, T.: Tight bounds for online class-constrained packing. Theoret. Comput. Sci. **321**(1), 103–123 (2004)
28. Shachnai, H., Tamir, T.: Polynomial time approximation schemes for class-constrained packing problems. J. Sched. **4**(6), 313–338 (2001)
29. Ullman, J.D.: The performance of a memory allocation algorithm. Technical report 100, Princeton University, Princeton, NJ (1971)
30. van Vliet, A.: An improved lower bound for online bin packing algorithms. Inf. Process. Lett. **43**(5), 277–284 (1992)
31. Xavier, E.C., Miyazawa, F.K.: The class constrained bin packing problem with applications to video-on-demand. Theoret. Comput. Sci. **393**(1–3), 240–259 (2008)
32. Yao, A.C.C.: New algorithms for bin packing. J. ACM **27**, 207–227 (1980)

The Online Multicommodity Connected Facility Location Problem

Mário César San Felice[1]([envelope]) [iD], Cristina G. Fernandes[2] [iD],
and Carla Negri Lintzmayer[3] [iD]

[1] Department of Computing, Federal University of São Carlos, São Carlos, Brazil
felice@ufscar.br
[2] Department of Computer Science, University of São Paulo, São Paulo, Brazil
cris@ime.usp.br
[3] Center for Mathematics, Computation and Cognition, Federal University of ABC,
Santo André, Brazil
carla.negri@ufabc.edu.br

Abstract. Grandoni and Rothvoß introduced the Multicommodity Connected Facility Location problem, a generalization of the Connected Facility Location problem which arises from a combination of the Facility Location and the Steiner Forest problems through the rent-or-buy model. We consider the online version of this problem and present a randomized algorithm that is $O(\log^2 n)$-competitive, where n is the number of given client pairs. Our algorithm combines the sample-and-augment framework of Gupta, Kumar, Pál, and Roughgarden with previous algorithms for the Online Prize-Collecting Facility Location and the Online Steiner Forest problems. Also, for the special case of the problem with edge scale factor equals 1, we show that a variant of our algorithm is deterministic and $O(\log n)$-competitive. Finally, we speculate on the possibility of finding a $O(\log n)$-competitive algorithm for the general case and the difficulties to achieve such ratio.

Keywords: Online algorithms · Competitive analysis
Facility Location · Steiner Forest · Rent-or-buy problems
Randomized algorithms

1 Introduction

In the Multicommodity Connected Facility Location problem (MCFL), one is given pairs of clients that must be connected through a path which may use parts of an infrastructure network. This network, consisting of a set of open facilities and a forest spanning them, can be seen as a subway network with

M. C. San Felice—Partial support CAPES PNPD 1522390, CNPq 456792/2014-7, FAPESP 2013/03447-6, and FAPESP 2017/11382-2.
C. G. Fernandes—Partial support CNPq 308116/2016-0, 456792/2014-7, and FAPESP 2013/03447-6.
C. N. Lintzmayer—Supported by FAPESP 2016/14132-4.

R. Solis-Oba and R. Fleischer (Eds.): WAOA 2017, LNCS 10787, pp. 118–131, 2018.
https://doi.org/10.1007/978-3-319-89441-6_10

facilities corresponding to subway stations. A pair of clients might represent a starting and ending point of a user's trip. The path chosen to take the user from her starting point to her ending point might use parts of the subway network, as long as entering in and exiting at subway stations, and is preferably as short as possible. The facility network is expensive to build, but once built, can be used by all users to shorten their trips.

We initiate the study of the online version of the MCFL, a combination of the Facility Location and the Steiner Forest problem through the rent-or-buy model which, as the example above illustrates, is of interest. In this problem, pairs of clients are given online, and one has to decide whether to open more facilities, how to connect them to other facilities, and also how to connect the pairs of clients. There is a cost to open each facility and a scale factor that measures how much more expensive it is to build the infrastructure network than to let users walk that way. The cost of a solution is then the sum of the cost to open the chosen facilities, the cost of the edges in the facility spanning forest multiplied by the scale factor, and the cost of the edges not in the facility network used by each client pair. In the next paragraphs we describe some related problems.

The Online Facility Location problem (OFL) is an infrastructure building problem in which one must decide where to place facilities and to which facility to connect each client. The clients are revealed one at a time and each one needs to be connected to a facility before the next client arrives. The goal is to minimize the cost of opening facilities plus the cost of connecting clients. There are algorithms for the OFL with competitive ratios $O\left(\frac{\log n}{\log \log n}\right)$ [7,13] and $O(\log n)$ [6,8,15], where n is the number of clients. The former one is asymptotically optimal, as the competitive ratio of any algorithm for the OFL is $\Omega\left(\frac{\log n}{\log \log n}\right)$ [7]. The Online Prize-Collecting Facility Location problem (OPFL) is a generalization of the OFL in which each client has a penalty cost and one may decide to pay the client's penalty instead of connecting it to a facility. Elmachtoub and Levi [3] and San Felice et al. [4] independently showed deterministic $O(\log n)$-competitive algorithms for the OPFL.

The Online Steiner Tree problem (OST) is a network design problem in which one must choose a tree that connects all terminals. The terminals are revealed one at a time and each one needs to be connected to the current tree before the next one arrives. The goal is to minimize the cost of the edges in the tree. There are $O(\log n)$-competitive algorithms for the OST [11], where n is the number of terminals, and these are asymptotically optimal as the competitive ratio of any algorithm for the OST is $\Omega(\log n)$ [11]. The Online Steiner Forest problem (OSF) is a generalization of the OST in which one receives client pairs and must build a forest that connects each client pair. Berman and Coulston [2] showed a deterministic $O(\log n)$-competitive algorithm for the OSF, where n is the number of given client pairs. The Online Multicommodity Rent-or-Buy problem (OMRoB) is a generalization of the OSF in which one is also given client pairs and may rent edges, by paying their regular cost, or buy edges, by paying M times their cost, to build a two-layer network that connects each pair. Rented edges are used only by the current pair, while bought edges are used by the current pair and eventually by subsequent client pairs. Umboh [16] and Awerbuch et al. [1] respectively showed a deterministic and a randomized $O(\log n)$-competitive algorithm for the OMRoB, where n is the number of given client pairs.

The Online Connected Facility Location problem (OCFL) arises from a combination of the OFL and the OST with the rent-or-buy model. A solution is also a two-layer network: a more expensive layer (the bought edges) connects the installed facilities, and a cheaper layer (the rented edges) connects clients to the facilities. One must decide where to place facilities, keep them connected to a given root, and connect each arriving client to a facility. An edge scale factor M is given and edges used to connect facilities to the root cost M times their cost, while edges used to connect clients to facilities have their normal cost. Umboh [16] and San Felice et al. [5] respectively showed a deterministic and a randomized $O(\log n)$-competitive algorithm for the OCFL, where n is the number of clients.

Our Contributions. We consider the Online Multicommodity Connected Facility Location problem (OMCFL), a generalization of both the OFL and the OMRoB in which one receives client pairs and must build a two-layer network that connects each given pair. The expensive layer of the network (the bought edges) is a forest connecting installed facilities, while the second layer (the rented edges) is used to connect the client pairs, eventually using the facility spanning forest. The client pair connection in the chosen two-layer network may have subpaths in the facility forest as long as these start and end at open facilities. An edge scale factor M is also given and edges in the facility layer of the network cost M times the regular edge cost, while edges used only to connect client pairs have their regular cost.

We give a randomized $O(\log^2 n)$-competitive algorithm for the OMCFL, where n is the number of client pairs, inspired on the algorithm by Grandoni and Rothvoß [9] for the MCFL. It combines the sample-and-augment framework of Gupta et al. [10] with above mentioned algorithms for the OPFL [4] and for the OSF [2]. Also, in the special case in which the edge scale factor M is equal to 1, we show how to modify our algorithm to deterministically achieve an $O(\log n)$-competitive ratio. Note that asymptotically this is optimal, since the $\Omega(\log n)$ lower bound for the OST applies to the OMCFL. This special case is closely related to the Steiner Tree-Star problem, addressed by Khuller and Zhu [12] to capture the scenario of private line data network design, in which we need to pay an extra amount for each "switch", i.e., installed facility.

2   Notations and Definitions

Each of the problems mentioned above has an offline version, which we refer to by omitting the O from the acronym of the problem. For instance, the offline version of the OMCFL is denoted by MCFL.

In this paper, let $G = (V, E)$ be a complete graph, $d : V \times V \to \mathbb{Q}^+$ be a symmetric function that respects the triangle inequality, referred to as *distance* on V, and $f : V \to \mathbb{Q}^+$ be a *facility opening cost function*. If H is a subgraph of G we denote by $V(H)$ and $E(H)$, respectively, the sets of vertices and edges of H. If $E' \subseteq E$, then we denote by $G[E']$ the subgraph induced by the edges in E'.

A *client-penalty duo* in G is a pair $(s, \pi(s))$ in which $s \in V$ is called *client*, and $\pi(s)$ is a nonnegative rational called *penalty*. We use D to denote a sequence of client-penalty duos. An instance of the OPFL is a quadruple (G, d, f, D) and a solution consists of an assignment ϕ of clients to elements of V or to null. Such an assignment induces a set F^ϕ of the elements in V that are in the image of ϕ, called *facilities*, and it also induces a subsequence D^ϕ of D consisting of the pairs in D whose client was assigned to null. The cost of such a solution for the OPFL is the sum of the opening cost for F^ϕ, the penalties for clients in D^ϕ, and the distance between each client not in D^ϕ and its assigned facility.

A *client pair* in G is a pair $p = (s, t)$ in which $s, t \in V$. We use P to denote a sequence of client pairs. A spanning forest H in G connects the pairs in P if there is no pair in P whose elements are in different components of H. An instance of the OSF is a triple (G, d, P) and a solution is a spanning forest H in G that connects all pairs in P. The cost of H is the sum of $d(e)$ for all edges e in H.

We denote the empty sequence by λ. Given a sequence S and an element p, we denote by $S + p$ the sequence obtained from appending p to the end of S. For instance, if $Q = ((1, 2), (3, 4))$, $p = (1, 3)$, and $P = Q + p$, then $P = ((1, 2), (3, 4), (1, 3))$.

3   The OMCFL Problem

An instance of the OMCFL is a complete graph $G = (V, E)$, a distance d on V, a facility opening cost function f, a cost scaling factor M, and a sequence P of client pairs in G. The offline part of the instance of the OMCFL is the tuple (G, d, f, M) and we leave implicit that $G = (V, E)$. The online part of the instance is the sequence P.

An online algorithm for the OMCFL must keep a set $E^b \subseteq E$ of *bought edges*, a set $F \subseteq V$ of *open facilities*, and sets $E_p^r \subseteq E$ of *rented edges* for each client pair p received that connects $p = (s, t)$ using E^b and F. We say that a set E_p^r *connects p using E^b and F* if s and t are in the same component of $G[E^b \cup E_p^r]$ and there is an (s, t)-path L_p in $G[E^b \cup E_p^r]$ for which all maximal subpaths of L_p in $G[E_p^r]$ have their extremes in $F \cup \{s, t\}$. In other words, we can connect s and t using rented and bought edges, provided that the common vertex between a rented edge and a bought edge which are adjacent in the (s, t)-path corresponds to an open facility.

The solution produced by the algorithm consists of F, E^b, and the sets E_p^r for each $p \in P$. The goal is to find a solution $(F, E^b, (E_p^r)_{p \in P})$ which minimizes the sum of the opening cost for F, with the cost of edges in E^b times M, plus the cost of edges in E_p^r for each p in P.

4   The OMCFL Algorithm

In this section, we describe an algorithm for the OMCFL, inspired on the algorithm for the MCFL due to Grandoni and Rothvoß [9]. For that, let $\mathrm{ALG_{OPFL}}$ be the deterministic $O(\log n)$-competitive algorithm for the OPFL proposed by San Felice et al. [4], with n being the length of the given client-penalty duo sequence, and let $\mathrm{ALG_{OSF}}$ be the deterministic $O(\log n)$-competitive algorithm for the OSF proposed by Berman and Coulston [2], with n being the length of the given client pair sequence.

The general idea of our algorithm for the OMCFL, which is presented in pseudocode as Algorithm 1, is to mark each arriving pair $p = (s, t)$ with probability $1/M$. If p is unmarked, then the algorithm can only rent edges to connect s and t. Otherwise, it opens the facilities assigned to s and t by $\mathrm{ALG_{OPFL}}$, uses $\mathrm{ALG_{OSF}}$ to decide which edges to buy, and completes the (s, t)-path by renting edges.

Input: (G, d, f, M) ▷ offline part of the input
1 $P \leftarrow \lambda; \quad D \leftarrow \lambda; \quad Q \leftarrow \lambda;$ ▷ λ is the empty sequence
2 $F \leftarrow \emptyset; \quad E^b \leftarrow \emptyset; \quad d' \leftarrow d; \quad E' \leftarrow \emptyset;$
3 **while** *a new pair $p = (s, t)$ arrives* **do**
4 $\quad$ $\pi_p \leftarrow \mathrm{dist}(G, d', s, t)/2;$
5 $\quad$ send (s, π_p) and (t, π_p) to $\mathrm{ALG_{OPFL}}(G, d, f, D)$ obtaining $\phi(s)$ and $\phi(t)$;
6 $\quad$ $D \leftarrow D + (s, \pi_p) + (t, \pi_p);$ ▷ append the two pairs to D
7 $\quad$ **if** $\phi(s) \neq$ null *and* $\phi(t) \neq$ null **then**
8 $\quad\quad$ mark p with probability $\frac{1}{M}$;
9 $\quad\quad$ **if** *p is marked* **then**
10 $\quad\quad\quad$ send $(\phi(s), \phi(t))$ to $\mathrm{ALG_{OSF}}(G, d, Q)$ obtaining an edge set $E_p^b;$
11 $\quad\quad\quad$ $Q \leftarrow Q + (\phi(s), \phi(t));$ ▷ append a new pair to Q;
12 $\quad\quad\quad$ $F \leftarrow F \cup \{\phi(s), \phi(t)\};$ ▷ open facilities at $\phi(s)$ and $\phi(t)$
13 $\quad\quad\quad$ $E^b \leftarrow E^b \cup E_p^b;$ ▷ buy the edges to connect $\phi(s)$ to $\phi(t)$
14 $\quad\quad\quad$ **for** $x, y \in F$ *in the same component of $G[E^b]$* **do**
15 $\quad\quad\quad\quad$ $d'(x, y) \leftarrow 0; \quad E' \leftarrow E' \cup \{xy\};$
16 $\quad$ consider an (s, t)-shortest path in G with costs $d';$
17 $\quad$ let E_p^r be the edges of this path except for those in $E';$
18 $\quad$ $P \leftarrow P + p;$
19 **return** $(F, E^b, (E_p^r)_{p \in P});$

Algorithm 1. Algorithm for the OMCFL problem.

Formally, let (G, d, f, M) be the offline part of the input for the OMCFL algorithm, and recall that it receives one client pair at a time. The algorithm stores the sequence P of pairs given so far, the set $F \subseteq V$ of open facilities, the set $E^b \subseteq E$ of bought edges, and sets $E_p^r \subseteq E$ of rented edges per given pair p. It also stores a sequence D of client-penalty duos corresponding to P, used by $\mathrm{ALG_{OPFL}}$. To set the penalty of each client in D and to make the choice of E_p^r for each pair p, the algorithm keeps a function $d' : V \times V \to \mathbb{Q}^+$ such that $d'(x, y) = 0$ if x, $y \in F$ are in the same component of the current graph $G[E^b]$, otherwise d' coincides with d. We define $\text{dist}(G, d', s, t)$ as the length of a shortest (s, t)-path in G with edge costs given by d'. The algorithm also keeps a set $E' \subseteq E$ of the edges in G where d and d' differ. Facilities, when opened by the algorithm, are opened in pairs, and Q is the sequence of such pairs, used by ALG_{OSF}.

The algorithm runs ALG_{OPFL} with (G, d, f) and the sequence D, given online, to decide when to randomly choose and mark a client pair. If a pair is marked, the facilities to which its clients were connected by ALG_{OPFL} are open, d' and E' are updated accordingly, and these two facilities form a pair that is fed to ALG_{OSF}, which is running with (G, d) and Q, given online. The algorithm buys the edges added to the solution that ALG_{OSF} is constructing. For every client pair p, the set E_p^r is then the edge set of a shortest path connecting p in G with costs d', excluding the edges in E'.

Observe that the initialization of d' and E' at Line 2, and their updates in Lines 14-15 assure they have the correct value during the algorithm. Using this, for every p, the set $E_p^r \subseteq E$ indeed connects p using E^b and F, and the algorithm produces a solution for the OMCFL.

5   Competitive Analysis of the Algorithm

In this section we analyze the competitivity of the OMCFL algorithm, given in Algorithm 1, which computes a solution $(F, E^b, (E_p^r)_{p \in P})$ when given instance (G, d, f, M, P). The cost of such solution is

$$\text{ALG}_{\text{OMCFL}}(P) = \sum_{i \in F} f(i) + M \sum_{e \in E^b} d(e) + \sum_{p \in P} \sum_{e \in E_p^r} d(e),$$

where we denote the first term in the right side by $O(P)$, the second term by $B(P)$, and the last term by $R(P)$. Note that $\text{ALG}_{\text{OMCFL}}(P)$, $O(P)$, $B(P)$, and $R(P)$ are random variables.

Similarly, consider an MCFL offline optimal solution $(F^*, E^{b*}, (E_p^{r*})_{p \in P})$ with which we are comparing. The cost of such solution is

$$\text{OPT}_{\text{MCFL}}(P) = \sum_{i \in F^*} f(i) + M \sum_{e \in E^{b*}} d(e) + \sum_{p \in P} \sum_{e \in E_p^{r*}} d(e),$$

where we denote the first term in the right side by $O^*(P)$, the second term by $B^*(P)$, and the last term by $R^*(P)$. Note that these three terms depend on the particular optimal solution we fixed, even though $\text{OPT}_{\text{MCFL}}(P)$ does not.

The next theorem is the main result of this section.

Theorem 1. *For any input (G, d, f, M, P) of the OMCFL and $n = |P|$, we have*

$$\mathbf{E}[\text{ALG}_{\text{OMCFL}}(P)] = O(\log^2 n)\, \text{OPT}_{\text{MCFL}}(P).$$

First we state an auxiliary lemma. Recall that D is the sequence of client-penalty duos obtained from the client pairs in P by the OMCFL algorithm.

Lemma 1. $\mathrm{OPT}_{\mathrm{PFL}}(D) \leq \mathrm{OPT}_{\mathrm{MCFL}}(P)$.

Proof. Consider an optimal MCFL solution $(F^*, E^{b^*}, (E_p^{r^*})_{p \in P})$ for the instance (G, d, f, M, P), with cost $O^*(P) + B^*(P) + R^*(P)$. Let us describe a solution for the PFL instance (G, d, f, D). Take $F' = F^*$ and let D' be the subsequence of D that contains the two client-penalty duos coming from each pair $p = (s, t) \in P$ such that s and t are in the same component of $G[E_p^{r^*}]$. As argued in Sect. 2, such pair (F', D') induces an assignment ϕ which is a solution for this PFL instance. Let us show that the cost of this solution is at most $O^*(P) + R^*(P)$.

Clearly the facility opening cost is $O^*(P)$. The penalty cost for the two clients in D' coming from a pair p is equal to $2\pi_p = \mathrm{dist}(G, d', s, t) \leq \sum_{e \in E_p^{r^*}} d(e)$. On the other hand, for the two clients not in D', coming from a pair $p = (s, t)$ for which s and t are not in the same component of $G[E_p^{r^*}]$, the connection cost is at most $\sum_{e \in E_p^{r^*}} d(e)$, because the edges in $E_p^{r^*}$ connect s and t to facilities in F^*. Thus, $\mathrm{OPT}_{\mathrm{PFL}}(D) \leq O^*(P) + R^*(P)$. $\square$

Let ϕ be the solution produced by $\mathrm{ALG}_{\mathrm{OPFL}}$ for instance (G, d, f, D). The cost of such solution, denoted by $\mathrm{ALG}_{\mathrm{OPFL}}(D)$, is the sum of the opening cost for F^ϕ, the penalties for clients in D^ϕ, and the distance between each client not in D^ϕ and its assigned facility. More precisely,

$$\mathrm{ALG}_{\mathrm{OPFL}}(D) = \sum_{v \in F^\phi} f(v) + \sum_{j \in D^\phi} \pi(j) + \sum_{j \notin D^\phi} d(j, \phi(j)),$$

where we denote the first term in the right side by $O'(D)$, the second term by $\Pi(D)$, and the last term by $C'(D)$. Also, let $\mathrm{OPT}_{\mathrm{PFL}}(D)$ denote the minimum cost of a solution for (G, d, f, D).

Now we bound the OMCFL facility opening cost by comparing it to the OPFL opening cost.

Lemma 2. $O(P) \leq O'(D)$.

Proof. Since the set F of facilities of our algorithm is a subset of the set of facilities open by $\mathrm{ALG}_{\mathrm{OPFL}}$, the lemma follows. $\square$

Let H be the solution produced by $\mathrm{ALG}_{\mathrm{OSF}}$ for instance (G, d, Q). The cost of such solution, denoted by $\mathrm{ALG}_{\mathrm{OSF}}(Q)$, is the sum of $d(e)$ for all edges e in H. That is,

$$\mathrm{ALG}_{\mathrm{OSF}}(Q) = \sum_{e \in E(H)} d(e).$$

Also, we denote the minimum cost of a solution for (G, d, Q) by $\mathrm{OPT}_{\mathrm{SF}}(Q)$.

Next we prove a bound on the expectation of the OMCFL buying cost $B(P)$. First we group the pairs in P according to the way they were treated by the algorithm. One group is the set P^π of pairs that failed in the test of Line 7, that is, at least one of the clients in the pair paid penalty in ALG_{OPFL}. The second group is the set P^m of pairs that were marked in Line 8 and the third group is the set P^u of the remaining pairs, that is, pairs for which Line 8 was executed, but they were unmarked.

Lemma 3. $\mathbf{E}[B(P)] = O(\log^2 n)\,\text{OPT}_{\text{MCFL}}(P)$, *where* $n = |P|$.

Proof. Note that E^b is the set of edges in the solution constructed by ALG_{OSF} for (G, d, Q) and $|Q| \leq n$. Therefore

$$B(P) \leq M\,\text{ALG}_{\text{OSF}}(Q) = M\,O(\log n)\,\text{OPT}_{\text{SF}}(Q). \tag{1}$$

Next we relate $\text{OPT}_{\text{SF}}(Q)$ to $\text{OPT}_{\text{MCFL}}(P)$ by describing how to build a feasible SF solution for instance (G, d, Q) from the optimal MCFL solution $(F^*, E^{b^*}, (E_p^{r^*})_{p \in P})$. The feasible SF solution consists of the edge set E^{b^*}, the edge sets $E_p^{r^*}$, and the edges $(s, \phi(s))$ and $(t, \phi(t))$ for each $p = (s, t) \in P^m$. Since Q consists of exactly the pairs $(\phi(s), \phi(t))$ for $p = (s, t) \in P^m$, the edge set of this solution indeed connects all pairs in Q, as each set $E_p^{r^*}$ connects p using E^{b^*}. Thus, we have

$$\text{OPT}_{\text{SF}}(Q) \leq \frac{B^*(P)}{M} + \sum_{p \in P^m} \sum_{e \in E_p^{r^*}} d(e) + \sum_{p = (s,t) \in P^m} (d(s, \phi(s)) + d(t, \phi(t))).$$

Note that

$$\mathbf{E}\left[\sum_{p \in P^m} \sum_{e \in E_p^{r^*}} d(e)\right] \leq \frac{1}{M} \sum_{p \in P} \sum_{e \in E_p^{r^*}} d(e) = \frac{R^*(P)}{M}.$$

Similarly, we have

$$\mathbf{E}\left[\sum_{(s,t) \in P^m} (d(s, \phi(s)) + d(t, \phi(t)))\right] = \frac{1}{M} \sum_{(s,t) \in P^m \cup P^u} (d(s, \phi(s)) + d(t, \phi(t)))$$

$$\leq \frac{C'(D)}{M}.$$

Thus, we have

$$\mathbf{E}[\text{OPT}_{\text{SF}}(Q)] \leq \frac{B^*(P) + R^*(P) + C'(D)}{M}.$$

Going back to (1) and taking the expectation, we deduce

$$\mathbf{E}[B(P)] = M\,O(\log n)\,\mathbf{E}[\text{OPT}_{\text{SF}}(Q)]$$

$$= M\,O(\log n)\left(\frac{B^*(P) + R^*(P) + C'(D)}{M}\right)$$

$$= O(\log n)\,(B^*(P) + R^*(P) + \text{ALG}_{\text{OPFL}}(D))$$

$$= O(\log n)\,(\text{OPT}_{\text{MCFL}}(P) + O(\log n)\text{OPT}_{\text{PFL}}(D))$$

$$= O(\log^2 n)\,\text{OPT}_{\text{MCFL}}(P), \tag{2}$$

where (2) follows from Lemma 1. $\qquad\square$

To bound the renting cost, we split it into three terms according to the grouping of the pairs. Namely, $R(P) = R^\pi(P) + R^m(P) + R^u(P)$, where $R^\pi(P)$, $R^m(P)$, and $R^u(P)$ are the renting costs for the pairs in P^π, P^m, and P^u respectively. Each of the next lemmas bounds one of these terms.

Next we bound the first part of the OMCFL renting cost by comparing it to the OPFL penalty cost.

Lemma 4. $R^\pi(P) \leq 2\Pi(D)$.

Proof. For each pair $p = (s,t) \in P^\pi$, as the test in Line 7 failed for p, Lines 8-15 were not executed in the iteration in which p was treated and d' did not change in this iteration. Therefore, $\sum_{e \in E_p^r} d(e) = \mathrm{dist}(G, d', s, t) = 2\pi_p$. Also, s or t paid the penalty π_p in $\mathrm{ALG_{OPFL}}$. Thus, we have

$$R^\pi(P) = \sum_{p \in P^\pi} \sum_{e \in E_p^r} d(e) = \sum_{p \in P^\pi} 2\pi_p \leq 2\Pi(D).$$

$\square$

In what follows we bound the second part of the OMCFL renting cost by comparing it to the OPFL connection cost.

Lemma 5. $R^m(P) \leq C'(D)$.

Proof. For each pair $p = (s,t) \in P^m$, Lines 10–15 were executed. Thus, the renting cost $\sum_{e \in E_p^r} d(e)$ is exactly the connection cost that is paid by $\mathrm{ALG_{OPFL}}$ to connect s to $\phi(s)$ and t to $\phi(t)$. Hence, we have

$$R^m(P) = \sum_{p \in P^m} \sum_{e \in E_p^r} d(e) = \sum_{p=(s,t) \in P^m} (d(s, \phi(s)) + d(t, \phi(t))) \leq C'(D).$$

$\square$

Finally, we bound the expectation of the last part of the OMCFL renting cost.

Lemma 6. $\mathbf{E}[R^u(P)] = O(\log^2 n)\, \mathrm{OPT_{MCFL}}(P)$, *where* $n = |P|$.

Proof. In this lemma we consider only pairs that are not in P^π. The main idea of this proof is to bound the expected renting cost of the unmarked pairs by comparing it with the expected buying cost.

For each pair $p = (s,t)$, consider a shortest (s,t)-path in G with costs d' computed at the beginning of the iteration in which p is treated. Let S_p be the edges of this path that are not in E'. Thus, we may assume $S_p = E_p^r$ if $p \notin P^m$. Let Z_p be the edge set that would be obtained if one sent $(\phi(s), \phi(t))$ to $\mathrm{ALG_{OSF}}(G, d, Q)$ just after Line 5. Thus, $Z_p = E_p^b$ if $p \in P^m$. Let $d(S_p) = \sum_{e \in S_p} d(e)$ and $d(Z_p) = \sum_{e \in Z_p} d(e)$. Note that $d(S_p)$ and $d(Z_p)$ are random variables. Also,

$$d(S_p) \leq d(Z_p) + d(s, \phi(s)) + d(t, \phi(t)), \tag{3}$$

because Z_p with E' and the edges $(s, \phi(s))$ and $(t, \phi(t))$ contain an (s,t)-path in G with costs d', while S_p with some edges from E' correspond to a shortest (s,t)-path in G with costs d'. Recall that $d'(e) = 0$ for every edge $e \in E'$ and $d'(e) = d(e)$ for every edge $e \notin E'$.

Note that $E_p^b = Z_p$ if $p \in P^m$ and $E_p^b = \emptyset$ otherwise. Similarly, let $E_p^u = S_p$ if $p \in P^u$ and $E_p^u = \emptyset$ otherwise. For random variables X and Y, the expression $\mathbf{E}[X \mid Y]$ is the random variable that takes on the value $\mathbf{E}[X \mid Y = y]$ when $Y = y$ [14, Definition 2.7]. Using this notation, for a pair $p \notin P^\pi$, we have

$$\mathbf{E}\left[\sum_{e \in E_p^u} d(e) \mid d(S_p), d(Z_p)\right] = \frac{M-1}{M} d(S_p) \leq d(S_p),$$

and

$$\mathbf{E}\left[\sum_{e \in E_p^b} M d(e) \mid d(S_p), d(Z_p)\right] = \frac{1}{M} M d(Z_p) = d(Z_p).$$

Since the previous inequalities hold for any values of $d(S_p)$ and $d(Z_p)$, they hold without the conditionals. Thus, using (3), we deduce

$$\mathbf{E}\left[\sum_{e \in E_p^u} d(e)\right] \leq \mathbf{E}\left[\sum_{e \in E_p^b} M d(e)\right] + d(s, \phi(s)) + d(t, \phi(t)).$$

Recall that $R^u(P) = \sum_{p \in P^u} \sum_{e \in E_p^r} d(e) = \sum_{p \notin P^\pi} \sum_{e \in E_p^u} d(e)$ as $E_p^u = \emptyset$ for $p \notin P^u$, and that $B(P) = M \sum_{p \in P^m} \sum_{e \in E_p^b} d(e) = M \sum_{p \notin P^\pi} \sum_{e \in E_p^b} d(e)$ as $E_p^b = \emptyset$ for $p \notin P^m$. Therefore, summing the previous inequality over all pairs not in P^π, applying Lemma 3 we have

$$\mathbf{E}[R^u(P)] \leq \mathbf{E}[B(P)] + C'(D) = O(\log^2 n) \operatorname{OPT}_{\mathrm{MCFL}}(P),$$

as $C'(D) \leq \operatorname{ALG}_{\mathrm{OPFL}}(D) = O(\log n) \operatorname{OPT}_{\mathrm{PFL}}(D) = O(\log n) \operatorname{OPT}_{\mathrm{MCFL}}(D)$ by Lemma 1. □

We conclude this section with the proof of Theorem 1.

Proof (of Theorem 1). Using Lemmas 2, 3, 4, 5, and 6, we have

$$\begin{aligned}
\mathbf{E}[\operatorname{ALG}_{\mathrm{OMCFL}}(P)] &= \mathbf{E}[O(P)] + \mathbf{E}[B(P)] + \mathbf{E}[R(P)] \\
&= \mathbf{E}[O(P)] + \mathbf{E}[B(P)] + \mathbf{E}[R^\pi(P) + R^m(P) + R^u(P)] \\
&\leq O'(D) + O(\log^2 n) \operatorname{OPT}_{\mathrm{MCFL}}(P) \\
&\quad + 2\Pi(D) + C'(D) + O(\log^2 n) \operatorname{OPT}_{\mathrm{MCFL}}(P) \\
&= O(\log^2 n) \operatorname{OPT}_{\mathrm{MCFL}}(P),
\end{aligned}$$

because $O'(D) + 2\Pi(D) + C'(D) \leq 2 \operatorname{ALG}_{\mathrm{OPFL}}(D) = O(\log n) \operatorname{OPT}_{\mathrm{MCFL}}(D)$ by Lemma 1. □

6   The OMCFL in the Special Case Where $M = 1$

In this section we focus on the special case of OMCFL with $M = 1$, which is closely related to the Steiner Tree-Star problem, addressed by Khuller and Zhu [12]. We consider a modification of Algorithm 1, called Algorithm 2, in which we buy edges in a slightly different way. We show that, while the bound for Algorithm 1 does not improve in this special case, Algorithm 2 achieves $O(\log n)$ competitive ratio for it, which is asymptotically best possible.

In Algorithm 2 we keep Q equal to the sequence of pairs in P^m and we replace Lines 10 and 11 of Algorithm 1 by

10 send (s, t) to $\mathrm{ALG_{OSF}}(G, d, Q)$ obtaining an edge set E_p^b;
10' add to E_p^b the edges $(s, \phi(s))$ and $(t, \phi(t))$;
11 $Q \leftarrow Q + (s, t)$;

In the following analysis of Algorithm 2 we use the same notation used in the analysis of Algorithm 1.

Note that, when $M = 1$, the behavior of Algorithm 2 is deterministic, and all pairs for which $\phi(s)$ and $\phi(t)$ are not null end up being marked. Therefore, when we split the OMCFL renting cost, there is no need to consider $R^u(P)$, i.e.,

$$R(P) = R^\pi(P) + R^m(P).$$

Also, the upper bounds $O(P) \leq O'(D)$, $R^\pi(P) \leq 2\Pi(D)$, and $R^m(P) \leq C'(D)$ respectively from Lemmas 2, 4, and 5 still apply. Thus, it only remains to present a better bound for the buying cost $B(P)$ of Algorithm 2.

Lemma 7. *In the special case of OMCFL in which $M = 1$ and for $n = |P|$, we have*

$$B(P) = O(\log n)\mathrm{OPT_{MCFL}}(P).$$

Proof. We divide $B(P)$ into $B^{\mathrm{ext}}(P)$ and $B^{\mathrm{core}}(P)$. The former is the buying cost for $(s, \phi(s))$ and $(t, \phi(t))$, while the latter is the buying cost to connect s to t, for all pairs $p = (s, t) \in P^m$. Thus, we have

$$B^{\mathrm{ext}}(P) \leq \sum_{p=(s,t)\in P^m} (d(s, \phi(s)) + d(t, \phi(t))) \leq C'(D) \leq \mathrm{ALG_{OPFL}}(D),$$

and

$$B^{\mathrm{core}}(P) = \mathrm{ALG_{OSF}}(Q) = O(\log n)\,\mathrm{OPT_{SF}}(Q).$$

Consider an MCFL offline optimal solution $(F^*, E^{b^*}, (E_p^{r^*})_{p\in P})$. Note that the edges in E^{b^*} together with the edges in $E_p^{r^*}$, for each pair $p \in P^m$, connect all the pairs in P^m, which composes a feasible SF solution for the instance (G, d, Q). Thus, we have

$$\mathrm{OPT_{SF}}(Q) \leq B^*(P) + \sum_{p\in P^m} \sum_{e\in E_p^{r^*}} d(e) \leq B^*(P) + R^*(P).$$

Putting everything together, we have

$$\begin{aligned} B(P) &= B^{\text{ext}}(P) + B^{\text{core}}(P) \\ &\leq \text{ALG}_{\text{OPFL}}(D) + O(\log n)\, \text{OPT}_{\text{SF}}(Q) \\ &= O(\log n)\, \text{OPT}_{\text{PFL}}(D) + O(\log n)\, (B^*(P) + R^*(P)) \\ &= O(\log n)\, \text{OPT}_{\text{MCFL}}(P), \end{aligned} \tag{4}$$

where (4) follows due to Lemma 1.   □

We denote the cost of Algorithm 2 by $\text{ALG2}_{\text{OMCFL}}$. The next theorem is the main result of this section.

Theorem 2. *In the special case of OMCFL in which $M = 1$ and for $n = |P|$, we have*

$$\text{ALG2}_{\text{OMCFL}}(P) = O(\log n)\, \text{OPT}_{\text{MCFL}}(P).$$

Proof. Using Lemmas 1, 2, 4, 5, and 7, we have

$$\begin{aligned} \text{ALG2}_{\text{OMCFL}}(P) &= O(P) + B(P) + R^{\pi}(P) + R^m(P) \\ &\leq O'(D) + O(\log n)\, \text{OPT}_{\text{MCFL}}(P) + 2\Pi(D) + C'(D) \\ &= O(\log n)\, \text{OPT}_{\text{MCFL}}(P). \end{aligned}$$

□

7   Final Remarks

In this paper we proposed the Online Multicommodity Connected Facility Location problem and gave a randomized $O(\log^2 n)$-competitive algorithm. We also gave a modified version of the algorithm which is $O(\log n)$-competitive in the special case of OMCFL with $M = 1$.

On a high level, the idea of our main theorem's proof is to bound the cost of the more expensive network layer, which connects the facilities, and then use it to bound the expected renting cost of the unmarked clients. When we applied this idea to the Online Connected Facility Location problem (OCFL) in [5], we were able to achieve a tighter competitive ratio (i) by building the more expensive network layer in a cheaper way, which is similar to the one we used in our algorithm for the OMCFL special case, and (ii) by showing that if a certain amount of clients rented a path nearby an inactive facility, it was expected that the facility would be opened and a path from it would be bought into the more expensive network layer.

Two main difficulties arised when we tried to apply these ideas to the OMCFL analysis. The first, related to (i), is that although we are able to build the more expensive network layer in a cheaper way, it seems harder to compare its cost to the renting cost of the unmarked clients, since the facilities opened by the algorithm and the ones opened by the offline optimal solution may be in different forest components, while in the OCFL there was only one component.

The second difficulty, related to (ii), is that in the OCFL a path from a client always goes to the root, thus entering only once in the more expensive network layer, while in the OMCFL an (s,t)-path may pass through several connected components of such layer. It is somewhat easy to bound the renting cost of the edges in the path extremes, since it relates to the facility connection cost, albeit not directly in the case of unmarked clients. However, the renting cost of the edges in the middle of the path is harder to bound.

Future work includes finding a $O(\log n)$-competitive algorithm for OMCFL with $M > 1$. We believe that, combining some of the ideas in this paper with the approach from Umboh [16], which uses hierarchic decompositions of trees, may help to overcome the difficulties and find a deterministic competitive algorithm. Also, we are interested in finding competitive algorithms for other two-layer network design problems.

Finally, Algorithm 1 with $M = 1$ is a $O(\log^2 n)$-competitive algorithm for a natural online and multicommodity generalization of the Steiner Tree-Star problem [12]. As far as we know, it is an interesting open question if there exists a $O(\log n)$-competitive algorithm for this related problem.

References

1. Awerbuch, B., Azar, Y., Bartal, Y.: On-line generalized Steiner problem. Theor. Comput. Sci. **324**(2–3), 313–324 (2004)
2. Berman, P., Coulston, C.: On-line algorithms for Steiner tree problems (extended abstract). In: Proceedings of the 29th Annual ACM Symposium on Theory of Computing (STOC), pp. 344–353 (1997)
3. Elmachtoub, A.N., Levi, R.: From cost sharing mechanisms to online selection problems. Math. Oper. Res. **40**(3), 542–557 (2014)
4. San Felice, M.C., Cheung, S., Lee, O., Williamson, D.P.: The online prize-collecting facility location problem. In: Proceedings of the VIII Latin-American Algorithms, Graphs and Optimization Symposium (LAGOS), volume 50 of Electronic Notes in Discrete Mathematics, pp. 151–156 (2015)
5. San Felice, M.C., Williamson, D.P., Lee, O.: A randomized $O(\log n)$-competitive algorithm for the online connected facility location problem. Algorithmica **76**(4), 1139–1157 (2016)
6. Fotakis, D.: A primal-dual algorithm for online non-uniform facility location. J. Discrete Algorithms **5**(1), 141–148 (2007)
7. Fotakis, D.: On the competitive ratio for online facility location. Algorithmica **50**(1), 1–57 (2008)
8. Fotakis, D.: Online and incremental algorithms for facility location. SIGACT News **42**(1), 97–131 (2011)
9. Grandoni, F., Rothvoß, T.: Approximation algorithms for single and multi-commodity connected facility location. In: Günlük, O., Woeginger, G.J. (eds.) IPCO 2011. LNCS, vol. 6655, pp. 248–260. Springer, Heidelberg (2011). https://doi.org/10.1007/978-3-642-20807-2_20
10. Gupta, A., Kumar, A., Pál, M., Roughgarden, T.: Approximation via cost sharing: simpler and better approximation algorithms for network design. J. ACM **54**(3) (2007). Article 11

11. Imase, M., Waxman, B.M.: Dynamic Steiner tree problem. SIAM J. Discrete Math. **4**(3), 369–384 (1991)
12. Khuller, S., Zhu, A.: The general Steiner tree-star problem. Inf. Process. Lett. **84**(4), 215–220 (2002)
13. Meyerson, A.: Online facility location. In: Proceedings of the 42nd Annual IEEE Symposium on Foundations of Computer Science (FOCS), pp. 426–431 (2001)
14. Mitzenmacher, M., Upfal, E.: Probability and Computing - Randomized Algorithms and Probabilistic Analysis. Cambridge University Press, New York (2005)
15. Nagarajan, C., Williamson, D.P.: Offline and online facility leasing. Discrete Optim. **10**(4), 361–370 (2013)
16. Umboh, S.: Online network design algorithms via hierarchical decompositions. In: Proceedings of the 26th Annual ACM-SIAM Symposium on Discrete Algorithms (SODA), pp. 1373–1387 (2015)

A Match in Time Saves Nine: Deterministic Online Matching with Delays

Marcin Bienkowski$^{(\boxtimes)}$, Artur Kraska, and Paweł Schmidt

Institute of Computer Science, University of Wrocław, Wrocław, Poland
{marcin.bienkowski,artur.kraska,pawel.schmidt}@cs.uni.wroc.pl

Abstract. We consider the problem of online Min-cost Perfect Matching with Delays (MPMD) introduced by Emek et al. (STOC 2016). In this problem, an even number of requests appear in a metric space at different times and the goal of an online algorithm is to match them in pairs. In contrast to traditional online matching problems, in MPMD all requests appear online and an algorithm can match any pair of requests, but such decision may be delayed (e.g., to find a better match). The cost is the sum of matching distances and the introduced delays.

We present the first deterministic online algorithm for this problem. Its competitive ratio is $O(m^{\log_2 5.5}) = O(m^{2.46})$, where $2m$ is the number of requests. In particular, the bound does not depend on other parameters of the metric, such as its aspect ratio. Unlike previous (randomized) solutions for the MPMD problem, our algorithm does not need to know the metric space in advance and it does not require the space to be finite.

Keywords: Online matching · Delays · Rent-or-buy
Competitive analysis

1 Introduction

In this paper, we give a deterministic online algorithm for the problem of Min-cost Perfect Matching with Delays (MPMD) [6,26]. For an informal description, imagine that there are players who are logging in real time into a gaming website, each wanting to play chess against another human player. The system pairs the players according to their known capabilities, such as playing strength and a decision with whom to match a given player can be delayed until a reasonable match is found. That is, the website tries to simultaneously minimize two objectives: the waiting times of players and their dissimilarity, i.e., each player would like to play with another one with comparable capabilities. An algorithm running the website has to work online, without the knowledge about future player arrivals and make its decision irrevocably: once two players are paired, they remain paired forever.

Supported by Polish National Science Centre grant 2016/22/E/ST6/00499.

1.1 Problem Definition

More formally, in the MPMD problem there is a metric space $\mathcal{X}$ with a distance function dist : $\mathcal{X} \times \mathcal{X} \to \mathbb{R}$, both known from the beginning to an online algorithm. The online part of the input is a sequence of $2m$ requests $\{(p_i, t_i)\}_{i=1}^{2m}$, where point $p_i \in \mathcal{X}$ corresponds to a player in our informal description above and t_i is the time of its arrival satisfying $t_1 \leq t_2 \leq \ldots \leq t_{2m}$. The integer m is not known a priori to the online algorithm. At any time τ, the online algorithm may decide to match any pair of requests (p_i, t_i) and (p_j, t_j) that have already arrived ($\tau \geq t_i$ and $\tau \geq t_j$) and have not been matched yet. The cost incurred by such *matching edge* is $\mathrm{dist}(p_i, p_j) + (\tau - t_i) + (\tau - t_j)$, i.e., the sum of the *connection cost* and the *waiting costs* of these two requests.

The goal is to eventually match all requests and minimize the total cost. We use a typical yardstick to measure the performance of an online algorithm: the competitive ratio [14], defined as the maximum, over all inputs, of the ratios between the cost of the online algorithm and the cost of an optimal offline solution OPT that knows the entire input sequence in advance.

1.2 Previous Work

The MPMD problem was introduced by Emek et al. [26], who presented a randomized $O(\log^2 n + \log \Delta)$-competitive algorithm. There, n is the number of points in the metric space $\mathcal{X}$ and Δ is its aspect ratio (the ratio between the largest and the smallest distance in $\mathcal{X}$). The competitive ratio was subsequently improved by Azar et al. [6] to $O(\log n)$. They showed that the competitive ratio of any randomized algorithm is at least $\Omega(\sqrt{\log n})$. The currently best lower bound of $\Omega(\log n / \log \log n)$ for randomized solutions was given by Ashlagi et al. [3].

So far, the construction of a competitive *deterministic* algorithm for general metric spaces remained an open problem. It was hypothesized that the competitive ratio achievable by deterministic algorithms might be superpolynomial in n (cf. Sect. 5 of [6]). Deterministic algorithms were known only for simple spaces: Azar et al. [6] gave an $O(\text{height})$-competitive algorithm for trees and Emek et al. [27] constructed a 3-competitive deterministic solution for two-point metrics (the latter competitive ratio is best possible).

1.3 Our Contribution

In this paper, we give the first deterministic algorithm for an arbitrary metric space, whose competitive ratio is $O(m^{\log_2 5.5}) = O(m^{2.46})$, where $2m$ is the number of requests. While previous solutions to the MPMD problem [6,26] required $\mathcal{X}$ to be finite and known a priori (to approximate it first by a random HST tree [28] or a random HST tree with reduced height [9]), our solution works even when $\mathcal{X}$ is revealed in online manner. That is, we require only that, together with any request r, the online algorithm learns the distances from r to all previous, not yet matched requests.

We note that m is uncomparable with n (the number of different points in the metric space $\mathcal{X}$) and their relation depends on the application. For instance, in the traditional player ranking systems, such as Elo rating [25] used for chess, $\mathcal{X}$ is just a finite set of integers (ratings) and typically there are many players with the same rating (multiple requests appear at the same point of $\mathcal{X}$). For such setting, our algorithm is admittedly outperformed by the deterministic algorithm for trees by Azar et al. [6] (which is $O(n)$-competitive for such metric). On the other hand, there is an emerging trend for representing players in online games as multi-dimensional vectors [4,19,21]: some real-time strategy games and first-person shooters started to characterize players by their rank, reflex, planning, offensive and defensive skills, number of actions per minute, network latency, etc. For such cases, $\mathcal{X}$ is a subset of $\mathbb{R}^k$ with ℓ^1 metric, and thus $m \ll n$.

Our online algorithm ALG uses a simple, local, semi-greedy scheme to find a suitable matching pair. In the analysis, we fix a final perfect matching of OPT and observe what happens when we gradually add matching edges that ALG creates during its execution. That is, we trace the evolution of alternating paths and cycles in time. To bound the cost of ALG, we charge the cost of an edge that ALG is adding against the cost of already existing matching edges from the same alternating path. Interestingly, our charging argument on alternating cycles bears some resemblance to the analyses of algorithms for the problems that are not directly related to MPMD: online metric (bipartite) matching on line metrics [2] and offline greedy matching [44].

1.4   Related Work

Originally, matching problems have been studied in variants where delaying decisions was not permitted. The setting most similar to the MPMD problem is called online *metric bipartite matching*. It involves m *offline points* given to an algorithm at the beginning and m *requests* presented in online manner that need to be matched (immediately after their arrival) to offline points. Both the points and the requests lie in a common metric space and the goal is to minimize the weight of a perfect matching created by the algorithm. For general metric spaces, the best randomized solution is $O(\log m)$-competitive [8,30,41], and the deterministic algorithms achieve the optimal competitive ratio of $2m - 1$ [31,36]. Interestingly, even for line metrics [2,29,37], the best known deterministic algorithm attains a competitive ratio that is polynomial in m [2].

In comparison, in the MPMD problem considered in this paper, all $2m$ requests appear in online manner (the requests need not be matched immediately after arrival), m is not known to an algorithm, and we allow to match any pair of them. That said, there is also a bipartite variant of the MPMD problem, in which all requests appear online, but m of them are negative and m are positive. An algorithm may then only match pairs of requests of different polarities [3,5].

The MPMD problem can be cast as augmenting min-cost perfect matching with a time axis, allowing the algorithm to delay its decisions, but penalizing the delays. There are many other problems that use this paradigm:

most notably the ski-rental problem and its continuous counterpart, the spin-block problem [33], where a purchase decision can be delayed until renting cost becomes sufficiently large. Such rent-or-buy (wait-or-act) trade-offs are also found in other areas, for example in aggregating messages in computer networks [1,12,24,32,35,43], in aggregating orders in supply-chain management [10,11,15,16,18,20] or in some scheduling variants [7].

Finally, there is a vast amount of work devoted to other online matching variants, where offline points and online requests are connected by graph edges and the goal is to maximize the weight or the cardinality of the produced matching. These types of matching problems have been studied since the seminal work of Karp et al. [34] and are motivated by applications to online auctions [13,17,22,23,34,38,40,42]. They were also studied under stochastic assumptions on the input, see, e.g., a survey by Mehta [39].

2 Algorithm

We will identify requests with the points at which they arrive. To this end, we assume that all requested points are different, but we allow distances between different metric points to be zero. For any request p, we denote the time of its arrival by $\mathsf{atime}(p)$.

Our algorithm is parameterized with real numbers $\alpha > 0$ and $\beta > 1$, whose exact values will be optimized later. For any request p, we define its waiting time at time $\tau \geq \mathsf{atime}(p)$ as

$$\mathsf{wait}_\tau(p) = \tau - \mathsf{atime}(p)$$

and its budget at time τ as

$$\mathsf{budget}_\tau(p) = \alpha \cdot \mathsf{wait}_\tau(p).$$

Our online algorithm ALG matches two requests p and q at time τ as soon as the following two conditions are satisfied.

- *Budget sufficiency*: $\mathsf{budget}_\tau(p) + \mathsf{budget}_\tau(q) \geq \mathsf{dist}(p, q)$.
- *Budget balance*: $\mathsf{budget}_\tau(p) \leq \beta \cdot \mathsf{budget}_\tau(q)$ and $\mathsf{budget}_\tau(q) \leq \beta \cdot \mathsf{budget}_\tau(p)$.

Note that the budget balance condition is equivalent to relations on waiting times, i.e., $\mathsf{wait}_\tau(p) \leq \beta \cdot \mathsf{wait}_\tau(q)$ and $\mathsf{wait}_\tau(q) \leq \beta \cdot \mathsf{wait}_\tau(p)$.

If the conditions above are met simultaneously for many point pairs, we break ties arbitrarily, and process them in any order. Note that at the time when p and q become matched, the sum of their budgets may exceed $\mathsf{dist}(p, q)$. For example, this occurs when q appears at time strictly larger than $\mathsf{atime}(p) + \mathsf{dist}(p, q)$: they are then matched by ALG as soon as the budget balance condition becomes true.

The observation below follows immediately by the definition of ALG.

Observation 1. *Fix time τ and two requests p and q, such that $\mathsf{atime}(p) \leq \tau$ and $\mathsf{atime}(q) \leq \tau$. Assume that neither p nor q has been matched by ALG strictly before time τ. Then exactly one of the following conditions holds:*

- $\alpha \cdot (\mathsf{wait}_\tau(p) + \mathsf{wait}_\tau(q)) \leq \mathsf{dist}(p, q)$,
- $\alpha \cdot (\mathsf{wait}_\tau(p) + \mathsf{wait}_\tau(q)) > \mathsf{dist}(p, q)$ *and* $\mathsf{wait}_\tau(p) \geq \beta \cdot \mathsf{wait}_\tau(q)$,
- $\alpha \cdot (\mathsf{wait}_\tau(p) + \mathsf{wait}_\tau(q)) > \mathsf{dist}(p, q)$ *and* $\mathsf{wait}_\tau(q) \geq \beta \cdot \mathsf{wait}_\tau(p)$.

3   Analysis

To analyze the performance of ALG, we look at matchings generated by ALG and by an optimal offline algorithm OPT. If points p and q were matched at time τ by ALG, then we say that ALG creates a (matching) edge $e = (p, q)$. Its cost is

$$\text{cost}_{\text{ALG}}(e) = \text{cost}_{\text{ALG}}(p, q) = \text{dist}(p, q) + \text{wait}_\tau(p) + \text{wait}_\tau(q).$$

We call e an ALG-edge. The cost_{OPT} of an edge in the solution of OPT (an OPT-edge) is defined analogously. In an optimal solution, however, the matching time is always equal to the arrival time of the later of two matched requests.

We consider a dynamically changing graph consisting of requested points, OPT-edges and ALG-edges. For the analysis, we assume that it changes in the following way: all requested points and all OPT-edges are present in the graph from the beginning, but the ALG-edges are added to the graph in m steps, in the order they are created by ALG.

At all times, the matching edges present in the graph form alternating paths or cycles (i.e., paths or cycles whose edges are interleaved ALG-edges and OPT-edges). Furthermore, any node-maximal alternating path starts and ends with OPT-edges. Assume now that a matching edge e created by ALG is added to the graph. It may either connect the ends of two different alternating paths, thus creating a single longer alternating path or connect the ends of one alternating path, generating an alternating cycle. In the former case, we call edge e *non-final*, in the latter case—*final*. Note that at the end of the ALG execution, when m ALG-edges are added, the graph contains only alternating cycles.

We extend the notion of cost to alternating path and cycles. For any cycle C, $\text{cost}(C)$ is simply the sum of costs of its edges: the cost of an OPT-edge on such cycle is the cost paid by OPT and the cost of an ALG-edge is that of ALG. We also define $\text{cost}_{\text{OPT}}(C)$, $\text{cost}_{\text{ALG}}(C)$ and $\text{cost}_{\text{ALG-NF}}(C)$ as the costs of OPT-edges, ALG-edges and non-final ALG-edges on cycle C, respectively. Clearly, $\text{cost}_{\text{ALG}}(C) + \text{cost}_{\text{OPT}}(C) = \text{cost}(C)$. We define the same notions for alternating paths; as a path P does not contain final ALG-edges, $\text{cost}_{\text{ALG-NF}}(P) = \text{cost}_{\text{ALG}}(P)$.

An alternating path is called κ-*step maximal alternating path* if it exists in the graph after ALG matched κ pairs and it cannot be extended, i.e., it ends with two requests that are not yet matched by the first κ ALG-edges.

3.1   Tree Construction

To facilitate the analysis, along with the graph, we create a dynamically changing forest F of binary trees, where each leaf of F corresponds to an OPT-edge and each internal (non-leaf) node of F to a non-final ALG-edge (and vice versa). After ALG matched κ pairs, each subtree of F corresponds to a κ-step maximal alternating path or to an alternating cycle. More precisely, at the beginning, F consists of m single nodes representing OPT-edges. Afterwards, whenever an ALG-edge is created, we perform the following operation on F.

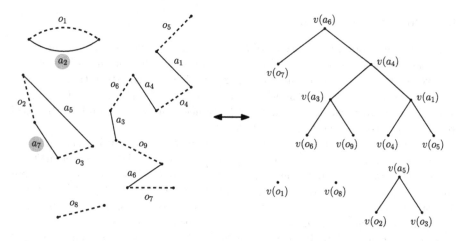

Fig. 1. The left side contains an example graph consisting of all OPT-edges $o_1, o_2, \ldots, o_9$ (dashed lines) and the first $\kappa = 7$ ALG-edges $a_1, a_2, \ldots, a_7$ (solid lines). ALG-edges are numbered in the order they were created and added to the graph. Shaded ALG-edges (a_2 and a_7) are final, the remaining ones are non-final. The right side depicts the corresponding forest F: leaves of F represent OPT-edges and non-leaf nodes of F correspond to non-final ALG-edges. Trees rooted at nodes $v(o_1)$ and $v(o_5)$ represent alternating cycles and those rooted at nodes $v(a_6)$ and $v(o_8)$ represent alternating paths in the graph.

- When a non-final ALG-edge $e = (p, q)$ is added to the graph, we look at the two alternating paths P and Q that end with p and q, respectively. We take the corresponding trees $T(P)$ and $T(Q)$ of F. We add a node $v(e)$ (representing edge e) to F and make $T(P)$ and $T(Q)$ its subtrees.
- When a final ALG-edge $e = (p, q)$ is added to the graph, it turns an alternating path P into an alternating cycle C. We then simply say that the tree $T(P)$ that corresponded to P, now corresponds to C.

An example of the graph and the associated forest F is presented in Fig. 1.

For any tree node w, we define its weight $\mathsf{weight}(w)$ as the cost of the corresponding matching edge, i.e., the cost of an OPT-edge for a leaf and the cost of a non-final ALG-edge for a non-leaf node. For any node w, by T_w we denote the tree rooted at w. We extend the notion of weight in a natural manner to all subtrees of F. In these terms, the weight of a tree T in F is equal to the total cost of the corresponding alternating path. (If T represents an alternating cycle C, then its weight is equal to the cost of C minus the cost of the final ALG-edge from C.)

Note that we consistently used terms "points" and "edges" for objects that ALG and OPT are operating on in the metric space $\mathcal{X}$. On the other hand, the term "nodes" will always refer to tree nodes in F and we will not use the term "edge" to denote an edge in F.

3.2   Outline of the Analysis

Our approach to bounding the cost of ALG is now as follows. We look at the forest F at the end of ALG execution. The corresponding graph contains only alternating cycles. The cost of non-final ALG-edges is then, by the definition, equal to the total weight of internal (non-leaf) nodes of F, while the cost of OPT-edges is equal to the total weight of leaves of F. Hence, our goal is to relate the total weight of any tree to the weight of its leaves.

The central piece of our analysis is showing that for any internal node w with children u and v, it holds that $\mathsf{weight}(w) \leq \xi \cdot \min\{\mathsf{weight}(T_u), \mathsf{weight}(T_v)\}$, where ξ is a constant depending on parameters α and β (see Corollary 4). Using this relation, we will bound the total weight of any tree by $O(m^{\log_2 (\xi+2)-1})$ times the total weight of its leaves. This implies the same bound on the ratio between non-final ALG-edges and OPT-edges on each alternating cycle.

Finally, we show that the cost of final ALG-edges incurs at most an additional constant factor in the total cost of ALG.

3.3   Cost of Non-final ALG-Edges

As described in Sect. 3.1, when ALG adds a κ-th ALG-edge e to the graph, and this edge is non-final, e joins two $(\kappa - 1)$-step maximal alternating paths P and Q. We will bound $\mathsf{cost}_{\mathrm{ALG}}(e)$ by a constant (depending on α and β) times $\min\{\mathsf{cost}(P), \mathsf{cost}(Q)\}$. We start with bounding the waiting cost of ALG related to one endpoint of e.

Lemma 2. *Let $e = (p, q)$ be the κ-th ALG-edge added at time τ, such that e is non-final. Let $P = (a_1, a_2, \ldots, a_\ell)$ be the $(\kappa - 1)$-step maximal alternating path ending at $p = a_1$. Then, $\mathsf{wait}_\tau(p) \leq \max\{\alpha^{-1}, \beta/(\beta - 1)\} \cdot \mathsf{cost}(P)$.*

Proof. First we lower-bound the cost of an alternating path P. We look at any edge (a_i, a_{i+1}) from P. Its cost (no matter whether paid by ALG or OPT) is certainly larger than $\mathsf{dist}(a_i, a_{i+1}) + |\mathsf{atime}(a_i) - \mathsf{atime}(a_{i+1})|$. Therefore, using triangle inequality (on distances and times), we obtain

$$\mathsf{cost}(P) \geq \sum_{i=1}^{\ell-1} (\mathsf{dist}(a_i, a_{i+1}) + |\mathsf{atime}(a_i) - \mathsf{atime}(a_{i+1})|)$$

$$\geq \mathsf{dist}(a_1, a_\ell) + |\mathsf{atime}(a_1) - \mathsf{atime}(a_\ell)|. \tag{1}$$

Therefore, in our proof we will simply bound $\mathsf{wait}_\tau(p) = \mathsf{wait}_\tau(a_1)$ using either $\mathsf{dist}(a_1, a_\ell)$ or $|\mathsf{atime}(a_1) - \mathsf{atime}(a_\ell)|$.

Recall that ALG matches a_1 at time τ. Consider the state of a_ℓ at time τ. If a_ℓ has not been presented to ALG yet ($\mathsf{atime}(a_\ell) > \tau$), then $\mathsf{wait}_\tau(a_1) = \tau - \mathsf{atime}(a_1) < \mathsf{atime}(a_\ell) - \mathsf{atime}(a_1) < \beta/(\beta - 1) \cdot (\mathsf{atime}(a_\ell) - \mathsf{atime}(a_1))$, and the lemma follows.

In the remaining part of the proof, we assume that a_ℓ was already presented to the algorithm ($\mathsf{atime}(a_\ell) \leq \tau$). As P is a $(\kappa - 1)$-step maximal alternating path, a_ℓ is not matched by ALG right after ALG creates $(\kappa-1)$-th matching edge. The earliest time when a_ℓ may become matched is when ALG creates the next, κ-th matching edge, i.e., at time τ. Therefore a_ℓ is not matched before time τ.

Now observe that there must be a reason for which requests a_1 and a_ℓ have not been matched with each other before time τ. Roughly speaking, either the sum of budgets of requests a_1 and a_ℓ does not suffice to cover the cost of $\mathsf{dist}(a_1, a_\ell)$ or one of them waits significantly longer than the other. Formally, we apply Observation 1 to the pair (a_1, a_ℓ) obtaining three possible cases. In each of the cases we bound $\mathsf{wait}_\tau(a_1)$ appropriately.

Case 1 (insufficient budgets). If $\alpha \cdot (\mathsf{wait}_\tau(a_1) + \mathsf{wait}_\tau(a_\ell)) \leq \mathsf{dist}(a_1, a_\ell)$, then by non-negativity of $\mathsf{wait}_\tau(a_\ell)$, it follows that $\mathsf{wait}_\tau(a_1) \leq \alpha^{-1} \cdot \mathsf{dist}(a_1, a_\ell)$.

Case 2 (a_1 waited much longer than a_ℓ). If $\alpha \cdot (\mathsf{wait}_\tau(a_1) + \mathsf{wait}_\tau(a_\ell)) > \mathsf{dist}(a_1, a_\ell)$ and $\mathsf{wait}_\tau(a_1) \geq \beta \cdot \mathsf{wait}_\tau(a_\ell)$, then $\mathsf{atime}(a_\ell) - \mathsf{atime}(a_1) = \mathsf{wait}_\tau(a_1) - \mathsf{wait}_\tau(a_\ell) \geq (1 - 1/\beta) \cdot \mathsf{wait}_\tau(a_1)$. Therefore, $\mathsf{wait}_\tau(a_1) \leq \beta/(\beta - 1) \cdot |\mathsf{atime}(a_1) - \mathsf{atime}(a_\ell)|$.

Case 3 (a_ℓ waited much longer than a_1). If $\alpha \cdot (\mathsf{wait}_\tau(a_1) + \mathsf{wait}_\tau(a_\ell)) > \mathsf{dist}(a_1, a_\ell)$ and $\mathsf{wait}_\tau(a_\ell) \geq \beta \cdot \mathsf{wait}_\tau(a_1)$, then $\mathsf{atime}(a_1) - \mathsf{atime}(a_\ell) = \mathsf{wait}_\tau(a_\ell) - \mathsf{wait}_\tau(a_1) \geq (\beta - 1) \cdot \mathsf{wait}_\tau(a_1)$. Thus, $\mathsf{wait}_\tau(a_1) \leq 1/(\beta - 1) \cdot |\mathsf{atime}(a_1) - \mathsf{atime}(a_\ell)| < \beta/(\beta - 1) \cdot |\mathsf{atime}(a_1) - \mathsf{atime}(a_\ell)|$. □

Lemma 3. *Let $e = (p, q)$ be the κ-th ALG-edge, such that e is non-final. Let $P = (a_1, a_2, \ldots, a_\ell)$ and $Q = (b_1, b_2, \ldots, b_{\ell'})$ be the $(\kappa - 1)$-step maximal alternating path ending at $p = a_1$ and $q = b_1$, respectively. Then,*

$$cost_{\text{ALG}}(e) \leq (1 + \alpha) \cdot (\beta + 1) \cdot \max\{\alpha^{-1}, \beta/(\beta - 1)\} \cdot \min\{cost(P), cost(Q)\}.$$

Proof. Let τ be the time when p is matched with q by ALG. Using the definition of $cost_{\text{ALG}}$, we obtain

$$
\begin{aligned}
\mathsf{cost}_{\text{ALG}}(p, q) &= \mathsf{dist}(p, q) + \mathsf{wait}_\tau(p) + \mathsf{wait}_\tau(q) \\
&\leq \mathsf{budget}_\tau(p) + \mathsf{budget}_\tau(q) + \mathsf{wait}_\tau(p) + \mathsf{wait}_\tau(q) \\
&= (1 + \alpha) \cdot (\mathsf{wait}_\tau(p) + \mathsf{wait}_\tau(q)) \\
&\leq (1 + \alpha) \cdot (\beta + 1) \cdot \min\{\mathsf{wait}_\tau(p), \mathsf{wait}_\tau(q)\}. \quad (2)
\end{aligned}
$$

The first inequality follows by the budget sufficiency condition of ALG and the second one by the budget balance condition.

By Lemma 2, $\mathsf{wait}_\tau(p) \leq \max\{\alpha^{-1}, \beta/(\beta - 1)\} \cdot \mathsf{cost}(P)$ and $\mathsf{wait}_\tau(q) \leq \max\{\alpha^{-1}, \beta/(\beta - 1)\} \cdot \mathsf{cost}(Q)$, which combined with (2) immediately yield the lemma. □

Recall now the iterative construction of the forest F from Sect. 3.1: whenever a non-final matching edge e created by ALG joins two alternating paths P and Q, we add a new node w to F, such that $\mathsf{weight}(w) = \mathsf{cost}_{\text{ALG}}(e)$ and make trees

$T(P)$ and $T(Q)$ its children. These trees correspond to paths P and Q, and satisfy $\mathsf{weight}(T(P)) = \mathsf{cost}(P)$ and $\mathsf{weight}(T(Q)) = \mathsf{cost}(Q)$. Therefore, Lemma 3 immediately implies the following equivalent relation on tree weights.

Corollary 4. *Let w be an internal node of the forest F whose children are u and v. Then,*

$$\mathsf{weight}(w) \leq (1+\alpha)\cdot(\beta+1)\cdot\max\left\{\alpha^{-1}, \beta/(\beta-1)\right\}\cdot\min\left\{\mathsf{weight}(T_u), \mathsf{weight}(T_v)\right\}.$$

This relation can be used to express the total weight of a tree of F in terms of the total weight of its leaves. The proof of the following technical lemma is deferred to Sect. 4. Here, we present how to use it to bound the cost of ALG on non-final edges of a single alternating cycle.

Lemma 5. *Let T be a weighted full binary tree and $\xi \geq 0$ be any constant. Assume that for each internal node w with children u and v, their weights satisfy $\mathsf{weight}(w) \leq \xi \cdot \min\{\mathsf{weight}(T_u), \mathsf{weight}(T_v)\}$. Then,*

$$\mathsf{weight}(T) \leq (\xi + 2) \cdot |L(T)|^{\log_2(\xi/2+1)} \cdot \mathsf{weight}(L(T)),$$

where $L(T)$ is the set of leaves of T and $\mathsf{weight}(L(T))$ is their total weight.

Lemma 6. *Let C be an alternating cycle obtained from combining matchings of ALG and OPT. Then $\mathsf{cost}_{\mathrm{ALG-NF}}(C) \leq (\xi+2)\cdot m^{\log_2(\xi/2+1)} \cdot \mathsf{cost}_{\mathrm{OPT}}(C)$, where $\xi = (1+\alpha) \cdot (\beta+1) \cdot \max\{\alpha^{-1}, \beta/(\beta-1)\}$.*

Proof. As described in Sect. 3.1, C is associated with a tree T from forest F, such that OPT-edges of C correspond to the set of leaves of T (denoted $L(T)$) and non-final ALG-edges of C correspond to internal (non-leaf) nodes of T. Hence, $\mathsf{cost}_{\mathrm{OPT}}(C) = \mathsf{weight}(L(T))$ and $\mathsf{cost}_{\mathrm{ALG-NF}}(C) + \mathsf{cost}_{\mathrm{OPT}}(C) = \mathsf{weight}(T)$.

By Corollary 4, the weight of any internal tree node w with children u, v satisfies $\mathsf{weight}(w) \leq \xi \cdot \min\{\mathsf{weight}(T_u), \mathsf{weight}(T_v)\}$. Therefore, we may apply Lemma 5 to tree T, obtaining $\mathsf{weight}(T) \leq (\xi+2)\cdot|L(T)|^{\log_2(\xi/2+1)}\cdot\mathsf{weight}(L(T))$, and thus

$$\begin{aligned}
\mathsf{cost}_{\mathrm{ALG-NF}}(C) \leq \mathsf{weight}(T) &\leq (\xi + 2) \cdot |L(T)|^{\log_2(\xi/2+1)} \cdot \mathsf{weight}(L(T)) \\
&\leq (\xi + 2) \cdot m^{\log_2(\xi/2+1)} \cdot \mathsf{weight}(L(T)) \\
&= (\xi + 2) \cdot m^{\log_2(\xi/2+1)} \cdot \mathsf{cost}_{\mathrm{OPT}}(C).
\end{aligned}$$

The last inequality follows as $|L(T)|$, the number of T leaves, is equal to the number of OPT-edges on cycle C, which is clearly at most m. □

3.4 Cost of Final ALG-Edges

In the previous section, we derived a bound on the cost of all non-final ALG-edges. The following lemma shows that the cost of final ALG-edges contribute at most a constant factor to the competitive ratio.

Lemma 7. *Let e be a final* ALG*-edge matched at time τ and C be the alternating cycle containing e. Then* $cost_{ALG}(e) \leq (1 + \alpha) \cdot \max\{\alpha^{-1}, (\beta + 1)/(\beta - 1)\} \cdot$ $(cost_{ALG-NF}(C) + cost_{OPT}(C))$.

Proof. Fix a final ALG-edge $e = (p, q)$, where $\mathsf{atime}(q) \geq \mathsf{atime}(p)$. By the budget sufficiency condition of ALG,

$$\mathsf{cost}_{ALG}(e) \leq (1 + \alpha) \cdot (\mathsf{wait}_\tau(p) + \mathsf{wait}_\tau(q)). \tag{3}$$

Our goal now is to bound $\mathsf{wait}_\tau(p) + \mathsf{wait}_\tau(q)$ in terms of $\mathsf{dist}(p, q)$ or $\mathsf{atime}(q) - \mathsf{atime}(p)$. Observe that whenever ALG matches two requests, the budget sufficiency condition of ALG or one of the inequalities of the budget balance condition is satisfied with equality. We apply this observation to the pair (p, q).

- If the budget sufficiency condition holds with equality, then $\alpha \cdot (\mathsf{wait}_\tau(p) + \mathsf{wait}_\tau(q)) = \mathsf{dist}(p, q)$, and therefore $\mathsf{wait}_\tau(p) + \mathsf{wait}_\tau(q) = \alpha^{-1} \cdot \mathsf{dist}(p, q)$.
- If the budget balance condition holds with equality, $\beta \cdot \mathsf{wait}_\tau(q) = \mathsf{wait}_\tau(p)$. Then,

$$\begin{aligned}
(\beta - 1) \cdot (\mathsf{wait}_\tau(p) + \mathsf{wait}_\tau(q)) &= (\beta - 1) \cdot (\beta + 1) \cdot \mathsf{wait}_\tau(q) \\
&= (\beta + 1) \cdot (\mathsf{wait}_\tau(p) - \mathsf{wait}_\tau(q)) \\
&= (\beta + 1) \cdot (\mathsf{atime}(q) - \mathsf{atime}(p)).
\end{aligned}$$

Hence, in either case it holds that

$$\mathsf{wait}_\tau(p) + \mathsf{wait}_\tau(q) \leq \max\left\{\alpha^{-1}, \frac{\beta + 1}{\beta - 1}\right\} \cdot (\mathsf{dist}(p, q) + |\mathsf{atime}(q) - \mathsf{atime}(p)|). \tag{4}$$

Finally, we bound $\mathsf{dist}(p, q) + |\mathsf{atime}(q) - \mathsf{atime}(p)|$ in terms of costs of other edges of C. These edges form a path $P = (a_1, a_2, \ldots, a_\ell)$, where $a_1 = p$ and $a_\ell = q$. By the triangle inequality applied to distances and time differences (in the same way as in (1)), we obtain that

$$\mathsf{dist}(p, q) + |\mathsf{atime}(q) - \mathsf{atime}(p)| \leq \mathsf{cost}(P) = \mathsf{cost}_{ALG-NF}(C) + \mathsf{cost}_{OPT}(C). \tag{5}$$

The lemma follows immediately by combining (3), (4) and (5). □

3.5 The Competitive Ratio

Finally, we optimize constants α and β used throughout the previous sections and bound the competitiveness of ALG.

Theorem 8. *For $\beta = 2$ and $\alpha = 1/2$, the competitive ratio of algorithm* ALG *is $O(m^{\log_2 5.5}) = O(m^{2.46})$, where $2m$ is the number of requests in the input sequence.*

Proof. The union of matchings constructed by ALG and OPT can be split into a set $\mathcal{C}$ of disjoint cycles. It is sufficient to show that we have the desired performance guarantee on each cycle from $\mathcal{C}$.

Fix a cycle $C \in \mathcal{C}$. Let $e = (p, q)$ be the final ALG-edge of C. By Lemma 7, $\mathrm{cost_{ALG}}(e) \leq 4.5 \cdot (\mathrm{cost_{ALG-NF}}(C) + \mathrm{cost_{OPT}}(C))$. Therefore, the competitive ratio of ALG is at most

$$\frac{\mathrm{cost_{ALG}}(C)}{\mathrm{cost_{OPT}}(C)} \leq \frac{5.5 \cdot \mathrm{cost_{ALG-NF}}(C) + 4.5 \cdot \mathrm{cost_{OPT}}(C)}{\mathrm{cost_{OPT}}(C)}$$

$$\leq O\left(m^{\log_2 5.5}\right) = O\left(m^{2.46}\right),$$

where the second inequality follows by Lemma 6. $\square$

4   Relating Weights in Trees (Proof of Lemma 5)

We start with the following technical claim that will facilitate the inductive proof of Lemma 5.

Lemma 9. *Fix any constant $\xi \geq 0$ and let $f(a) = a^{\log_2(\xi+2)}$. Then,*

$$\xi \cdot \min\{f(x), f(y)\} + f(x) + f(y) \leq f(x + y)$$

for all $x, y \geq 0$.

Proof. Fix any $z \geq 0$ and let $g_z(a) = (\xi + 1) \cdot f(a) + f(z - a)$. We observe that $g_z(0) = f(z)$ and $g_z(z/2) = (\xi + 1) \cdot f(z/2) + f(z/2) = (\xi + 2) \cdot (z/2)^{\log_2(\xi+2)} = z^{\log_2(\xi+2)} = f(z)$. Moreover, the function g_z is convex in the interval $[0, z]$ as it is a sum of two convex functions. As $g_z(0) = g_z(z/2) = f(z)$, by convexity, $g_z(a) \leq f(z)$ for any $a \in [0, z/2]$.

To prove the lemma, assume without loss of generality that $x \leq y$. By the monotonicity, $f(x) \leq f(y)$, and therefore

$$\xi \cdot \min\{f(x), f(y)\} + f(x) + f(y) = (\xi + 1) \cdot f(x) + f((x + y) - x)$$

$$= g_{x+y}(x)$$

$$\leq f(x + y).$$

The last inequality follows as $x \leq (x + y)/2$. $\square$

Proof (of Lemma 5). We scale weights of all nodes, so that the average weight of each leaf is 1, i.e., we define a scaled weight function ws as

$$\mathsf{ws}(w) = \mathsf{weight}(w) \cdot \frac{|\mathsf{L}(T)|}{\mathsf{weight}(\mathsf{L}(T))}.$$

Note that ws also satisfies $\mathsf{ws}(w) \leq \xi \cdot \min\{\mathsf{ws}(T_u), \mathsf{ws}(T_v)\}$. Moreover, since we scaled all weights in the similar way, $\mathsf{ws}(T)/\mathsf{ws}(\mathsf{L}(T)) = \mathsf{weight}(T)/\mathsf{weight}(\mathsf{L}(T))$, and hence to show the lemma, it suffices to bound the term $\mathsf{ws}(T)/\mathsf{ws}(\mathsf{L}(T))$.

For any node $w \in T$ and the corresponding subtree T_w rooted at w, we define $\mathsf{size}(T_w) = \mathsf{ws}(\mathsf{L}(T_w)) + |\mathsf{L}(T_w)|$. We inductively show that for any node of $w \in T$, it holds that

$$\mathsf{ws}(T_w) \leq \mathsf{size}(T_w)^{\log_2(\xi+2)}. \tag{6}$$

For the induction basis, assume that w is a leaf of T. Then,

$$\mathsf{ws}(T_w) = \mathsf{ws}(\mathsf{L}(T_w)) \leq \mathsf{size}(T_w) \leq \mathsf{size}(T_w)^{\log_2(\xi+2)},$$

where the last inequality follows as $\mathsf{size}(T_w) \geq |\mathsf{L}(T_w)| = 1$ and $\xi \geq 0$.

For the inductive step, let w be a non-leaf node of T and let u and v be its children. Then,

$$\begin{aligned}
\mathsf{ws}(T_w) &= \mathsf{ws}(T_u) + \mathsf{ws}(T_v) + \mathsf{ws}(w) \\
&\leq \mathsf{ws}(T_u) + \mathsf{ws}(T_v) + \xi \cdot \min\left\{ \mathsf{ws}(T_u), \mathsf{ws}(T_v) \right\} \\
&\leq \mathsf{size}(T_u)^{\log_2(\xi+2)} + \mathsf{size}(T_v)^{\log_2(\xi+2)} \\
&\quad + \xi \cdot \min\left\{ \mathsf{size}(T_u)^{\log_2(\xi+2)}, \mathsf{size}(T_v)^{\log_2(\xi+2)} \right\} \\
&\leq (\mathsf{size}(T_u) + \mathsf{size}(T_v))^{\log_2(\xi+2)} \\
&= \mathsf{size}(T_w)^{\log_2(\xi+2)}.
\end{aligned}$$

The first inequality follows by the lemma assumption and the second one by the inductive assumptions for T_u and T_v. The last inequality is a consequence of Lemma 9 and the final equality follows by the additivity of function size.

Recall that we scaled weights so that $\mathsf{ws}(\mathsf{L}(T)) = |\mathsf{L}(T)|$. Therefore, applying (6) to the whole tree T yields $\mathsf{ws}(T) \leq (\mathsf{ws}(\mathsf{L}(T)) + |\mathsf{L}(T)|)^{\log_2(\xi+2)} = (2 \cdot |\mathsf{L}(T)|)^{\log_2(\xi+2)} = (\xi+2) \cdot |\mathsf{L}(T)|^{\log_2(\xi+2)}$. Hence,

$$\frac{\mathsf{weight}(T)}{\mathsf{weight}(\mathsf{L}(T))} = \frac{\mathsf{ws}(T)}{\mathsf{ws}(\mathsf{L}(T))} \leq \frac{(\xi+2) \cdot |\mathsf{L}(T)|^{\log_2(\xi+2)}}{|\mathsf{L}(T)|} = (\xi+2) \cdot |\mathsf{L}(T)|^{\log_2(\xi/2+1)},$$

which concludes the proof.   □

5   Conclusions

We showed a deterministic algorithm ALG for the MPMD problem whose competitive ratio is $O(m^{\log_2 5.5})$. The currently best lower bound (holding even for randomized solutions) is $\Omega(\log n / \log \log n)$ [3]. A natural research direction would be to narrow this gap.

It is not known whether the analysis of our algorithm is tight. However, one can show that its competitive ratio is at least $\Omega(m^{\log_2 1.5}) = \Omega(m^{0.58})$. To this end, assume that all requests arrive at the same time. For such input, OPT does not pay for delays and simply returns the min-cost perfect matching. On the other hand, ALG computes the same matching as a greedy routine (i.e., it greedily connects two nearest, not yet matched requests). Hence, even if we neglect the delay costs of ALG, its competitive ratio would be at least the approximation ratio of the greedy algorithm for min-cost perfect matching. The latter was shown to be $\Theta(m^{\log_2 1.5})$ by Reingold and Tarjan [44].

The reasoning above indicates an inherent difficulty of the problem. In order to beat the $\Omega(m^{\log_2 1.5})$ barrier, an online algorithm has to handle settings when all requests are given simultaneously more effectively. In particular, for such and similar input instances it has to employ a non-local and non-greedy policy of choosing requests to match.

References

1. Albers, S., Bals, H.: Dynamic TCP acknowledgment: penalizing long delays. SIAM J. Discret. Math. **19**(4), 938–951 (2005)
2. Antoniadis, A., Barcelo, N., Nugent, M., Pruhs, K., Scquizzato, M.: A $o(n)$-competitive deterministic algorithm for online matching on a line. In: Bampis, E., Svensson, O. (eds.) WAOA 2014. LNCS, vol. 8952, pp. 11–22. Springer, Cham (2015). https://doi.org/10.1007/978-3-319-18263-6_2
3. Ashlagi, I., Azar, Y., Charikar, M., Chiplunkar, A., Geri, O., Kaplan, H., Makhijani, R.M., Wang, Y., Wattenhofer, R.: Min-cost bipartite perfect matching with delays. In: Proceedings of the 20th International Workshop on Approximation Algorithms for Combinatorial Optimization (APPROX), pp. 1:1–1:20 (2017)
4. Avontuur, T., Spronck, P., van Zaanen, M.: Player skill modeling in Starcraft II. In: Proceedings of the 9th AAAI Conference on Artificial Intelligence and Interactive Digital Entertainment, AIIDE-2013 (2013)
5. Azar, Y., Chiplunkar, A., Kaplan, H.: Polylogarithmic bounds on the competitiveness of min-cost (bipartite) perfect matching with delays (2016). https://arxiv.org/abs/1610.05155
6. Azar, Y., Chiplunkar, A., Kaplan, H.: Polylogarithmic bounds on the competitiveness of min-cost perfect matching with delays. In: Proceedings of the 28th ACM-SIAM Symposium on Discrete Algorithms (SODA), pp. 1051–1061 (2017)
7. Azar, Y., Epstein, A., Jeż, Ł., Vardi, A.: Make-to-order integrated scheduling and distribution. In: Proceedings of the 27th ACM-SIAM Symposium on Discrete Algorithms (SODA), pp. 140–154 (2016)
8. Bansal, N., Buchbinder, N., Gupta, A., Naor, J.: A randomized $O(\log^2 k)$-competitive algorithm for metric bipartite matching. Algorithmica **68**(2), 390–403 (2014)
9. Bansal, N., Buchbinder, N., Mądry, A., Naor, J.: A polylogarithmic-competitive algorithm for the k-server problem. J. ACM **62**(5), 40:1–40:49 (2015)
10. Bienkowski, M., Böhm, M., Byrka, J., Chrobak, M., Dürr, C., Folwarczný, L., Jeż, Ł., Sgall, J., Thang, N.K., Veselý, P.: Online algorithms for multi-level aggregation. In: Proceedings of the 24th European Symposium on Algorithms (ESA), pp. 12:1–12:17 (2016)
11. Bienkowski, M., Byrka, J., Chrobak, M., Jeż, Ł., Nogneng, D., Sgall, J.: Better approximation bounds for the joint replenishment problem. In: Proceedings of the 25th ACM-SIAM Symposium on Discrete Algorithms (SODA), pp. 42–54 (2014)
12. Bienkowski, M., Byrka, J., Chrobak, M., Jeż, Ł., Sgall, J., Stachowiak, G.: Online control message aggregation in chain networks. In: Dehne, F., Solis-Oba, R., Sack, J.-R. (eds.) WADS 2013. LNCS, vol. 8037, pp. 133–145. Springer, Heidelberg (2013). https://doi.org/10.1007/978-3-642-40104-6_12

13. Birnbaum, B., Mathieu, C.: On-line bipartite matching made simple. SIGACT News **39**(1), 80–87 (2008)
14. Borodin, A., El-Yaniv, R.: Online Computation and Competitive Analysis. Cambridge University Press, Cambridge (1998)
15. Brito, C., Koutsoupias, E., Vaya, S.: Competitive analysis of organization networks or multicast acknowledgement: how much to wait? Algorithmica **64**(4), 584–605 (2012)
16. Buchbinder, N., Feldman, M., Naor, J.S., Talmon, O.: O(depth)-competitive algorithm for online multi-level aggregation. In: Proceedings of the 28th ACM-SIAM Symposium on Discrete Algorithms (SODA), pp. 1235–1244 (2017)
17. Buchbinder, N., Jain, K., Naor, J.S.: Online primal-dual algorithms for maximizing ad-auctions revenue. In: Arge, L., Hoffmann, M., Welzl, E. (eds.) ESA 2007. LNCS, vol. 4698, pp. 253–264. Springer, Heidelberg (2007). https://doi.org/10.1007/978-3-540-75520-3_24
18. Buchbinder, N., Kimbrel, T., Levi, R., Makarychev, K., Sviridenko, M.: Online make-to-order joint replenishment model: primal dual competitive algorithms. In: Proceedings of the 19th ACM-SIAM Symposium on Discrete Algorithms (SODA), pp. 952–961 (2008)
19. Chen, Z., Sun, Y., El-Nasr, M.S., Nguyen, T.D.: Player skill decomposition in multiplayer online battle arenas (2017). http://arxiv.org/abs/1702.06253
20. Chrobak, M.: Online aggregation problems. SIGACT News **45**(1), 91–102 (2014)
21. Delalleau, O., Contal, E., Thibodeau-Laufer, E., Ferrari, R.C., Bengio, Y., Zhang, F.: Beyond skill rating: advanced matchmaking in Ghost Recon Online. IEEE Trans. Comput. Intel. AI Games **4**(3), 167–177 (2012)
22. Devanur, N.R., Jain, K.: Online matching with concave returns. In: Proceedings 44th ACM Symposium on Theory of Computing (STOC), pp. 137–144 (2012)
23. Devanur, N.R., Jain, K., Kleinberg, R.D.: Randomized primal-dual analysis of RANKING for online bipartite matching. In: Proceedings of the 24th ACM-SIAM Symposium on Discrete Algorithms (SODA), pp. 101–107 (2013)
24. Dooly, D.R., Goldman, S.A., Scott, S.D.: On-line analysis of the TCP acknowledgment delay problem. J. ACM **48**(2), 243–273 (2001)
25. Elo, A.E.: The Rating of Chessplayers, Past and Present. Arco Publishing, New York (1978)
26. Emek, Y., Kutten, S., Wattenhofer, R.: Online matching: haste makes waste! In: Proceedings of the 48th ACM Symposium on Theory of Computing (STOC), pp. 333–344 (2016)
27. Emek, Y., Shapiro, Y., Wang, Y.: Minimum cost perfect matching with delays for two sources. In: Fotakis, D., Pagourtzis, A., Paschos, V.T. (eds.) CIAC 2017. LNCS, vol. 10236, pp. 209–221. Springer, Cham (2017). https://doi.org/10.1007/978-3-319-57586-5_18
28. Fakcharoenphol, J., Rao, S., Talwar, K.: A tight bound on approximating arbitrary metrics by tree metrics. J. Comput. Syst. Sci. **69**(3), 485–497 (2004)
29. Fuchs, B., Hochstättler, W., Kern, W.: Online matching on a line. Theor. Comput. Sci. **332**(1–3), 251–264 (2005)
30. Gupta, A., Lewi, K.: The online metric matching problem for doubling metrics. In: Czumaj, A., Mehlhorn, K., Pitts, A., Wattenhofer, R. (eds.) ICALP 2012, Part I. LNCS, vol. 7391, pp. 424–435. Springer, Heidelberg (2012). https://doi.org/10.1007/978-3-642-31594-7_36
31. Kalyanasundaram, B., Pruhs, K.: Online weighted matching. J. Algorithms **14**(3), 478–488 (1993)

32. Karlin, A.R., Kenyon, C., Randall, D.: Dynamic TCP acknowledgement and other stories about e/(e − 1). Algorithmica **36**(3), 209–224 (2003)
33. Karlin, A.R., Manasse, M.S., McGeoch, L.A., Owicki, S.: Competitive randomized algorithms for non-uniform problems. Algorithmica **11**(6), 542–571 (1994)
34. Karp, R.M., Vazirani, U.V., Vazirani, V.V.: An optimal algorithm for on-line bipartite matching. In: Proceedings of the 22nd ACM Symposium on Theory of Computing (STOC), pp. 352–358 (1990)
35. Khanna, S., Naor, J.S., Raz, D.: Control message aggregation in group communication protocols. In: Widmayer, P., Eidenbenz, S., Triguero, F., Morales, R., Conejo, R., Hennessy, M. (eds.) ICALP 2002. LNCS, vol. 2380, pp. 135–146. Springer, Heidelberg (2002). https://doi.org/10.1007/3-540-45465-9_13
36. Khuller, S., Mitchell, S.G., Vazirani, V.V.: On-line algorithms for weighted bipartite matching and stable marriages. Theor. Comput. Sci. **127**(2), 255–267 (1994)
37. Koutsoupias, E., Nanavati, A.: The online matching problem on a line. In: Solis-Oba, R., Jansen, K. (eds.) WAOA 2003. LNCS, vol. 2909, pp. 179–191. Springer, Heidelberg (2004). https://doi.org/10.1007/978-3-540-24592-6_14
38. Mahdian, M., Yan, Q.: Online bipartite matching with random arrivals: an approach based on strongly factor-revealing LPs. In: Proceedings of the 43rd ACM Symposium on Theory of Computing (STOC), pp. 597–606 (2011)
39. Mehta, A.: Online matching and ad allocation. Found. Trends Theor. Comput. Sci. **8**(4), 265–368 (2013)
40. Mehta, A., Saberi, A., Vazirani, U.V., Vazirani, V.V.: Adwords and generalized online matching. J. ACM **54**(5), 265–368 (2007)
41. Meyerson, A., Nanavati, A., Poplawski, L.J.: Randomized online algorithms for minimum metric bipartite matching. In: Proceedings of the 7th ACM-SIAM Symposium on Discrete Algorithms (SODA), pp. 954–959 (2006)
42. Naor, J., Wajc, D.: Near-optimum online ad allocation for targeted advertising. In: Proceedings of the 16th ACM Conference on Economics and Computation (EC), pp. 131–148 (2015)
43. Pignolet, Y.A., Schmid, S., Wattenhofer, R.: Tight bounds for delay-sensitive aggregation. Discret. Math. Theor. Comput. Sci. **12**(1), 39–58 (2010)
44. Reingold, E.M., Tarjan, R.E.: On a greedy heuristic for complete matching. SIAM J. Comput. **10**(4), 676–681 (1981)

Online Packing of Rectangular Items
into Square Bins

Janusz Januszewski and Łukasz Zielonka[✉]

Institute of Mathematics and Physics, UTP University of Science and Technology,
Al. Prof. S. Kaliskiego 7, 85-789 Bydgoszcz, Poland
{januszew,Lukasz.Zielonka}@utp.edu.pl

Abstract. Any list of rectangular items of total area not greater than
0.2837 can be packed online into the unit square (90°-rotations are
allowed). Furthermore, we describe a 4.84-competitive 1-space bounded
2-dimensional bin packing algorithm and present the lower bound of
3.246 for the competitive ratio.

Keywords: Online Algorithms · Competitive analysis
Two dimensional bin packing

1 Introduction

In *online* packing problems, items arrive one by one. Each item that has arrived
is packed into a bin and cannot be moved henceforth (inside a bin as well as
between bins). Each item must be packed without knowing any information
about future items.

In *t-space bounded* model, t bins are available for packing, i.e., each item can
be packed online only into one of t active bins. If it is impossible to pack an item
into any active bin, then we close one of the current active bins and open a new
active bin to pack that item. If an active bin has been closed, it is never used
again. We will consider the following problem of $2D$ bin packing. Each item
is a rectangle of side lengths not greater than 1. Moreover, arriving items are
packed into square bins of size 1×1 and 90°-rotations are allowed.

Given a sequence S of items, denote by $A(S)$ the number of bins used by algorithm A to pack items from S. Furthermore, denote by $OPT(S)$ the minimum
possible number of bins used to pack items from S with the most efficient offline
method. By the asymptotic competitive ratio for algorithm A we mean:

$$R_A^\infty = \lim_{n\to\infty} \sup_S \left\{ \frac{A(S)}{OPT(S)} \,|\, OPT(S) = n \right\}.$$

1.1 Related Work

Fifty years ago Moon and Moser [19] proved that any collection of squares with
total area not greater than $1/2$ can be packed offline into a unit square, which

© Springer International Publishing AG, part of Springer Nature 2018
R. Solis-Oba and R. Fleischer (Eds.): WAOA 2017, LNCS 10787, pp. 147–163, 2018.
https://doi.org/10.1007/978-3-319-89441-6_12

is tight. Moser [18] conjectured also that a similar result may be obtained for packing rectangles of side lengths not greater than 1 (this has been confirmed in [12]).

The online version of this problem remains open. The best known upper bound for both square and rectangle packing is 1/2 (two squares of side lengths greater than 1/2 cannot be packed into the unit square). In the case of online packing squares the following lower bounds were presented: 5/16 in [13], 1/3 in [11], 11/32 in [8] and 3/8 (see [1]). Moreover, at the 12th International Workshop on Approximation and Online Algorithms (WAOA 2014), Brubach [2] used a multidirectional shelf approach and some others techniques to obtain the new bound 2/5. Brubach also stated that the results obtained for squares may be extended for rectangles. However, he has not given any explicit lower bound. Although there are many papers concerning online packing rectangles into strips or into rectangular containers, it seems that the only lower bound (equal to 0.1464) for packing into the unit square was given twenty years ago by Lassak [16]. The aim of this note is to improve this bound.

The one-dimensional case of the space bounded bin packing problem has been extensively studied and the best possible algorithms are known: the Next-Fit algorithm [15] for the one-space bounded model and the Harmonic algorithm [17] when the number of active bins goes to infinity. The questions concerning t-space bounded d-dimensional packing ($d \geq 2$) have been studied in a number of papers. For large number of active bins, Epstein and van Stee [7] presented a $(\Pi_\infty)^d$-competitive space bounded algorithm, where $\Pi_\infty \approx 1.69103$ is the competitive ratio of the one-dimensional algorithm Harmonic. A 8.48-competitive packing strategy of packing of rectangular items with only one active bin was described in [5]. In [3], a 1-space bounded 2-dimensional bin packing algorithm with the competitive ratio 5.155 was given and the lower bound of 3 was proved. In [4], a 5.06-competitive 1-space bounded method was presented and the lower bound of 3.167 was described. We improve both upper and lower bounds given in [4]. The cases with two and three active bins were considered in [9,10,14,20].

1.2 Our Results

In Sect. 2 we describe the $3TP$-algorithm (three types packing algorithm). This algorithm will be used in Sect. 3 for online packing into a single bin. We improve the upper bound of 0.1464 given in [16]. By using the $3TP(h)$-algorithm for $h = 5/36$ we show that any sequence of rectangles of sides of length not greater than 1 with total area not greater than 0.2837 can be packed online into the unit square.

In Sect. 4 we use the $3TP(h)$-algorithm, with $h = 4/9$, to obtain the new upper bound of 4.84 for the competitive ratio for 1-space bounded model of packing of rectangular items.

Furthermore, we show in Sect. 5 that no 1-space bounded algorithm can achieve the competitive ratio less than 3.246.

2   Packing Algorithm

In the packing algorithm presented by Lassak in [16] every incoming rectangular item Q_i with height h_i (h_i is not greater than width of Q_i) is packed into a *proper layer* for it. The layer is either a rectangle with size $1 \times (0.4757\ldots)^k/2$ provided $(0.4757\ldots)^{k+1}/2 < h_i \leq (0.4757\ldots)^k/2$, where k is a positive integer or a rectangle with size $1 \times h_i$ provided $h_i > (0.4757\ldots)/2$. Our method of packing rectangular items into a bin is an improvement of techniques described in [3,4,6]. Chin et al. distinguish three types of items according to their widths: A-items with width greater than $1/2$, B-items with width between $1/8$ and $1/2$ and C-items with width smaller than or equal to $1/8$. In this packing method C-items are packed into a row (container) divided by three subrows. While A-items are packed using a top-down approach (as high as it is possible), B-items and containers are packed using a bottom-up approach (as low as it is possible) in two vertical strips. Our packing algorithm is a modification of the $2BR$-algorithm presented in [14], where two bins were active. We use the LR-method described in [14] for packing C-items (in this paper C-item is a rectangle with width smaller than $1/9$).

2.1   Types of Items

Let $P_1, P_2, \ldots$ be a sequence of rectangular items. Denote by w_i the width and by h_i the height of P_i. We will also write $P_i = (w_i, h_i)$. Since $90°$-rotations are allowed, we may assume that $w_i \geq h_i$. Let us add, that some items (called C-items) will be rotated before packing. Items are divied into three types:

- if $w_i \geq 1/3$, then P_i is an A-item;
- if $1/9 \leq w_i < 1/3$, then P_i is a B-item;
- if $w_i < 1/9$, then P_i is a C-item.

2.2   Description of LR-Method

Any C-item is rotated before packing. Therefore the height w_i of any packed C-item is not smaller than its width h_i. C-items are classified into classes $C_1, C_2, C_3, \ldots$. Put

$$c_1 = 1/9,$$
$$a = (5 - \sqrt{21})/2 \approx 0.2087 \quad \text{and}$$
$$c_{k+1} = ac_k.$$

We say that a C-item P_i

- belongs to class C_{2k-1} if $c_k/2 < w_i \leq c_k$;
- belongs to class C_{2k} if $c_{k+1} = ac_k < w_i \leq c_k/2$.

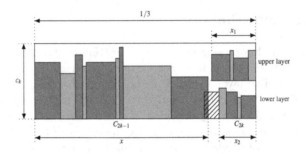

Fig. 1. Packing C-items (after rotating by $90°$) into a c_k-container

By a c_k-*container* (*c-container*) we mean a rectangle of height c_k and width $1/3$.

Now we describe how to pack C-items into c-containers. Each item belonging to class C_{2k-1} or C_{2k} ($k \geq 1$) is packed (after rotating by $90°$) into a c_k-container (see Fig. 1). Items from class C_{2k-1} are packed by the LR-method from left to right along the bottom of the container. The container can be horizontally divided into two layers (upper and lower) of height $c_k/2$. Each item from class C_{2k} is packed by the LR-method into one of these layers from right to left. It is packed into the layer which is less filled along the bottom of the layer.

2.3 $3TP(h)$-Algorithm

Let $0 \leq h \leq 1$ and let $P_1, P_2, \ldots$ be a sequence of rectangular items. We describe how to pack any item $P_i = (w_i, h_i)$ into the active bin by $3TP(h)$-algorithm (see Fig. 2). Denote by L_1, L_2 and L_3 the vertical strips of width $1/3$ on the left hand side, in the middle, and on the right hand side of the active bin, respectively.

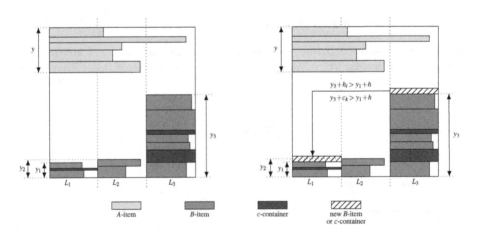

Fig. 2. Packing items into a bin

Furthermore, assume that the sum of heights of packed A-items is $y = y(i)$ and that the sum of heights of all B-items and all c-containers packed in L_j equals $y_j = y_j(i)$, for $j = 1, 2, 3$.

Description of $3TP(h)$-algorithm

1. If P_i is an A-item, then it is packed along the left side of the bin using a top-down approach, i.e., as high as it is possible.
2. Assume that P_i is a B-item.
   (a) If $y_3 + h_i \leq h + \min(y_1, y_2)$, then P_i is packed into L_3 along the left side using a bottom-up approach, i.e., as low as it is possible.
   (b) If $y_3 + h_i > h + \min(y_1, y_2)$ and if $y_1 < y_2$, then P_i is packed into L_1 along the left side as low as it is possible.
   (c) If $y_3 + h_i > h + \min(y_1, y_2)$ and if $y_1 \geq y_2$, then P_i is packed into L_2 along the left side as low as it is possible.
3. Assume that P_i is a C-item from class C_{2k-1} or C_{2k}. If there is a c_k-container into which P_i can be packed, then P_i is packed into this c_k-container by the LR-method. Otherwise, a new c_k-container is created. We can treat this container as a B-item of height c_k. This c_k-container is packed in the place described by Rule 2. Next, P_i is packed into this container by the LR-method.
4. If P_i cannot be packed into the bin by using this strategy, then we close the active bin and open a new active bin to pack this item.

2.4 Density of Packing of C-Items

Consider the packing of C-items. If there is more than one c_k-container, the last c_k-container (for any $k \geq 1$) could be almost empty and the non-last ones are almost full. The total height of last c_k-containers is smaller than

$$\sum_{k \geq 1} c_k = \sum_{k \geq 1} a^{k-1} c_1 = \frac{c_1}{1 - a} = \frac{\frac{1}{9}}{1 - \frac{5 - \sqrt{21}}{2}} = \frac{3 + \sqrt{21}}{54} \approx 0.14$$

Now we analyze the occupation ratio of non-last c_k-containers. In [14] copies of C-items (in a homothety of ratio $3/4$) were packed into copies of c-containers (in a homothety of ratio $3/4$). Since affine transformations preserve ratios of areas, the presented below result is a consequence of Lemma 1 of [14].

Lemma 1. *If m equals the sum of heights of all non-last c_k-containers ($k \in \{1, 2, \dots\}$) in a closed bin, then the total area of C-items packed into this bin is not smaller than $m/9$.*

Let us add, that $a \approx 0.2087$ is chosen as the smallest number such that the total area of packed C-items is not smaller than $1/3$ times the area of non-last c-containers.

3   Packing into a Single Bin

We improve the upper bound of 0.1464 presented in [16] by using the $3TP(h)$-algorithm for $h = 5/36 \approx 0.139$.

Theorem 1. *Any sequence of rectangles of sides of length not greater than 1 with total area not greater than $95/324 - \sqrt{21}/486 \approx 0.28378$ can be packed online into the unit square.*

Proof. Let S be a sequence of rectangles of sides of length not greater than 1. Furthermore, let the total area of rectangles in S be smaller than or equal to

$$\sigma_o^+ = 95/324 - \sqrt{21}/486 \approx 0.28378.$$

We pack the rectangles into the unit square I by using the $3TP(h)$-algorithm for $h = 5/36 \approx 0.139$. Contrary to the statement assume that rectangles from S cannot be packed into I. We show that this leads to a contradiction. Suppose that $P_u = (w_u, h_u)$ is the first item that cannot be packed into I by $3TP(5/36)$-algorithm.

Let

$$\lambda = \frac{1}{9} \sum_{k \geq 1} c_k = \frac{3 + \sqrt{21}}{486} \approx 0.0156$$

and let

$$d = y_3 - \min(y_1, y_2).$$

Let us clarify, that this is the value of d at the moment P_u arrives, i.e., $d = y_3(u) - \min(y_1(u), y_2(u))$. Obviously, $d \leq 5/36$ for the $3TP(5/36)$-algorithm.

Furthermore, denote by σ^+ the sum of areas of rectangles $P_1, \ldots, P_u$. To prove Theorem 1 it suffices to show that the total area of rectangles $P_1, \ldots, P_u$ is greater than σ_o^+, i.e., that

$$\sigma^+ > \sigma_o^+,$$

which is a contradiction.

Clearly, the total area of A-items is not smaller than $1/3$ times the sum of heights of these items. The total area of B-items is not smaller than $1/9$ times the sum of heights of these items. Denote by m_+ the sum of heights of all c-containers packed in the bin. The total area of c-containers equals $m_+/3$. However, the last c-containers can be almost empty. By Lemma 1 we know that the sum of areas of all packed C-items is not smaller than $\frac{1}{9}(m_+ - \sum_{k \geq 1} c_k) = \frac{1}{9}m_+ - \lambda$. B-items as well as c-containers are packed in three vertical strips. Thus the total area of B-items and C-items packed into the bin is greater than $\frac{1}{9}(y_1 + y_2 + y_3) - \lambda$. In particular, the sum of areas of packed B-items and C-items is greater than $\frac{1}{3}\min(y_1, y_2, y_3) - \lambda$.

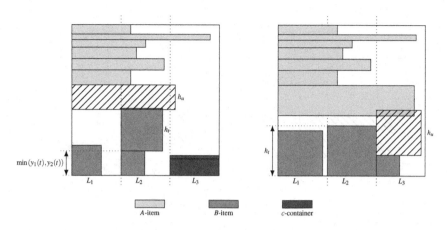

Fig. 3. $y_3 < \min(y_1, y_2)$.

Consider few cases depending on the size of $P_u = (w_u, h_u)$. Cases 1b (for $c_k = 1/9$ and either h_u close to 0 or $h_u = 1/3$) and 2b (for h_u equal to 1/18) give the worst case result of $\sigma_o^+ \approx 0.28378$.

Case 1: P_u is an A-item.
Subcase 1a: $w_u > 2/3$. If I contains only A-items, then

$$\sigma^+ > \frac{1}{3}(1 - h_u) + w_u h_u > \frac{1}{3}(1 - h_u) + \frac{2}{3}h_u > \frac{1}{3} > \sigma_o^+.$$

Otherwise, either a B-item or a c-container makes it is impossible to pack P_u.

First consider the case, when there is a B-item P_t such that the distance between its top and the bottom of the bin equals $\max(y_1, y_2, y_3)$.

If $P_t \subset L_1 \cup L_2$, then by Rule 2b and 2c of the description of $3TP(5/36)$-algorithm, $y_3 - \min(y_1(t), y_2(t)) > 5/36 - h_t$ (clearly, $y_3 = y_3(u) = y_3(t)$ in this case). On Fig. 3 (left) P_t is the second B-item contained in L_2. Figure 3 (right) illustrates the case when $w_u < 1/3$; however, if we exchange P_u to a rectangle of width $w_u > 2/3$, then P_t is the B-item contained in L_2. The total area of packed A-items is not smaller than $\frac{1}{3}(1 - h_u - h_t - \min(y_1(t), y_2(t)))$. The total area of packed B-items and C-items minus the area of P_t is not smaller than $\frac{1}{9}\min(y_1(t), y_2(t)) + \frac{1}{9}\min(y_1(t), y_2(t)) + \frac{1}{9}y_3 - \lambda$. Consequently,

$$\sigma^+ > \frac{1}{3}\big(1 - h_u - h_t - \min(y_1(t), y_2(t))\big) + 2 \cdot \frac{1}{9}\min(y_1(t), y_2(t))$$
$$+ \frac{1}{9}\Big(\frac{5}{36} - h_t + \min(y_1(t), y_2(t))\Big) - \lambda + w_t h_t + w_u h_u$$
$$> \frac{1}{3}(1 - h_u - h_t) + \frac{1}{9}\Big(\frac{5}{36} - h_t\Big) - \lambda + h_t^2 + \frac{2}{3}h_u$$
$$> \frac{1}{3}(1 - h_t) + \frac{1}{9}\Big(\frac{5}{36} - h_t\Big) - \lambda + h_t^2 = \Big(h_t - \frac{2}{9}\Big)^2 + \frac{97}{324} - \lambda \geq \frac{97}{324} - \lambda = \sigma_o^+.$$

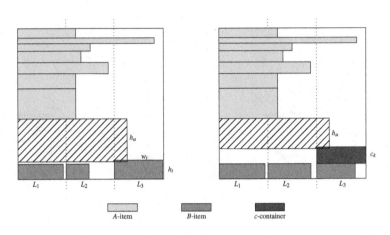

Fig. 4. $w_u > 2/3$.

If $P_t \subset L_3$ and, at the same time, $y_3 - h_t \leq \min(y_1, y_2)$ (see Fig. 4, left), then

$$\sigma^+ > \frac{1}{3}(1 - h_u - h_t) + w_t h_t + w_u h_u - \lambda \geq \frac{1}{3}(1 - h_u - h_t) + h_t^2 + \frac{2}{3}h_u - \lambda$$

$$> h_t^2 - \frac{1}{3}h_t + \frac{1}{3} - \lambda = \left(h_t - \frac{1}{6}\right)^2 + \frac{11}{36} - \lambda > 0.289 > \sigma_o^+.$$

If $P_t \subset L_3$ and, at the same time, $y_3 - h_t > \min(y_1, y_2)$ (see Fig. 5, left), then

$$\sigma^+ > \frac{1}{3}(1 - d - h_u) + \frac{1}{9}d + w_u h_u - \lambda \geq \frac{1}{3} - \frac{2}{9}d + \frac{1}{3}h_u - \lambda$$

$$> \frac{1}{3} - \frac{2}{9} \cdot \frac{5}{36} - \lambda > 0.286 > \sigma_o^+.$$

Now consider the case, when there is a c_k-container K such that the distance between its top and the bottom of the bin equals $\max(y_1, y_2, y_3)$.

Denote by l the distance between the bottom of K and the bottom of I.

If $K \subset L_1 \cup L_2$, then

$$\sigma^+ > \frac{1}{3}\left(1 - h_u - c_k - l\right) + 2 \cdot \frac{1}{9}l + \frac{1}{9}\left(\frac{5}{36} - c_k + l\right) - \lambda + \frac{1}{9}c_k + w_u h_u$$

$$> \frac{1}{3} - \frac{1}{3}c_k + \frac{1}{9} \cdot \frac{5}{36} - \lambda > \frac{1}{3} - \frac{1}{3} \cdot \frac{1}{9} + \frac{1}{9} \cdot \frac{5}{36} - \lambda > \sigma_o^+.$$

If $K \subset L_3$ and $y_3 - c_k \leq \min(y_1, y_2)$ (see Fig. 4, right), then

$$\sigma^+ > \frac{1}{3}(1 - h_u - c_k) + \frac{1}{9}c_k + w_u h_u - \lambda > \frac{1}{3} \cdot \frac{8}{9} + \frac{1}{9} \cdot \frac{1}{9} - \lambda > 0.293 > \sigma_o^+.$$

If $K \subset L_3$ and $y_3 - c_k > \min(y_1, y_2)$ (as on Fig. 5, right), then

$$\sigma^+ > \frac{1}{3}(1 - d - h_u) + \frac{1}{9}d + w_u h_u - \lambda > \frac{1}{3} - \frac{2}{9} \cdot \frac{5}{36} - \lambda > 0.286 > \sigma_o^+.$$

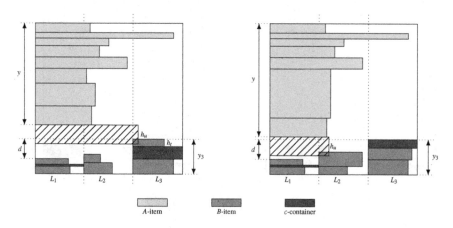

Fig. 5. Case 1

Subcase 1b: $w_u \leq 2/3$ and $y_1 + y_2 > 0$ (see Fig. 5, right). Obviously, $w_u \geq 1/3$. If there is a B-item P_t in $L_1 \cup L_2$ such that the distance between its top and the bottom of the bin equals $\max(y_1, y_2)$, then

$$\sigma^+ > \frac{1}{3}(1 - h_u - h_t) + h_t^2 + \frac{1}{9}\left(\frac{5}{36} - h_t\right) + w_u h_u - \lambda$$

$$= \left(h_t - \frac{2}{9}\right)^2 + \frac{97}{324} + h_u\left(w_u - \frac{1}{3}\right) - \lambda \geq \frac{97}{324} - \lambda = \sigma_o^+.$$

If there is a c_k-container in $L_1 \cup L_2$ such that the distance between its top and the bottom of the bin equals $\max(y_1, y_2)$, then

$$\sigma^+ > \frac{1}{3}(1 - h_u - c_k) + \frac{1}{9}\left(\frac{5}{36} - c_k\right) + w_u h_u - \lambda$$

$$\geq \frac{1}{3}\left(1 - \frac{1}{9}\right) + \frac{1}{9}\left(\frac{5}{36} - \frac{1}{9}\right) + h_u\left(w_u - \frac{1}{3}\right) - \lambda \geq \frac{97}{324} - \lambda = \sigma_o^+.$$

Subcase 1c: $w_u \leq 2/3$ and $y_1 + y_2 = 0$. We have

$$\sigma^+ > \frac{1}{3}(1 - h_u) + \frac{1}{3}h_u = \frac{1}{3} > \sigma_o^+.$$

Case 2: P_u is a B-item.
Subcase 2a: $y_3 \leq \min(y_1, y_2)$ and $d + h_u \leq 5/36$, i.e., we try to pack P_u into L_3 (see Fig. 3, right). We have

$$\sigma^+ > \frac{1}{3}(1 - h_u) + h_u^2 - \lambda = \left(h_u - \frac{1}{6}\right)^2 + \frac{11}{36} - \lambda > 0.289 > \sigma_o^+.$$

Subcase 2b: $y_3 > \min(y_1, y_2)$ and $d + h_u \leq 5/36$, i.e., we try to pack P_u into L_3 (see Fig. 6, left). Since $0 < d \leq 5/36 - h_u$, we get

$$\sigma^+ > \frac{1}{3}(1 - d - h_u) + \frac{1}{9}d + h_u^2 - \lambda = \frac{1}{3} - \frac{2}{9}d + h_u^2 - \frac{1}{3}h_u - \lambda$$

$$\geq \frac{1}{3} - \frac{2}{9}\left(\frac{5}{36} - h_u\right) + h_u^2 - \frac{1}{3}h_u - \lambda = \left(h_u - \frac{1}{18}\right)^2 + \frac{97}{324} - \lambda \geq \frac{97}{324} - \lambda = \sigma_o^+.$$

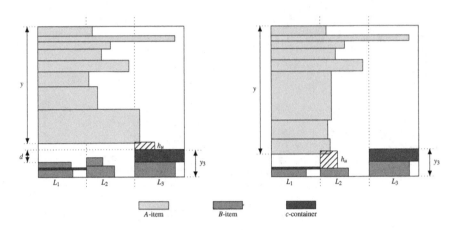

Fig. 6. Case 2.

Subcase 2c: $y_3 > \min(y_1, y_2)$ *and* $d + h_u > 5/36$, *i.e., we try to pack* P_u *into* $L_1 \cup L_2$ (*see Fig. 6, right*). *We have*

$$\sigma^+ > \frac{1}{3}(1 - h_u) + h_u^2 - \lambda = \left(h_u - \frac{1}{6}\right)^2 + \frac{11}{36} - \lambda > 0.289 > \sigma_o^+.$$

Subcase 2d: $y_3 \leq \min(y_1, y_2)$ *and* $d + h_u > 5/36$, *i.e., we try to pack* P_u *into* $L_1 \cup L_2$. *Clearly,* $y_3 - \min(y_1, y_2) > \frac{5}{36} - h_u$. *The total area of packed* A-*items is not smaller than* $\frac{1}{3}(1 - h_u - \min(y_1, y_2))$. *The total area of packed* B-*items and* C-*items is not smaller than* $\frac{1}{9}\min(y_1, y_2) + \frac{1}{9}\min(y_1, y_2) + \frac{1}{9}y_3 - \lambda$. *Consequently,*

$$\sigma^+ > \frac{1}{3}(1 - h_u - \min(y_1, y_2)) + 2 \cdot \frac{1}{9}\min(y_1, y_2)$$
$$+ \frac{1}{9}\left(\frac{5}{36} - h_u + \min(y_1, y_2)\right) - \lambda + w_u h_u$$
$$\geq \frac{1}{3}(1 - h_u) + \frac{1}{9}\left(\frac{5}{36} - h_u\right) - \lambda + h_u^2$$
$$= \left(h_u - \frac{2}{9}\right)^2 + \frac{97}{324} - \lambda \geq \frac{97}{324} - \lambda = \sigma_o^+.$$

Case 3: P_u *is a* C-*item from class* C_{2k-1} *or* C_{2k}.

If $c_k = 1/9$, then there is no almost empty c_1-container in I. This implies that the sum of heights of almost empty c_k-containers is smaller than $\sum_{k \geq 2} c_k$. By the proof of Lemma 1 of [14] we deduce that the total area of C-items packed into I plus the area of P_u is not smaller than $\mu/9$, where μ is equal to the sum of heights of all c_1-containers and all non-last c_k-containers for $k \geq 2$.

Put

$$\lambda^- = \frac{1}{9}\sum_{k \geq 2} c_k = \frac{1}{9}\left(\sum_{k \geq 1} c_k - \frac{1}{9}\right) = \lambda - \frac{1}{81}.$$

Subcase 3a: $y_3 \leq \min(y_1, y_2)$, *i.e., we try to create in* L_3 *a new* c_k-*container.*
If $c_k = 1/9$, then

$$\sigma^+ > \frac{1}{3}(1 - c_k) - \lambda^- = \frac{1}{3} \cdot \frac{8}{9} + \frac{1}{81} - \lambda > 0.293 > \sigma_0^+.$$

If $c_k \leq a/9 = (5 - \sqrt{21})/18$, then

$$\sigma^+ > \frac{1}{3}\left(1 - \frac{a}{9}\right) - \lambda > 0.31 > \sigma_0^+.$$

Subcase 3b: $y_3 > \min(y_1, y_2)$ *and* $d + c_k \leq 5/36$, *i.e., we try to create in* L_3 *a new* c_k-*container.*
If $c_k = 1/9$, then

$$\sigma^+ > \frac{1}{3}(1 - d - c_k) + \frac{1}{9}d - \lambda^- = \frac{1}{3} - \frac{2}{9}d - \frac{1}{3}c_k - \lambda^-$$

$$\geq \frac{1}{3} - \frac{2}{9}\left(\frac{5}{36} - c_k\right) - \frac{1}{3}c_k - \lambda^- = \frac{49}{162} - \frac{1}{9}c_k - \lambda^-$$

$$= \frac{49}{162} - \frac{1}{9} \cdot \frac{1}{9} - \lambda + \frac{1}{81} > 0.286 > \sigma_0^+.$$

If $c_k \leq a/9 = (5 - \sqrt{21})/18$, then

$$\sigma^+ > \frac{1}{3}(1 - d - c_k) + \frac{1}{9}d - \lambda \geq \frac{49}{162} - \frac{1}{9}c_k - \lambda \geq \frac{49}{162} - \frac{1}{9} \cdot \frac{a}{9} - \lambda > 0.284 > \sigma_0^+.$$

Subcase 3c: $y_3 > \min(y_1, y_2)$ *and* $d + c_k > 5/36$, *i.e., we try to create in* $L_1 \cup L_2$ *a new* c_k-*container.*
If $c_k = 1/9$, then

$$\sigma^+ > \frac{1}{3}\left(1 - \frac{1}{9}\right) - \lambda^- > 0.293 > \sigma_0^+.$$

If $c_k \leq a/9 = (5 - \sqrt{21})/18$, then

$$\sigma^+ > \frac{1}{3}\left(1 - \frac{a}{9}\right) - \lambda > 0.31 > \sigma_0^+.$$

$\square$

4   Upper Bound

We improve the upper bound of 5.06 given in [4]. We show that the $3TP(4/9)$-algorithm is 4.84-competitive for 1-space bounded 2-dimensional bin packing.

For a given sequence of items, suppose that the number of closed bins used by packing strategy $3TP(4/9)$ is n. Let σ_A^j, σ_B^j and σ_C^j denote the total area of A-, B- and C- items packed into the j-th bin, respectively. The average occupation for all closed bins is

$$\frac{1}{n}\sum_{j=1}^{n}(\sigma_A^j + \sigma_B^j + \sigma_C^j).$$

Consider the packing configuration of the j-th bin presented on Fig. 2. We have

$$o_A^j \geq \frac{y}{3},$$

$$o_B^j \geq \frac{1}{9}\left(y_1 + y_2 + y_3 - \sum_{k \geq 1} c_k - m\right),$$

$$o_C^j \geq \frac{m}{9},$$

where m is the total height of non-last c-containers and $\sum_{k \geq 1} c_k$ is greater than the sum of heights of last containers. The inequality $o_C^j \geq m/9$ follows by Lemma 1.

Let $P_{u(j)} = (w_{u(j)}, h_{u(j)})$ be the first item that cannot be packed into the j-th bin. We will also write u instead of $u(j)$, for short. By using an amortized analysis, if $P_{u(j)}$ is either an A-item or a B-item that cannot be packed into the active bin, then this item will contribute some area to the just recently closed bin. If P_u is an A-item, than its contribution in the newly created bin is $h_u/3$ and the remaining area of this item is equal to $h_u w_u - h_u/3$, which will be contribute to the previous bin. If P_u is a B-item, than its contribution in the newly created bin is $h_u/9$ and the remaining area of this item is equal to $h_u w_u - h_u/9$.

If $P_u = P_{u(j)}$ is an A-item, put $q_j = h_u w_u - h_u/3$. If P_u is a B-item, put $q_j = h_u w_u - h_u/9$. If P_u is a C-item, put $q_j = 0$. Moreover, let $q_0 = 0$ and $q_n = 0$. Clearly, we have

$$\frac{1}{n}\sum_{j=1}^{n}(o_A^j + o_B^j + o_C^j) = \frac{1}{n}\sum_{j=1}^{n}(o_A^j + o_B^j + o_C^j + q_j - q_{j-1}).$$

Put

$$\sigma_0 = \frac{105 - \sqrt{21}}{486} \approx 0.2066,$$

$$\sigma_j = o_A^j + o_B^j + o_C^j + q_j - q_{j-1},$$

$$\lambda = \frac{1}{9}\sum_{k \geq 1} c_k = \frac{3 + \sqrt{21}}{486} \approx 0.0156$$

and

$$d = y_3 - \min(y_1, y_2).$$

Note that $d \leq 4/9$ for $3TP(4/9)$-algorithm.

To prove Theorem 2 it suffices to show that

$$\sigma_j \geq \sigma_0 \approx 0.2066 \approx (4.84)^{-1}.$$

Theorem 2. *The asymptotic competitive ratio of the 1-space bounded 2-dimensional bin packing algorithm $3TP(4/9)$ is equal to $81(105+\sqrt{21})/1834 \approx 4.84$.*

Proof. As in the proof of Theorem 1 consider three cases depending on the size of P_u.

Case 1: $P_u = (w_u, h_u)$ is an A-item.

Subcase 1a: $w_u > 2/3$. Put $d_u = \max(y_1, y_2, y_3) - \min(y_1, y_2, y_3)$. Clearly, $0 \leq d_u \leq 4/9$. By $q_j = h_u w_u - h_u/3 > h_u/3$ (similarly as on Fig. 5, left, where $d_u = d$)

$$\sigma_j = \sigma_A^j + \sigma_B^j + \sigma_C^j + q_j - q_{j-1} \geq \frac{1}{3}(1 - d_u) + \frac{1}{9}\left(d_u - \sum_{k \geq 1} c_k\right) = \frac{1}{3} - \frac{2}{9}d_u - \lambda$$

$$\geq \frac{1}{3} - \frac{2}{9} \cdot \frac{4}{9} - \lambda > \sigma_0.$$

Subcase 1b: $w_u \leq 2/3$ and $y_1 + y_2 > 0$. Clearly, $q_j = h_u w_u - h_u/3 \geq h_u^2 - h_u/3$. Moreover, at least one B-item or c-container was packed into $L_1 \cup L_2$.

If there is a B-item P_t in L_2 such that the distance between its top and the bottom of the bin equals $\max(y_1, y_2)$ (similarly as on Fig. 5, right), then, by Rule 2 of the description of $3TP(4/9)$-algorithm, we have $y_3 + h_t > 4/9 + \min(y_1, y_2 - h_t)$. As a consequence, $y_3 - \min(y_1, y_2 - h_t) > 4/9 - h_t$ and

$$\sigma_j \geq \frac{1}{3}(1 - h_u - h_t) + q_j + h_t w_t + \frac{1}{9}\left(\frac{4}{9} - h_t\right) - \lambda$$

$$\geq \frac{1}{3}(1 - h_u - h_t) + h_u^2 - \frac{1}{3}h_u + h_t^2 + \frac{1}{9}\left(\frac{4}{9} - h_t\right) - \lambda$$

$$= \left(h_u - \frac{1}{3}\right)^2 + \left(h_t - \frac{2}{9}\right)^2 + \frac{2}{9} - \lambda \geq \frac{2}{9} - \lambda = \sigma_0.$$

Similarly, if there is a B-item in L_1 such that the distance between its top and the bottom of the bin equals $\max(y_1, y_2)$, then $\sigma_j \geq \sigma_0$.

If there is a c_k-container in $L_1 \cup L_2$ such that the distance between its top and the bottom of the bin equals $\max(y_1, y_2)$, then

$$\sigma_j \geq \frac{1}{3}(1 - h_u - c_k) + h_u^2 - \frac{1}{3}h_u + \frac{1}{9}\left(\frac{4}{9} - c_k\right) - \lambda$$

$$= \left(h_u - \frac{1}{3}\right)^2 - \frac{4}{9}c_k + \frac{22}{81} - \lambda \geq -\frac{4}{9} \cdot \frac{1}{9} + \frac{22}{81} - \lambda = \frac{2}{9} - \lambda = \sigma_0.$$

Subcase 1c: $w_u \leq 2/3$ and $y_1 + y_2 = 0$. We have

$$\sigma_j \geq \frac{1}{3}(1 - h_u) + q_j = \frac{1}{3}(1 - h_u) + h_u^2 - \frac{1}{3}h_u = \left(h_u - \frac{1}{3}\right)^2 + \frac{2}{9} > \sigma_0.$$

Case 2: P_u is a B-item.

Subcase 2a: $d + h_u \leq 4/9$, i.e., we try to pack P_u into L_3. If $d < 0$, then

$$\sigma_j \geq \frac{1}{3}(1 - h_u) - \lambda \geq \frac{1}{3} \cdot \frac{2}{3} - \lambda = \sigma_0.$$

If $d \geq 0$ by $q_j = h_u w_u - h_u/9 \geq h_u^2 - h_u/9$ and $d \leq 4/9 - h_u$ we get

$$
\sigma_j \geq \frac{1}{3}\left(1 - d - h_u\right) + \frac{1}{9}d + q_j - \lambda \geq \frac{1}{3} - \frac{2}{9}d + h_u^2 - \frac{4}{9}h_u - \lambda
$$
$$
\geq \frac{1}{3} - \frac{2}{9}\left(\frac{4}{9} - h_u\right) + h_u^2 - \frac{4}{9}h_u - \lambda = \left(h_u - \frac{1}{9}\right)^2 + \frac{2}{9} - \lambda \geq \frac{2}{9} - \lambda = \sigma_0.
$$

Subcase 2b: $d + h_u > 4/9$, i.e., we try to pack P_u into $L_1 \cup L_2$ (similarly as on Fig. 6, right). We have

$$
\sigma_j \geq \frac{1}{3}(1 - h_u) - \lambda \geq \frac{1}{3} \cdot \frac{2}{3} - \lambda = \sigma_0.
$$

Case 3: P_u is a C-item from class C_{2k-1} or C_{2k}.
Subcase 3a: $d + c_k \leq 4/9$, i.e., we try to create in L_3 a new c_k-container. If $d < 0$, then

$$
\sigma_j \geq \frac{1}{3}(1 - c_k) - \lambda > \sigma_0.
$$

If $d \geq 0$, then by $c_k \leq 1/9$ and $d \leq 4/9 - c_k$ we obtain

$$
\sigma_j \geq \frac{1}{3}\left(1 - d - c_k\right) + \frac{1}{9}d - \lambda = \frac{1}{3} - \frac{2}{9}d - \frac{1}{3}c_k - \lambda
$$
$$
\geq \frac{1}{3} - \frac{2}{9}\left(\frac{4}{9} - c_k\right) - \frac{1}{3}c_k - \lambda = \frac{19}{81} - \frac{1}{9}c_k - \lambda \geq \frac{2}{9} - \lambda = \sigma_0.
$$

Subcase 3b: $d + c_k > 4/9$, i.e., we try to create in $L_1 \cup L_2$ a new c_k-container. We have

$$
\sigma_j \geq \frac{1}{3}(1 - c_k) - \lambda \geq \frac{8}{27} - \lambda > \sigma_0.
$$

Since $\sigma_j \geq \sigma_0 \approx 0.2066$, it follows that the average occupation in each bin is not smaller than σ_0. Consequently, the asymptotic competitive ratio of this packing strategy is not greater than $\sigma_0^{-1} \approx 4.84$. $\qquad\square$

Remark. It is easy to verify that the asymptotic competitive ratio of $3TP(h)$-algorithm is greater than $81(105 + \sqrt{21})/1834 \approx 4.84$ provided $h \neq 4/9$. If $h < 4/9$, then the worst case is Subcase 1b. If $h > 4/9$, then the worst case is Subcase 2a (for $d \geq 0$).

5   Lower Bound

We will improve slightly the lower bound of 3.167 given in [4]. We show that no online algorithm can achieve the competitive ratio less than 3.246. In the proof we use the method described in [4] with additional types of items.

Theorem 3. *The asymptotic competitive ratio of any 1-space bounded 2-dimensional bin packing algorithm is greater than 3.246.*

Proof. Let $n \geq 11$ be an integer and let $\varepsilon < 1/(5418(n+1))$. Denote by $\lceil l \rceil$ the smallest integer greater than or equal to l. Consider the following sequence S of items:

$$X_1, X_2, \ldots, X_{2n}, A_1, B_1, A_2, B_2, \ldots, A_n, B_n, T_1, \ldots, T_n,$$
$$U_1, \ldots, U_5, V_1, U_6, \ldots, U_{10}, V_2, U_{11}, \ldots, U_{5 \cdot \lceil n/5 \rceil}, V_{\lceil n/5 \rceil}, W_{\lceil n/5 \rceil + 1}, \ldots, W_n.$$

The first $2n$ square items have side lengths close to $1/2$:

$$X_{2i-1} = \big(1/2 + i\varepsilon, 1/2 + i\varepsilon\big), \quad X_{2i} = \big(1/2 - i\varepsilon + \varepsilon, 1/2 - i\varepsilon + \varepsilon\big).$$

Clearly, the sum of side lengths on any two consecutive X-items is greater than 1 and, consequently, these items cannot be packed into one bin. Thus, at least $2n$ bins are used for packing items $X_1, \ldots, X_{2n}$. On the other hand, it is possible to pack X_{2i-1} and X_{2i+2} into one bin. The reason is that $\frac{1}{2} + i\varepsilon + \frac{1}{2} - (i+1)\varepsilon + \varepsilon = 1$.

The next $2n$ items in the sequence have one side of length close to $1/3$:

$$A_i = \big(2/3 + i\varepsilon, 1/3 + i\varepsilon, \big), \quad B_i = \big(1/3 - i\varepsilon + \varepsilon, 1/3 + i\varepsilon - \varepsilon\big).$$

Items B_i, A_{i+1} and B_{i+1} can be packed into one bin while items A_i, B_i and A_{i+1} cannot. Since no two items A_i and A_{i+1} (with the corresponding item B_i) can be packed into one bin in any online algorithm, at least n bins are used for packing these items. Obviously, X_{2n}, A_1 and B_1 can be packed online into one bin, but it does not matter to verify the asymptotic ratio. In the offline strategy, it is possible to pack $X_{2i-1}, X_{2i+2}, A_i, B_{i+1}$ into one bin (see Fig. 7, left).

The next n items in S have one side of length close to $1/43$: $T_i = \big(1, 1/43 + \varepsilon\big)$. Any bin may contain at most 42 such items. Thus, at least $n/42$ bins are used for packing these items. On the other hand, items $X_{2i-1}, X_{2i+2}, A_i, B_{i+1}, T_i$ can be packed into one bin.

The sequence S contains also $6 \cdot \lceil n/5 \rceil$ items of type U and V with one side of length close to $1/7$:

$$U_i = \big(6/7 - \lceil i/5 \rceil \cdot \varepsilon, 1/7 + \lceil i/5 \rceil \cdot \varepsilon + 2\varepsilon\big), \quad V_i = \big(1/7 + i\varepsilon + 2\varepsilon, 1/7 + i\varepsilon + 2\varepsilon\big).$$

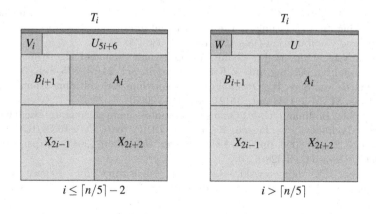

Fig. 7. Optimal packing

It is easy to verify that no six consecutive items of type U (with the corresponding item of type V) can be packed into one bin. Consequently, at least $\lceil n/5 \rceil$ bins are used for packing these items.

At least $(n - \lceil n/5 \rceil)/36$ bins are used to pack $n - \lceil n/5 \rceil$ items $W_i = (1/7 + \varepsilon, 1/7 + \varepsilon)$. On the other hand, it is possible to pack X_{2i-1}, X_{2i+2}, $A_i, B_{i+1}, V_i, U_{5i+6}$ into one bin for $i = 1, \ldots, \lceil n/5 \rceil - 2$ (see Fig. 7, left). Moreover, any item of type W and any item of type U can be packed into one bin together with items $X_{2i-1}, X_{2i+2}, A_i, B_{i+1}$ (see Fig. 7, right).

By the above consideration we conclude that there exist integers c_1 and c_2 such that for any n it is possible to pack items from S into $n + c_1$ bins in the optimal offline strategy. However, there is no online algorithm which can pack all items from S into

$$2n + n + n/42 + n/5 + 4n/(5 \cdot 36) - c_2$$

bins. This implies that no online algorithm can achieve the competitive ratio less than $3 + 1/42 + 1/5 + 4/(5 \cdot 36) = 409/126 \approx 3.246$. □

References

1. Brubach, B.: Improved online square-into-square packing. arxiv.org/pdf/1401.5583.pdf
2. Brubach, B.: Improved bound for online square-into-square packing. In: Bampis, E., Svensson, O. (eds.) WAOA 2014. LNCS, vol. 8952, pp. 47–58. Springer, Cham (2015). https://doi.org/10.1007/978-3-319-18263-6_5
3. Zhang, Y., Chen, J., Chin, F.Y.L., Han, X., Ting, H.-F., Tsin, Y.H.: Improved online algorithms for 1-space bounded 2-dimensional bin packing. In: Cheong, O., Chwa, K.-Y., Park, K. (eds.) ISAAC 2010. LNCS, vol. 6507, pp. 242–253. Springer, Heidelberg (2010). https://doi.org/10.1007/978-3-642-17514-5_21
4. Chin, F.Y.L., Han, X., Poon, C.K., Ting, H.-F., Tsin, Y.H., Ye, D., Zhang, Y.: Online algorithms for 1-space bounded 2-dimensional bin packing and square packing. Theor. Comput. Sci. **554**, 135–149 (2014)
5. Chin, F.Y.L., Ting, H.-F., Zhang, Y.: 1-space bounded algorithms for 2-dimensional bin packing. In: Proceedings of the 20th Annual International Symposium on Algorithms and Computation (ISAAC), pp. 321–330 (2009)
6. Chin, F.Y.L., Ting, H.-F., Zhang, Y.: One-space bounded algorithms for two-dimensional bin packing. Int. J. Found. Comput. Sci. **21**(6), 875–891 (2010)
7. Epstein, L., van Stee, R.: Optimal online algorithms for multidimensional packing problems. SIAM J. Comput. **35**(2), 431–448 (2005)
8. Fekete, S.P., Hoffmann, H.-F.: Online square-into-square packing. In: Raghavendra, P., Raskhodnikova, S., Jansen, K., Rolim, J.D.P. (eds.) APPROX/RANDOM - 2013. LNCS, vol. 8096, pp. 126–141. Springer, Heidelberg (2013). https://doi.org/10.1007/978-3-642-40328-6_10
9. Grzegorek, P., Januszewski, J.: Online algorithms for 3-space bounded 2-dimensional bin packing and square packing. Rom. J. Inf. Sci. Technol. **17**(2), 189–202 (2014)
10. Grzegorek, P., Januszewski, J.: A note on one-space bounded square packing. Inf. Process. Lett. **115**, 872–876 (2015)

11. Han, X., Iwama, K., Zhang, G.: Online removable square packing. Theory Comput. Syst. **43**(1), 38–55 (2008)
12. Januszewski, J.: Packing rectangles into the unit square. Geom. Dedicata **8**, 13–18 (2000)
13. Januszewski, J., Lassak, M.: On-line packing sequences of cubes in the unit cube. Geom. Dedicata **67**, 285–293 (1997)
14. Januszewski, J., Zielonka, L.: Improved online algorithms for 2-space bounded 2-dimensional bin packing. Int. J. Found. Comput. Sci. **27**(4), 407–429 (2016)
15. Johnson, D.S.: Fast algorithms for bin packing. J. Comput. Syst. Sci. **8**, 272–314 (1974)
16. Lassak, M.: On-line potato-sack algorithm efficient for packing into small boxes. Periodica Math. Hung. **34**, 105–110 (1997)
17. Lee, C.C., Lee, D.T.: A simple on-line bin packing algorithm. J. ACM **32**, 562–572 (1985)
18. Moser, L.: Poorly formulated unsolved problems of combinatorial geometry. Mimeographed (1966)
19. Moon, J., Moser, L.: Some packing and covering theorems. Colloq. Math. **17**, 103–110 (1967)
20. Zhao, X., Shen, H.: On-line algorithms for 2-space bounded cube and hypercube packing. Tsinghua Sci. Technol. **20**(3), 255–263 (2015)

A Tight Lower Bound for Online Convex Optimization with Switching Costs

Antonios Antoniadis[1] and Kevin Schewior[1,2(✉)]

[1] Max-Planck-Institut für Informatik, Saarbrücken, Germany
{aantonia,schewior}@mpi-inf.mpg.de
[2] Universidad de Chile, Santiago, Chile

Abstract. We investigate online convex optimization with switching costs (OCO; Lin et al., INFOCOM 2011), a natural online problem arising when rightsizing data centers: A server initially located at p_0 on the real line is presented with an online sequence of non-negative convex functions $f_1, f_2, \ldots, f_n : \mathbb{R} \to \mathbb{R}_+$. In response to each function f_i, the server moves to a new position p_i on the real line, resulting in cost $|p_i - p_{i-1}| + f_i(p_i)$. The total cost is the sum of costs of all steps. One is interested in designing competitive algorithms.

In this paper, we solve the problem in the classical sense: We give a lower bound of 2 on the competitive ratio of any possibly randomized online algorithm, matching the competitive ratio of previously known deterministic online algorithms (Andrew et al., COLT 2013/arXiv 2015; Bansal et al., APPROX 2015). It has been previously conjectured that $(2 - \epsilon)$-competitive algorithms exist for some $\epsilon > 0$ (Bansal et al., APPROX 2015).

1 Introduction

We consider a fundamental online problem that models the adaptation of a single parameter to an environment that becomes known in an online fashion. Here, costs are incurred both by changing the parameter and by the convex functions of the parameter that model the environment and are revealed online. More specifically, in online convex optimization with switching costs (OCO) a server is initially located at the origin of the real line (modeling the range of parameters) and receives an online sequence of non-negative convex functions. As an answer to each such function, the server may move to a new point on the real line, paying the moved distance and the function value at the endpoint. The goal is to be competitive with respect to the total costs.

The problem and its variants have received considerable attention in the recent literature [1,4,12–15], also because it has applications in the right-sizing

A. Antoniadis—Supported by the Deutsche Forschungsgemeinschaft (DFG, German Research Foundation) under AN 1262/1-1.
K. Schewior—Supported by CONICYT grant PCI PII 20150140 and the Millennium Nucleus Information and Coordination in Networks ICM/FIC RC130003.

© Springer International Publishing AG, part of Springer Nature 2018
R. Solis-Oba and R. Fleischer (Eds.): WAOA 2017, LNCS 10787, pp. 164–175, 2018.
https://doi.org/10.1007/978-3-319-89441-6_13

of data centers [13]. Here, the parameter models the number of processors that are turned on, and load that is to be processed arrives online. Turning processors on or off, that is, changing the parameter, incurs (energy) cost. In addition, the energy required for handling the load in the most energy-efficient way is a convex function of the number of processors that are turned on.

There are two deterministic 2-competitive algorithms known, due to Andrew et al. [1,2] and Bansal et al. [4]. The best known lower bound on the competitive ratio of any (possibly randomized) algorithm is 1.86 [4]. We close the gap by improving the lower bound to 2.

Further Related Work. The problem was introduced by Lin et al. in 2011 [13]. They showed that the offline version of the problem can be modeled as a convex program and gave a 3-competitive online algorithm. Andrew et al. subsequently claimed a 2-competitive algorithm [1], but the argument contained a flaw that was fixed later [2]. Independently, Bansal et al. [4] gave another 2-competitive algorithm. They also showed that any randomized algorithm can be easily derandomized without loss in the competitive ratio. Furthermore they gave a 3-competitive memoryless algorithm and showed that this is best possible for memoryless algorithms.

On the lower-bound side, there is straightforward reduction from the randomized ski-rental problem [10,11] to the OCO problem without loss in the competitive ratio, implying a lower bound of $e/(e-1) \approx 1.58$ on the achievable competitive ratio. This was later improved to 1.86 [4]. We will discuss known lower bounds in more detail below.

The problem is also interesting because it is a not too restrictive special case of the very general online problem of metrical task systems (MTS) [7]. Here, the real line is replaced by any metrical space, and the functions arriving online are arbitrary (possibly non-convex) functions. For deterministic algorithms and size k of the base set of the metrical space, a competitive ratio of $2k-1$ can be achieved, and this is tight [7]. For randomized algorithms and input sequences of length n, there is a lower bound of $\Omega(\log n / \log \log n)$ [5,6] and an upper bound of $\mathcal{O}(\log^2 n \log \log n)$ [8] on the achievable competitive ratio.

A number of other problems related to MTS and OCO have been considered. In the following we list known results in decreasing order of generality. MTS on space $\mathbb{R}^d$ with convex functions, that is, the natural generalization of OCO to higher dimensions, has been considered as convex-function chasing [3]. It has been claimed in the same paper that the problem can be reduced to so-called halfspace chasing (defined below) at the cost of a constant factor in the competitive ratio, but there is a gap in the proof. There is no other work on general convex-function chasing when $d \geq 2$.

The restriction of convex-function chasing to functions that either take ∞ (or some large value) or 0 as values is usually referred to as convex-body chasing, as introduced by Friedman and Linial [9] in 1993. This problem can be essentially reduced to halfspace chasing, that is, the problem in which, for each function, the preimage of ∞ forms a halfspace of $\mathbb{R}^d$ [9]. The only result known on halfspace chasing is an $\mathcal{O}(1)$-competitive algorithm for $d=2$ [9]. For affine-subspace chasing, in which the preimage of each function is an affine subspace of $\mathbb{R}^d$, Antoniadis et al. [3] give a $2^{\mathcal{O}(d)}$-competitive algorithm.

The OCO problem has also been considered within the online-learning setting. Here, the basic difference is that the optimum is assumed to be stationary, and one cares about additive additional cost (regret). It was shown that sublinear regret is achievable but not simultaneously with a constant competitive ratio [1].

Previous Lower Bounds and Our Contribution. As mentioned earlier, the randomized ski-rental problem can be reduced to OCO without loss in the competitive ratio. This is because a ski-rental instance with k days, rental cost ϵ, and buying cost 1 can be translated to a sequence of k (decreasing) functions of the form $f^{\mathrm{dec}}(x) = \epsilon \cdot |1 - x|$. Moving to position $p \in [0, 1]$ then corresponds to buying skis with probability p, and the costs are exactly the same in both problems. Recall that, in the limit for $\epsilon \to 0$, the unique optimal online algorithm moves over from 0 to 1 at exponentially increasing speed, resulting in competitive ratio $e/(e-1)$ [10].

The lower bound of 1.86 [4] relies on realizing that this algorithm moves over to 1 too fast to have competitive ratio $e/(e-1)$ in case a number of (increasing) functions $f^{\mathrm{inc}}(x) = \epsilon \cdot |x|$ are added at the end of the sequence. In fact, the lower bound is obtained by analyzing the optimal algorithm for sequences that start with an arbitrary number of decreasing functions followed by a large number of increasing functions. The resulting speed of moving over to 1 as long as only decreasing functions are presented is still increasing. Note however that, given the algorithm is at position p when the first increasing function is presented, the best lower bound on its additional cost we can obtain is p if we only consider this special type of instances.

Although it was conjectured [4] that 2 is not the optimal competitive ratio for OCO, it is quite obvious by the above insight that the previously known lower bound is not tight, but also that one lacked techniques to analyze arbitrary instances consisting only of functions of the type f^{dec} or f^{inc}. In the following we call such instances *bitonic*. The contribution of this paper is a technique for analyzing bitonic instances as well as the tight lower bound for OCO obtained by this analysis.

In our analysis, we consider an algorithm ALG^* that moves over to 1 at constant speed as long as only decreasing functions are presented. More precisely, ALG^* moves a distance of $\epsilon/2$ towards the minimum of each line that is presented (if it is not at that minimum, in which case it does not move). On bitonic instances, ALG^* coincides with the 2-competitive algorithm from [4]. In fact, we implicitly show that ALG^* is the optimal algorithm for bitonic instances.

For the proof of the lower bound of 2 for some given online algorithm ALG, we adversarially construct an instance on which (i) ALG^* has cost that is (asymptotically) by a $2 - \epsilon/2$ factor larger than the optimal cost and (ii) the cost of ALG is at least that of ALG^*. The adversarial strategy is fairly simple: Whenever the position p_i of ALG is to the right of the position p_i^* of ALG^*, we present an increasing function; otherwise we present a decreasing one. If ALG^* reaches either 0 or 1 in the second round or later, we present a large number of the type of function that is free for ALG^*. We show that (i) and (ii) are true after finitely many steps.

The proof of (i) relies on distinguishing the case that ALG^* reaches 0 or 1 (in the second round or later) and the one in which it does not. In the first case the $2 - \epsilon/2$ bound is easily seen to be exact. In the case in which ALG^* never reaches 0 or 1, the bound only holds asymptotically: For n rounds, we show that the cost of OPT is $n \cdot \epsilon/2 + \mathcal{O}(1)$ while that of ALG^* is $n \cdot (2 - \epsilon/2) \cdot \epsilon/2 - \mathcal{O}(1)$.

To conclude the proof by showing (ii), we define *phases* such that the cost of ALG are at least as large as that of ALG^* within each phase. Whenever ALG "overtakes" ALG^*, we let a phase end, and the claim follows from the fact that ALG has been paying higher function cost throughout the phase and, by overtaking, also higher movement cost. If ALG^* has reached 0 or 1, we argue that we can let the phase (and instance) end after a finite amount of additional functions.

Overview of this Paper. We give preliminary definitions and results in Sect. 2. In Sect. 3, we give the lower bound. Here, we first describe ALG^* and give the adversarial instance. We then bound the costs of OPT, ALG^*, and finally ALG. We give concluding remarks in Sect. 4.

2   Preliminaries

Problem Definition. Online convex optimization with switching cost (OCO) is formally defined as follows. Let $p_0 = 0$. In *round* $i = 1, \ldots, n$, a convex function $f_i : \mathbb{R} \to \mathbb{R}_+$ arrives, and the required output as an answer is some point $p_i \in \mathbb{R}$. The cost in round i is the *movement cost* $|p_{i-1} - p_i|$ plus the *function cost* $f_i(p_i)$. The total cost is the sum over the costs in all rounds. For a given algorithm ALG or the offline optimum OPT and a given instance I, we refer to the total cost on I by $\mathrm{cost}(\mathsf{ALG}, I)$ or $\mathrm{cost}(\mathsf{OPT}, I)$, respectively.

An algorithm is called an online algorithm if its choices in round i solely depend on the input up to round i (and possibly randomization). We call a (possibly randomized) algorithm *c-competitive* if $\mathbb{E}[\mathrm{cost}(\mathsf{ALG}, I)] \leq c \cdot \mathrm{cost}(\mathsf{OPT}, I)$ for all instances I.

Notation. The (to the algorithm unknown) number of rounds of some input instance I is referred to as $|I|$. In our description of the lower bound, p_i will be the position of an arbitrary algorithm ALG (whose competitive ratio we lower bound) and p_i^* that of the auxiliary algorithm ALG^* defined below.

Randomized and Deterministic Algorithms. To be able to restrict to designing a lower bound for deterministic algorithms, we use the following lemma due to [4].

Lemma 1 (Bansal et al. [4]). *If there exists a c-competitive randomized algorithm for online convex optimization with switching costs, then there also exists a c-competitive deterministic algorithm.*

This lemma simply follows by derandomizing any randomized algorithm by, at all times, moving to the expected location of the randomized algorithm. Convexity of the input functions and Jensen's inequality shows that the competitive ratio cannot become larger.

3   The Lower Bound

The goal of this section is to show the main result of this paper, as stated in the following theorem.

Theorem 1. *For any $\delta > 0$, there does not exist a $(2-\delta)$-competitive algorithm for online convex optimization with switching costs.*

To prove the theorem, we start by restricting ourselves to what we call *bitonic instances* which contain only two distinct functions. We then define an auxiliary algorithm ALG* that runs on any such bitonic instance. With the help of ALG* we are able to define an adversarial input sequence I for any online algorithm ALG. The function that arrives in the next round in this adversarial input sequence I depends only on the current positions p_i and p_i^* of ALG and ALG*, respectively. Our lower bound then follows by directly comparing algorithms ALG and ALG* to each other, as well as ALG* to a specific simple algorithm OPT' whose cost is upper bounding the cost of the optimal offline solution OPT.

Although we do not need these facts and do not prove them directly, OPT' can be thought of as the optimal offline algorithm and ALG* as the optimal online algorithm. The intuition then is that whenever the state of ALG differs from that of the optimal online algorithm ALG* it gets "punished" by the adversary.

3.1   Bitonic Input Instances

For some fixed $\epsilon \in (0, 2\delta)$ with $2/\epsilon$ integer, let f^{inc} and f^{dec} be non-negative convex functions with

$$f^{\mathrm{inc}}(x) = \epsilon \cdot x$$
$$\text{and } f^{\mathrm{dec}}(x) = \epsilon \cdot (1-x) \text{ for } x \in [0,1],$$

e.g., $f^{\mathrm{inc}}(x) = \epsilon \cdot |x|$ and $f^{\mathrm{dec}}(x) = \epsilon \cdot |1-x|$. The function superscripts refer to *increasing* and *decreasing*. Our lower bound will only use these functions. We define the corresponding class of instances.

Definition 1. *An instance is called* bitonic *if it only consists of f^{inc} and f^{dec}.*

To get some intuition about bitonic instances, note that moving beyond 0 or 1 from inside the interval $[0, 1]$ will only incur more cost since both f^{inc} and f^{dec} are non-decreasing in these directions.

Observation 1. For bitonic instances, if there is a c-competitive online algorithm, there is also a c-competitive online algorithm that never moves to points outside $[0, 1]$.

For the remainder of the paper, we will only consider bitonic instances and, hence by the observation, (deterministic) algorithms that only move to points in $[0, 1]$.

3.2   An Adversarial Strategy for Bitonic Instances

As already mentioned, the definition of the adversarial strategy depends on the state of algorithm ALG^* in each round. We therefore start by formally defining ALG^*.

Algorithm ALG^*:
If $f_i = f^{\mathrm{dec}}$, move to $p_i := \min\{p_{i-1} + \epsilon/2, 1\}$;
if $f_i = f^{\mathrm{inc}}$, move to $p_i := \max\{0, p_{i-1} - \epsilon/2\}$.

We note that ALG^* is a version of the 2-competitive algorithm from [4] restricted to bitonic instances.

Consider now an arbitrary given algorithm ALG. We define an *adversarial sequence* which is based on comparing, in each round, the points p_i and p_i^* at which ALG and ALG^* reside, respectively.

An Adversarial Input Sequence:
The function arriving in the first round is f^{dec}.
For any subsequent round i we distinguish between three cases:

- If $p_{i-1} = p_{i-1}^*$, then $f_i := f_{i-1}$;
- If $p_{i-1} < p_{i-1}^*$, then $f_i := f^{\mathrm{dec}}$;
- if $p_{i-1} > p_{i-1}^*$, then $f_i := f^{\mathrm{inc}}$.

The instance ends whenever one of the following termination conditions becomes true for $i > 1$:

(i) $p_i^* = 0$ for the previous $4/\epsilon$ rounds,
(ii) $p_i^* = 1$ for the previous $4/\epsilon$ rounds,
(iii) $\frac{i \cdot (2-\epsilon/2) \cdot \epsilon/2 - 6}{i \cdot \epsilon/2 + 1} > 2 - \delta$
       and $((f_i = f^{\mathrm{dec}}$ and $p_i \geq p_i^*)$ or $(f_i = f^{\mathrm{inc}}$ and $p_i \leq p_i^*))$.

It is easy to verify that the adversarial input sequence is well defined and will eventually terminate: If Conditions (i) and (ii) never occur, then ALG^* never reaches 0 or 1 (if it did, f^{inc} or f^{inc}, respectively, would get repeated and Condition (i) or (ii), respectively, would eventually become true). For ALG^* to never reach 0 or 1, however, ALG needs to "overtake" (implies second part of Condition (iii)) ALG^* at least every $2/\epsilon$ steps. It can also be seen easily that, by our choice of ϵ, the first part of Condition (iii) becomes true for large enough i, so eventually Condition (iii) will occur.

3.3 The Offline Algorithm OPT′

In this subsection we define and analyze the offline algorithm OPT′ which we will use to upper bound the cost of the actual optimal offline solution OPT.

Algorithm OPT′:
Let $\#f^{\text{dec}}$ and $\#f^{\text{inc}}$ be the numbers of times that each corresponding function appears in the input instance I.
If $\#f^{\text{inc}} \geq \#f^{\text{dec}}$, then stay at $p^{\text{OPT}'} = p_0 = 0$ throughout I.
Else, move to $p^{\text{OPT}'} = 1$ in response to the first function, and stay there throughout I.

In the following lemma we analyze the cost of the offline algorithm OPT′ on an adversarial sequence I as defined above. Note that this upper bounds the cost of OPT.

Lemma 2. *Consider an adversarial input instance I as described above on which* ALG* *ends at position* $q\epsilon/2$, $q \in [0, 2/\epsilon] \cap \mathbb{N}$. *Further, let* $s \leq 4/\epsilon$ *be the number of rounds that* ALG* *was in position 0 or 1 during I. Then we have* $\text{cost}(\text{OPT}', I) = (|I| - s + q) \cdot \frac{\epsilon}{2}$ *when I terminates due to Condition (iii), and* $\text{cost}(\text{OPT}', I) = (|I| - 4/\epsilon) \cdot \frac{\epsilon}{2}$ *when I terminates due to Conditions (i) or (ii).*

Proof. If I terminates due to Condition (i), then we have $p^{\text{OPT}'} = 0$, and OPT′ pays ϵ for each f^{dec} in I and 0 for each f^{inc} in I. Furthermore, since I terminated due to Condition (i) we have that $\#f^{\text{inc}} = \#f^{\text{dec}} + 4/\epsilon$ and in turn

$$\#f^{\text{dec}} = \frac{|I| - \frac{4}{\epsilon}}{2}.$$

So in total for this case we have

$$\text{cost}(\text{OPT}', I) = \#f^{\text{dec}} \cdot \epsilon = \left(|I| - \frac{4}{\epsilon}\right) \cdot \frac{\epsilon}{2}.$$

If I terminates due to Condition (ii), then we have $p^{\text{OPT}'} = 1$, and OPT′ pays 1 in movement cost, and ϵ and 0 for each f^{inc} and f^{dec} in I, respectively.

$$\#f^{\text{dec}} = \#f^{\text{inc}} + \frac{4}{\epsilon} + \frac{2}{\epsilon} = \#f^{\text{inc}} + \frac{6}{\epsilon},$$

and in turn

$$\#f^{\text{inc}} = \frac{|I| - \frac{6}{\epsilon}}{2}.$$

Therefore in total

$$\text{cost}(\text{OPT}', I) = 1 + \#f^{\text{inc}} \cdot \epsilon = 1 + \frac{|I| - \frac{6}{\epsilon}}{2} \cdot \epsilon = \left(|I| - \frac{4}{\epsilon}\right) \cdot \frac{\epsilon}{2}.$$

Finally, if I terminates due to Condition (iii), then we distinguish between two cases:

Case 1: If $p^{OPT'} = 0$, then OPT' pays ϵ for each f^{dec} in I and 0 for each f^{inc} in I. Furthermore, it must be the case that $q \leq 1/\epsilon$, and if $q = 0$ then it might be the case that $s > 0$. Hence it holds that $\#f^{dec} = \#f^{inc} + q - s$. We therefore have in total:

$$\text{cost}(\text{OPT}', I) = \#f^{dec} \cdot \epsilon = (|I| + q - s) \cdot \frac{\epsilon}{2},$$

as claimed. Note that in this case OPT' has no movement cost.

Case 2: On the other hand, if $p^{OPT'} = 1$, then OPT' has a movement cost of 1, a function cost of ϵ for each f^{inc} in I, and a function cost of 0 for each f^{dec} in I. Furthermore, since ALG^* has stoped at $q \cdot (1 - 2/\epsilon)$ and $\#f^{dec} = \#f^{inc} + (2/\epsilon - q) + s$, we have

$$\begin{aligned}
\text{cost}(\text{OPT}', I) &= \#f^{inc} \cdot \epsilon + 1 \\
&= \left(\frac{|I| + q - s}{2} - \frac{1}{\epsilon} \right) \cdot \epsilon + 1 \\
&= (|I| + q - s) \cdot \frac{\epsilon}{2},
\end{aligned}$$

similarly to the computation above.   □

This concludes the proof of the lemma.

3.4   Concluding the Proof

In order to conclude the proof of Theorem 1, we show that for any algorithm ALG and the adversarial instance I defined above it holds that

$$\text{cost}(\text{ALG}, I) \geq \text{cost}(\text{ALG}^*, I) \geq (2-\delta) \cdot \text{cost}(\text{OPT}', I) \geq (2-\delta) \cdot \text{cost}(\text{OPT}, I). \quad (1)$$

We note that the last inequality directly follows by the feasibility of OPT'; it therefore remains to show the first two inequalities. We first show the second inequality, which is implied by the following lemma.

Lemma 3. *For any online algorithm* ALG *and the adversarial instance defined above, we have*

$$\text{cost}(\text{ALG}^*, I) \geq (2 - \delta) \cdot \text{cost}(\text{OPT}', I).$$

Proof. We distinguish the condition due to which I was terminated.

Case 1: Termination Condition (i). By the definition of ALG^* and the adversarial strategy, since $p_i^* = 0$ it must be the case that $\#f^{inc} = \#f^{dec} + 4/\epsilon$. By Lemma 2, $\text{cost}(\text{OPT}, I) = \left(|I| - \frac{4}{\epsilon} \right) \cdot \frac{\epsilon}{2}$.

By the definition of ALG^* in every Round i it incurs a cost of $|p_i^* - p_{i-1}^*| + f(p_i^*)$, where $f(p_i^*)$ refers to either f^{inc} or f^{dec} – whatever arrived in round i. We note that exactly the last $4/\epsilon$ rounds we have that $p_i^* = p_{i-1}^*$. So

$$\mathrm{cost}(\mathsf{ALG}^*, I) = \left(|I| - \frac{4}{\epsilon}\right) \cdot \frac{\epsilon}{2} + \sum_{i=1}^{|I|} f(p_i^*).$$

In order to analyze the second term in the above expression, consider an edge $e_i = [i\epsilon/2, (1+i)\epsilon/2)$ for $i = 0, 1 \ldots 2/\epsilon - 1$ as the interval between any two consecutive points (multiples of $\epsilon/2$) visited by ALG^*. One can observe that since we started at 0, ended at 0 and always traversed edges of length $\epsilon/2$ either to the left or to the right, each such edge has length $\epsilon/2$ and furthermore, each such edge was traversed an equal number of times towards the left as towards the right. So consider an edge $e = [x, x + \epsilon/2)$ which was traversed $2k$ times in total, k times towards the left due to an f^{inc} and k times towards the right due to an f^{dec}. The total function cost with respect to edge e is therefore:

$$k \left(f^{\mathrm{inc}}(x) + f^{\mathrm{dec}}(x + \epsilon/2) \right)$$
$$= k \left(x\epsilon + \left(1 - (x + \epsilon/2)\,\epsilon\right) \right)$$
$$= k\epsilon - \epsilon^2/2.$$

We note that this expression is independent of x and therefore also independent of which edge e we are considering. Thus if we sum over all edges, and add the moving cost we have in total,

$$\mathrm{cost}(\mathsf{ALG}^*, I) = \left(|I| - \frac{4}{\epsilon}\right)\epsilon - \left(|I| - \frac{4}{\epsilon}\right)\frac{\epsilon^2}{4}$$

The competitive ratio of ALG^* therefore is at least $2 - \epsilon/2 \geq 2 - \delta$.

Case 2: Termination Condition (ii). Since $p_i^* = 1$, by the definition of ALG^*, it must be the case in I that $\#f^{\mathrm{dec}} = \#f^{\mathrm{inc}} + \frac{2}{\epsilon} + \frac{4}{\epsilon} = \#f^{\mathrm{inc}} + \frac{6}{\epsilon}$. By Lemma 2, $\mathrm{cost}(\mathsf{OPT}, I) = (|I| - 4/\epsilon)\frac{\epsilon}{2}$.

Similarly to Case 1, we have:

$$\mathrm{cost}(\mathsf{ALG}^*, I) = \left(|I| - \frac{4}{\epsilon}\right) \cdot \frac{\epsilon}{2} + \sum_{i=1}^{|I|} f(p_i^*).$$

If we again consider edges as in the proof of Case 1 above, then it is easy to see, that under ALG^* each edge is traversed exactly once more from left to right than it is from right to left. So the total function cost is the same as in the case above for a total of $|I| - 2/\epsilon - 4/\epsilon$ rounds,

$$\frac{\epsilon^2}{2} \sum_{i=1}^{2/\epsilon-1} i = 1 - \epsilon/2,$$

for the remaining $2/\epsilon$ non-zero cost rounds. This gives in total

$$\text{cost}(\text{ALG}^*, I) = \left(|I| - \frac{4}{\epsilon}\right)\frac{\epsilon}{2} + \left(|I| - \frac{6}{\epsilon}\right)\frac{\epsilon}{2} - \left(|I| - \frac{6}{\epsilon}\right)\frac{\epsilon^2}{4} + 1 - \frac{\epsilon}{2}$$

$$= \left(|I| - \frac{4}{\epsilon}\right)\epsilon - \left(|I| - \frac{4}{\epsilon}\right)\frac{\epsilon^2}{4}.$$

Thus the competitive ratio follows in this case exactly as in Case 1.

Case 3: Termination Condition (iii). Lemma 2 implies $\text{cost}(\text{OPT}', I) = (|I| + q - s) \cdot \epsilon/2 \le |I|\epsilon/2 + 1$. On the other hand we have that

$$\text{cost}(\text{ALG}^*, I) \ge \left(|I| - q - s\right)\epsilon - \left(|I| - q - s\right)\frac{\epsilon^2}{4} \ge |I|\epsilon - |I|\frac{\epsilon^2}{2} - 6,$$

which gives the desired competitive ratio for this case as well because of the first part of Condition (iii).

This concludes the proof of the lemma. □

We finally show the following lemma, which implies the first inequality of (1).

Lemma 4. *For any online algorithm* ALG *and the adversarial instance defined above, we have*

$$\text{cost}(\text{ALG}, I) \ge \text{cost}(\text{ALG}^*, I).$$

Proof. The idea underlying this proof is to divide the complete input sequence into subsequences called *phases*. Phases are contiguous subsequences of maximum size such that only one distinct function occurs in them. We will show that, in each phase, the cost of ALG is at least the cost of ALG*. To do so, consider some phase lasting from Round ℓ to Round r, and assume w.l.o.g. that it only consists of functions of the type f^{dec}. We distinguish two cases as to the reason why the phases has ended.

Case 1: The phase ended because the instance ended due to Condition (i) or (ii). Then by definition the last $4/\epsilon$ functions were either all f^{inc} or all f^{dec} respectively. For each round during the phase in which ALG* had positive function cost, ALG had at least the same function cost. So it suffices to compare the movement cost of ALG* to the movement cost of ALG and the remaining function cost of ALG. So assume that ALG* had movement cost of d^*, and ALG had movement cost of $d < d^*$. It follows that $|p_i - p_i^*| \ge d^* - d$, for each i of the last $4/\epsilon$ rounds during this phase. So during these last $4/\epsilon$ rounds, ALG had an extra function cost of at least $\epsilon \cdot (d^* - d)$ thus giving a total cost of at least $d^* - d$.

Case 2: Otherwise. Note that we have $p_i \le p_i^*$ for all $i \in \{\ell - 1, \ell, \ldots, r - 1\}$ and $p_r = p_r^* + x$ for some $x \ge 0$. This means that, in this phase, ALG has had movement cost of at least x more than ALG*. Furthermore $f^{\text{dec}}(p_i) \ge f^{\text{dec}}(p_i^*)$ for $i \in \{\ell, \ldots r - 1\}$, and $f^{\text{dec}}(p_r) = f^{\text{dec}}(p_r^* + x) = (1 - p_r^* - x)\epsilon = (1 - p_r^*)\epsilon - x\epsilon \ge f^{\text{dec}}(p_r^*) - x$.

This concludes the proof of the lemma. □

We note that we do not use the exact step size of ALG* in the proof of the lemma. Indeed, the lemma would still hold if we redefined ALG* to still move towards the minimizer of each function but with a different step size. Even though the emerging variant of ALG* has a competitive ratio strictly larger than 2, this does not show a general lower bound strictly larger than 2. The reason is that the performance of the variant of ALG* on the emerging adversarial instance is *smaller* than that of ALG*, meaning we would only be able to show a weaker version of Lemma 3.

4   Conclusion

With the present problem being solved in the classical sense, it would be interesting to look at its generalization to higher dimensions. Even for two dimensions, it is an open problem whether $\mathcal{O}(1)$-competitive algorithms exist.

References

1. Andrew, L.H., Barman, S., Ligett, K., Lin, M., Meyerson, A., Roytman, A., Wierman, A.: A tale of two metrics: simultaneous bounds on competitiveness and regret. In: Conference on Learning Theory (COLT), pp. 741–763 (2013)
2. Andrew, L.H., Barman, S., Ligett, K., Lin, M., Meyerson, A., Roytman, A., Wierman, A.: A tale of two metrics: simultaneous bounds on competitiveness and regret. CoRR, abs/1508.03769 (2015)
3. Antoniadis, A., Barcelo, N., Nugent, M., Pruhs, K., Schewior, K., Scquizzato, M.: Chasing convex bodies and functions. In: Kranakis, E., Navarro, G., Chávez, E. (eds.) LATIN 2016. LNCS, vol. 9644, pp. 68–81. Springer, Heidelberg (2016). https://doi.org/10.1007/978-3-662-49529-2_6
4. Bansal, N., Gupta, A., Krishnaswamy, R., Pruhs, K., Schewior, K., Stein, C.: A 2-competitive algorithm for online convex optimization with switching costs. In: Workshop on Approximation Algorithms for Combinatorial Optimization Problems (APPROX), pp. 96–109 (2015)
5. Bartal, Y., Bollobás, B., Mendel, M.: Ramsey-type theorems for metric spaces with applications to online problems. J. Comput. Syst. Sci. 72(5), 890–921 (2006)
6. Bartal, Y., Linial, N., Mendel, M., Naor, A.: On metric Ramsey-type phenomena. In: ACM Symposium on Theory of Computing (STOC), pp. 463–472 (2003)
7. Borodin, A., Linial, N., Saks, M.E.: An optimal on-line algorithm for metrical task system. J. ACM 39(4), 745–763 (1992)
8. Fiat, A., Mendel, M.: Better algorithms for unfair metrical task systems and applications. SIAM J. Comput. 32(6), 1403–1422 (2003)
9. Friedman, J., Linial, N.: On convex body chasing. Discret. Comput. Geom. 9, 293–321 (1993)
10. Karlin, A.R., Manasse, M.S., McGeoch, L.A., Owicki, S.S.: Competitive randomized algorithms for nonuniform problems. Algorithmica 11(6), 542–571 (1994)
11. Karlin, A.R., Manasse, M.S., Rudolph, L., Sleator, D.D.: Competitive snoopy caching. Algorithmica 3, 77–119 (1988)
12. Lin, M., Liu, Z., Wierman, A., Andrew, L.H.: Online algorithms for geographical load balancing. In: International Green Computing Conference (IGCC), pp. 1–10 (2012)

13. Lin, M., Wierman, A., Andrew, L.H., Thereska, E.: Dynamic right-sizing for power-proportional data centers. IEEE/ACM Trans. Netw. **21**(5), 1378–1391 (2013)
14. Liu, Z., Lin, M., Wierman, A., Low, S.H., Andrew, L.H.: Greening geographical load balancing. IEEE/ACM Trans. Netw. **23**(2), 657–671 (2015)
15. Wang, K., Lin, M., Ciucu, F., Wierman, A., Lin, C.: Characterizing the impact of the workload on the value of dynamic resizing in data centers. Perform. Eval. **85–86**, 1–18 (2015)

A k-Median Based Online Algorithm for the Stochastic k-Server Problem

Abhijin Adiga[1], Alexander D. Friedman[2], and Sharath Raghvendra[3(✉)]

[1] Biocomplexity Institute of Virginia Tech, Blacksburg, USA
abhijin@bi.vt.edu
[2] Department of Mathematics, Virginia Tech, Blacksburg, USA
adfriedm@vt.edu
[3] Department of Computer Science, Virginia Tech, Blacksburg, USA
sharathr@vt.edu

Abstract. We consider the k-server problem in the random arrival model and present a simple k-median based algorithm for this problem. Let $\sigma = \langle r_1, \ldots, r_n \rangle$ be the sequence of requests. Our algorithm will batch the requests into $\log n$ groups where the $(i+1)^{st}$ group contains requests $\langle r_{2^i+1}, \ldots, r_{2^{i+1}} \rangle$. To process the requests of group $i+1$, our algorithm will place the k servers at the k-median centers of the first 2^i requests. When a new request of this group arrives, the algorithm will simply assign the server associated with the nearest k-median center to serve it. We show that this simple algorithm, in the random arrival model, has a competitive ratio of at most $O(\alpha)$ and an additive cost of $O(\Delta k \log n)$, where Δ is the diameter of the requests and α is a lower bound on the competitive ratio of any online algorithm in this model.

For our analysis, we use the following fact: In the random arrival model, the expected cost of serving the next request is minimized when servers are located at the k-median of the requests that have not yet arrived (unprocessed requests). But our algorithm instead uses the k-median of the requests seen so far as a proxy. Using existing analysis techniques, we obtain only a large bound on the difference between k-median of the unprocessed requests and that of the processed ones. In particular, in addition to $O(\alpha)$ times the optimal cost, for some $\epsilon > 0$, existing analysis techniques will also give an additive cost of ϵn for serving n requests. We present a new analysis to show that when the number of processed and unprocessed requests are of comparable sizes, the cost of serving n requests incurs only an additive cost of $k\Delta$ (independent of n and significantly better than the previous methods). We then apply this bound for serving each of the $\log n$ groups and obtain an overall bound which is $O(\alpha)$ times the optimal cost with an additive error of $O(k\Delta \log n)$.

Keywords: k-server problem · Random arrival model
k-median clustering

A. D. Friedman and S. Raghvendra were supported under grant NSF-CCF 1464276.
A. Adiga was supported by the DTRA CNIMS Contract HDTRA1-11-D-0016-0001.

R. Solis-Oba and R. Fleischer (Eds.): WAOA 2017, LNCS 10787, pp. 176–189, 2018.
https://doi.org/10.1007/978-3-319-89441-6_14

1   Introduction

In this era of instant gratification, consumers demand access to goods and services in real-time. Business ventures, therefore, have to schedule their delivery of goods and services often without the complete knowledge of the future requests or their order of arrival. To assist with this, we need robust and competitive *online algorithms* that immediately and irrevocably allocate resources to requests in real-time with minimal cost.

In this paper, we consider the celebrated *k-server problem*. Initially, we are given a set of locations $\mathcal{K}$ for the k servers. These are locations in a discrete metric space. Requests are generated in the same metric space by an adversary and are revealed one at a time. When a request $r \in \sigma$ is revealed, we have to immediately move one of the k servers to serve this request and incur a cost equal to the distance between their locations. The objective is to design an algorithm that allocates servers to requests that is competitive with respect to the minimum-cost offline solution.

For any algorithm $\mathcal{A}$, initial configuration of servers $\mathcal{K}$, and a sequence of requests σ, let $w_{\mathcal{A}}(\sigma, \mathcal{K})$ be the cost incurred when the algorithm $\mathcal{A}$ assigns the requests to servers. Let $w_{\text{OPT}}(\sigma, \mathcal{K})$ be the smallest cost solution generated by an offline algorithm that has complete knowledge of the request sequence σ and assigns servers to requests based on their arrival order. We say that $\mathcal{A}$ is α-competitive if, for a constant $\Phi_0 \geq 0$, the cost incurred by our algorithm satisfies,

$$w_{\mathcal{A}}(\sigma, \mathcal{K}) \leq \alpha w_{\text{OPT}}(\sigma, \mathcal{K}) + \Phi_0$$

for any request set and their arrival order.

In the *adversarial model*, there is an adversary who knows the server locations and the assignments made by the algorithm and generates a sequence to maximize α. In the *random arrival model* [1], the adversary chooses the locations of the requests in σ before the algorithm executes but their arrival order is a permutation chosen uniformly at random from the set of all possible permutations of the requests. In practical situations, it may be useful to assume that the requests are arriving i.i.d. from a known or an unknown distribution $\mathcal{D}$. Under these (known and unknown) models, the adversary is weaker than in the random arrival model and therefore, the competitive ratio for the random arrival model is an upper bound on its competitive ratio in the known and the unknown distribution models; see [2] for an algorithm in these models. We refer to the k-server problem under the known and unknown distribution as well as the random arrival model as the *stochastic k-server problem*.

For the stochastic k-server problem, the competitive ratio is expressed with respect to the expected costs. More specifically, $\mathcal{A}$ is α-competitive if,

$$\mathbb{E}[w_{\mathcal{A}}(\sigma, \mathcal{K})] \leq \alpha \mathbb{E}[w_{\text{OPT}}(\sigma, \mathcal{K})] + \Phi_0.$$

Previous Work. The k-server problem is central to the theory of online algorithms. The problem was first posted by Manasse *et al.* [3]. In the adversarial model, the best-known deterministic algorithm for this problem is the $(2k-1)$-competitive work function algorithm [4]. It is known that no deterministic algorithm can achieve a competitive ratio better than k and is conjectured that in fact there is a k-competitive algorithm for this problem. This conjecture is popularly called the *k-server conjecture*.

Bansal *et al.* [5] presented an $O(\log^{O(1)} n \log k)$-competitive randomized algorithm for the k-server problem in the *oblivious adversary model*. This model is similar to the adversarial model; however, the adversary does not have any knowledge of the random choices made by the algorithm. There is also an online algorithm [6] for the closely related online metric matching problem. This algorithm is known to achieve an optimal competitive ratio of $2H_n - 1$ in the random arrival model. To the best of our knowledge, the stochastic k-server problem has not been studied before.

In the stochastic models, we can view any initial set of requests to be a sample that is chosen uniformly at random from σ. In this paper, using this sample, we approximate the k-median of the remaining requests leading to an improved algorithm for the stochastic k-server problem. Independent and uniform random samples have been used before in the context of sub-linear time algorithms for the k-median problem; see [7,8]. In [7], it has been shown that a random sample of size $\tilde{O}(\Delta/\epsilon^2)$ can be used to approximate the k-median within a constant factor of the optimal k-median with an additional additive cost of ϵn. Meyerson *et al.* [8] show that if all the optimal k-median clusters are dense ($\geq \epsilon n/k$), then a very small random sample of size $\approx k/\epsilon$ can be used to approximate the k-median within a constant factor. In this paper, we will present and analyze a k-median based deterministic algorithm for the stochastic k-server problem.

Our Results. First, we present a simple algorithm which we refer to as the zoned algorithm for the k-server problem in the known distribution model, i.e., the request locations are i.i.d from a distribution $\mathcal{D}$ on the discrete metric space. The zoned algorithm will associate one server to each of the centers of the k-median of $\mathcal{D}$ and when a request arrives, the algorithm simply assigns the server associated with the closest k-median center to this request. The cost of serving any request is lower bounded by the average k-median cost of the distribution $\mathcal{D}$. Using triangle inequality, we can bound the cost incurred by zoned algorithm by twice the cost incurred by any optimal online algorithm for this problem.

Next, for the unknown distribution model and the random arrival model, we present an adaptive version of the zoned algorithm. Let $\sigma = \langle r_1, \ldots, r_n \rangle$ be the request sequence. Our algorithm will batch the requests into $\log n$ groups where the $(i+1)^{st}$ group (denoted by σ_{i+1}) contains requests $\langle r_{2^i+1}, \ldots, r_{2^{i+1}} \rangle$. To process requests group $(i+1)$, we apply the zoned algorithm using the k-median centers of the first 2^i requests.

In the random arrival model, the first t requests is a uniformly chosen random subset of size t. Using existing bounds [7], a random subset of size $\tilde{O}(\Delta/\epsilon^2)$ can be used to estimate the average k-median cost within a constant factor with an additional additive cost of ϵ. Unfortunately, despite having a large random subset $(\sigma_0 \cup \sigma_1, \ldots \cup \sigma_{\log n - 1})$ of $n/2$ requests (i.e., $\epsilon \approx \sqrt{\Delta/n}$) to estimate the k-median of $n/2$ requests of $\sigma_{\log n}$, we can only bound the average k-median cost within a constant factor with an additional additive cost of $\epsilon n/2 \approx \sqrt{\Delta n}$. Lower bounds on uniform random sample based estimation of k median seem to suggest that there is very little scope for improving this analysis for small-sized random subsets; see [7] for details on the upper and lower bound. In our case, since the sample size is large $(= n/2)$, we present a different analysis to show that the k-median of this large random subset is a good proxy for the k-median of σ_{i+1}. Using this analysis, we show that the overall cost incurred in serving requests of σ_{i+1} has an additive cost of only $k\Delta$ (independent of n) and the total additive cost over all the $\log n$ groups to be $O(k\Delta \log n)$ leading to the following theorem (in Sect. 4.2):

Theorem 1. *Let σ be a sequence of n requests from a discrete metric space (X, d). For any $\alpha > 1.5$, the expected cost of the adaptive zoned algorithm (Algorithm 2) for serving σ is upper bounded as follows.*

$$\mathbb{E}[w_\mathcal{A}(\sigma, \mathcal{K})] \leq 2\alpha n \, \mathrm{medavg}(\sigma) + \left(\frac{8}{e} \left(\frac{2\alpha + 1}{2\alpha - 3} \right)^2 + 1 \right) k\Delta \log n + \Phi_0.$$

Here $\mathrm{medavg}(\sigma)$ is the average k-median cost of all the requests in σ and is formally defined in Sect. 2.

In the random arrival model, the cost of serving the ith request can be lower bounded by the average k-median cost of the remaining (unprocessed) $n - i + 1$ requests. We show (in Sect. 4.3) that the lower bound on the cost of serving all the requests can still be related to the average k-median cost $\mathrm{medavg}(\sigma)$ within an additive cost of $O(k\Delta \log n)$.

Theorem 2. *Let (X, d) be any discrete metric space and σ be a multi-set of n points from X. Let $\mathcal{A}$ be any online algorithm to serve σ under the random arrival model, with initial configuration of servers $\mathcal{K}$. Let Δ be the diameter of X and for any $0 < \delta < 1$, the expected cost of $\mathcal{A}$ is*

$$\mathbb{E}[w_\mathcal{A}(\sigma, \mathcal{K})] \geq \frac{n-1}{2} \delta \, \mathrm{medavg}(\sigma) - \frac{2\delta}{(1 - \delta)^2} (k + 2)\Delta \log n.$$

Combining the two theorems, it follows that the adaptive zoned algorithm performs within a constant factor of the cost incurred by the best online algorithm along with an additional additive cost of $O(k\Delta \log n)$.

In Sect. 2, we present the basic terminology required for our algorithm and its analysis. In Sect. 3, we present the zoned algorithm for the known distribution model along with its analysis. In Sect. 4, we present the adaptive zoned algorithm in the random arrival model. We present our analysis of the upper bound of the cost in Sect. 4.2 and lower bound of the cost in Sect. 4.3.

2   Preliminaries

Let P be a multi-set of n points in a given discrete metric space (X, d). For any point $p \in P$ and a set $K \subset X$, we define $d(p, K)$ to be the distance of p to its nearest neighbor in K. We define the distance of the set P to the set K denoted by $d(P, K)$.

$$d(P, K) = \sum_{p \in P} d(p, K).$$

The average distance of P from K, denoted as $d_{\text{avg}}(P, K)$ is $d_{\text{avg}}(P, K) = \frac{1}{|P|} \sum_{x \in P} d(x, K)$. We define the k-median of P to be a set of k points $K^* \subseteq X$, given by

$$K^* = \arg\min_{K \subset X, |K| = k} d(P, K).$$

We refer to K^* as the k-median centers of P. The cost of the k-median K^* denoted by $\text{med}(P)$ is $\text{med}(P) = d(P, K^*)$, and the average cost of this k-median, denoted by $\text{medavg}(P)$, is given by

$$\text{medavg}(P) = \frac{\text{med}(P)}{|P|}.$$

In several instances, we denote the k-median of a set A by K^A.

The definition of k-median K^* extends easily to the case where we are given a probability distribution $\mathcal{D}(\cdot)$ on the discrete metric space X:

$$K^* = \arg\min_{K \subset X, |K| = k} \sum_{x \in X} \mathcal{D}(x) d(x, K),$$

and let $\text{medavg}(\mathcal{D}, X) = \sum_{x \in X} \mathcal{D}(x) d(x, K^*)$.

Theorem 3 (Chernoff bounds). *Suppose $X_1, \ldots, X_n$ are independent random binary variables, X denotes their sum, and $\mu = \mathbb{E}[X]$. Then*

$$\mathbb{P}[X \geq (1 + \delta)\mu] \leq e^{-\delta^2 \mu / 3}, \quad 0 < \delta < 1, \tag{1a}$$

$$\mathbb{P}[X \geq (1 + \delta)\mu] \leq e^{-\delta \mu / 3}, \quad 1 < \delta, \tag{1b}$$

$$\mathbb{P}[X \leq (1 - \delta)\mu] \leq e^{-\delta^2 \mu / 2}, \quad 0 < \delta < 1. \tag{1c}$$

3   Zoned Algorithm

We begin by introducing the zoned algorithm for the k-server problem in the known distribution (Algorithm 1) and the random arrival model (Algorithm 2). The core idea behind the algorithm is that the discrete metric space (X, d) can be partitioned into k zones each with a single server. For any request in a given zone, the corresponding server of this zone will serve it. For the known distribution, the partition is induced by the k-median of the distribution and is presented below.

Algorithm 1. Zoned algorithm for a known distribution

Data: Metric space (X, d), probability distribution $\mathcal{D}$, a sequence of requests $\sigma = (r_1, \ldots, r_n)$, k-servers $(\kappa_1, \ldots, \kappa_k)$ and their initial positions

Result: Sequence of servers assigned in an online manner

1 Compute the k-median centers K^* of the distribution $\mathcal{D}$;
2 $\phi : K^* \to (\kappa_1, \ldots, \kappa_k)$, minimum-cost bipartite matching from K^* to the initial locations of k-servers;
3 **for** r *in* σ **do**
4 $\quad$ Find k-median center c with distance $d(r, K^*)$;
5 $\quad$ Move server $\phi(c)$ to r;
6 **end**

The following lemma bounds the cost of any online algorithm with the k-median cost.

Lemma 1. *Given a discrete metric space (X, d), and any request sequence σ of n locations chosen i.i.d. from a known distribution $\mathcal{D}$ on X, the expected cost of any online algorithm $\mathbb{A}$ is*

$$\mathbb{E}[w_{\mathbb{A}}(\sigma, \mathcal{K})] \geq n \cdot \mathrm{medavg}(\mathcal{D}, X).$$

Proof. When a request arrives the algorithm must assign a server to it, so the cost of the algorithm must be at least the distance from the closest server to the request. The expected cost is therefore bounded below by the expected distance of a request to its nearest server. This is minimized if the servers have the configuration of the k-median centers K^* with the expected distance of $\mathrm{medavg}(\mathcal{D}, X)$. The result then follows by linearity of expectation.

Theorem 4. *The zoned algorithm $\mathcal{A}$ has an expected cost that is at most twice the cost incurred by any optimal online algorithm in the known distribution model.*

Proof. Let the initial configuration of the k servers be $\mathcal{K}$. For every request $r \in \sigma$, let $k \in K^*$ be its closest center. The zoned algorithm will move the server $\phi(k)$ to the request point r. By the triangle inequality this distance is less than if we had moved the server to c first, and then to r after. Every request under such a modification therefore incurs at most two costs, movement from c and movement to c. The expected distance of any request to its closest center is $\mathrm{medavg}(\mathcal{D}, X)$, so using the modification, by linearity of expectation, and by Lemma 1:

$$\mathbb{E}[w_{\mathcal{A}}(\sigma, \mathcal{K})] \leq 2n \cdot \mathrm{medavg}(\mathcal{D}, X) + \Phi_0 \leq 2w_{\mathbb{A}}(\sigma, \mathcal{K}) + \Phi_0$$

for n requests, any online algorithm $\mathbb{A}$, and where Φ_0 is the cost of the matching of $\mathcal{K}$ and K^*.

4 Random Arrival Model

4.1 Adaptive Zoned Algorithm

We present and analyze a slightly modified zoned algorithm for the random arrival model. We partition the request sequence σ into $\log n$ groups $\sigma = \sigma_0 ^\frown \sigma_1, \ldots, ^\frown \sigma_{\log n}$ where $|\sigma_0| = 1$ and $|\sigma_i| = 2^{i-1}$. For the request in σ_i we apply the zoned algorithm by using the k-median of the requests in $\sigma_0 \cup \sigma_1, \ldots \cup \sigma_{i-1}$. Note that after serving requests in σ_i, we need to recompute the k-median of all the requests seen so far and move the k-servers to these locations (implicitly through the mapping ϕ). This results in a reconfiguration cost which is bounded by $O(k\Delta \log n)$. The algorithm is presented next.

Algorithm 2. Adaptive zoned algorithm $\mathbb{A}$ that runs in the Random Arrival Model

---

Data: Metric (X, d), a sequence of
requests $\sigma = (r_1, \ldots, r_n)$, k-server $(\kappa_1, \ldots, \kappa_k)$ and their initial positions

Result: Sequence of servers assigned in an online manner

1   $1 \to i$;
2   $1 \to j$;
3   $K^{(1)}$ is the initial positions of the k-servers $(\kappa_1, \ldots, \kappa_k)$;
4   ϕ is the map of locations in $K^{(1)}$ to the servers that are located there;
5   **for** r *in* σ **do**
6   Find center $c \in K^{(j)}$ that minimizes $d(r, K^{(j)})$;
7   Move server $\phi(c)$ to σ;
8   **if** $i = 2^{j-1}$ **then**
9   Compute the k-median centers $K^{(j)}$ on $\{r_1, \ldots, r_i\}$;
10      $\phi : K^{(j)} \to (\kappa_1, \ldots, \kappa_k)$, minimum-cost bipartite matching from $K^{(j)}$ to the servers at their current locations;
11      $j \leftarrow j + 1$;
12   **end**
13     $i \leftarrow i + 1$;
14 **end**

4.2 Upper Bound

We use the following simple lemma in bounding the cost of the adaptive zoned algorithm.

Lemma 2. *Let P be any set of n points and let Q be a subset of P that is chosen uniformly at random from all possible subsets of size t. Then,*

$$\mathbb{E}[d(Q, K^P)] = t \cdot \mathrm{medavg}(P).$$

Proof. Let $\mathcal{S}_t$ be the set of all subsets of P with cardinality exactly t. The number of such subsets is given by $|\mathcal{S}_t| = \binom{n}{t}$. Therefore, the expected value of $d(Q, K^P)$ can be bounded by

$$\mathbb{E}[d(Q, K^P)] = \sum_{Q \in \mathcal{S}_t} \frac{1}{\binom{n}{t}} \sum_{q \in Q} d(q, K^P).$$

Every point $q \in P$ appears in exactly $\binom{n-1}{t-1}$ subsets of $\mathcal{S}_t$. Therefore, we can rewrite the expected value as

$$\mathbb{E}[d(Q, K^P)] = \frac{\binom{n-1}{t-1}}{\binom{n}{t}} \sum_{q \in P} d(q, K^P) = \frac{t}{n} \sum_{q \in P} d(q, K^P) = t \cdot \mathrm{medavg}(P).$$

Lemma 3. *Let P be a set of n points with diameter Δ where n is a power of 2. For a random permutation of the points of P, let A and B correspond to the first $n/2$ and the last $n/2$ points of this permutation. Let K^P be the k-median centers of P. Given that A has been observed and K^A is the k-median centers of A, and for any $\alpha > 1.5$, the expected value of distance $d(B, K^A)$ is at most $\alpha n \, \mathrm{medavg}(P) + \frac{4}{e} \left(\frac{2\alpha+1}{2\alpha-3} \right)^2 k\Delta.$*

Proof. Since A is the first $n/2$ points of a random permutation of P, A can be considered to be a subset chosen uniformly at random from the set of all subsets of size $n/2$. For the optimal k-median centers of A, namely K^A, we will bound the expected cost of $d(B, K^A)$. To help with the analysis, we partition P into k clusters based on assigning each point $p \in P$ to its closest k-median center in K^P (the optimal k-median of P). Let C_j^P denote the jth cluster with $k_j^P \in K^P$ as its median center and let $s_j = |C_j^P|$ be the size of this cluster. Let $A_j = A \cap C_j^P$ and $B_j = B \cap C_j^P$. By using triangle inequality (as in [8]), we can bound the distance of median center of cluster C_j^P, i.e., k_j^P to its closest median center in K^A

$$d(k_j^P, K^A) \le \min_{x \in A_j} \left(d(k_j^P, x) + d(x, K^A) \right) \le \frac{1}{|A_j|} \sum_{x \in A_j} \left(d(k_j^P, x) + d(x, K^A) \right).$$

The last inequality follows from the fact that the minimum of a set of numbers is bounded from above by its average. Using this bound, we can bound the distance from any point $y \in C_j^P$ to its closest median center in K^A by

$$d(y, K^A) \le d(y, k_j^P) + d(k_j^P, K^A) \le d(y, k_j^P) + \frac{1}{|A_j|} \sum_{x \in A_j} (d(k_j^P, x) + d(x, K^A)).$$

Therefore, the total distance of points in B_j to their closest center in K^A is bounded by

$$d(B_j, K^A) \le \sum_{y \in B_j} d(y, k_j^P) + \frac{|B_j|}{|A_j|} \sum_{x \in A_j} \left(d(k_j^P, x) + d(x, K^A) \right). \tag{2}$$

Next, we will bound the expected value of $d(B_j, K^A)$. Note that the expected value $\mathbb{E}[|A_j|] = \frac{s_j}{2} = \mathbb{E}[|B_j|]$. Consider the event $\mathcal{E}$: $(|A_j| > (1-\delta)\mathbb{E}[|A_j|])$, where $0 < \delta < 1$. Applying Chernoff bound (Theorem 3, inequality (1c)), event $\mathcal{E}$ occurs with probability at least $1 - \exp\left(-\frac{\delta^2 s_j}{4}\right)$. When event $\mathcal{E}$ occurs,

$$\frac{|B_j|}{|A_j|} = \frac{s_j - |A_j|}{|A_j|} < \frac{s_j(1 - \frac{1-\delta}{2})}{(1-\delta)\frac{s_j}{2}} = \frac{1+\delta}{1-\delta}.$$

and we can bound $d(B_j, K^A)$ by

$$\sum_{y \in B_j} d(y, k_j^P) + \frac{1+\delta}{1-\delta} \sum_{x \in A_j} \left(d(k_j^P, x) + d(x, K^A)\right).$$

When $\mathcal{E}$ does not occur (which has a probability of at most $\exp\left(-\frac{\delta^2 s_j}{4}\right)$), i.e., $|A_j| \le (1-\delta)\mathbb{E}[|A_j|]$, we use a trivial upper bound of $d(B_j, K^A) \le s_j \Delta$. Applying this, we have the following upper bound for $\mathbb{E}[d(B_j, K^A)]$, where the expectation is over all possible permutations of P.

$$\mathbb{E}[d(B_j, K^A)] = \Pr(\mathcal{E})\mathbb{E}[d(B_j, K^A) \mid \mathcal{E}] + \Pr(\overline{\mathcal{E}})\mathbb{E}[d(B_j, K^A) \mid \overline{\mathcal{E}}]$$

$$< \mathbb{E}\left[\sum_{y \in B_j} d(y, k_j^P) + \frac{1+\delta}{1-\delta} \sum_{x \in A_j} \left(d(k_j^P, x) + d(x, K^A)\right)\right]$$

$$+ \exp\left(-\frac{\delta^2 s_j}{4}\right) s_j \Delta. \tag{3}$$

We set $\delta := 2\sqrt{\frac{\log(s_j/\tau)}{s_j}}$. When $s_j \ge \tau$, we can reduce the last term in (3) to $\exp\left(-\frac{\delta^2 s_j}{4}\right) s_j \Delta \le \tau \Delta$. When $s_j < \tau$, we can simply bound $d(B_j, K^A) \le |B_j|\Delta < \tau \Delta$. Also, we note that $\frac{1+\delta}{1-\delta}$ is a monotonically increasing function of δ for $0 < \delta < 1$. For $x > 0$, $\delta = 2\sqrt{\frac{\log(s_j/\tau)}{s_j}}$ attains the maximum value of $\frac{2}{\sqrt{e\tau}}$ at $s_j = e\tau$. Therefore, $\frac{1+\delta}{1-\delta} \le \frac{\sqrt{e\tau}+2}{\sqrt{e\tau}-2}$. Using these bounds, we rewrite (3) as

$$\mathbb{E}[d(B_j, K^A)] \le \mathbb{E}\left[\sum_{y \in B_j} d(y, k_j^P) + \frac{\sqrt{e\tau}+2}{\sqrt{e\tau}-2} \sum_{x \in A_j} \left(d(k_j^P, x) + d(x, K^A)\right)\right] + \tau \Delta.$$

Summing over all clusters,

$$\mathbb{E}[d(B, K^A)] \le \mathbb{E}\left[\sum_{j=1}^{k}\sum_{y \in B_j} d(y, k_j^P) + \frac{\sqrt{e\tau}+2}{\sqrt{e\tau}-2} \sum_{j=1}^{k}\sum_{x \in A_j} \left(d(k_j^P, x) + d(x, K^A)\right)\right]$$

$$+ k\tau\Delta$$

$$\le \mathbb{E}\left[\sum_{y \in B} d(y, K^P) + \frac{\sqrt{e\tau}+2}{\sqrt{e\tau}-2} \sum_{x \in A} \left(d(K^P, x) + d(x, K^A)\right)\right] + k\tau\Delta.$$

Since $\sum_{x \in A} d(x, K^A) \le \sum_{x \in A} d(x, K^P)$, we have

$$\mathbb{E}[d(B, K^A)] \le \mathbb{E}\left[\sum_{y \in B} d(y, K^P) + 2\frac{\sqrt{e\tau} + 2}{\sqrt{e\tau} - 2} \sum_{x \in A} d(x, K^P)\right] + k\tau\Delta.$$

Since A and B are subsets chosen uniformly at random from all possible subsets of P of size $n/2$, by Lemma 2, we have $\mathbb{E}\left[\sum_{y \in B} d(y, K^P)\right] = |B| \operatorname{medavg}(P)$ and $\mathbb{E}\left[\sum_{y \in A} d(y, K^P)\right] = |A| \operatorname{medavg}(P)$. Therefore,

$$\mathbb{E}[d(B, K^A)] \le |B| \operatorname{medavg}(P) + 2\frac{\sqrt{e\tau} + 2}{\sqrt{e\tau} - 2}|A| \operatorname{medavg}(P) + k\tau\Delta$$

and,

$$\mathbb{E}[d(B, K^A)] \le \left(\frac{\sqrt{e\tau} + 2}{\sqrt{e\tau} - 2} + \frac{1}{2}\right) n \operatorname{medavg}(P) + k\tau\Delta.$$

Setting $\alpha = \frac{\sqrt{e\tau}+2}{\sqrt{e\tau}-2} + \frac{1}{2}$ we arrive at the final expression.

Proof (Proof of Theorem 1). Let $\sigma = \sigma^{(0)} \frown \sigma^{(1)} \frown \sigma^{(2)} \frown \cdots$, where $\frown$ denotes concatenation, so that $|\sigma^{(0)}| = 1$ and $|\sigma^{(i)}| = 2^{i-1}$. Let $T^{(i)} = \sigma^{(0)} \frown \sigma^{(1)} \frown \cdots \frown \sigma^{(i)}$ and $K^{(i)}$ denote the k-median centers of $T^{(i)}$. Here, we use the terms multi-set and sequence interchangeably. By algorithm $\mathcal{A}$, each sequence $\sigma^{(i)}$ is served by $K^{(i-1)}$. Noting that $T^{(i)} = T^{(i-1)} \frown \sigma^{(i)}$ and $|\sigma^{(i)}| = |T^{(i-1)}| = 2^{i-1}$, we apply Lemma 3 with $P = T^{(i)}$, $A = T^{(i-1)}$ and $B = \sigma^{(i)}$. The expected distance $d(\sigma^{(i)}, K^{(i-1)}) \le \alpha |T^{(i)}| \operatorname{medavg}(T^{(i)}) + \frac{4}{e}\left(\frac{2\alpha+1}{2\alpha-3}\right)^2 k\Delta$, and the cost to serve $\sigma^{(i)}$ is at most twice this cost. In addition, for each $\sigma^{(i)}$, the cost to move the severs from their position at the end of $\sigma^{(i-1)}$ to $K^{(i)}$ is at most $k\Delta$. Therefore, $\mathbb{E}[w_{\mathcal{A}}(\sigma, \mathcal{K})]$ can be bounded as follows.

$$\mathbb{E}[w_{\mathcal{A}}(\sigma, \mathcal{K})] \le 2\sum_i \alpha |T^{(i)}| \operatorname{medavg}(T^{(i)}) + \sum_i \left(\frac{8}{e}\left(\frac{2\alpha+1}{2\alpha-3}\right)^2 k\Delta + k\Delta\right) + \Phi_0.$$

Since, $\operatorname{medavg}(T^{(i)}) = \frac{1}{|T^{(i)}|}\sum_{x \in T^{(i)}} d(x, K^{(i)}) \le \frac{1}{|T^{(i)}|}\sum_{x \in T^{(i)}} d(x, K^*)$,

$$\mathbb{E}[w_{\mathcal{A}}(\sigma, \mathcal{K})] \le 2\alpha \sum_i \sum_{x \in T^{(i)}} d(x, K^*) + \left(\frac{8}{e}\left(\frac{2\alpha+1}{2\alpha-3}\right)^2 + 1\right) k\Delta \log n + \Phi_0$$

$$= 2\alpha \sum_{x \in \sigma} d(x, K^*) + \left(\frac{8}{e}\left(\frac{2\alpha+1}{2\alpha-3}\right)^2 + 1\right) k\Delta \log n + \Phi_0$$

$$= 2\alpha n \operatorname{medavg}(\sigma) + \left(\frac{8}{e}\left(\frac{2\alpha+1}{2\alpha-3}\right)^2 + 1\right) k\Delta \log n + \Phi_0.$$

Hence proved.

4.3   Lower Bound

Let S_j denote the set of all possible subsets of σ of size j. The expected value of medavg(A) for a set A of size m chosen uniformly at random from S_m is denoted by medavg$_{\mathbb{E}}(m)$, and defined as

$$\text{medavg}_{\mathbb{E}}(m) = \frac{1}{|S_m|} \sum_{A \in S_m} \text{medavg}(A).$$

We will first prove the following lemmas which bounds the cost of any online algorithm from below by $\sum_{i=1}^{j} \text{medavg}_{\mathbb{E}}(i)$.

Lemma 4. $\mathbb{E}[w_A(\sigma, \mathcal{K})] \geq \sum_{i=1}^{n} \text{medavg}_{\mathbb{E}}(i)$.

Proof. Supposing $i - 1$ requests have been served, let A denote the subset of S containing the points yet to be served, i.e., $|A| = n - i + 1$. Let K be the current configuration of the servers. Since every element of A has the same probability of being picked, for any k-set K, the expected distance of the ith request from K is

$$\frac{1}{|A|} \sum_{x \in A} d(x, K) \geq \text{medavg}(A).$$

Since every set $A \in S_{n-i+1}$ is equiprobable, the cost of serving the ith request is medavg$_{\mathbb{E}}(n - i + 1)$ and the result follows.

Lemma 5. *For any* $m > m'$, medavg$_{\mathbb{E}}(m) \geq$ medavg$_{\mathbb{E}}(m')$.

Proof. Let $A \in S_m$ and $B \subset A$ of size m'. Noting that every element of A occurs in exactly $\binom{m-1}{m'-1}$ subsets B, we have

$$\text{medavg}(A) = \frac{1}{|A|} \sum_{x \in A} d(x, K^A) = \frac{1}{m} \frac{1}{\binom{m-1}{m'-1}} \sum_{B \subset A, \, |B| = m'} d(x, K^A)$$

$$\geq \frac{1}{m} \frac{1}{\binom{m-1}{m'-1}} \sum_{B \subset A, \, |B| = m'} d(x, K^B)$$

$$= \frac{m'}{m} \frac{1}{\binom{m-1}{m'-1}} \sum_{B \subset A, \, |B| = m'} \text{medavg}(B).$$

Since, every B is a subset of exactly $\binom{n-m'}{m-m'}$ sets of S_m,

$$\sum_{A \in S_m} \text{medavg}(A) \geq \binom{n-m'}{m-m'} \frac{m'}{m} \frac{1}{\binom{m-1}{m'-1}} \sum_{B \in S_{m'}} \text{medavg}(B)$$

$$= \frac{|S_m|}{|S_{m'}|} \sum_{B \in S_{m'}} \text{medavg}(B).$$

Lemma 6. $\mathbb{E}[w_A(\sigma, \mathcal{K})] \geq \lceil \frac{n}{2} \rceil \, \mathrm{medavg}_{\mathbb{E}}(\lceil \frac{n}{2} \rceil)$.

Proof. From Lemma 5, for any $m > \frac{n}{2}$, $\mathrm{medavg}_{\mathbb{E}}(m) \geq \mathrm{medavg}_{\mathbb{E}}(\lceil \frac{n}{2} \rceil)$. From Lemma 4,

$$\mathbb{E}[w_A(\sigma, \mathcal{K})] \geq \sum_{i=1}^{n} \mathrm{medavg}_{\mathbb{E}}(i) \geq \sum_{i=1}^{\lceil \frac{n}{2} \rceil} \mathrm{medavg}_{\mathbb{E}}(n - i + 1) \geq \left\lceil \frac{n}{2} \right\rceil \mathrm{medavg}_{\mathbb{E}}\left(\left\lceil \frac{n}{2} \right\rceil\right).$$

Following Meyerson *et al.* [8], for any k-set K of σ, let $\beta(K, b) = (B_1, B_2, \ldots, B_b)$ denote the partition of σ induced by K as follows: We order points of σ by their distance to the closest point in K (low to high), and divide this sequence into b bins each containing equal number of points $\frac{n}{b}$. Henceforth, let $b = \frac{n(1-\delta)^2}{4(k+2)\log n}$ for $0 < \delta < 1$.

Lemma 7. *Let A be any random subset of σ of size $\lceil \frac{n}{2} \rceil$. With probability at least $1 - \frac{1}{n}$, every bin of $\beta(K^A, b)$ has at most $\frac{\delta n}{2b}$ points from A.*

Proof. We will show that with high probability the above statement is satisfied for all k-sets K. Since K^A is one among these sets, the bound follows. Let K be any k-set and $\beta(K, b)$ be the induced partition. The expected number of points of A belonging to each bin is $\frac{n}{2b}$. Using Chernoff bound (1c), for $0 < \delta < 1$, the probability that a bin has at most $\frac{\delta n}{2b}$ points is at most $e^{-\frac{n(1-\delta)^2}{4b}}$. Since there are at most n^k sets K and b bins, applying union bound the proof follows.

Proof (Proof of Theorem 2). Let $A \subset \sigma$ of size $\frac{n}{2}$, and $\beta(K^A, b)$ be the partition of σ induced by K^A. Let $d_j^{\min}$ and $d_j^{\max}$ denote the minimum and maximum distances respectively of points from bin B_j. From Lemma 7, with probability at least $1 - \frac{1}{n}$, every bin of $\beta(K^A, b)$ has at most $\frac{\delta n}{2b}$ points of A. Therefore,

$$\sum_{x \in A} d(x, K^A) \geq \sum_{j=1}^{b} \frac{\delta n}{2b} d_j^{\min} \geq \frac{\delta}{2} \sum_{j=1}^{b-1} \frac{n}{b} d_j^{\max}.$$

The last inequality follows from the fact that $d_j^{\min} \geq d_{j-1}^{\max}$. For each j, since $d_j^{\max}$ is the maximum assigned distance, $\frac{n}{b} d_j^{\max} \geq \sum_{y \in B_j} d(y, K^A)$. Finally, we note that $d_b^{\max} \leq \Delta$, the diameter. Combining,

$$\sum_{x \in A} d(x, K^A) \geq \frac{\delta}{2} \sum_{j=1}^{b-1} \sum_{y \in B_j} d(y, K^A) + \frac{\delta}{2} \sum_{y \in B_b} d(y, K^A) - \frac{\delta n}{2b} \Delta$$

$$= \frac{\delta}{2} \sum_{j=1}^{b} \sum_{y \in B_j} d(y, K^A) - \frac{\delta n}{2b} \Delta$$

$$= \frac{\delta}{2} \sum_{y \in \sigma} d(y, K^A) - \frac{\delta n}{2b} \Delta.$$

Therefore, with probability $(1 - 1/n)$

$$\text{medavg}(A) \geq \frac{\delta}{n} \sum_{y \in \sigma} d(y, K^A) - \frac{\delta \Delta}{b} \geq \delta \, \text{medavg}(\sigma) - \frac{\delta \Delta}{b}.$$

Hence,

$$\text{medavg}_\mathbb{E}\left(\left\lceil \frac{n}{2} \right\rceil\right) \geq \left(1 - \frac{1}{n}\right)\left(\frac{\delta}{n} \sum_{y \in \sigma} d(y, K^A) - \frac{\delta \Delta}{b}\right) + \frac{1}{n} \cdot 0$$

From Lemma 6 and the choice of $b = \frac{n(1-\delta)^2}{4(k+2)\log n}$,

$$\begin{aligned}
\mathbb{E}[w_A(\sigma, \mathcal{K})] &\geq \left\lceil \frac{n}{2} \right\rceil \text{medavg}_\mathbb{E}\left(\left\lceil \frac{n}{2} \right\rceil\right) \geq \left\lceil \frac{n}{2} \right\rceil \left(1 - \frac{1}{n}\right)\left(\delta \, \text{medavg}(\sigma) - \frac{\delta \Delta}{b}\right) \\
&\geq \frac{n-1}{2}\left(\delta \, \text{medavg}(\sigma) - \frac{\delta \Delta}{b}\right) \\
&= \frac{n-1}{2}\delta \, \text{medavg}(\sigma) - \frac{2\delta}{(1-\delta)^2}(k+2)\Delta \log n.
\end{aligned}$$

Hence proved.

5   Conclusion

In this paper, we presented and analyzed a simple k-median based algorithm for the stochastic k-server problem. Our result is based on proving a new and sharper approximation bound of the k-median of a large random subset of a point set with respect to the k-median of entire point set.

In the random arrival model, the cost of serving the next request is lower bounded by the average k-median cost of the requests that have not yet been processed. Clearly, the k servers cannot always be in the optimal k-median configuration even for the best online algorithm. Therefore, it is conceivable that one can prove a stronger lower bound. In particular, can we prove a stronger lower or upper bound and reduce the additive error from $O(k\Delta \log n)$ to $O(k\Delta)$ (completely independent of n)?

References

1. Mahdian, M., Yan, Q.: Online bipartite matching with random arrivals: an approach based on strongly factor-revealing LPs. In: Proceedings of the 43rd Annual ACM Symposium on Theory of Computing, STOC 2011, pp. 597–606 (2011)
2. Karande, C., Mehta, A., Tripathi, P.: Online bipartite matching with unknown distributions. In: Proceedings of the Forty-Third Annual ACM Symposium on Theory of Computing, STOC 2011, pp. 587–596. ACM, New York (2011)
3. Manasse, M.S., McGeoch, L.A., Sleator, D.D.: Competitive algorithms for server problems. J. Algorithms 11(2), 208–230 (1990)

4. Koutsoupias, E., Papadimitriou, C.H.: On the k-server conjecture. J. ACM **42**(5), 971–983 (1995)
5. Bansal, N., Buchbinder, N., Madry, A., Naor, J.: A polylogarithmic-competitive algorithm for the k-server problem. In: Proceedings of the IEEE 52nd Annual Symposium on Foundations of Computer Science (FOCS), pp. 267–276, October 2011
6. Raghvendra, S.: A Robust and optimal online algorithm for minimum metric bipartite matching. In: Approximation, Randomization, and Combinatorial Optimization. Algorithms and Techniques, APPROX/RANDOM 2016, vol. 60, pp. 18:1–18:16 (2016)
7. Czumaj, A., Sohler, C.: Sublinear-time approximation algorithms for clustering via random sampling. Random Struct. Algorithms **30**(1–2), 226–256 (2007)
8. Meyerson, A., O'callaghan, L., Plotkin, S.: A k-median algorithm with running time independent of data size. Mach. Learn. **56**(1–3), 61–87 (2004)

On Packet Scheduling with Adversarial Jamming and Speedup

Martin Böhm[1], Łukasz Jeż[2], Jiří Sgall[1], and Pavel Veselý[1(✉)]

[1] Computer Science Institute of Charles University, Prague, Czech Republic
{bohm,sgall,vesely}@iuuk.mff.cuni.cz
[2] Institute of Computer Science, University of Wrocław, Wrocław, Poland
lje@cs.uni.wroc.pl

Abstract. In Packet Scheduling with Adversarial Jamming packets of arbitrary sizes arrive over time to be transmitted over a channel in which instantaneous jamming errors occur at times chosen by the adversary and not known to the algorithm. The transmission taking place at the time of jamming is corrupt, and the algorithm learns this fact immediately. An online algorithm maximizes the total size of packets it successfully transmits and the goal is to develop an algorithm with the lowest possible asymptotic competitive ratio, where the additive constant may depend on packet sizes.

Our main contribution is a universal algorithm that works for any speedup and packet sizes and, unlike previous algorithms for the problem, it does not need to know these properties in advance. We show that this algorithm guarantees 1-competitiveness with speedup 4, making it the first known algorithm to maintain 1-competitiveness with a moderate speedup in the general setting of arbitrary packet sizes. We also prove a lower bound of $\phi + 1 \approx 2.618$ on the speedup of any 1-competitive deterministic algorithm, showing that our algorithm is close to the optimum. Additionally, we formulate a general framework for analyzing our algorithm locally and use it to show upper bounds on its competitive ratio for speedups in $[1, 4)$ and for several special cases, recovering some previously known results, each of which had a dedicated proof. In particular, our algorithm is 3-competitive without speedup, matching the algorithm and the lower bound of Jurdzinski *et al.* [7]. We use this framework also for the case of divisible packet sizes in which the size of a packet divides the size of any larger packet, to show that a slight modification of our algorithm is 1-competitive with speedup 2 and it achieves the optimal competitive ratio of 2 without speedup, again matching the algorithm and the lower bound of [7].

M. Böhm, J. Sgall and P. Veselý—Supported by the project 17-09142S of GA ČR and by the project 634217 of GAUK.

L. Jeż—Supported by Polish National Science Center grant 2016/21/D/ST6/02402.

R. Solis-Oba and R. Fleischer (Eds.): WAOA 2017, LNCS 10787, pp. 190–206, 2018.
https://doi.org/10.1007/978-3-319-89441-6_15

1   Introduction

We study an online packet scheduling model recently introduced by Anta *et al.* [1] and extended by Jurdzinski *et al.* [7]. In our model, packets of arbitrary sizes arrive over time. The algorithm schedules any packet of its choice at any time, but cannot interrupt its subsequent transmission; in the scheduling jargon, there is a single machine and no preemptions. There are, however, instantaneous *jamming errors* or *faults* at times chosen by the adversary, which are not known to the algorithm. A transmission taking place at the time of jamming is corrupt, and the algorithm learns this fact immediately. The packet whose transmission failed can be retransmitted immediately or at any later time.

The objective is to maximize the total size of packets successfully transmitted. In particular, the goal is to develop an online algorithm with the lowest possible competitive ratio. We focus on algorithms with resource augmentation, namely on online algorithms that transmit packets $s \geq 1$ times faster than the offline optimum solution they are compared against; such algorithm is often said to be speed-s, running at speed s, or having a speedup of s. We note that this notion has been introduced for a similar job scheduling problem [8], in which speedup is required for constant ratio. In contrast, our problem allows constant ratio with no speedup, so we consider the competitive ratio as a function of it. This deviates from previous work, which focused on the case with no speedup or on the speedup sufficient for ratio 1, ignoring intermediate cases.

The distinguishing feature of the model is that the number of different packet sizes is a constant and that these packet sizes are considered to be constants. Thus, the additive term in the definition of the competitive ratio may depend on the number and values of the packet sizes. (It is easy to observe that, at speed 1, without such additive term, it is impossible to attain constant competitive ratio that does not depend on packet sizes, even for only two sizes.)

1.1   Previous Results

The model was introduced by Anta *et al.* [1], who resolved it for two packet sizes: If $\gamma > 1$ denotes the ratio of the two sizes, then the optimum ratio for deterministic algorithms is $(\gamma + \lfloor \gamma \rfloor)/\lfloor \gamma \rfloor$, which is always in the range $[2, 3)$. Anta *et al.* [1] note that their lower bound strategy applies to randomized algorithms as well, but their argument only applies to the adaptive adversary model. To our best knowledge, randomized algorithms for the more common oblivious adversary model were never considered in this setting. Anta *et al.* [1] demonstrate the necessity of instantaneous error feedback by proving that discovering errors upon completed transmission rules out constant competitive ratio. They also provide improved results for a stochastic online setting.

These results were extended by Jurdzinski *et al.* [7], who proved that the optimum ratio for the case of multiple (though fixed) packet sizes is given by the same formula for the two packet sizes which maximize it. Moreover, Jurdzinski *et al.* [7] gave further results for *divisible* packet sizes, i.e., instances in which every packet size divides every larger packet size, for which the optimum ratio is 2. Namely, they generalized the 2-competitive algorithm to a setting with f independent transmission channels (or machines in scheduling parlance); jamming errors on each channel are independent, and a packet can only be transmitted on one channel at a time. Furthermore, for the single channel divisible case, they proved that speed 2 is sufficient for 1-competitiveness in the resource augmentation setting.

1.2   Our Results

Our main new results concern the analysis of the non-divisible case with speedup, where we show that speed 4 is sufficient for a competitive ratio of 1 (Sect. 4). However, our major contribution is a uniform algorithm that works well in every setting, together with a uniform analysis framework (Sect. 3). This contrasts with the results of Jurdzinski et al. [7], where each upper bound was attained by a dedicated algorithm with independently crafted analysis. Moreover, unlike those algorithms, which all require prior knowledge of all possible packet sizes and speedup, in the sense that there are different algorithms for speeds 1 and 2 (only divisible case), with none for intermediate speeds, our algorithm is oblivious in that it requires no such information. Furthermore, our algorithm is more appealing as it is significantly simpler and "work-conserving" or "busy", i.e., transmitting some packet whenever there is one pending, which is desirable in practice. See Sect. 2 for the description of our algorithm.

To recover the 1-competitiveness for divisible instances at speed 2, we have to modify our algorithm slightly, as otherwise we can guarantee this ratio only at speed 2.5. This is to be expected, as divisible instances are a very special case. On the other hand, we prove that our original algorithm is 1-competitive on far broader class of "well-separated" instances at sufficient speed: If the ratio between two successive packet sizes (in their sorted list) is no smaller than $\alpha \geq 1$, our algorithm is 1-competitive if its speed is at least S_α, which is a non-increasing function of α such that $S_1 = 4$ and $\lim_{\alpha \to \infty} S_\alpha = 2$; see Sect. 3.2 for the precise definition of S_α.

In Sect. 5 we complement these results with two lower bounds on speed that is necessary to achieve 1-competitiveness. The first one proves that even for two divisible packet sizes, speed 2 is required to attain 1-competitiveness, establishing optimality of our modified algorithm and that of Jurdzinski et al. [7] for the divisible case. The second lower bound strengthens the previous construction by showing that for non-divisible instances with more packet sizes, speed $\phi + 1 \approx 2.618$ is required for 1-competitiveness. Both results hold even if all packets are released together.

The analyses of our algorithm are mostly tight as (a) on general instances, the algorithm is no better than $(1 + 2/s)$-competitive for $s < 2$ and no better than $4/s$-competitive for $s \in [2, 4)$, (b) on divisible instances, it is no better than $4/3$-competitive for $s < 2.5$, and (c) it is at least 2-competitive for $s < 2$, even for two divisible packet sizes. Due to space limitations, we omit some proofs and the analysis of the algorithm for the divisible case.

1.3   Related Results

We first mention a few results concerning the model with adversarial jamming. Anta *et al.* [3] consider essentially the same problem but with a different objective: minimizing the number and/or the total size of pending (i.e., not yet transmitted) packets. They focus on multiple channels (or machines, as therein the problem is formulated as job scheduling), and investigate what speedup is necessary and sufficient for competitiveness. Note that for 1-competitiveness, minimizing the total size of pending packets is equivalent to our objective. In Sects. 4, 5, 6 and 7 they obtain a tight bound of 2 on speedup necessary for 1-competitiveness for two packet sizes, although in their lower bound of 2 they do not release all packets together, unlike we do in our lower bound constructions. In Sect. 8 they claim to have a 1-competitive algorithm with speedup $7/2$, but the proof is incorrect; even an example showing that our algorithm is no better than $4/s$-competitive for $s \in [2, 4)$ works the same way for their algorithm.

Georgiou *et al.* [6] consider the same problem without any speedup, restricted to jobs of unit size, under various efficiency and correctness measures, again focusing on multiple machines/channels. This article is motivated by distributed computation, and thus distinguishes between different information models and ways in which the machines communicate. In yet another work, Anta *et al.* [2] consider latency on top of the previous two objectives, and study the competitiveness of popular scheduling algorithms for all these measures. Garncarek *et al.* [5] consider "synchronized" parallel channels that all suffer errors at the same time. This work distinguishes between "regular" jamming errors and "crashes", which also cause the algorithm's state to reset, losing any information stored about the past events and proves that for two packet sizes the optimum ratio tends to $4/3$ for the former and to $\phi = (\sqrt{5} + 1)/2 \approx 1.618$ for the latter as the number f of channels tends to infinity.

As mentioned before, Kalyanasundaram and Pruhs [8] introduced resource augmentation. Among other results they proved that a constant competitive ratio is possible with a constant speedup for a preemptive variant of real-time scheduling where each job has a release time, deadline, processing time and a weight and the objective is to maximize the weight of jobs completed by their deadlines on a single machine. Subsequently resource augmentation was applied in various scenarios. Of the most relevant for us are those that considered algorithms with speedup that are 1-competitive, i.e., as good as the optimum. We mention two models that still contain interesting open problems.

For real-time scheduling, Phillips *et al.* [12] considered the underloaded case in which there exists a schedule that completes all the jobs. It is easy to see that on a single machine, the earliest deadline first (EDF) algorithm is then an optimal online algorithm. Phillips *et al.* [12] proved that EDF on m machines is 1-competitive with speedup $2 - 1/m$. (Here the weights are not relevant.) Intriguingly, finding a 1-competitive algorithm with minimal speedup is wide open: It is known that speedup at least $6/5$ is necessary, it has been conjectured that speedup $e/(e - 1)$ is sufficient, but the best upper bound proven is $2 - 2/(m + 1)$ from [11]. See Schewior [13] for more on this problem.

Later these results were extended to real-time scheduling of overloaded systems, where for uniform density (i.e., weight equal to processing time) Lam *et al.* [10] have shown that a variant of EDF with admission control is 1-competitive with speedup 2 on a single machine and with speedup 3 on more machines. For non-uniform densities, the necessary speedup is a constant if each job is tight (its deadline equals its release time plus its processing time) [9]. Without this restriction it is no longer constant, depending on the ratio ξ of the maximum and minimum weight. It is known that it is at least $\Omega(\log \log \xi)$ and at most $O(\log \xi)$ [4, 10]; as far as we are aware, closing this gap is still an open problem.

2   Algorithms, Preliminaries, Notations

The general idea of the algorithm is that after each error, we start by transmitting packets of small sizes, only increasing the size of packets after a sufficiently long period of uninterrupted transmissions. It turns out that the right tradeoff is to transmit a packet only if it would have been transmitted successfully if started just after the last error. It is also crucial that the initial packet after each error has the right size, namely to ignore small packet sizes if the total size of not transmitted packets of those sizes is small compared to a larger packet that can be transmitted. This guarantees that if no error occurs, all currently pending packets with size equal to or larger than the size of the initial packet are eventually transmitted before the algorithm starts a smaller packet.

We start by some notations. We assume there are k distinct non-zero packet sizes denoted by ℓ_i and ordered so that $\ell_1 < \cdots < \ell_k$. For convenience, we define $\ell_0 = 0$. We say that the packet sizes are divisible if ℓ_i divides ℓ_{i+1} for all $i = 1, \ldots, k-1$. For a packet p, let $\ell(p)$ denote the size of p. For a set of packets A, let $\ell(A)$ denote the total size of all the packets in A.

During the run of an algorithm, at time t, a packet is pending if it is released before or at t, not completed before or at t and not started before t and still running. At time t, if no packet is running, the algorithm may start any pending packet. As a convention of our model, if a fault (jamming error) happens at time t and this is the completion time of a previously scheduled packet, this packet is considered completed. Also, at the fault time, the algorithm may start a new packet.

Let $L_{ALG}(i, Y)$ denote the total size of packets of size ℓ_i completed by an algorithm ALG during a time interval Y. Similarly, $L_{ALG}(\geq i, Y)$ (resp. $L_{ALG}(< i, Y)$) denotes the total size of packets of size at least ℓ_i (resp. less than ℓ_i) completed by an algorithm ALG during a time interval Y; formally we define $L_{ALG}(\geq i, Y) = \sum_{j=i}^{k} L_{ALG}(j, Y)$ and $L_{ALG}(< i, Y) = \sum_{j=1}^{i-1} L_{ALG}(j, Y)$. We use the notation $L_{ALG}(Y)$ with a single parameter to denote the size $L_{ALG}(\geq 1, Y)$ of packets of all sizes completed by ALG during Y and the notation L_{ALG} without parameters to denote the size of all packets of all sizes completed by ALG at any time.

By convention, the schedule starts at time 0 and ends at time T, which is a part of the instance unknown to an online algorithm until it is reached. (This is similar to the times of jamming errors, one can also alternatively say that after T the errors are so frequent that no packet is completed.) Algorithm ALG is called R-competitive, if there exists a constant C, possibly dependent on k and $\ell_1, \ldots, \ell_k$, such that for any instance and its optimal schedule OPT we have $L_{OPT}((0,T]) \leq R \cdot L_{ALG}((0,T]) + C$. We remark that in our analyses we show only a crude bounds on C.

We denote the algorithm ALG with speedup $s \geq 1$ by $ALG(s)$. The meaning is that in $ALG(s)$, packets of size L need time L/s to process. In the resource-augmentation variant, we are mainly interested in finding the smallest s such that $ALG(s)$ is 1-competitive, compared to OPT that runs at speed 1.

2.1   Algorithms $MAIN$ and DIV

We divide the run of the algorithm into phases. Each phase starts by an invocation of Step (2) and ends either by a fault or by an invocation of Step (4). The periods of idle time, i.e., waiting in Step (1), do not belong to any phase. Throughout the algorithm, t denotes the current time. The time t_B denotes the last invocation of Step (2), which is the start of the current phase. We set $\mathrm{rel}(t) = s(t - t_B)$. Since the algorithm does not insert unnecessary idle time, $\mathrm{rel}(t)$ denotes the amount of transmitted packets in the current phase. (Note that we use $\mathrm{rel}(t)$ only when there is no packet running at time t, so there is no partially executed packet.) Thus $\mathrm{rel}(t)$ can be thought as a measure of time relative to the start of the current phase (scaled by the speed of the algorithm). Note also that the algorithm can evaluate $\mathrm{rel}(t)$ without knowing the speedup, as it can simply observe the total size of the transmitted packets.

We now give a common description of the two algorithms, $MAIN$ for general packet sizes and DIV for divisible sizes. They differ only in Step (3), where DIV enforces an additional *divisibility condition*. Let $P^{<i}$ denote the set of pending packets of sizes $\ell_1, \ldots, \ell_{i-1}$ at any given time.

(1) If no packet is available, stay idle until the next release time.
(2) Let i be the maximal $i \leq k$ such that there is a pending packet of size ℓ_i and $\ell(P^{<i}) < \ell_i$. Schedule a packet of size ℓ_i.
(3) Choose the maximum i such that (i) there is a pending packet of size ℓ_i, (ii) $\ell_i \leq \mathrm{rel}(t)$ and (iii)*in case of DIV*,ℓ_i *divides rel* (t).
    Schedule a packet of size ℓ_i. Repeat Step (3) as long as such i exists.
(4) If no packet satisfies the condition in Step (3), go to Step 1.

We first note that the algorithm is well-defined, i.e., that it is always able to choose a packet in Step (2) if it has any packets pending, and that if it succeeds in sending it, the length of thus started phase can be related to the total size of the packets completed in it.

Lemma 1. *Let ALG be either $MAIN$ or DIV. In Step (2), ALG always chooses some packet if it has any pending. Moreover, if ALG completes the first packet in the phase, then $L_{ALG(s)}((t_B, t_E]) > s(t_E - t_B)/2$, where t_B denotes the start of the phase and t_E its end (by a fault or Step (4)).*

Proof. For the first property, note that the packet of the smallest size is eligible. For the second property, note that there is no idle time in the phase, and that only the last packet chosen by ALG in the phase may not complete due to a jam. By the condition in Step (3), the size of this jammed packet is no larger than the total size of all the packets ALG previously completed in this phase, which yields the bound.

The following lemma shows a crucial property of the algorithm, namely that if packets of size ℓ_i are pending, the algorithm schedules packets of size at least ℓ_i most of the time. Its proof also explains the reasons behind our choice of the first packet in a phase in Step (2) of the algorithm. We give the statement and proof of the lemma for $MAIN$. For DIV, the first part of the lemma still holds, while the divisibility condition makes the proof a bit more tedious; on the other hand divisibility of packet sizes ensures that the bound in the second part of the lemma can be improved.

Lemma 2. *Let u be a start of a phase in $MAIN(s)$ and $t = u + \ell_i/s$.*

(i) *If a packet of size ℓ_i is pending at time u and no fault occurs in (u, t), then the phase does not end before t.*

(ii) *Suppose that $v > u$ is such that any time in $[u, v)$ a packet of size ℓ_i is pending and no fault occurs. Then the phase does not end in (u, v) and $L_{MAIN(s)}(< i, (u, v]) < \ell_i + \ell_{i-1}$. (Recall that $\ell_0 = 0$.)*

Proof

(i) Suppose for a contradiction that the phase started at u ends at time $t' < t$. We have $\mathrm{rel}(t') < \mathrm{rel}(t) = \ell_i$. Let ℓ_j be the smallest packet size among the packets pending at t'. As there is no fault, the reason for a new phase has to be that $\mathrm{rel}(t') < \ell_j$, and thus Step (3) did not choose a packet to be scheduled. Also note that any packet scheduled before t' was completed. This implies, first, that there is a pending packet of size ℓ_i, as there was one at time u and there was insufficient time to complete it, so j is well-defined and $j \le i$. Second, all packets of sizes smaller than ℓ_j pending at u were completed before t', so their total size is at most $\mathrm{rel}(t') < \ell_j$. However, this contradicts the fact that the phase started by a packet smaller than ℓ_j at time u, as a pending packet of the smallest size equal to or larger than ℓ_j satisfied the condition in Step (2) at time u. (Note that it is possible that no packet of size ℓ_j is pending at u; however, at least a packet of size ℓ_i is pending.)

(ii) By (i), the phase that started at u does not end before time t if no fault happens. A packet of size ℓ_i is always pending by the assumption of the lemma, and it is always a valid choice of a packet in Step (3) from time t on. Thus, the phase that started at u does not end in (u, v), and moreover only packets of sizes at least ℓ_i are started in $[t, v)$. It follows that packets of sizes smaller than ℓ_i are started only before time t and their total size is thus less than $\mathrm{rel}(t) + \ell_{i-1} = \ell_i + \ell_{i-1}$. The lemma follows.

3   Local Analysis and Results

In this section we formulate a general method for analyzing our algorithms by comparing locally within each phase the size of "large" packets completed by the algorithm and by the adversary. This method simplifies a complicated induction used in [7]; we easily obtain the same upper bounds of 2 and 3 on competitiveness for divisible and unrestricted packet sizes respectively at no speedup, as well as several new results for the non-divisible cases. In Sect. 4, we use a more complex charging scheme to obtain our main result.

For the analysis, let ALG denote one of the algorithms $MAIN$ and DIV, and let $s \geq 1$ be the speedup. We fix an instance and its schedules for $ALG(s)$ and OPT.

3.1   Critical Times and Master Theorem

The common scheme is the following. We carefully define a sequence of critical times $C_k \leq C_{k-1} \leq \cdots \leq C_1 \leq C_0$, where $C_0 = T$ is the end of the schedule, satisfying two properties: (1) the algorithm has completed almost all pending packets of size ℓ_i released before C_i and (2) in $(C_i, C_{i-1}]$, a packet of size ℓ_i is always pending. Properties (1) and (2) allow us to relate $L_{OPT}(i, (0, C_i])$ and $L_{OPT}(\geq i, (C_i, C_{i-1}])$ respectively to their "ALG counterparts". As each packet completed by OPT belongs to exactly one of these sets, summing the bounds gives the desired results. These two facts together imply R-competitiveness of the algorithm for appropriate R and speed s.

We first define the notion of i-good times so that they satisfy property (1), and then choose the critical times among their suprema so that those satisfy property (2) as well.

Definition 1. *Let ALG be one of the algorithms $MAIN$ and DIV, and let $s \geq 1$ be the speedup. For $i = 1, \ldots k$, time t is called i-good if one of the following conditions holds:*

(i) At time t, algorithm $ALG(s)$ starts a new phase by scheduling a packet of size larger than ℓ_i, or
(ii) at time t, no packet of size ℓ_i is pending for $ALG(s)$, or
(iii) $t = 0$.

We define critical times $C_0, C_1, \ldots, C_k$ iteratively as follows:

- $C_0 = T$, *i.e., it is the end of the schedule.*
- *For $i = 1, \ldots, k$, C_i is the supremum of i-good times t such that $t \leq C_{i-1}$.*

Note that all C_i's are defined and $C_i \geq 0$, as time $t = 0$ is i-good for all i.

The choice of C_i implies that each C_i is of one of the three types (the types are not disjoint):

- C_i is i-good and a phase starts at C_i (this includes $C_i = 0$),
- C_i is i-good and $C_i = C_{i-1}$, or

– there exists a packet of size ℓ_i pending at C_i, however, any such packet was
  released at C_i.

If the first two options do not apply, then the last one is the only remaining
possibility (as otherwise some time in a non-empty interval $(C_i, C_{i-1}]$ would be
i-good); in this case, C_i is not i-good but only the supremum of i-good times.

First we bound the total size of packets of size ℓ_i completed before C_i; the
proof actually only uses the fact that each C_i is the supremum of i-good times
and justifies the definition above.

Lemma 3. *Let ALG be one of the algorithms MAIN and DIV, and let $s \geq 1$
be the speedup. Then, for any i, $L_{OPT}(i, (0, C_i]) \leq L_{ALG(s)}(i, (0, C_i]) + \ell_k$.*

Proof. If C_i is i-good and satisfies condition (i) in Definition 1, then by the
description of Step (2) of the algorithms, the total size of pending packets of
size ℓ_i is less than the size of the scheduled packet, which is at most ℓ_k and the
lemma follows.

In all the remaining cases it holds that $ALG(s)$ has completed all the jobs of
size ℓ_i released before C_i. Thus the inequality holds trivially even without the
additive term.

Our remaining goal is to bound $L_{OPT}(\geq i, (C_i, C_{i-1}])$. We divide $(C_i, C_{i-1}]$
into i-segments by the faults. We prove the bounds separately for each i-segment.
One important fact is that for the first i-segment we use only a loose bound,
as we can use the additive constant. The critical part is then the bound for
i-segments started by a fault, this part determines the competitive ratio and
is different for each case. We summarize the general method by the following
definition and master theorem.

Definition 2. *The interval $(u, v]$ is called an initial i-segment if $u = C_i$ and v
is either C_{i-1} or the first time of a fault after u, whichever comes first.*

*The interval $(u, v]$ is called a proper i-segment if $u \in (C_i, C_{i-1})$ is a time of
a fault and v is either C_{i-1} or the first time of a fault after u, whichever comes
first.*

Note that there is no i-segment if $C_{i-1} = C_i$.

Theorem 1 (Master Theorem). *Let ALG denote one of the algorithms DIV
and MAIN, and let $s \geq 1$ be the speedup. Suppose that $R \geq 1$ satisfies that for
each $i = 1, \ldots, k$ and each proper i-segment $(u, v]$ with $v - u \geq \ell_i$ we have*

$$(R - 1)L_{ALG(s)}((u, v]) + L_{ALG(s)}(\geq i, (u, v]) \geq L_{OPT}(\geq i, (u, v]). \qquad (1)$$

Suppose furthermore that for the initial i-segment $(u, v]$ we have

$$L_{ALG(s)}(\geq i, (u, v]) > s(v - u) - 4\ell_k. \qquad (2)$$

Then $ALG(s)$ is R-competitive.

Proof. First note that for a proper i-segment $(u, v]$, u is a fault time. Thus if $v - u < \ell_i$, then $L_{OPT}(\geq i, (u, v]) = 0$ and (1) is trivial. It follows that (1) holds even without the assumption $v - u \geq \ell_i$.

Now consider an initial i-segment $(u, v]$. We have $L_{OPT}(\geq i, (u, v]) \leq \ell_k + v - u$, as at most a single packet started before u can be completed. Combining this with (2) and using $s \geq 1$, we get $L_{ALG(s)}(\geq i, (u, v]) > s(v - u) - 4\ell_k \geq v - u - 4\ell_k \geq L_{OPT}(\geq i, (u, v]) - 5\ell_k$.

Summing this with (1) for all proper i-intervals and using $R \geq 1$ we get

$$(R - 1)\, L_{ALG(s)}((C_i, C_{i-1}]) + L_{ALG(s)}(\geq i, (C_i, C_{i-1}]) + 5\ell_k$$
$$\geq\ L_{OPT}(\geq i, (C_i, C_{i-1}]). \tag{3}$$

Note that for $C_i = C_{i-1}$, the Eq. (3) holds trivially.

To complete the proof of the theorem, note that each completed job in the optimum contributes to exactly one among the $2k$ terms $L_{OPT}(\geq i, (C_i, C_{i-1}])$ and $L_{OPT}(i, (0, C_i])$; similarly for $L_{ALG(s)}$. Thus by summing both (3) and Lemma 3 for all $i = 1, \ldots, k$ we obtain

$$L_{OPT} = \sum_{i=1}^{k} L_{OPT}\left(\geq i, (C_i, C_{i-1}]\right) + \sum_{i=1}^{k} L_{OPT}(i, (0, C_i])$$

$$\leq \sum_{i=1}^{k}(R - 1)L_{ALG(s)}((C_i, C_{i-1}]) + \sum_{i=1}^{k}\left(L_{ALG(s)}(\geq i, (C_i, C_{i-1}]) + 5\ell_k\right)$$

$$+ \sum_{i=1}^{k}\left(L_{ALG(s)}(i, (0, C_i]) + \ell_k\right)$$

$$\leq (R - 1)L_{ALG(s)} + L_{ALG(s)} + 6k\ell_k\ =\ R \cdot L_{ALG(s)} + 6k\ell_k.$$

The theorem follows.

3.2   Analysis of *MAIN*

The first part of the following lemma implies the condition (2) for the initial i-segments in all cases. The second part of the lemma is the base of the analysis of a proper i-segment, which is different in each situation.

Lemma 4

(i) *If $(u, v]$ is an initial i-segment then $L_{MAIN(s)}(\geq i, (u, v]) > s(v - u) - 4\ell_k$.*

(ii) *If $(u, v]$ is a proper i-segment and $s(v - u) \geq \ell_i$ then $L_{MAIN(s)}((u, v]) > s(v - u)/2$ and $L_{MAIN(s)}(\geq i, (u, v]) > s(v - u)/2 - \ell_i - \ell_{i-1}$. (Recall that $\ell_0 = 0$.)*

Proof

(i) Let $(u, v]$ be an initial i-segment. If the phase that starts at u or contains u ends before v, let u' be its end; otherwise let $u' = u$. We have $u' \leq u + \ell_i/s$, as otherwise any packet of size ℓ_i, pending throughout the i-segment by definition, would be an eligible choice in Step (3) of the algorithm, and the phase would not end before v. Using Lemma 2(ii), we have $L_{MAIN(s)}(< i, (u', v]) < \ell_i + \ell_{i-1} < 2\ell_k$. Since at most one packet at the end of the segment is unfinished, we have $L_{MAIN(s)}(\geq i, (u, v]) \geq L_{MAIN(s)}(\geq i, (u', v]) > s(v - u') - 3\ell_k > s(v - u) - 4\ell_k$.

(ii) Let $(u, v]$ be a proper i-segment. Thus u is a start of a phase that contains at least the whole interval $(u, v]$ by Lemma 2(ii). By the definition of C_i, u is not i-good, so the phase starts by a packet of size at most ℓ_i. If $v - u \geq \ell_i$ then the first packet finishes (as $s \geq 1$) and thus $L_{MAIN(s)}((u, v]) > s(v - u)/2$ by Lemma 1. The total size of completed packets smaller than ℓ_i is at most $\ell_i + \ell_{i-1}$ by Lemma 2 (ii), and thus $L_{MAIN(s)}(\geq i, (u, v]) > s(v - u)/2 - \ell_i - \ell_{i-1}$.

The next theorem gives a tradeoff of the competitive ratio of $MAIN(s)$ and the speedup s using our local analysis. In Theorem 5 we give a non-local analysis of $MAIN(s)$ and show that the algorithm is 1-competitive for $s \geq 4$. However, for $s = 1$ our local analysis is tight as already the lower bound from [1] shows that no algorithm is better than 3-competitive (for packet sizes 1 and $2 - \varepsilon$).

Theorem 2. *For $s \in [1, 4)$, $MAIN(s)$ is R_s-competitive where $R_s = 1 + 2/s$.*

Proof. Lemma 4(i) implies the condition (2) for the initial i-segments. We now prove (1) for any proper i-segment $(u, v]$ with $v - u \geq \ell_i$ and appropriate R. The bound then follows by the Master Theorem.

Since there is a fault at time u, we have $L_{OPT}(\geq i, (u, v]) \leq v - u$. For $s \in [1, 4)$, by Lemma 4(ii) we get $(2/s) \cdot L_{MAIN(s)}((u, v]) > v - u \geq L_{OPT}(\geq i, (u, v])$, which implies (1) for $R = 1 + 2/s$.

We can obtain better bounds on the speedup necessary for 1-competitiveness if the packet sizes are sufficiently different. Namely, we call the packet sizes $\ell_1, \ldots, \ell_k$ α-separated if $\ell_i \geq \alpha \ell_{i-1}$ holds for $i = 2, \ldots, k$. We start by analyzing $MAIN(s)$ for the case of divisible packet sizes, which is a special case of 2-separated packet sizes. The proofs of the following two theorems are omitted due to space limitations.

Theorem 3. *If the packet sizes are divisible, then $MAIN(s)$ is 1-competitive for $s \geq 2.5$.*

Next, we show that for α-separated packet sizes, $MAIN(S_\alpha)$ is 1-competitive for the following S_α. We define

$$\alpha_0 = \frac{1}{2} + \frac{1}{6}\sqrt{33} \approx 1.46, \text{ which is the positive root of } 3\alpha^2 - 3\alpha - 2.$$

$$\alpha_1 = \frac{3 + \sqrt{17}}{4} \approx 1.78, \text{ which is the positive root of } 2\alpha^2 - 3\alpha - 1.$$

$$S_\alpha = \begin{cases} \dfrac{4\alpha + 2}{\alpha^2} & \text{for } \alpha \in [1, \alpha_0], \\ 3 + \dfrac{1}{\alpha} & \text{for } \alpha \in [\alpha_0, \alpha_1), \text{ and} \\ 2 + \dfrac{2}{\alpha} & \text{for } \alpha \geq \alpha_1. \end{cases}$$

Note that S_α is decreasing in α, with a single discontinuity at α_1. We have $S_1 = 6$ and $S_2 = 3$, i.e., $MAIN(3)$ is 1-competitive for 2-separated packet sizes, which includes the case of divisible packet sizes studied above. The limit of S_α for $\alpha \to +\infty$ is 2. For $\alpha < (1 + \sqrt{3})/2 \approx 1.366$, we get $S_\alpha > 4$, while Theorem 5 shows that $MAIN(s)$ is 1-competitive for $s \geq 4$; the weaker result of Theorem 4 below reflect the limits of the local analysis.

Theorem 4. *Let $\alpha > 1$. If the packet sizes are α-separated, then $MAIN(s)$ is 1-competitive for any $s \geq S_\alpha$.*

4   *MAIN* with Speed 4

In this section we prove that speed 4 is sufficient for *MAIN* to be 1-competitive. We remark that speed 4 is also necessary for our algorithm.

Intuition. For $s \geq 4$ we have that if at the start of a phase $MAIN(s)$ has a packet of size ℓ_i pending and the phase has length at least ℓ_i, then $MAIN(s)$ completes a packet of size at least ℓ_i. To show this, assume that the phase starts at time t. Then the first packet p of size at least ℓ_i is started before time $t + 2\ell_i/s$ by Lemma 2(ii) and by the condition in Step (3) it has size smaller than $2\ell_i$. Thus it completes before time $t + 4\ell_i/s \leq t + \ell_i$, which is before the end of the phase. This property does not hold for $s < 4$. It is important in our proof, as it shows that if the optimal schedule completes a job of some size, and such job is pending for $MAIN(s)$, then $MAIN(s)$ completes a job of the same size or larger. However, this is not sufficient to complete the proof by a local (phase-by-phase) analysis similar to the previous section, as the next example shows.

Assume that at the beginning, we release N packets of size 1, N packets of size $1.5 - 2\varepsilon$, one packet of size $3 - 2\varepsilon$ and a sufficient number of packets of size $1 - \varepsilon$, for a small $\varepsilon > 0$. Our focus is on packets of size at least 1. Supposing $s = 4$ we have the following phases:

- First, there are N phases of length 1. In each phase the optimum completes a packet of size 1, while among packets of size at least 1, $MAIN(s)$ completes a packet of size $1.5 - 2\varepsilon$, as it starts packets of sizes $1 - \varepsilon$, $1 - \varepsilon$, $1.5 - 2\varepsilon$, $3 - 2\varepsilon$, in this order, and the last packet is jammed.
- Then there are N phases of length $1.5 - 2\varepsilon$ where the optimum completes a packet of size $1.5 - 2\varepsilon$ while among packets of size at least 1, the algorithm completes only a single packet of size 1, as it starts packets of sizes $1 - \varepsilon$, $1 - \varepsilon$, 1, $3 - 2\varepsilon$, in this order. The last packet is jammed, since for $s = 4$ the phase must have length at least $1.5 - \varepsilon$ to complete it.

In phases of the second type, the algorithm does not complete more (in terms of total size) packets of size at least 1 than the optimum. Nevertheless, in our example, packets of size $1.5 - 2\varepsilon$ were already finished by the algorithm, and this is a general rule. The novelty in our proof is a complex charging argument that exploits such subtle interaction between phases.

Outline of the proof. We define critical times C_i' similarly as before, but without the condition that they should be ordered, i.e., C_i' is simply the supremum of i-good times; consequently, either $C_i' \leq C_{i-1}'$ or $C_i' > C_{i-1}'$ may hold. Our plan is to define, for each i, charges between packets of size ℓ_i up to time C_i', since nearly each packet of size ℓ_i completed by the adversary before C_i' must be completed by the algorithm before C_i'. Each such charge from an adversary's packet leads to an algorithm's packet of the same size and each algorithm's packet receives at most one such charge; we thus call these charges 1-to-1 charges.

We want to charge the remaining adversary's packets of size ℓ_i (scheduled after C_i') to some algorithm's packets in the same phase. For technical reasons it is actually more convenient to partition the schedule not into phases but into blocks inbetween successive faults. A block can contain several phases of the algorithm separated by an execution of Step (4); however, in the most important and tight part of the analysis the blocks coincide with phases.

After the critical time C_i', packets of size ℓ_i are always pending for the algorithm, and thus (as we observed above) the algorithm schedules a packet of size at least ℓ_i in blocks in which the adversary completes a packet of size ℓ_i. In the crucial lemma of the proof, based on these observations and their refinements, we show that we can assign remaining adversary's packets (i.e., those without 1-to-1 charges) to algorithm's packets in the same block so that for each algorithm's packet q the total size of packets assigned to it is at most $\ell(q)$. However, we cannot use this assignment directly to charge the remaining packets, as some of the algorithm's big packets may receive 1-to-1 charges, and in this case the analysis needs to handle the interaction of different blocks. This very issue can be seen even in our introductory example.

To deal with this, we process blocks in the order of time from the beginning to the end of the schedule, simultaneously completing the charging to the packets in the current block of $MAIN(s)$ schedule and possibly modifying ADV (the schedule of the adversary) in the future blocks. In fact, in the assignment described above, we include not only the packets in ADV without 1-to-1 charge but also packets in ADV with a 1-to-1 charge to a later block. After creating the assignment, if we have a packet q in ALG that receives a 1-to-1 charge from a packet p in a later block of ADV, we remove p from ADV in that later block and replace it there by the packets assigned to q (that are guaranteed to be of smaller total size than p). After these swaps, the 1-to-1 charges together with the assignment form a valid charging that charges the remaining not swapped packets in ADV in this block together with the removed packets from the later blocks in ADV to the packets of $MAIN(s)$ in the current block. This charging is now independent of the other blocks, so we can continue with the next block.

Furthermore, to analyze blocks in which the algorithm does not complete any packet, we shift the targets of the 1-to-1 charges backward in time by accounting for the first few packets of each size completed by the adversary at the beginning of the schedule and after C_i' in the additive constant of the competitive ratio. The formal proof of the theorem is omitted due to space limitations.

Theorem 5. *$MAIN(s)$ is 1-competitive for $s \geq 4$.*

5   Lower Bounds

In this section we study lower bounds on the speed necessary to achieve 1-competitiveness. Due to space limitations we only sketch the proofs.

We start with a lower bound of 2 which holds even for the divisible case. It follows that our algorithm DIV and the algorithm in Jurdzinski et al. [7] are optimal. Note that this lower bound follows from results of Anta et al. [3] by a similar construction, although the packets in their construction are not released together.

Theorem 6. *There is no 1-competitive deterministic online algorithm running with speed $s < 2$, even if packets have sizes only 1 and ℓ for $\ell > 2s/(2 - s)$ and all of them are released at time 0.*

Proof sketch. For a contradiction, consider an algorithm ALG running with speed $s < 2$ that is claimed to be 1-competitive with additive constant A where A may depend on ℓ. At time 0 the adversary releases $N_1 > A/\ell$ packets of size ℓ and N_0 packets of size 1 where $N_0 \gg N_1$ is large enough. These are all packets in the instance.

The adversary's strategy works by blocks where a block is a time interval between two faults, and the first block begins at time 0. The adversary ensures that in each such block ALG completes no packet of size ℓ and moreover ADV either completes a packet of size ℓ or completes more packets of size 1 than ALG.

Let t be the time of the last fault; initially $t = 0$. Let $\tau \geq t$ be the time when ALG starts the first packet ℓ after t; we set $\tau = \infty$ if it does not happen. Note that we use here that ALG is deterministic. In a block beginning at time t, the adversary proceeds according to the first case below that applies.

(D1) If ADV has less than $2\ell/s$ pending packets of size 1, then the end of the schedule is at t.

(D2) If ADV has all packets ℓ completed, then it stops the current process and issues a fault at times $t+1, t+2, \ldots$ Between each two consecutive faults after t it completes one packet of size 1 and it continues issuing faults until it has no pending packet of size 1. Then there is the end of the schedule. Clearly, ALG may complete only packets of size 1 after t as $\ell > 2s/(2 - s) > s$.

(D3) If $\tau \geq t + \ell/s - 2$, then the next fault is at time $t + \ell$. In the current block, the adversary completes a packet ℓ. ALG completes at most $s\ell$ packets of size 1 and then it possibly starts a packet ℓ at τ (if $\tau < t + \ell$) which is jammed, since it would be completed at $\tau + \ell/s$ and a calculation shows that $\tau + \ell/s > t + \ell$; here we use $s < 2$ and $\ell > 2s/(2 - s)$.

(D4) Otherwise, if $\tau < t + \ell/s - 2$, then the next fault is at time $\tau + \ell/s - \varepsilon$ for a small enough $\varepsilon > 0$. In the current block, ADV completes as many packets of size 1 as it can, that is $\lfloor \tau + \ell/s - \varepsilon - t \rfloor$; note that by Case (D1), ADV has enough pending packets of size 1. The algorithm does not complete the packet of size ℓ started at τ, because it would finish at $\tau + \ell/s$.

First notice that the process above ends, since in each block the adversary completes a packet. We now show $L_{ADV} > L_{ALG} + A$ which contradicts the claimed 1-competitiveness of ALG.

If the adversary's strategy ends in Case (D2), then ADV has all packets ℓ completed and then it schedules all 1's, thus $L_{ADV} = N_1 \cdot \ell + N_0 > A + N_0$. However, ALG does not complete any packet ℓ and hence $L_{ALG} \leq N_0$ which concludes this case.

Otherwise, the adversary's strategy ends in Case (D1). A calculation shows that in a block created in Case (D4), ADV finishes more 1's than ALG, thus the gain of ADV is at least 1 in this block. In each block created in Case (D3) ALG completes at most $s\ell$ packets of size 1 and ADV finishes one packet of size ℓ, thus there at most N_1 such blocks and in each such block $(u, v]$ it holds $L_{ADV}((u, v]) - L_{ALG}((u, v]) \geq (1 - s) \cdot \ell$. Since ADV has only a few packets of size 1 pending at the end, we get that the total gain of ADV in blocks created in Case (D4) is much bigger than the gain of ALG in blocks created in Case (D3) for a large enough N_0. Hence $L_{ADV} > L_{ALG} + A$, showing that the algorithm cannot be 1-competitive in this case.

Our main lower bound of $\phi + 1 = \phi^2 \approx 2.618$ generalizes the construction of Theorem 6 for more packet sizes, which are no longer divisible. Still, we make no use of release times.

Theorem 7. *There is no 1-competitive deterministic online algorithm running with speed $s < \phi + 1$, even if all packets are released at time 0.*

Proof sketch. The adversary chooses $\varepsilon > 0$ small enough and $k \in \mathbb{N}$ large enough. For convenience, the smallest size in the instance is ε instead of 1. There will be $k + 1$ packet sizes in the instance, namely $\ell_0 = \varepsilon$, and $\ell_i = \phi^{i-1}$ for $i = 1, \ldots, k$.

Suppose for a contradiction that there is an algorithm ALG running with speed $s < \phi + 1$ that is claimed to be 1-competitive with additive constant A where A may depend on ℓ_i's, in particular on ε and k. The adversary issues N_i packets of size ℓ_i at time 0, for $i = 0, \ldots, k$; N_i's are chosen large enough so that $N_0 \gg N_1 \gg \cdots \gg N_k$. These are all packets in the instance.

The adversary's strategy works again by blocks. In a block beginning at t, the adversary proceeds as follows, using the first case below that applies:

- If the number of packets of size ε pending for ADV is small, then the schedule ends. We show that in blocks in which ADV schedules packets of size ε, its total size of completed packets is larger than that of ALG. Then for N_0 large enough, the same holds overall, i.e., total size of completed packets in the schedule of ADV is much larger than that in the schedule of ALG.
- If ADV has no pending packet of size ℓ_i for some $i \geq 1$, then it stops the current process and continues with a different strategy in which it tries to schedule all packets of size less than ℓ_i, while preventing the algorithm from completing any packet of size at least ℓ_i. If ALG starts a packet of size at least ℓ_i too early in a block, ADV schedules packets of size ε and gains in terms of total size. Otherwise, ADV is able to complete any packet of size less than ℓ_i. After completing all packets of size less than ℓ_i, ADV schedules all

packets of size ε. We show that ALG has completed overall only a bounded number of packets of size at least ℓ_i, in particular, it scheduled much less than N_i packets of size ℓ_i. Then using $N_i \gg N_j$ for $i < j$ we get that ADV schedules much more than ALG in terms of total size of completed packets.

- If ALG starts the first packet p of size ℓ_1 before time $t + (\phi - 1)/s$, or it starts the first packet q of size at least ℓ_2 before time $t + 1/s$ (note that only one of the cases occurs), ADV issues a fault so that p, resp. q is not completed by ALG. ADV schedules as many packets of size ε as possible and a calculation shows that ADV gains in terms of total size of completed packets for a small enough ε. This case ensures that a packet of size ℓ_1, or a packet of size at least ℓ_2 is not started too early.
- If ALG starts the first packet p of size at least ℓ_{i+1} before the first packet of size ℓ_i for an integer $i \geq 1$ as small as possible, then ADV issues a fault so that p is not finished by ALG. ADV schedules a packet of size ℓ_i and we show that the block is sufficiently long to complete it. Intuitively, this case means that ALG skips a packet of size ℓ_i and schedules a larger packet earlier. We remark that this case is tight only for $i \leq 2$.
- Otherwise, the next fault occurs at $t + \ell_k$ and ADV schedules a packet of size ℓ_k in this block. We show that ALG cannot complete a packet of size ℓ_k in this block, provided that none of the previous cases applied. This case is tight; here we use the definition of the sequence ℓ_i.

The process above eventually ends, since in each block ADV schedules a packet. In both cases of the end, it holds $L_{ADV} > L_{ALG} + A$, contradicting the claimed 1-competitiveness.

References

1. Fernández Anta, A., Georgiou, C., Kowalski, D.R., Widmer, J., Zavou, E.: Measuring the impact of adversarial errors on packet scheduling strategies. J. Sched. **19**(2), 135–152 (2016)
2. Fernández Anta, A., Georgiou, C., Kowalski, D.R., Zavou, E.: Competitive analysis of task scheduling algorithms on a fault-prone machine and the impact of resource augmentation. In: Pop, F., Potop-Butucaru, M. (eds.) ARMS-CC 2015. LNCS, vol. 9438, pp. 1–16. Springer, Cham (2015). https://doi.org/10.1007/978-3-319-28448-4_1
3. Fernández Anta, A., Georgiou, C., Kowalski, D.R., Zavou, E.: Online parallel scheduling of non-uniform tasks. Theor. Comput. Sci. **590**, 129–146 (2015)
4. Chrobak, M., Epstein, L., Noga, J., Sgall, J., van Stee, R., Tichý, T., Vakhania, N.: Preemptive scheduling in overloaded systems. J. Comput. Syst. Sci. **67**, 183–197 (2003)
5. Garncarek, P., Jurdziński, T., Loryś, K.: Fault-tolerant online packet scheduling on parallel channels. In: 2017 IEEE International Parallel and Distributed Processing Symposium (IPDPS), pp. 347–356, May 2017
6. Georgiou, C., Kowalski, D.R.: On the competitiveness of scheduling dynamically injected tasks on processes prone to crashes and restarts. J. Parallel Distrib. Comput. **84**, 94–107 (2015)

7. Jurdzinski, T., Kowalski, D.R., Lorys, K.: Online packet scheduling under adversarial jamming. In: Bampis, E., Svensson, O. (eds.) WAOA 2014. LNCS, vol. 8952, pp. 193–206. Springer, Cham (2015). https://doi.org/10.1007/978-3-319-18263-6_17
8. Kalyanasundaram, B., Pruhs, K.: Speed is as powerful as clairvoyance. J. ACM **47**(4), 617–643 (2000)
9. Koo, C.-Y., Lam, T.W., Ngan, T.-W., Sadakane, K., To, K.-K.: On-line scheduling with tight deadlines. Theor. Comput. Sci. **295**, 251–261 (2003)
10. Lam, T.W., Ngan, T.-W., To, T.-T.: Performance guarantee for EDF under overload. J. Algorithms **52**, 193–206 (2004)
11. Lam, T.W., To, K.-K.: Trade-offs between speed and processor in hard-deadline scheduling. In Proceedings of the 10th Annual ACM-SIAM Symposium on Discrete Algorithms (SODA), pp. 623–632. ACM/SIAM (1999)
12. Phillips, C.A., Stein, C., Torng, E., Wein, J.: Optimal time-critical scheduling via resource augmentation. Algorithmica **32**, 163–200 (2002)
13. Schewior, K.: Deadline scheduling and convex-body chasing. Ph.D dissertation, TU Berlin (2016)

Non-clairvoyant Scheduling to Minimize Max Flow Time on a Machine with Setup Times

Alexander Mäcker[(✉)], Manuel Malatyali, Friedhelm Meyer auf der Heide, and Sören Riechers

Computer Science Department, Heinz Nixdorf Institute, Paderborn University, Paderborn, Germany
{alexander.maecker,manuel.malatyali,fmadh,soeren.riechers}@upb.de

Abstract. Consider a problem in which n jobs that are classified into k types arrive over time at their release times and are to be scheduled on a single machine so as to minimize the maximum flow time. The machine requires a setup taking s time units whenever it switches from processing jobs of one type to jobs of a different type. We consider the problem as an online problem where each job is only known to the scheduler as soon as it arrives and where the processing time of a job only becomes known upon its completion (non-clairvoyance).

We are interested in the potential of simple "greedy-like" algorithms. We analyze a modification of the FIFO strategy and show its competitiveness to be $\Theta(\sqrt{n})$, which is optimal for the considered class of algorithms. For $k = 2$ types it achieves a constant competitiveness. Our main insight is obtained by an analysis of the smoothed competitiveness. If processing times p_j are independently perturbed to $\hat{p}_j = (1 + X_j)p_j$, we obtain a competitiveness of $O(\sigma^{-2} \log^2 n)$ when X_j is drawn from a uniform or a (truncated) normal distribution with standard deviation σ. The result proves that bad instances are fragile and "practically" one might expect a much better performance than given by the $\Omega(\sqrt{n})$-bound.

Keywords: Scheduling · Flow time · Setup times · Smoothed analysis

1 Introduction

Consider a scheduling problem in which there is a single machine for processing jobs arriving over time. Each job is defined by a release time at which it arrives, a size describing the time required to process it and it belongs to exactly one of k types. Whenever the machine switches from processing jobs of one type to jobs of a different type, a setup needs to take place for the reconfiguration of the machine. During a setup the machine cannot process any workload. A natural

This work was partially supported by the German Research Foundation (DFG) within the Collaborative Research Centre "On-The-Fly Computing" (SFB 901).

© Springer International Publishing AG, part of Springer Nature 2018
R. Solis-Oba and R. Fleischer (Eds.): WAOA 2017, LNCS 10787, pp. 207–222, 2018.
https://doi.org/10.1007/978-3-319-89441-6_16

objective in such a model is the minimization of the time each job remains in the system. This objective was introduced in [7] as maximum flow time, defined as the maximum time a job spends in the system, that is, the time between the arrival of a job and its completion. It describes the quality of service as, for example, perceived by users and aims at schedules being responsive to each job. In settings in which user interaction is present or processing times may depend on other further inputs not known upon the arrival of a job, it is also natural to assume the concept of non-clairvoyance, as introduced in [20].

There are several applications for a model with job types and setup times mentioned in the literature. Examples are settings in which a server has to answer requests of different types, depending on the data to be loaded into memory and to be accessed by the server [9,22]; manufacturing systems, in which machines need to be reconfigured or cleaned during the manufacturing of different customer orders [21]; or a setting where an intersection of two streets is equipped with traffic lights and where setup times describe the time drivers need for start-up once they see green light [12].

In this paper, we study the potential of "greedy-like" online algorithms in terms of their (smoothed) competitiveness. The formal model and notions are given in Sect. 2. In Sect. 4 we analyze the competitiveness of "greedy-like" algorithms and show matching upper and lower bounds of $\Theta(\sqrt{n})$, where the bound is achieved by a simple modification of the *First In First Out* (FIFO) strategy. For the special case of $k = 2$ types, the competitiveness improves to $O(1)$. Our main result is an analysis of the smoothed competitiveness of this algorithm in Sect. 5, which is shown to be $O(\sigma^{-2} \log^2 n)$ where σ denotes the standard deviation of the underlying smoothing distribution. It shows worst case instances to be fragile against random noise and that, except on some pathological instances, the algorithm achieves a much better performance than suggested by the worst case bound on the competitiveness.

Proofs omitted throughout the paper can be found in the full version [17].

2   Model and Notions

We consider a scheduling problem in which n jobs, partitioned into k types, are to be scheduled on a single machine. Each job j has a size (processing time) $p_j \in \mathbb{R}_{\geq 1}$, a release time $r_j \in \mathbb{R}_{\geq 0}$ and a parameter τ_j defining the type it belongs to. The machine can process at most one job at a time. Whenever it switches from processing jobs of one type to a different one and before the first job is processed, a setup taking constant s time units needs to take place during which the machine cannot be used for processing. The goal is to compute a non-preemptive schedule, in which each job runs to completion without interruption once it is started, that minimizes the maximum flow time $F := \max_{1 \leq j \leq n} F_j$ where F_j is the amount of time job j spends in the system. That is, a job j arriving at r_j, started in a schedule at t_j and completing its processing at $c_j := t_j + p_j$ has a flow time $F_j := c_j - r_j$.

Given a schedule, a *batch* is a sequence of jobs, all of a common type τ, that are processed in a contiguous interval without any intermediate setup.

For a batch B, we use $\tau(B)$ to denote the common type τ of B's jobs and $w(B) := \sum_{j \in B} p_j$ to denote its workload. We refer to setup times and idle times as *overhead* and overhead is associated to a job j if it directly precedes j in the schedule. For an interval $I = [a, b]$ we also use $l(I) := a$ and $r(I) := b$ and $w(I) := \sum_{j:r_j \in I} p_j$ to denote the workload released in interval I.

Non-clairvoyant Greedy-Like Online Algorithms. We consider our problem in an *online* setting where jobs arrive over time at their release times and are not known to the scheduler in advance. Upon arrival the scheduler gets to know a job together with its type but does not learn about its processing time, which is only known upon its completion (*non-clairvoyance*) [20]. We are interested in the potential of conceptually simple and efficient *greedy-like* algorithms. For classical combinatorial offline problems, the concept of greedy-like algorithms has been formalized by Borodin et al. in [8] by *priority algorithms*. We adopt this concept and for our online problem we define greedy-like algorithms to work as follows: When a job completes (and when the first job arrives), the algorithm determines a total ordering of *all possible* jobs without looking at the actual instance. It then chooses (among already arrived yet unscheduled jobs) the next job to be scheduled by looking at the instance and selecting the job coming first according to this ordering.

Quality Measure. To analyze the quality of online algorithms, we facilitate competitive analysis. It compares solutions of the online algorithm to solutions of an optimal offline algorithm which knows the complete instance in advance. Precisely, an algorithm ALG is called *c-competitive* if, on any instance $\mathcal{I}$, $F(\mathcal{I}) \leq c \cdot F^*(\mathcal{I})$, where $F(\mathcal{I})$ and $F^*(\mathcal{I})$ denote the flow time of ALG and an optimal (clairvoyant) offline solution on instance $\mathcal{I}$, respectively.

Although competitive analysis is the standard measure for analyzing online algorithms, it is often criticized to be overly pessimistic. That is, a single or a few pathological and very rarely occurring instances can significantly degrade the quality with respect to this measure. To overcome this, various alternative measures have been proposed in the past (e.g. see [11, 14, 15]). One approach introduced in [6] is *smoothed competitiveness*. Here the idea is to slightly perturb instances dictated by an adversary by some random noise and then analyze the expected competitiveness, where expectation is taken with respect to the random perturbation. Formally, if input instance $\mathcal{I}$ is smoothed according to some probability function f and if we use $N(\mathcal{I})$ to denote the instances that can be obtained by smoothing $\mathcal{I}$ according to f, the smoothed competitiveness c_{smooth} is defined as $c_{\text{smooth}} := \sup_{\mathcal{I}} \mathbb{E}_{\hat{\mathcal{I}} \xleftarrow{f} N(\mathcal{I})} \left[\frac{F(\hat{\mathcal{I}})}{F^*(\hat{\mathcal{I}})} \right]$. We will smoothen instances by randomly perturbing processing times. We assume the adversary to be *oblivious* with respect to perturbations. That is, the adversary constructs the instance based on the knowledge of the algorithm and f (so that $\mathcal{I}$ is defined at the beginning and is not a random variable).

3 Related Work

The problem supposedly closest related to ours is presented in a paper by Divakaran and Saks [10]. They consider the clairvoyant variant in which the processing time of each job is known upon arrival. Additionally, they allow the setup time to be dependent on the type. For this problem, they provide an $O(1)$-competitive online algorithm. Also, they show that the offline problem is NP-hard in case the number k of types is part of the input. In case k is a constant, it was known before that the problem can be solved optimally in polynomial time by a dynamic program proposed by Monma and Potts in [19]. When all release times are identical, then the offline problem reduces to the classical makespan minimization problem with setup times. It has been considered for m parallel machines and it is known [16] to be solvable by an FPTAS if m is assumed to be constant; here an (F)PTAS is an approximation algorithm that finds a schedule with makespan at most by a factor of $(1 + \varepsilon)$ larger than the optimum in time polynomial in the input size (and $\frac{1}{\varepsilon}$). For variable m, Jansen and Land propose a PTAS for this problem in [13]. Since in general a large body of literature for scheduling with setup considerations has evolved over time, the interested reader is referred to the surveys by Allahverdi et al. [1–3].

Our model can also be seen as a generalization of classical models without setup times. In this case, it is known that FIFO is optimal for minimizing maximum flow time on a single machine. On m parallel machines FIFO achieves a competitiveness of $3 - 2/m$ (in the (non-)preemptive case) as shown by Mastrolilli in [18]. Further results include algorithms for (un-)related machines with speed augmentation given by Anand et al. in [4] and for related machines proposed by Bansal and Cloostermans in [5].

The concept of smoothed analysis has so far, although considered as an interesting alternative to classical competitiveness (e.g. [11,14,15]), only been applied to two problems. In [6], Bechetti et al. study the Multilevel Feedback Algorithm for minimizing total flow time on parallel machines when preemption is allowed and non-clairvoyance is assumed. They consider a smoothing model in which initial processing times are integers from the interval $[1, 2^K]$ and they are perturbed by replacing the k least significant bits by a random number from $[1, 2^k]$. They prove a smoothed competitiveness of $O((2^k/\sigma)^3 + (2^k/\sigma)^2 2^{K-k})$, where σ denotes the standard deviation of the underlying distribution. This, for example, becomes $O(2^{K-k})$ for the uniform distribution. This result significantly improves upon the lower bounds of $\Omega(2^K)$ and $\Omega(n^{\frac{1}{3}})$ known for the classical competitiveness of deterministic algorithms [20]. In [23], Schäfer and Sivadasan apply smoothed competitive analysis to metrical task systems (a general framework for online problems covering, for example, the paging and the k-server problem). While any deterministic online algorithm is (on any graph with n nodes) $\Omega(n)$-competitive, the authors, amongst others, prove a sublinear smoothed competitiveness on graphs fulfilling certain structural properties. Finally, a notion similar to smoothed competitiveness has been applied by Scharbrodt, Schickinger and Steger in [24]. They consider the problem of minimizing the total completion time on parallel machines and analyze the Shortest Expected Processing Time

First strategy. While it is $\Omega(n)$-competitive, they prove an expected competitiveness, defined as $\mathbb{E}\left[\frac{\text{ALG}}{\text{OPT}}\right]$, of $O(1)$ if processing times are drawn from a gamma distribution.

4   A Non-clairvoyant Online Algorithm

In this section, we present a simple greedy-like algorithm and analyze its competitiveness. The idea of the algorithm BALANCE, as presented in Algorithm 1, is to find a tradeoff between preferring jobs with early release times and jobs that are of the type the machine is currently configured for. This is achieved by the following idea: Whenever the machine is about to idle at some time t, BALANCE checks whether there is a job j available that is of the same type τ_j as the machine is currently configured for, denoted by $active(t)$. If this is the case and if there is no job j' with a "much smaller" release time than j, job j is assigned to the machine. The decision whether a release time is "much smaller" is taken based on a parameter λ, called *balance parameter*. This balance parameter is grown over time based on the maximum flow time encountered so far and, at any time, is of the form α^q, for some $q \in \mathbb{N}$ which is increased over time and some constant α determined later.[1] Note that BALANCE is a greedy-like algorithm by using the adjusted release times for determining the ordering of jobs.

Algorithm 1. Description of BALANCE

---
(1) Let $\lambda = \alpha$.   ▷ for some constant α
(2) If the machine idles at time t,
       process available job with smallest **adjusted release time** $\bar{r}_j(t)$

$$\bar{r}_j(t) := \begin{cases} r_j & \text{if } \tau_j = active(t) \\ r_j + \lambda & \text{else} \end{cases}$$

     after doing a setup if necessary.
     To break a tie, prefer job j with $\tau_j = active(t)$.
(3) As soon as a job j completes with $F_j \geq \alpha\lambda$, set $\lambda := \alpha\lambda$.

4.1   Basic Properties of Balance

The following two properties follow from the definition of BALANCE and relate the release times and flow times of consecutive jobs, respectively. For a job j, let $\lambda(j)$ denote the value of λ when j was scheduled.

Proposition 1. *Consider two jobs j_1 and j_2. If $\tau_{j_1} = \tau_{j_2}$, both jobs are processed according to FIFO. Otherwise, if j_2 is processed after j_1 in a schedule of* BALANCE, $r_{j_2} \geq r_{j_1} - \lambda(j_1)$.

---
[1] A variant of this algorithm with a fixed λ and hence without Step (3) has been previously mentioned in [10] as an algorithm with $\Omega(n)$-competitiveness for the clairvoyant variant of our problem.

Proof. The first statement directly follows from the definition of the algorithm. Consider the statement for two jobs j_1 and j_2 with $\tau_{j_1} \neq \tau_{j_2}$. Let t be the point in time at which j_1 is assigned to the machine. If active$(t) \neq \tau_{j_1}$ and active$(t) \neq \tau_{j_2}$, it follows $r_{j_2} \geq r_{j_1}$. If active$(t) = \tau_{j_1}$, then because j_1 is preferred over j_2, we have $r_{j_1} = \bar{r}_{j_1}(t) \leq \bar{r}_{j_2}(t) = r_{j_2} + \lambda(j_1)$. Finally, if active$(t) = \tau_{j_2}$ we know by the fact that j_1 is preferred that $r_{j_1} + \lambda(j_1) = \bar{r}_{j_1}(t) < \bar{r}_{j_2}(t) = r_{j_2}$, which proves the proposition. $\qquad\square$

Proposition 2. *Consider two jobs j_1 and j_2. If j_1 is processed before j_2 and no job is processed in between, then $F_{j_2} \leq F_{j_1} + p_{j_2} + s + \lambda(j_1)$.*

4.2   Competitiveness

We carefully define specific subschedules of a given schedule S of BALANCE, which we will heavily use throughout our analysis of the (smoothed) competitiveness. Given $\alpha^q \geq F^*$, $q \in \mathbb{N}_0$, let S_{α^q} be the subschedule of S that starts with the first job j with $\lambda(j) = \alpha^q$ and ends with the last job j' with $\lambda(j') = \alpha^q$. For a fixed δ, let $S_{\alpha^q}^{\delta}$ be the suffix of S_{α^q} such that the first job in $S_{\alpha^q}^{\delta}$ is the last one in S_{α^q} with the following properties: (1) It has a flow time of at most $(\alpha - \delta)\alpha^q$, and (2) it starts a batch. (We will prove in Lemma 1 that $S_{\alpha^q}^{\delta}$ always exists.) Without loss of generality, let $j_1, \ldots, j_m$ be the jobs in $S_{\alpha^q}^{\delta}$ such that they are sorted by their starting times, $t_1 < t_2 < \ldots < t_m$. Let $B_1, \ldots, B_\ell$ be the batches in $S_{\alpha^q}^{\delta}$.

The main idea of Lemma 1 is to show that, in case a flow time of $F > \alpha^{q+1}$ is reached, the interval $[r_{j_1}, r_{j_m}]$ is in a sense dense: Workload plus setup times in $S_{\alpha^q}^{\delta}$ is at least by $\delta\alpha^q$ larger than the length of this interval. Intuitively, this holds as otherwise the difference in the flow times F_{j_1} and F_{j_m} could not be as high as $\delta\alpha^q$, which, however, needs to hold by the definition of $S_{\alpha^q}^{\delta}$. Additionally, the flow time of all jobs is shown to be lower bounded by $3\alpha^q$. Roughly speaking, this holds due to the following observation: If a fixed job has a flow time below $3\alpha^q$, then the job starting the next batch can, on the one hand, not have a much smaller release time (by definition of the algorithm). On the other hand, it will therefore not be started much later, leading to the fact that the flow time cannot be too much larger than $3\alpha^q$ (and in particular, is below $(\alpha - \delta)\alpha^q$ for sufficiently small δ).

Lemma 1. *Let $\alpha^q \geq F^*$ and $\delta \leq \alpha - 10$. Then $S_{\alpha^q}^{\delta}$ always exists and all jobs in $S_{\alpha^q}^{\delta}$ have a flow time of at least $3\alpha^q$. Also, if $F > \alpha^{q+1}$, it holds $\sum_{i=1}^{\ell} w(B_i) + r_{j_1} - r_{j_m} \geq \delta\alpha^q - (\ell - 1)s$.*

Proof. We first prove that a job with the two properties starting off $S_{\alpha^q}^{\delta}$ exists. Let $\tilde{j}_1, \ldots$ be the jobs in S_{α^q}. Consider the last job $\tilde{j}_0$ processed directly before $\tilde{j}_1$. By Proposition 2 we have $F_{\tilde{j}_1} \leq F_{\tilde{j}_0} + p_{\tilde{j}_1} + s + \alpha^{q-1} \leq \alpha^q + p_{\tilde{j}_0} + s + \alpha^{q-1} + p_{\tilde{j}_1} + s + \alpha^{q-1} < 6\alpha^q$. Among jobs in S_{α^q} that have a different type than $\tilde{j}_1$, consider the job $\tilde{j}_i$ with the lowest starting time. We show that it is a candidate for starting $S_{\alpha^q}^{\delta}$, implying that $S_{\alpha^q}^{\delta}$ exists. Property (2) directly follows by construction. For the flow time of $\tilde{j}_i$, we know that only jobs of the same type as $\tilde{j}_1$ are scheduled between $\tilde{j}_1$ and $\tilde{j}_i$. This implies that jobs $\tilde{j}_2, \ldots, \tilde{j}_{i-1}$ are released in the interval $[r_{\tilde{j}_1}, r_{\tilde{j}_i} + \alpha^q]$. The interval can contain a workload of at most $r_{\tilde{j}_i} + \alpha^q - r_{\tilde{j}_1} + F^*$ (see also Proposition 3), hence the flow time of job $\tilde{j}_i$ is at most $F_{\tilde{j}_1} = F_{\tilde{j}_i} \leq (r_{\tilde{j}_1} + F_{\tilde{j}_1} + (r_{\tilde{j}_i} + \alpha^q - r_{\tilde{j}_1} + F^*) + s + p_{\tilde{j}_i}) - r_{\tilde{j}_i} \leq 9\alpha^q + p_{\tilde{j}_i} = 9\alpha^q + p_{\tilde{j}_1} \leq (\alpha - \delta)\alpha^q$. Property (1) and the existence of $S_{\alpha^q}^\delta$ follow. Since $t_{j_1} = c_{j_1} - p_{j_1}$ and $F_{j_1} = c_{j_1} - r_{j_1}$, we also have $t_{j_1} \leq r_{j_1} + 9\alpha^q \leq r_{j_1} + (\alpha - \delta)\alpha^q (*)$.

We now show that during $S_{\alpha^q}^\delta$, the machine does not idle and each job in $S_{\alpha^q}^\delta$ has a flow time of at least $3\alpha^q$. Assume this is not the case. Denote by t the last time in $S_{\alpha^q}^\delta$ where either an idle period ends or a job with a flow time of less than $3\alpha^q$ completes. We denote the jobs scheduled after t by $\hat{j}_1, \ldots$ and the first job of the first batch started at or after t by $\hat{j}_i$. Similar to above, all jobs $\hat{j}_1, \ldots, \hat{j}_i$ are released in the interval $[t - 3\alpha^q, r_{\hat{j}_i} + \alpha^q]$. The overall workload of these jobs is at most $r_{\hat{j}_i} + 4\alpha^q - t + F^* \leq r_{\hat{j}_i} + 5\alpha^q - t$. Job $\hat{j}_i$ is thus finished by $t + (r_{\hat{j}_i} + 5\alpha^q - t) + s \leq r_{\hat{j}_i} + 6\alpha^q$. This is a contradiction to $F_{\hat{j}_i} > (\alpha - \delta)\alpha^q$.

Finally, since there are no idle times and by (*), for the last job j_m of $S_{\alpha^q}^\delta$ we have $F_{j_m} \leq r_{j_1} + (\alpha - \delta)\alpha^q + \sum_{i=1}^{\ell} w(B_i) + (\ell - 1)s - r_{j_m}$. By the assumption that $F > \alpha^{q+1}$ and the definition of j_m to be the first job with flow time at least α^{q+1}, we obtain the desired result. $\qquad \square$

We will also make use of Corollary 1, which follows from the proof of Lemma 1.

Corollary 1. *The statement of Lemma 1 also holds if $S_{\alpha^q}^\delta$ is replaced by $S_{\alpha^q}^\delta(j)$ for any job $j \in S_{\alpha^q}^\delta$ with $F_j \leq (\alpha - \delta)\alpha^q$, where $S_{\alpha^q}^\delta(j)$ is the suffix of $S_{\alpha^q}^\delta$ starting with job j.*

Next we give simple lower bounds for the optimal flow time F^*. Besides the direct lower bound $F^* \geq \max\{s, p_{max}\}$, where $p_{max} := \max_{1 \leq j \leq n} p_j$, we can also prove a bound as given in Proposition 3. For a given interval I, let $\mathrm{overhead}_{\mathrm{OPT}}(I)$ be the overhead in OPT between the jobs j_1 and j_2 released in I and being processed first and last in OPT, respectively. Precisely, $j_1 := \operatorname{argmin}_{j:r_j \in I} t_j$ and $j_2 := \operatorname{argmax}_{j:r_j \in I} t_j$.

Proposition 3. *As lower bounds for F^* we have $F^* \geq \max\{s, p_{max}\}$ as well as $F^* \geq \max_I \{w(I) + \mathrm{overhead}_{\mathrm{OPT}}(I) - |I|\}$.*

Proof. We have $F^* \geq c_{j_2} - r_{j_2}$. On the other hand, $c_{j_2} \geq r_{j_1} + \mathrm{overhead}_{\mathrm{OPT}}(I) + w(I)$. Thus, $F^* \geq \mathrm{overhead}_{\mathrm{OPT}}(I) + w(I) + l(I) - r(I) = w(I) + \mathrm{overhead}_{\mathrm{OPT}}(I) - |I|$. $\qquad \square$

Combining Lemma 1 and Proposition 3, we easily obtain that the competitiveness can essentially be bounded by the difference in the number of setups OPT and BALANCE perform on those jobs which are part of $S_{\alpha^q}^\delta$. Let $I(S_{\alpha^q}^\delta)$ be the interval in which all jobs belonging to $S_{\alpha^q}^\delta$ are released, $I(S_{\alpha^q}^\delta) := [\min_j \{r_j : j \in S_{\alpha^q}^\delta\}, \max_j \{r_j : j \in S_{\alpha^q}^\delta\}]$. We have the following bound.

Lemma 2. *Let $\alpha^{q+1} \leq F < \alpha^{q+2}$ and $3 \leq \delta \leq \alpha - 10$ and $\alpha^q \geq F^*$. It holds $F \leq \alpha^2 (\delta - 2)^{-1}(F^* + \mathrm{overhead}_{\mathrm{BALANCE}}(S_{\alpha^q}^\delta) - \mathrm{overhead}_{\mathrm{OPT}}(I(S_{\alpha^q}^\delta)))$.*

Throughout the rest of the paper, we assume that $\delta = 3$ and $\alpha = 13$ fulfilling the properties of Lemmas 1 and 2. Our goal now is to bound the competitiveness by upper bounding the difference of the overhead of OPT and BALANCE in $S_{\alpha^q}^{\delta}$ for some $\alpha^q = \Omega(\sqrt{n \cdot s \cdot p_{max}})$. In Lemma 4 we will see that to obtain a difference of $i \cdot s$, a workload of $\Omega(i \cdot \alpha^q)$ is required. Using this, we can then upper bound the competitiveness based on the overall workload of $O(n \cdot p_{max})$ available in a given instance in Theorem 1. Before we can prove Lemma 4 we need the following insight. Given $S_{\alpha^q}^{\delta}$ for some $q \in \mathbb{N}_0$ such that $\alpha^q \geq F^*$. Let $j_{\tau,i}$ be the first job of the i-th batch of some fixed type τ in $S_{\alpha^q}^{\delta}$. We show that the release times of jobs $j_{\tau,i}$ and $j_{\tau,i+1}$ differ by at least α^q. Intuitively, this holds due to the definition of the balance parameter and the fact that in $S_{\alpha^q}^{\delta}$ all jobs starting a batch have a flow time of at least $3\alpha^q$.

Lemma 3. *Given $S_{\alpha^q}^{\delta}$, it holds $r_{j_{\tau,i}} > r_{j_{\tau,i-1}} + \alpha^q$, for all $i \geq 2$ and all τ.*

Lemma 4. *Let B be a batch in OPT. If all jobs from B are part of $S_{\alpha^q}^{\delta}$, an overhead of at most $2s$ is associated to them in the schedule of BALANCE.*

Also, if the overhead associated to B in OPT is smaller than $2s$ and is $2s$ in the schedule of BALANCE, it needs to hold

1. *$w(B) \geq \alpha^q - F^* - s =: \bar{w}$ and*
2. *jobs of B with size at least $\bar{w}$ need to be released in an interval of length α^q.*

Proof. Assume to the contrary that BALANCE processes the jobs of B in three batches with $j_1, j_2, j_3 \in B$ being the jobs starting the first, the second and the third batch, respectively. Then there need to be two jobs i_1 and i_2 that are processed between the first and second and second and third such batch, respectively. Since j_2 is preferred over i_2 and by Lemma 3, we have $r_{i_2} \geq r_{j_2} \geq r_{j_1} + \alpha^q$. Also, since i_2 is preferred over j_3 and by Lemma 1, we have $r_{i_2} + \alpha^q \leq r_{j_3}$. Hence, OPT cannot process i_2 before nor after B (since then either j_1 or i_2 would have a flow time larger than F^*), which is a contradiction to the fact that B is a batch in OPT.

If BALANCE processes the jobs of B in two batches, let $j_1, j_2 \in B$ be the jobs starting the first batch and the second batch, respectively. We start with the case that $w(B) < \bar{w}$ and show a contradiction. We know that $r_{j_2} > r_{j_1} + \alpha^q$. Consider an optimal schedule. As j_1 cannot be started after $r_{j_1} + F^*$ and because OPT processes j_1 and j_2 in the same batch B, the processing of B needs to cover the interval $[r_{j_1} + F^*, r_{j_2}] \supseteq [r_{j_1} + F^*, r_{j_1} + \alpha^q]$. As $w(B) < \alpha^q - F^* - s$ this implies an additional overhead of at least s associated to B, contradicting our assumption.

Therefore, assume that $w(B) \geq \bar{w}$ but there is no interval of length α^q with jobs of B of size at least $\bar{w}$. We know that j_1 needs to be started not later than $r_{j_1} + F^*$. Also, the workload of jobs of B released until $r_{j_1} + \alpha^q$ is below $\bar{w}$. Hence, there needs to be a job in B released not before $r_{j_1} + \alpha^q$. This implies that the processing of B needs to cover the entire interval $[r_{j_1} + F^*, r_{j_1} + \alpha^q]$. However, this implies an additional overhead associated to B of at least s, contradicting our assumption. $\square$

We are now ready to bound the competitiveness of BALANCE.

Theorem 1. BALANCE *is* $O(\sqrt{n})$*-competitive. It holds* $F = O(F^* + \sqrt{np_{max}s})$.

For the case $k = 2$ we can even strengthen the statement of Lemma 3. Given $S_{\alpha^q}^\delta$, let job j_i be the first job of the i-th batch in $S_{\alpha^q}^\delta$ and note that $\tau_{j_i} = \tau_{j_{i+2}}$ as the batches form an alternating sequence of the two types. We have the following lemma.

Lemma 5. *Given* $S_{\alpha^q}^\delta$, *if* $k = 2$ *then it holds* $r_{j_i} > r_{j_{i-1}} + \alpha^q$, *for all* $i \geq 3$.

Based on this fact, we can show that OPT can essentially not process any jobs that belong to different batches in $S_{\alpha^q}^\delta$ in one batch. Hence, OPT performs roughly the same amount of setups as BALANCE does and we have the following theorem by Lemma 2.

Theorem 2. *If* $k = 2$, *then* BALANCE *is* $O(1)$*-competitive.*

To conclude this section, we show that the bound of $O(\sqrt{n})$ from Theorem 1 for the competitiveness of BALANCE is tight and that a lower bound of $\Omega(\sqrt{n})$ holds for any greedy-like algorithm as defined in Sect. 2. This also implies that the $\Omega(\sqrt{n})$ bound holds for BALANCE independent of how λ is chosen or increased (and even if done at random). The construction in the proof of Theorem 3 is a generalization of a worst-case instance given in [10].

Theorem 3. *Any greedy-like algorithm is* $\Omega(\sqrt{n})$*-competitive.*

5   Smoothed Competitive Analysis

In this section, we analyze the smoothed competitiveness of BALANCE. We consider the following *multiplicative smoothing model* from [6]. Let p_j be the processing time of a job j as specified by the adversary in instance $\mathcal{I}$. Then the perturbed instance $\hat{\mathcal{I}}$ is defined as $\mathcal{I}$ but with processing times $\hat{p}_j$ defined by $\hat{p}_j = (1 + X_j)p_j$ where X_j depends on the smoothing distribution. For $0 < \varepsilon < 1$ being a fixed parameter describing the amount of perturbation, we consider two smoothing distributions. In case of a *uniform smoothing distribution*, X_j is chosen uniformly at random from the interval $[-\varepsilon, \varepsilon]$. More formally, $X_j \sim \mathcal{U}(-\varepsilon, \varepsilon)$ where $\mathcal{U}(a, b)$ denotes the continuous uniform distribution with probability density function $f(x) = \frac{1}{b-a}$ for $a \leq x \leq b$ and $f(x) = 0$ otherwise. Hence, for $\hat{p}_j$ we have $\hat{p}_j \in [(1 - \varepsilon)p_j, (1 + \varepsilon)p_j]$. In case of a *normal smoothing distribution*, X_j is chosen from a normal distribution with expectation 0, standard deviation $\sigma = \frac{\varepsilon}{\sqrt{2.64}}$ and truncated at -1 and 1. More formally, $X_j \sim \mathcal{N}_{(-1,1)}(0, \sigma^2)$ where $\mathcal{N}_{(a,b)}(\mu, \sigma^2)$ denotes the truncated normal distribution with probability density function $f(x) = \frac{\phi(\frac{x-\mu}{\sigma})}{\sigma(\Phi(\frac{b-\mu}{\sigma}) - \Phi(\frac{a-\mu}{\sigma}))}$ for $a < x < b$ and $f(x) = 0$ otherwise. Here $\phi(\cdot)$ denotes the density function of the standard normal distribution and $\Phi(\cdot)$ the respective (cumulative) distribution function.

Our goal is to prove a smoothed competitiveness of $O(\varepsilon^{-2} s \log^2 n)$. We analyze the competitiveness by conditioning it on the flow time of OPT and its relation to the flow time of BALANCE. Let $\mathcal{E}^q_{\text{OPT}}$ be the event that $F^* \in [\alpha^q, \alpha^{q+1})$ and $\mathcal{E}^q_{\text{BALANCE}}$ be the event that $F > c_1 \alpha^{q+1} \varepsilon^{-2} s \log^2 n$ (for a constant value of c_1 determined by the analysis). Also, denote by $\bar{\mathcal{E}}^q_x$ the respective complementary events. Then for a fixed instance $\mathcal{I}$ we obtain

$$
\mathbb{E}_{\hat{\mathcal{I}} \leftarrow N(\mathcal{I})} \left[\frac{F(\hat{\mathcal{I}})}{F^*(\hat{\mathcal{I}})} \right] = \sum_{q \in \mathbb{N}} \mathbb{E} \left[\frac{F(\hat{\mathcal{I}})}{F^*(\hat{\mathcal{I}})} \mid \mathcal{E}^q_{\text{OPT}} \wedge \bar{\mathcal{E}}^q_{\text{BALANCE}} \right] \cdot \Pr[\mathcal{E}^q_{\text{OPT}} \wedge \bar{\mathcal{E}}^q_{\text{BALANCE}}]
$$

$$
+ \sum_{q = \lfloor \log_\alpha s \rfloor}^{\lceil \log_\alpha n \rceil} \mathbb{E} \left[\frac{F(\hat{\mathcal{I}})}{F^*(\hat{\mathcal{I}})} \mid \mathcal{E}^q_{\text{OPT}} \wedge \mathcal{E}^q_{\text{BALANCE}} \right] \cdot \Pr[\mathcal{E}^q_{\text{OPT}} \wedge \mathcal{E}^q_{\text{BALANCE}}]
$$

$$
+ \sum_{q > \lceil \log_\alpha n \rceil} \mathbb{E} \left[\frac{F(\hat{\mathcal{I}})}{F^*(\hat{\mathcal{I}})} \mid \mathcal{E}^q_{\text{OPT}} \wedge \mathcal{E}^q_{\text{BALANCE}} \right] \cdot \Pr[\mathcal{E}^q_{\text{OPT}} \wedge \mathcal{E}^q_{\text{BALANCE}}].
$$

Note that the first sum is by definition directly bounded by $O(\varepsilon^{-2} s \log^2 n)$ and the third one by $O(\sqrt{s})$ according to Theorem 1. Thus, we only have to analyze the second sum. We show that we can complement the upper bound on the ratio, which can be as high as $O(\sqrt{n})$ by Theorem 1, by $\Pr[\mathcal{E}^q_{\text{OPT}} \wedge \mathcal{E}^q_{\text{BALANCE}}] \leq 1/n$. From now on we consider an arbitrary but fixed $q \geq \lfloor \log_\alpha s \rfloor$, and in the following we analyze $\Pr[\mathcal{E}^q_{\text{OPT}} \wedge \mathcal{E}^q_{\text{BALANCE}}]$. Let $\Gamma := \alpha^{i-1}$ such that i is the largest integer with $\alpha^i \leq c_1 \varepsilon^{-2} \alpha^{q+1} s \log^2 n$. Thus we have $\Gamma \geq c_1 \alpha^{q-1} \varepsilon^{-2} s \log^2 n$.

On a high level, the idea of our proof is as follows: We first define a careful partitioning of the time horizon into consecutive intervals (Sect. 5.1). Depending on the amount of workload released in each such interval and an estimation of the amount of setups required for the respective jobs (Sect. 5.2), we then classify each of them to either be dense or sparse (Sect. 5.3). We distinguish two cases depending on the number of dense intervals in $\mathcal{I}$. If this number is sufficiently large, F^* is, with high probability (w.h.p.), not much smaller than F (Lemma 9). This holds as w.h.p. the perturbation increases the workload in a dense interval so that even these jobs cannot be scheduled with a low flow time by OPT. In case the number of dense intervals is small, the analysis is more involved. Intuitively, we can show that w.h.p. there is only a logarithmic number of intervals between any two consecutive sparse intervals in which the perturbation decreases the workload to a quite small amount. Between such sparse intervals the flow time cannot increase too much (even in the worst-case) and during a sparse interval BALANCE can catch up with the optimum: If taking a look at the flow time of the job completing at time t and continuing this consideration over time, we then obtain a sawtooth pattern always staying below a not too large bound for the flow time of BALANCE (Lemma 10).

5.1 Partitioning of Instance $\mathcal{I}$

We define a partitioning of the instance $\mathcal{I}$, on which our analysis of the smoothed competitiveness will be based. We partition the time interval $[r_{min}, r_{max}]$, where

---

Algorithm 2. Description of SETUPESTIMATE(I)

Construct a sequence $(j_1, j_2, \ldots, j_m)$ of all jobs released in I as follows:

(1) For $i = 1, 2, \ldots, m$ set j_i to be job $j \notin (j_1, \ldots j_{i-1})$ with smallest $\bar{r}_j$, where

$$\bar{r}_j := \begin{cases} r_j & \text{if } \tau_j = \tau_{j_{i-1}} \\ r_j + \Gamma & \text{else.} \end{cases}$$

   To break a tie, prefer job j with $\tau_j = \tau_{j_{i-1}}$.

(2) Let $N_s(I)$ be the number of values i such that $\tau_{j_i} \neq \tau_{j_{i-1}}$.

---

r_{min} and r_{max} are the smallest and largest release time, as follows: Let a *candidate interval* C be an interval such that $|C| = \Gamma$ and such that for some τ it holds $\sum_{j:r_j \in C, \tau_j = \tau} p_j \geq \Gamma/4$. Intuitively, a candidate interval C is an interval on which, in $\hat{\mathcal{I}}$, BALANCE possibly has to perform more setups than OPT does (which, if all jobs released in the interval belong to S_Γ^δ and under the assumption that $\mathcal{E}_{\text{OPT}}^q \wedge \mathcal{E}_{\text{BALANCE}}^q$ holds, according to Lemma 4 requires a workload of at least $\frac{\Gamma}{2}$ in $\hat{\mathcal{I}}$ and hence, at least $\frac{\Gamma}{4}$ in $\mathcal{I}$). Let C_1 be the first candidate interval C. For $i > 1$ let C_i be the first candidate interval C that does not overlap with C_{i-1}.

Now we consider groups of $\mu := \left\lceil \frac{\varepsilon^2 \Gamma}{c_2 s^2 \log^2 n} \right\rceil$ many consecutive candidate intervals C_i, for some constant c_2 determined by the further analysis. Precisely, these groups are defined as $I_1 = [r_{min}, r(C_\mu)], I_2 = (r(C_\mu), r(C_{2\mu})]$ and so on. In the rest of the paper we consistently use I_i to denote these intervals. Let $\bigcup_i I_i = [r_{min}, r_{max}]$ by (possibly) extending the last I_i so that its right endpoint is r_{max}. Although it worsens constants involved in the competitiveness, we use $\mu \leq \frac{2\varepsilon^2 \Gamma}{c_2 s^2 \log^2 n}$ for $c_1 \geq \alpha c_2$ for the sake of simplicity.

5.2 Estimation of Setups in I_i

Before we can now define intervals I_i to be dense or sparse, we need an estimate $N_s(I)$ on the number of setups OPT and BALANCE perform on jobs released in a given interval I. We require $N_s(I)$ to be a value uniquely determined by the instance $\mathcal{I}$ and hence, in particular not to be a random variable. This is essential for our analysis and avoids any computation of conditional probabilities. For the definition of $N_s(I)$ consider the construction by SETUPESTIMATE(I) in Algorithm 2. For a fixed interval I, it essentially mimics BALANCE in S_Γ^δ in the sense that Lemma 6 holds completely analogous to Lemma 4. Also, note that the construction is indeed invariant to job sizes and hence to perturbations. It should not be understood as an actual algorithm for computing a schedule, however, for ease of presentation we refer to the sequence constructed as if it was a schedule. Particularly, we say that it processes two jobs j_i and $j_{i'}$ with $\tau_{j_i} = \tau_{j_{i'}}$ in different batches if there is an i'' such that $i < i'' < i'$ with $\tau_{j_{i''}} \neq \tau_{j_i}$.

For two jobs j_1 and j_2 of a common type τ which start two batches in SETUPESTIMATE(I), $r_{j_2} \geq r_{j_1} + \Gamma$ holds. Hence, by the exact same line of arguments as in the proof of Lemma 4 we have the following lemma.

Lemma 6. *Assume $\mathcal{E}^q_{\mathrm{OPT}} \wedge \mathcal{E}^q_{\mathrm{BALANCE}}$ holds. Let B be a batch in OPT. Let I be such that $r_j \in I$ for all $j \in B$. An overhead of at most $2s$ is associated to B in SETUPESTIMATE(I).*

Also, if the overhead associated to B in OPT is smaller than $2s$ and is $2s$ in the schedule of SETUPESTIMATE(I), $w(B) \geq \Gamma - F^* - s =: \bar{w}$ needs to hold and jobs of B with size at least $\bar{w}$ need to be released in an interval of length Γ.

In the next two lemmas we show that $N_s(I_i)$ is indeed a good estimation of the number of setups OPT and BALANCE have to perform, respectively. Lemma 7 essentially follows by Lemma 6 together with the definition of I_i to consist of μ many candidate intervals. To prove Lemma 8 we exploit the fact that all jobs in S^δ_Γ have a flow time of at least 3Γ by Lemma 1 so that BALANCE and SETUPESTIMATE essentially behave in the same way.

Lemma 7. *Assume $\mathcal{E}^q_{\mathrm{OPT}} \wedge \mathcal{E}^q_{\mathrm{BALANCE}}$ holds and consider I_i for a fixed i. For the overhead of OPT it holds $\mathrm{overhead}_{\mathrm{OPT}}(I_i) \geq (N_s(I_i) - 6\mu)s$.*

Lemma 8. *Consider an interval I_i and suppose that all jobs from I_i belong to S^δ_Γ. Then it holds $\mathrm{overhead}_{\mathrm{BALANCE}}(I_i) \leq N_s(I_i)s$, where $\mathrm{overhead}_{\mathrm{BALANCE}}(I_i)$ denotes the overhead of BALANCE associated to jobs j with $r_j \in I_i$.*

5.3   Good and Bad Events

We are now ready to define good and bad events, which are outcomes of the perturbation of the job sizes that help the algorithm to achieve a small and help the adversary to achieve a high competitiveness, respectively. Let $w_{\mathcal{I}}(I_i) := \sum_{j:r_j \in I_i} p_j$ and $w_{\hat{\mathcal{I}}}(I_i) := \sum_{j:r_j \in I_i} \hat{p}_j$ denote the workload released in the interval I_i in instance $\mathcal{I}$ and $\hat{\mathcal{I}}$, respectively. We distinguish two kinds of intervals I_i and associate a good and a bad event to each of them. We call an interval I_i to be *dense* if $w_{\mathcal{I}}(I_i) + N_s(I_i)s \geq |I_i|$ and associate an event $\mathcal{D}^{\mathrm{good}}_i$ or $\mathcal{D}^{\mathrm{bad}}_i$ to I_i depending on whether $w_{\hat{\mathcal{I}}}(I_i) \geq w_{\mathcal{I}}(I_i) + \varepsilon^2 \Gamma/(18\sqrt{c_2}s \log n)$ holds or not. Symmetrically, we call an interval I_i to be *sparse* if $w_{\mathcal{I}}(I_i) + N_s(I_i)s < |I_i|$ and associate an event $\mathcal{S}^{\mathrm{good}}_i$ or $\mathcal{S}^{\mathrm{bad}}_i$ to I_i depending on whether $w_{\hat{\mathcal{I}}}(I_i) \leq w_{\mathcal{I}}(I_i) - \varepsilon^2 \Gamma/(18\sqrt{c_2}s \log n)$ holds or not.

We next show two lemmas which upper bound $\Pr[\mathcal{E}^q_{\mathrm{OPT}} \wedge \mathcal{E}^q_{\mathrm{BALANCE}}]$ by the probability of occurrences of good events. As we will see in Theorem 4 this is sufficient as we can prove the respective good events to happen with sufficiently large probability.

Lemma 9. $\Pr[\mathcal{E}^q_{\mathrm{OPT}} \wedge \mathcal{E}^q_{\mathrm{BALANCE}}] \leq \Pr[\text{no event } \mathcal{D}^{\mathrm{good}}_i \text{ happens}]$.

Proof. We show that if an event $\mathcal{D}^{\mathrm{good}}_i$ happens, then $\mathcal{E}^q_{\mathrm{OPT}}$ does not hold. Consider a dense interval I_i and assume an event $\mathcal{D}^{\mathrm{good}}_i$ occurs. Then we have by definition of dense intervals and the definition of event $\mathcal{D}_i^{\mathrm{good}}$ that $w_{\mathcal{I}}(I_i) + N_s(I_i)s \geq |I_i|$ and $w_{\hat{\mathcal{I}}}(I_i) \geq w_{\mathcal{I}}(I_i) + \varepsilon^2\Gamma/(18\sqrt{c_2}s\log n)$. Taken together, $w_{\hat{\mathcal{I}}}(I_i) + N_s(I_i)s \geq |I_i| + \varepsilon^2\Gamma/(18\sqrt{c_2}s\log n)$. On the other hand, together with Lemma 7 we then have $w_{\hat{\mathcal{I}}}(I_i) + \mathrm{overhead}_{\mathrm{OPT}}(I_i) \geq |I_i| + \varepsilon^2\Gamma/(18\sqrt{c_2}s\log n) - 6s \cdot \mu$. By Proposition 3 we have

$$F^* \geq w_{\hat{\mathcal{I}}}(I_i) + \mathrm{overhead}_{\mathrm{OPT}}(I_i) - |I_i| \geq \frac{\varepsilon^2\Gamma}{18\sqrt{c_2}s\log n} - 6s\left(\frac{2\varepsilon^2\Gamma}{c_2 s^2 \log^2 n}\right)$$
$$\geq \frac{\varepsilon^2\Gamma}{18\sqrt{c_2}s\log n}\left(1 - \frac{12 \cdot 18}{\sqrt{c_2}\log n}\right) \geq \frac{c_1 \alpha^{q-1}\log n}{18\sqrt{c_2}}\left(1 - \frac{12 \cdot 18}{\sqrt{c_2}\log n}\right) > \alpha^{q+1}$$

for sufficiently large $c_1 > c_2$ and n. Then $\mathcal{E}_{\mathrm{OPT}}^q$ does not hold. $\qquad\square$

In Theorem 4 we will see that the number N_D of dense intervals in $\mathcal{I}$ can be bounded by $N_D = 7\log n$ as otherwise the probability for event $\mathcal{E}_{\mathrm{OPT}}^q$ to hold is only $1/n$.

Thus, next we consider the case $N_D < 7\log n$. Consider the sequence of events associated to sparse intervals. A *run* of events $\mathcal{S}_i^{\mathrm{bad}}$ is a maximal subsequence such that no event $\mathcal{S}_i^{\mathrm{good}}$ happens within this subsequence.

Lemma 10. *If* $N_D < 7\log n, \Pr[\mathcal{E}_{\mathrm{OPT}}^q \wedge \mathcal{E}_{\mathrm{BALANCE}}^q] \leq \Pr[\exists\ \text{run of } \mathcal{S}_i^{\mathrm{bad}} \text{ of length} \geq 14\log n].$

Proof. We assume that all runs of events $\mathcal{S}_i^{\mathrm{bad}}$ are shorter than $14\log n$ and $\mathcal{E}_{\mathrm{OPT}}^q \wedge \mathcal{E}_{\mathrm{BALANCE}}^q$ holds and show a contradiction. From $\mathcal{E}_{\mathrm{BALANCE}}^q$ we can deduce by Lemma 1 that S_Γ^δ exists. Since we will use the following reasoning iteratively, let $S = S_\Gamma^\delta$. Using the terminology from Lemma 1, let $j_1, j_2, \ldots, j_m$ be the jobs in S and, as before, ℓ be the number of batches in S. By Lemma 1 it needs to hold $\sum_{i=1}^{\ell} w_{\hat{\mathcal{I}}}(B_i) + r_{j_1} - r_{j_m} \geq 3\Gamma - (\ell - 1)s$. Let $I_{\iota+1}$ be the first interval I_i such that $l(I_{\iota+1}) \geq r_{j_1}$. Let κ be chosen such that κ is the smallest integer where in $I_{\iota+\kappa}$ an event $\mathcal{S}_{\iota+\kappa}^{\mathrm{good}}$ occurs if κ exists and otherwise set κ such that $I_{\iota+\kappa}$ ends with r_{max}. Note that it holds $\kappa < 21\log n$ because of the assumption $N_D < 7\log n$ and the length of the longest run. Let $I_\iota = [\min_{1\leq i\leq m} r_{j_i}, l(I_{\iota+1}))$. We claim that all jobs belonging to $\bigcup_{i=0}^{\kappa} I_{\iota+i}$ need to have a flow time below $\alpha\Gamma$. Assume this is not the case. We have a contradiction as

$$F^* \geq w_{\hat{\mathcal{I}}}(I(S)) + \mathrm{overhead}_{\mathrm{OPT}}(I(S)) - |I(S)|$$
$$\geq \sum_{i=1}^{\ell} w_{\hat{\mathcal{I}}}(B_i) + \mathrm{overhead}_{\mathrm{OPT}}(I(S)) + r_{j_1} - r_{j_m} - 2\Gamma$$
$$\geq \sum_{i=1}^{\ell} w_{\hat{\mathcal{I}}}(B_i) + r_{j_1} - r_{j_m} - 2\Gamma + \mathrm{overhead}_{\mathrm{BALANCE}}(I(S)) - \frac{21\log n \cdot 12s\varepsilon^2\Gamma}{c_2 s^2 \log^2 n}$$
$$\geq \Gamma - \frac{252\log ns\varepsilon^2\Gamma}{c_2 s^2 \log^2 n} \geq \varepsilon^2\Gamma\left(\frac{1}{\varepsilon^2} - \frac{252}{c_2 s\log n}\right) \geq \frac{1}{2}\varepsilon^2 c_1\varepsilon^{-2}\alpha^{q-1}s\log^2 n$$
$$> \alpha^{q+1},$$

where we used Proposition 3 in the first inequality, the fact that $|I(S)| \leq (r_{j_m} - r_{j_1}) + 2\Gamma$ in the second, Lemmas 7 and 8 in the third, Lemma 1 in the fourth and in the remaining inequalities suitable values for $c_1 > c_2$ and the fact that $\Gamma \geq c_1 \varepsilon^{-2} \alpha^{q-1} s \log^2 n$. Observe that in case $r(I_{\iota+\kappa}) = r_{max}$, we are done as $\mathcal{E}_{\text{BALANCE}}^q$ cannot hold.

Otherwise, consider the situation directly before the first job $\tilde{j}$ with $r_{\tilde{j}} > r(I_{\iota+\kappa})$ is started. Denote the subschedule of S up to (not including) job $\tilde{j}$ by $\tilde{S}$. Let $\text{overhead}_{\text{BALANCE}}(I)$ be the overhead in S associated to jobs released in the interval I. Let $\text{overhead}_{\text{BALANCE}}(I, \tilde{S})$ and $\text{overhead}_{\text{BALANCE}}(I, \neg \tilde{S})$ be the overhead of jobs released in interval I and which are part and not part of $\tilde{S}$, respectively. Let $w_{\hat{\mathcal{I}}}(I, \tilde{S})$ and $w_{\hat{\mathcal{I}}}(I, \neg \tilde{S})$ be the workload of jobs released in interval I and which are part and not part of $\tilde{S}$, respectively. For brevity let

$$L = w_{\hat{\mathcal{I}}}([0, r_{j_1}), \tilde{S}) - w_{\hat{\mathcal{I}}}([r_{j_1}, r_{\tilde{j}}), \neg \tilde{S})$$
$$+ \text{overhead}_{\text{BALANCE}}([0, r_{j_1}), \tilde{S}) - \text{overhead}_{\text{BALANCE}}([r_{j_1}, r_{\tilde{j}}), \neg \tilde{S}).$$

We can then bound the workload and setups in $\tilde{S}$ by

$$w_{\hat{\mathcal{I}}}(\tilde{S}) + \text{overhead}_{\text{BALANCE}}(\tilde{S})$$
$$\leq w_{\hat{\mathcal{I}}}([r_{j_1}, l(I_{\iota+\kappa}))) + w_{\hat{\mathcal{I}}}(I_{\iota+\kappa}) + w_{\hat{\mathcal{I}}}([0, r_{j_1}), \tilde{S}) - w_{\hat{\mathcal{I}}}([r_{j_1}, r_{\tilde{j}}), \neg \tilde{S})$$
$$+ \text{overhead}_{\text{BALANCE}}([r_{j_1}, l(I_{\iota+\kappa}))) + \text{overhead}_{\text{BALANCE}}(I_{\iota+\kappa})$$
$$+ \text{overhead}_{\text{BALANCE}}([0, r_{j_1}), \tilde{S}) - \text{overhead}_{\text{BALANCE}}([r_{j_1}, r_{\tilde{j}}), \neg \tilde{S})$$
$$\leq l(I_{\iota+\kappa}) - r_{j_1} + F^* + 21 \log n 12 s \cdot \frac{\varepsilon^2 \Gamma}{c_2 s^2 \log^2 n} + |I_{\iota+\kappa}| - \frac{\varepsilon^2 \Gamma}{18 \sqrt{c_2} s \log n} + L$$
$$= r(I_{\iota+\kappa}) - r_{j_1} + F^* + \frac{\varepsilon^2 \Gamma}{s \log n 18 \sqrt{c_2}} \left(\frac{252 \cdot 18}{\sqrt{c_2}} - 1 \right) + L$$
$$\leq r(I_{\iota+\kappa}) - r_{j_1} + F^* + \frac{c_1 \alpha^{q-1} \log n}{18 \sqrt{c_2}} \left(\frac{252 \cdot 18}{\sqrt{c_2}} - 1 \right) + L$$
$$< r(I_{\iota+\kappa}) - r_{j_1} - 2F^* + L,$$

where we used Proposition 3 together with Lemma 7 and the fact that to $I_{\iota+\kappa}$ an event $S_{\iota+\kappa}^{\text{good}}$ is associated in the second inequality, the lower bound on Γ in the third inequality and suitable values for c_1 and c_2 in the last inequality. Then, job $\tilde{j}$ is started before $r_{j_1} + F_{j_1} + r(I_{\iota+\kappa}) - r_{j_1} - 2F^* + L + s$ and finished by $r(I_{\iota+\kappa}) + F_{j_1} + L$ with flow time $F_{\tilde{j}} \leq F_{j_1} + L$. For $S = S_\Gamma^\delta$ we have $L \leq -w_{\hat{\mathcal{I}}}([r_{j_1}, r_{\tilde{j}}), \neg \tilde{S}) - \text{overhead}_{\text{BALANCE}}([r_{j_1}, r_{\tilde{j}}), \neg \tilde{S})$ as no jobs with smaller release time than r_{j_1} can be part of S. Thus, $F_{\tilde{j}} < (\alpha - \delta)\Gamma - w_{\hat{\mathcal{I}}}([r_{j_1}, r_{\tilde{j}}), \neg \tilde{S}) - \text{overhead}_{\text{BALANCE}}([r_{j_1}, r_{\tilde{j}}), \neg \tilde{S})$. Now, applying the same arguments with $S = S_\Gamma^\delta(\tilde{j})$ and using Corollary 1 instead of Lemma 1, we find a further job with flow time at most $(\alpha - \delta)\Gamma$ (and all jobs processed before have flow time below $\alpha\Gamma$). Iterating this process we will eventually reach the end of the instance without finding a job with flow time at least $\alpha\Gamma$, contradicting that $\mathcal{E}_{\text{BALANCE}}^q$ holds. The formal (inductive) proof of this iterative process can be found in the full version. $\square$

To finally bound the probability of good events to happen, we need the following lemma.

Lemma 11. *Let J be a set of jobs and assume that processing times are perturbed according to a uniform or normal smoothing distribution. With probability at least $1/10$, $w_{\hat{I}}(J) \geq w_{\mathcal{I}}(J) + \frac{\varepsilon}{5}(\lfloor w_{\mathcal{I}}(J) \rfloor/3)^{0.5}$. Also, with probability at least $1/10$, $w_{\hat{I}}(J) \leq w_{\mathcal{I}}(J) - \frac{\varepsilon}{5}(\lfloor w_{\mathcal{I}}(J) \rfloor/3)^{0.5}$.*

Theorem 4. *The smoothed competitiveness of* BALANCE *is $O(\sigma^{-2} \log^2 n)$ when processing times p_j are perturbed independently at random to $\hat{p}_j = (1 + X_j)p_j$ where $X_j \sim \mathcal{U}(-\varepsilon, \varepsilon)$ or $X_j \sim \mathcal{N}_{(-1,1)}(0, \sigma^2)$.*

Proof. Recall that it only remains to prove $\Pr[\mathcal{E}_{\text{OPT}}^q \wedge \mathcal{E}_{\text{BALANCE}}^q] \leq 1/n$. First consider the case $N_D \geq 7 \log n$. For a fixed i we have $\Pr[\mathcal{D}_i^{\text{good}}] \geq \frac{1}{10}$ because of the following reasoning. According to Lemma 11, it holds $\Pr[w_{\hat{I}}(I_i) \geq w_{\mathcal{I}}(I_i) + \frac{\varepsilon}{5}(\lfloor w_{\mathcal{I}}(I_i) \rfloor/3)^{0.5}] \geq \frac{1}{10}$. By definition of I_i we can bound $\frac{\varepsilon}{5}(\lfloor w_{\mathcal{I}}(I_i) \rfloor/3)^{0.5} \geq \frac{\varepsilon}{5}(\frac{\mu \Gamma}{12})^{0.5} \geq \varepsilon^2 \frac{\Gamma}{18\sqrt{c_2 s} \log n}$ which implies $\Pr[\mathcal{D}_i^{\text{good}}] \geq \frac{1}{10}$. Because $N_D \geq 7 \log n$, the probability that no event $\mathcal{D}_i^{\text{good}}$ occurs is then upper bounded by $(1 - \frac{1}{10})^{7 \log n} \leq 1/n$ and so is $\Pr[\mathcal{E}_{\text{OPT}}^q \wedge \mathcal{E}_{\text{BALANCE}}^q]$ according to Lemma 9.

For the case $N_D < 7 \log n$ the same line of arguments gives $\Pr[\mathcal{S}_i^{\text{good}}] \geq 1/10$ for each sparse interval I_i. Hence, the probability for a run of events $\mathcal{S}_i^{\text{bad}}$ of length at least $14 \log n$ is at most $1/n$ and so is $\Pr[\mathcal{E}_{\text{OPT}}^q \wedge \mathcal{E}_{\text{BALANCE}}^q]$ by Lemma 10. □

To conclude, Theorem 4 shows a polylogarithmic smoothed competitiveness, which significantly improves upon the worst-case bound of $\Theta(\sqrt{n})$. It would be very interesting for future work to further investigate if it is possible to improve the smoothed analysis of BALANCE. Although we were not able to show such a result, it is quite possible that the actual smoothed competitiveness is independent of n. Though, proving such a result would probably require a different approach; the $\log n$ term in our result seems to be inherent to our analysis as it relies on the length of the longest run of bad events, each occurring with constant probability.

References

1. Allahverdi, A.: The third comprehensive survey on scheduling problems with setup times/costs. Eur. J. Oper. Res. **246**(2), 345–378 (2015)
2. Allahverdi, A., Gupta, J.N., Aldowaisan, T.: A review of scheduling research involving setup considerations. Omega **27**(2), 219–239 (1999)
3. Allahverdi, A., Ng, C.T., Cheng, T.C.E., Kovalyov, M.Y.: A survey of scheduling problems with setup times or costs. Eur. J. Oper. Res. **187**(3), 985–1032 (2008)
4. Anand, S., Bringmann, K., Friedrich, T., Garg, N., Kumar, A.: Minimizing maximum (weighted) flow-time on related and unrelated machines. Algorithmica **77**(2), 515–536 (2017)
5. Bansal, N., Cloostermans, B.: Minimizing maximum flow-time on related machines. Theory Comput. **12**(1), 1–14 (2016)

6. Becchetti, L., Leonardi, S., Marchetti-Spaccamela, A., Schäfer, G., Vredeveld, T.: Average-case and smoothed competitive analysis of the multilevel feedback algorithm. Math. Oper. Res. **31**(1), 85–108 (2006)
7. Bender, M.A., Chakrabarti, S., Muthukrishnan, S.: Flow and stretch metrics for scheduling continuous job streams. In: Proceedings of the 9th Annual ACM-SIAM Symposium on Discrete Algorithms (SODA 1998), pp. 270–279. ACM/SIAM (1998)
8. Borodin, A., Nielsen, M.N., Rackoff, C.: (Incremental) priority algorithms. Algorithmica **37**(4), 295–326 (2003)
9. Divakaran, S., Saks, M.: An online scheduling problem with job set-ups. Technical report, DIMACS Technical Report (2000)
10. Divakaran, S., Saks, M.E.: An online algorithm for a problem in scheduling with set-ups and release times. Algorithmica **60**(2), 301–315 (2011)
11. Hiller, B., Vredeveld, T.: Probabilistic alternatives for competitive analysis. Comput. Sci. - R&D **27**(3), 189–196 (2012)
12. Hofri, M., Ross, K.W.: On the optimal control of two queues with server setup times and its analysis. SIAM J. Comput. **16**(2), 399–420 (1987)
13. Jansen, K., Land, F.: Non-preemptive scheduling with setup times: a PTAS. In: Dutot, P.-F., Trystram, D. (eds.) Euro-Par 2016. LNCS, vol. 9833, pp. 159–170. Springer, Cham (2016). https://doi.org/10.1007/978-3-319-43659-3_12
14. Koutsoupias, E., Papadimitriou, C.H.: Beyond competitive analysis. SIAM J. Comput. **30**(1), 300–317 (2000)
15. López-Ortiz, A.: Alternative performance measures in online algorithms. In: Kao, M.-Y. (ed.) Encyclopedia of Algorithms, pp. 67–72. Springer, Heidelberg (2016). https://doi.org/10.1007/978-1-4939-2864-4_13
16. Mäcker, A., Malatyali, M., Meyer auf der Heide, F., Riechers, S.: Non-preemptive scheduling on machines with setup times. In: Dehne, F., Sack, J.-R., Stege, U. (eds.) WADS 2015. LNCS, vol. 9214, pp. 542–553. Springer, Cham (2015). https://doi.org/10.1007/978-3-319-21840-3_45
17. Mäcker, A., Malatyali, M., Meyer auf der Heide, F., Riechers, S.: Non-clairvoyant scheduling to minimize max flow time on a machine with setup times. CoRR (2017). 1709.05896
18. Mastrolilli, M.: Scheduling to minimize max flow time: offline and online algorithms. In: Lingas, A., Nilsson, B.J. (eds.) FCT 2003. LNCS, vol. 2751, pp. 49–60. Springer, Heidelberg (2003). https://doi.org/10.1007/978-3-540-45077-1_6
19. Monma, C.L., Potts, C.N.: On the complexity of scheduling with batch setup times. Oper. Res. **37**(5), 798–804 (1989)
20. Motwani, R., Phillips, S., Torng, E.: Non-clairvoyant scheduling. Theor. Comput. Sci. **130**(1), 17–47 (1994)
21. Potts, C.N.: Scheduling two job classes on a single machine. Comput. OR **18**(5), 411–415 (1991)
22. Sahney, V.K.: Single-server, two-machine sequencing with switching time. Oper. Res. **20**(1), 24–36 (1972)
23. Schäfer, G., Sivadasan, N.: Topology matters: smoothed competitiveness of metrical task systems. Theor. Comput. Sci. **341**(1–3), 216–246 (2005)
24. Scharbrodt, M., Schickinger, T., Steger, A.: A new average case analysis for completion time scheduling. J. ACM **53**(1), 121–146 (2006)

On-line Search in Two-Dimensional Environment

Dariusz Dereniowski$^{(\boxtimes)}$ⓘ and Dorota Urbańska

Faculty of Electronics, Telecommunications and Informatics,
Gdańsk University of Technology, 80-233 Gdańsk, Poland
deren@eti.pg.edu.pl, dorurban@student.pg.edu.pl

Abstract. We consider the following on-line pursuit-evasion problem. A team of mobile agents called searchers starts at an arbitrary node of an unknown network. Their goal is to execute a search strategy that guarantees capturing a fast and invisible intruder regardless of its movements using as few searchers as possible. As a way of modeling two-dimensional shapes, we restrict our attention to networks that are embedded into partial grids: nodes are placed on the plane at integer coordinates and only nodes at distance one can be adjacent. We give an on-line algorithm for the searchers that allows them to compute a connected and monotone strategy that guarantees searching any unknown partial grid with the use of $O(\sqrt{n})$ searchers, where n is the number of nodes in the grid. We prove also a lower bound of $\Omega(\sqrt{n}/\log n)$ in terms of achievable competitive ratio of any on-line algorithm.

Keywords: Connected search number · Distributed searching
On-line searching · Partial grid · Pursuit-evasion

1 Introduction

A team of mobile autonomous robots wants to search an area with the goal of finding a mobile intruder (or lost entity). The intruder is invisible, very fast and clever, which enforces robots to consider the worst case scenario for them since they want to have a search strategy that guarantees interception. The above problem is usually restated in discrete terms, naturally expressing the search game using graph-theoretic notation. Following the widely used terminology, the mobile entities performing the search are called *searchers*.

In this work we focus on the graph-theoretic problem statement, where the searchers operate in a given graph in which they move along edges. Moreover, what greatly influences algorithmic approaches is the assumption of whether the searchers know the graph in advance (off-line version), or whether the graph is unknown and the searchers learn its structure while conducting the search

Research partially supported by National Science Centre (Poland) grant number 2015/17/B/ST6/01887. The full version can be found at: https://arxiv.org/abs/1610.01458.

(on-line or distributed setting). We shortly review both approaches, giving later a formal statement of the problem we study in this work. In all cases we are interested in minimizing the number of searchers needed to clear the given network.[1] We omit discussion on a possible transition to a distributed model and we only note that it follows from our algorithm that it can be adopted to operate in distributed asynchronous setting, with searchers having local communication and polynomial memory. From the point of view of this work, the terms on-line and distributed are used exchangeably because we do not impose any communication, memory or synchronization restrictions on the agents (a more detailed discussion on this topic is provided in Sect. 2).

1.1 Related Work

Off-line Searching. The historically first studied graph searching model is called *edge search* [27,28], where the goal is to construct a search strategy that guarantees capturing a fast and invisible fugitive in a graph that is given as an input. A search strategy itself is a sequence of *moves*, where each move is one of the following: (i) placing a searcher on any graph node; (ii) removing a searcher from the node it occupies; (iii) sliding a searcher along an edge in order to clear it. We call a search strategy *monotone* if once an edge has been cleared, it must remain clean till the end of the search and *connected* if at any given time point the subgraph that is clean is connected. It is known that the edge search problem is monotone [23], while the connected search is not [32]. Moreover it turns out that each monotone edge search strategy can be converted (in polynomial time) into a monotone connected one by approximately 'doubling' the number of searchers [11]. Thus, for asymptotic results, like the one in this work, this gives another reason that justifies restricting attention to monotone connected search strategies. For connections between graph searching games and graph parameters, e.q., pathwidth, treewidth; see a survey [17] and since we adopt the connected searching problem in our on-line model, we point out to few recent works on the problem [2,3,9,10].

On-line Searching. In the distributed, or on-line, version of the problem it is assumed that the network is not known in advance to the searchers. In most cases, when designing distributed searching algorithms, the monotonicity requirement is adopted. (See [5] for an example how an optimal connected search strategy can be constructed in a distributed fashion when recontamination is allowed.) In this work we consider connected search strategies, which together with monotonicity allow us to assume that all searchers start at some node called the *homebase* and only moves of type (iii) are then made (see the definition of edge search above). During construction of a monotone strategy in an on-line way, there is naturally some 'cost' involved in terms of increased number of searchers required for guarding—this cost measured as the ratio of number of searchers that each on-line algorithm needs to use for some n-node graph and its monotone connected (off-line) search number is know to be $\Omega(n/\log n)$ [20].

---

[1] In this work the terms graph and network are used exchangeably.

Grid networks were studied in [8] where the searching model used the concept of temporal immunity: a node after cleaning remains protected (even if unguarded) against recontamination for a certain amount of time; see also [15,16]. For other distributed searching models and algorithms for specific network topologies see e.g. [14,18,26]. We note that our results may be of particular interest not only by providing theoretical insight into searching dynamics in distributed agent computations, but may also find applications in the field of robotics. Having in mind the vast literature on the subject we point the interested reader to a few references to recent works in this field [7,12,19,22,29–31].

1.2   Outline of this Work

The next section provides the notation used in this work and the problem statement. It is subdivided so that its first part defines the graph searching problem we study and the second part introduces the terminology related to the partial grid networks we consider. Section 3 gives a construction of a class of n-node networks such that each on-line algorithm uses $\Omega(\sqrt{n})$ searchers for some networks in the class which turns out to be $\Omega(\sqrt{n}/\log n)$ times more that an optimal off-line algorithm would use. This result serves as a lower bound in our further analysis. Section 4 describes an on-line algorithm that performs a guaranteed search in partial grids where it is assumed that the algorithm is given an upper bound n on the size of the network. We point out that this algorithm uses an on-line procedure from [6] as a subroutine that is called many times to clear selected parts of a grid and it can be seen as a generalization from a 'linear' graph structure studied in [6] to a 2-dimensional structure discussed in this work. Also, although both algorithms are conducted via some greedy rules which dictate how a search should 'expand' to unknown parts of the graph, the analysis of our algorithm is different from the one in [6]. We conclude that the final algorithm that receives no information on the underlying graph in advance uses $O(\sqrt{n})$ searchers. This result, stated in Theorem 4, is our main contribution. Due to space limitations, analysis of the algorithm and proofs of all lemmas and theorems are in the appendix and this extended abstract is intended to contain a detailed presentation of the algorithm with necessary notation and corresponding lower bound.

2   Definitions and Terminology

In this section we present the notation we use. We consider only simple undirected connected graphs. Given any graph $G = (V, E)$ and $X \subseteq V$, $G[X]$ is the subgraph of G induced by X: its node set is X and consists of all edges $\{u, v\}$ of G having both endpoints in X.

Problem Statement. A *connected k-search strategy* S for a network G is defined as follows. Initially, k searchers are placed on a node h of G, called the *homebase*. (We also say that S *starts at* h.) Then, S is a sequence of moves, where each *move* consists of selecting one searcher present at some node u and sliding the searcher along an edge $\{u, v\}$. Initially, all edges are *contaminated*. After each move of sliding a searcher along an edge $\{u, v\}$ it is declared to be *clean*. It becomes contaminated again (*recontaminated*) if at any time during execution of the strategy S at least one of its endpoints is not occupied by a searcher and is incident to a contaminated edge. If no recontamination happens in S, then S is called *monotone*. Regardless of whether the strategy is monotone or not, we require that the subgraph consisting of all clean edges is connected after each move of the search strategy. Finally, we require that after the last move of S all edges are clean. The minimum k such that there exists a node h and a (monotone) connected k-search strategy that starts at h is called the (*monotone*) *connected search number* of G and denoted by ($\mathtt{mcs}(G)$, respectively) $\mathtt{cs}(G)$.

We now state the on-line model we use. All searchers start at the homebase— a selected node of the network. The network itself is not known in advance to the searchers (except for the fact that the searchers may expect that the network is a partial grid). (We note here that our main algorithmic result will be obtained in two stages: first we describe an algorithm that as an input receives the upper bound n of the size of the network and then we use it to obtain our main result, an algorithm that works without any a priori information about the network.) We assume that the edges incident to each node are marked with unique labels (port numbers) and because only partial grids are considered in this work (see the definition below) we assume that labels naturally reflect all possible directions for each edge (i.e., left, right, up and down). A searcher present at a node is able to perceive these labels. The moves of searchers are performed by a centralized scheduler who has the following partial knowledge of the graph: it knows the subgraph induced by all nodes previously visited by searchers and also knows the edges incident to visited nodes and leading to unvisited nodes. The time is divided into steps, in each step the scheduler may for each searcher perform one of the two actions: either move a searcher from currently occupied node v to a neighbor of v, or leave the searcher at the currently occupied node.

Partial Grid Notation. Assume that a grid graph is embedded into two-dimensional Cartesian coordinate system with a horizontal x-axis and vertical y-axis, where there is a node at each point with integer coordinates and two nodes are adjacent if and only if the distance between them nodes equals one (in Euclidean metric). We define a partial grid $G = (V, E)$ with a set of nodes V and edges E, as a connected subgraph of such a grid graph. For convenience, the homebase is located in a $(0, 0)$ position. In order to refer to a node that corresponds to a point with coordinates (x, y) we write $v(x, y)$. In this work n denotes an upper bound of the number of nodes of a partial grid, such that $\sqrt{n}$ is an integer. Let us notice here that nodes at distance one in the grid are not necessarily neighbors in the graph. We note that some graphs can be embedded in various ways and for different embeddings our algorithm may perform differently.

Informally speaking, our algorithm will conduct a search by expanding the clean part of the graph from one 'checkpoint' to another. These checkpoints (defined formally later) will be subsets of nodes and their potential placements on the partial grid are dictated by a concept of a *frontier*. Take any $x = i\sqrt{n}$ for some integer i, $y = j\sqrt{n}$ for some integer j and take $i', j' \in \{0, 1\}, i' \neq j'$. Then, the line segment with endpoints (x, y) and $(x + \sqrt{n}i', y + \sqrt{n}j')$ is called a *frontier* and denoted by $F((x, y), (x + \sqrt{n}i', y + \sqrt{n}j'))$. Whenever the endpoints of a frontier are clear from the context or not important we will omit them. The set of all frontiers is denoted by $\mathcal{F}$. The subgraph induced by all nodes that belong to a frontier F is denoted by $G[F]$.

For $i \in \{1, \ldots, \sqrt{n}\}$ and some frontier $F = F((x, y), (x', y'))$, where $x \leq x'$ and $y \leq y'$, we define the *i-th rectangle of F*, denoted by $\mathcal{R}(F, i)$, as the rectangle with corner vertices $(x - i, y - i)$, $(x - i, y + i)$, $(x' + i, y' - i)$, $(x' + i, y' + i)$ if $y = y'$ and as the rectangle with corner vertices $(x - i, y - i)$, $(x + i, y - i)$, $(x' - i, y' + i)$, $(x' + i, y' + i)$ if $x = x'$.

The two above concepts, namely frontiers and rectangles, provide a template on how the search may 'progress'. However, due to the structure of a partial grid it may be possible that only certain nodes, but not all, that lie on a frontier are reached at some point of a search strategy. For this reason, our notation needs to be extended to subsets of nodes that lie on frontiers and the corresponding rectangles. Let $F = F((x, y), (x', y'))$ be some frontier. Any subset C of nodes of G that belong to F is called a *checkpoint*. The *0-th expansion of a checkpoint* C is C itself and is denoted by $C\langle 0 \rangle$. For $i \in \{1, \ldots, \sqrt{n}\}$ we define *an i-th expansion of C*, denoted by $C\langle i \rangle$, recursively as follows: the set $C\langle i \rangle$ consists of all nodes $v \notin C\langle 0 \rangle \cup C\langle 1 \rangle \cup \cdots \cup C\langle i - 1 \rangle$ for which there exists a node $u \in C\langle i - 1 \rangle$, such that there exists a path between v and u in the subgraph of G induced by nodes that lie on the rectangles $\mathcal{R}(F, 0), \mathcal{R}(F, 1), \ldots, \mathcal{R}(F, i)$. Define $C^+\langle i \rangle = C\langle 0 \rangle \cup \ldots \cup C\langle i \rangle$, $i \in \{0, \ldots, \sqrt{n}\}$. Informally, $C\langle i \rangle$ consists of only those nodes that belong to the rectangle $\mathcal{R}(F, i)$ that are connected to nodes of C by paths that lie 'inside' of $\mathcal{R}(F, i)$—this definition captures the behavior of searchers (in our algorithm) that guard the nodes of C and 'expand' from C in all directions: then possible nodes that belong to any of the rectangles $\mathcal{R}(F, 0), \mathcal{R}(F, 1), \ldots, \mathcal{R}(F, i)$ but do not belong to $C^+\langle i \rangle$ will not be reached by the searchers. Notice that not every node of the rectangle $\mathcal{R}(F, i)$ has to be guarded—only the ones that have neighbors outside $C^+\langle i \rangle$. See Fig. 1 for an exemplary checkpoint with its expansions.

3   Lower Bound

First note that a regular $\sqrt{n} \times \sqrt{n}$ grid requires $\Omega(\sqrt{n})$ searchers even in the offline setting [13], that is, when the network is known in advance and the searchers may decide on the location of the homebase. Therefore, our on-line algorithm is asymptotically optimal with respect to this worst case measure.

What we would also like to obtain is a lower bound expressed as a competitive ratio, which is defined as a maximized over all networks and all starting nodes ratio between number of searchers that a given algorithm uses and the search number that is optimal for a given network in an offline settings. In other words, for any on-line algorithm A, let $A(G, h)$ be the number of searchers that it uses to clean a network G starting from the homebase h and let $A(G) = \max_h A(G, h)$.

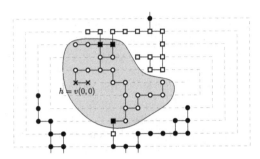

Fig. 1. Some expansions of a checkpoint C (here $\sqrt{n} = 9$); *crosses* denote $C = C\langle 0\rangle$, *gray area* covers nodes that belong to $C^+\langle 3\rangle$, *empty squares* denote nodes in $C\langle 4\rangle$ and *dark squares* denote the ones that need to be guarded when the gray area consists of the clean nodes. The horizontal dotted line that contains h is the considered frontier.

We aim at proving that for *each* on-line algorithm A there exists an n-node partial grid network G such that $A(G)/\mathtt{mcs}(G) = \Omega(\sqrt{n}/\log n)$. Due to space limitations we give a brief sketch of the proof and a formal analysis is postponed to the appendix.

Define a class of partial grids $\mathcal{L} = \bigcup_{l \geq 0} \mathcal{L}_l$, where $\mathcal{L}_l$ for $l \geq 0$ is defined recursively as follows. We take $\mathcal{L}_0$ to contain one network that is a single node located at $(0,0)$. Then, in order to describe how $\mathcal{L}_{l+1}$ is obtained from $\mathcal{L}_l$, $l \geq 0$, we introduce an operation of *extending* $G \in \mathcal{L}_l$ at i, for $i \in \{0, \ldots, l\}$. In this operation, first take G and add $l + 2$ new nodes located at coordinates: $(0, l+1), (1, l), \ldots, (j, l+1-j), \ldots, (l+1, 0)$. Call these coordinates the $(l+1)$-*th diagonal*. For each $j \in \{0, \ldots, i\}$ add an edge connecting the nodes $v(j, l-j)$ and $v(j, l-j+1)$, and for each $j \in \{i, \ldots, l\}$ add an edge connecting the nodes $v(j, l-j)$ and $v(j+1, l-j)$. Then, obtain $\mathcal{L}_{l+1}$ as follows: initially take $\mathcal{L}_{l+1}$ to be empty and then for each $G \in \mathcal{L}_l$ and for each $i \in \{0, \ldots, l\}$, obtain a network G' by extending G at i and add G' to $\mathcal{L}_{l+1}$. Notice here that a graph constructed this way is not only a partial grid, but also a tree. Intuitively (see appendix for a complete proof) whenever any algorithm reaches for the first time a node $v(i, j-i)$ in the j-th diagonal, the *adversary* decides to extend at i the network explored so far, thus always forcing the situation that the first node reached on a diagonal is of degree three.

Observe that each network G in $\mathcal{L}$ is a tree and therefore $\mathtt{mcs}(G) = O(\log(n))$, $n = |V(G)|$ [2,25]. This gives us the following.

Theorem 1. *For each on-line algorithm A computing a connected monotone search strategy there exists an n-node network G with homebase h such that $A(G, h)/\mathtt{mcs}(G) = \Omega(\sqrt{n}/\log n)$.*

4   The Algorithm

In this section we describe our algorithm that takes an upper bound on the size of the network as an input. We start below with the initialization performed at the beginning of the algorithm. Then, we give all pseudocodes ending with the main algorithm—Procedure GridSearching. After each move performed by searchers, each searcher that occupies a node that does not need to be guarded is said to be *free*. Each node that needs to be guarded is occupied by at least one searcher; if more searchers occupy such node then all of them except for one are also *free*. If, at some point, no node of the last expansion of some checkpoint needs to be guarded, then we say that the expansion is *empty*.

The initialization performed by the searchers is as follows. The initial checkpoint C_0 is the set of nodes of the connected component of $G[F((0,0),(\sqrt{n},0))]$ that contains h. Thus, initially $|C_0|$ searchers place themselves on all nodes of C_0 (note that the nodes of C_0 induce a path in G).

Procedure CleanExpansion. We start with an informal description of the procedure. When a new checkpoint C has been reached, our search strategy 'expands' from C by successively cleaning subgraphs $G[C^+\langle i\rangle]$ for $i \in \{1, \ldots, \sqrt{n}\}$. Once all nodes in $C^+\langle i-1\rangle$ are clean for some $0 < i \leq \sqrt{n}$, the transition to reaching the state in which all nodes in $C^+\langle i\rangle$ are clean requires cleaning all nodes of the i-th expansion of C. This is done by calling for every guarded node u from $C^+\langle i-1\rangle$ a special procedure (ModConnectedSearching, described below), which cleans nodes which belong to $C\langle i\rangle$ and 'has access' to them from u. Procedure CleanExpansion performs this job by using $O(\sqrt{n})$ searchers.

For cleaning all nodes of i-th expansion of C, provided that $G[C^+\langle i-1\rangle]$ is clean we will use a procedure from [6] (intuitively, we use the fact that our rectangles are of 'width' $2i+1$) which is more general and due to space limitations we directly rewrite it as we intend to use it.

Theorem 2 ([6]). *Let F be any frontier and let G' be any connected partial grid with nodes lying on rectangles $\mathcal{R}(F,0), \mathcal{R}(F,1), \ldots, \mathcal{R}(F,i)$, $i \geq 0$. There exists an on-line procedure ConnectedSearching that, starting at an arbitrarily chosen homebase in G', clears G' in a connected and monotone way using $6i+4$ searchers.*

Note that while using procedure ConnectedSearching, we will be cleaning a subgraph of $G[C^+\langle i\rangle]$ that is embedded into the entire partial grid and thus some nodes v of $G[C^+\langle i\rangle]$ have edges leading to neighbors that lie outside of $G[C^+\langle i\rangle]$. If such an edge is already clean, then no recontamination happens for the node v and moreover no searcher used by ConnectedSearching for the subgraph of $G[C^+\langle i\rangle]$ needs to stay at v. On the other hand, if such an edge is contaminated (and thus not reached yet by our search strategy), then v needs to be guarded and for that end we place an extra searcher on it that guards v during the remaining execution of ConnectedSearching. Note that in the latter case, the node v belongs to $\mathcal{R}(F,i)$, where F is the frontier that contains the nodes of C and therefore there exist $O(\sqrt{n})$ such nodes v. In other words, ConnectedSearching is called to clean a certain subgraph contained within $\mathcal{R}(F,i)$ and whenever a node on the rectangle $\mathcal{R}(F,i)$ has a contaminated edge leading outside of the rectangle $\mathcal{R}(F,i)$, then an extra searcher, not accommodated by ConnectedSearching

in Theorem 2, is introduced to be left behind to guard v. The modification of
`ConnectedSearching` that leaves behind a searcher on each such newly reached
node of $\mathcal{R}(F, i)$ will be denoted by `ModConnectedSearching`. Note that this procedure is invoked for every guarded node from $C^+\langle i-1 \rangle$ in order to clean $C^+\langle i \rangle$,
see Figs. 2 and 3 for examples.

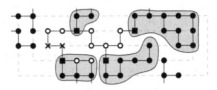

Fig. 2. Example of an execution of procedure `CleanExpansion`; *crosses* denote $C = C\langle 0 \rangle$, *empty circles* denote nodes that belong to $C^+\langle 1 \rangle$, *dark squares* denote the one
that belongs to $C^+\langle 1 \rangle$ and for which procedure `ModConnectedSearching` is invoked,
gray areas show nodes that will be cleaned in four calls of `ModConnectedSearching` in
order to clean $C\langle 2 \rangle$. Note that there are some nodes that belongs to $C^+\langle 1 \rangle$ and are
guarded at first, but after one of the calls of `ModConnectedSearching` there is no need
to guard them any more, so the procedure is not invoked for them.

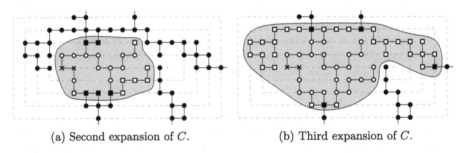

(a) Second expansion of C.   (b) Third expansion of C.

Fig. 3. Two expansions of some checkpoint C; *crosses* denote $C = C\langle 0 \rangle$, *empty circles* are nodes cleaned in previous expansions; *squares* are nodes explored in current
expansion; *dark circles* are nodes not reached yet by the searchers; and *dark squares*
are nodes that need to be guarded at the end of current expansion. *Gray areas* show
the clean part of the graph, i.e., $C^+\langle i \rangle$ for $i \in \{2, 3\}$.

It follows that it is enough to provide as an input to `ModConnectedSearching`:
a node in $C^+\langle i-1 \rangle$ that plays the role of homebase for `ModConnectedSearching`,
the frontier F and i. We note that each checkpoint used in our final algorithm is
obtained as follows: some frontier F is selected and then a checkpoint C is created
as some set of nodes that belong to F; thus we assume that with C is associated
such a unique frontier F. Thus, this approach guarantees us using at most $6i + 4$
searchers to clean $G[C\langle i \rangle]$ and, in addition to those, at most $2\sqrt{n} + 8i$ searchers
for guarding nodes laying on $\mathcal{R}(F, i)$. To summarize, we give a formal statement
of our procedure.

Procedure. `CleanExpansion`

Input: An expansion $C\langle i - 1\rangle$ with C contained in the frontier F, $i \geq 1$.
Result: Cleaning all nodes of $C\langle i\rangle$.

    while there exists a node $v \in C\langle i - 1\rangle$ with a contaminated neighbor u in $C\langle i\rangle$ **do**
        Place $6i + 4 + 2\sqrt{n} + 8i$ free searchers on v to be used by `ModConnectedSearching`.
        Call `ModConnectedSearching` for v as the homebase, frontier F and integer i.

Procedure `UpgradeCheckpoints`. By definition, if F is some frontier, then $\mathcal{R}(F, \sqrt{n})$ contains 10 frontiers. Thus, reaching the $\sqrt{n}$-th expansion $C\langle\sqrt{n}\rangle$ of a checkpoint of F provides a possibility of creating one new checkpoint for each of the above frontiers. Procedure `UpgradeCheckpoints`, which takes as an input $C\langle\sqrt{n}\rangle$ and a collection $\mathcal{C}$ of currently present checkpoints, generates these new checkpoints and adds them to $\mathcal{C}$ and removes C from $\mathcal{C}$. Also, if it happens that some newly constructed checkpoint belongs to the same frontier as some existing checkpoint in $\mathcal{C}$ and no expansion for the existing one has been performed yet, then both checkpoints are merged into one. Finally, any checkpoint in $\mathcal{C}$ whose lastly performed expansion is empty is removed from $\mathcal{C}$. We remark that only procedure `UpgradeCheckpoints` modifies the collection of checkpoints $\mathcal{C}$ and this procedure performs no cleaning moves.

Thus, to summarize, the 'lifetime' of a checkpoint is as follows. Once the 1-st expansion of C is performed, the checkpoint will remain in the collection $\mathcal{C}$ and possibly more expansions of C are made (in total at most $\sqrt{n}$ expansion are possible for each checkpoint). A checkpoint C may disappear from $\mathcal{C}$ in three ways:

– when C is in its 0-th expansion and another checkpoint C' appears in the same frontier (thus, C' is in its 0-th expansion) and then the nodes of C are added to C', or
– some expansion of C becomes empty (then C is not removed from $\mathcal{C}$ right away but during the subsequent call to `UpgradeCheckpoints`), or
– C reaches its $\sqrt{n}$-th expansion and procedure `UpgradeCheckpoints` is called for C (in which case C possibly 'gives birth' to new checkpoints during the execution of `UpgradeCheckpoints`).

Our algorithm maintains a collection $\mathcal{C}$ of currently used checkpoints.
Procedure `GridSearching`. We start with an informal description. The search strategy it produces is divided into phases. In each step of the algorithm, the checkpoint with the highest number of nodes that need to be guarded is being chosen and the next expansion is being made on it. When one of the checkpoints reaches its $\sqrt{n}$-th expansion, then the current phase ends and the procedure `UpgradeCheckpoints` is being invoked. Thus, the division of search strategy into phases is dictated by consecutive calls to procedure `UpgradeCheckpoints`. For an expansion C, in the pseudocode below we write $\delta(C)$ to refer to the set of nodes that belong to the last expansion of C and need to be guarded at a given point. We finish description of the algorithm by showing in Fig. 4 how our algorithm clears an exemplary partial grid network.

Procedure. UpgradeCheckpoints

Input: $C\langle\sqrt{n}\rangle$ and the collection of all checkpoints $\mathcal{C}$
Result: Updated collection $\mathcal{C}$

$\quad \mathcal{C} \leftarrow \mathcal{C} \setminus \{C\}$
$\quad \mathcal{C}_{new} \leftarrow \emptyset$
$\quad$**for each** frontier F on $\sqrt{n}$-th rectangle of the frontier containing C **do**
$\quad\quad$ Let C' consist of all guarded nodes in F.
$\quad\quad$ If $C' \neq \emptyset$, then $\mathcal{C}_{new} \leftarrow \mathcal{C}_{new} \cup \{C'\}$.
$\quad$**for each** C'' in $\mathcal{C}$ **do**
$\quad\quad$ **if** there exists $C' \in \mathcal{C}_{new}$ that is a subset of the same frontier as C'' **then**
$\quad\quad\quad$ **if** C'' is in 0-th expansion **then**
$\quad\quad\quad\quad$ $\mathcal{C} \leftarrow \mathcal{C} \setminus \{C''\}$
$\quad\quad\quad\quad$ Replace C' with $C'' \cup C'$ in $\mathcal{C}_{new}$.
$\quad \mathcal{C} \leftarrow \mathcal{C} \cup \mathcal{C}_{new}$
$\quad$**for each** C in $\mathcal{C}$ **do**
$\quad\quad$ **if** no node in the last expansion of C is guarded **then**
$\quad\quad\quad$ $\mathcal{C} \leftarrow \mathcal{C} \setminus \{C\}$

Procedure. GridSearching

Input: An integer n providing an upper bound on the size of the partial grid G.
Result: A monotone connected search strategy for G.
$\quad$ Perform the initialization (see the beginning of Sect. 4).
$\quad$**while** G is not clean **do**
$\quad\quad$ **while** no checkpoint has reached its $\sqrt{n}$-th expansion **do**
$\quad\quad\quad$ Let $C_{max} \in \mathcal{C}$ be such that $\delta(C_{max}) \geq \delta(C)$ for each $C \in \mathcal{C}$.
$\quad\quad\quad$ Let i be the number of expansions of C_{max} performed so far.
$\quad\quad\quad$ Invoke **CleanExpansion** for $C_{max}\langle i\rangle$.
$\quad\quad$ Invoke **UpgradeCheckpoints** for the $\sqrt{n}$-th expansion $C_{max}\langle\sqrt{n}\rangle$ and for $\mathcal{C}$.

We close this section, by giving the final results about the performance of our algorithm, which using a standard binary search (also referred as doubling) technique over n, can be turned into one that receives no information regarding the size of the underlying grid (the formal details are moved to the appendix due to space limitation).

Theorem 3. *Given an upper bound n of the size of the network as an input, the algorithm* GridSearching *clears in a connected and monotone way any unknown underlying partial grid network using $O(\sqrt{n})$ searchers.*

Theorem 4. *There exists a on-line algorithm that clears (starting at an arbitrary homebase) in a connected and monotone way any unknown underlying partial grid network using $O(\sqrt{n})$ searchers. The algorithm receives no prior information on the network.*

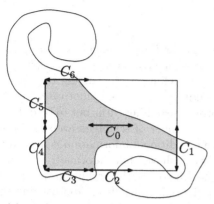

(a) At the end of the first phase C_0 (initial checkpoint) reaches its $\sqrt{n}$-th expansion. Procedure `UpgradeCheckpoints` creates 6 new checkpoints and removes C_0 from $\mathcal{C}$, i.e. $\mathcal{C} = \{C_1, C_2, C_3, C_4, C_5, C_6\}$.

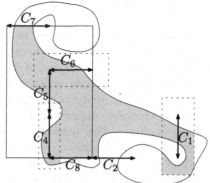

(b) $\sqrt{n}$-th expansion of C_5 ends the second phase. Checkpoints C_4 and C_6 are removed from $\mathcal{C}$ because (in our example) there is no need to guard any node on theirs expansions. C_3 is in 0-th expansion, so when a new checkpoint C_8 is created on the same frontier, C_3 is removed from $\mathcal{C}$; $\mathcal{C} = \{C_1, C_2, C_7, C_8\}$.

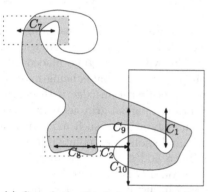

(c) C_1 ends the third phase. Notice that a new checkpoint C_9 emerged 'inside' already cleaned area by C_0; C_2 is removed from $\mathcal{C}$ even if $\sqrt{n}$-th expansion has not been reached but its last expansion has no nodes to be guarded; $\mathcal{C} = \{C_7, C_9, C_{10}\}$.

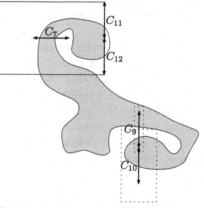

(d) C_7 ends fourth phase. Note that a new checkpoint C_{12} emerged on an edge of C_5's $\sqrt{n}$-th expansion; it was not be created in the second phase because then there was no access to the contaminated part; $\mathcal{C} = \{C_{11}, C_{12}\}$.

Fig. 4. Clearing an exemplary partial grid by procedure `GridSearching`; *gray areas* denote the clean part, *arrows* denote frontiers on which the marked checkpoints lie, *dotted rectangles* around checkpoints denote their current expansions and *solid rectangles* denote the $\sqrt{n}$-th expansions, which end phases.

5 Motivation

There exists a number of studies of graph searching problems in the graph-theoretic context. Much less is known for geometric scenarios. It turns out that [24] the geometric analogue of graph searching is challenging to analyze. The class of graphs we have selected to study in this work is motivated by the following arguments. First, on-line (monotone) searching turns out to be difficult in terms of achievable upper bound on the number of searchers even in simple topologies like trees. This suggest that some additional information is needed to perform on-line search efficiently and our work shows that, informally speaking, a two-dimensional sense of direction is enough to search a graph in asymptotically almost optimal way. Our second motivation comes from approaching the problem of geometric search by considering its discrete analogues, i.e., by modeling via graph theory. We give a short sketch to give an overview as the problem of modeling is out of scope of this work and we only refer to some recent works on the subject [1,4,21,24]. Consider a continuous search problem in which k searchers initially placed at the same location need to capture the fugitive hiding in an arbitrary polygon that possibly has holes. The polygon is not known a priori to the searchers. The fugitive is considered captured in time t when it is located at distance at most r from some searcher at time point t. (The distance r can be related to physical dimensions of searchers and/or their visibility range, etc.)

Consider the following transition from the above continuous searching problem of a polygon to a discrete one. Overlap the coordinate system with the polygon in such a way that the origin coincides with the original placement of the searchers. Then, place nodes on all points with coordinates being multiples of r and lying in the polygon. Connect two nodes with an edge if the edge is contained in the polygon. In this way we obtain a partial grid network. In this brief sketch we omit potential problems that may arise in such modeling, like obtaining disconnected networks or having 'blind spots', i.e., points in the polygon that cannot be cleared by using the above nodes and edges only. We say that a partial grid network G covers the polygon if G is connected and for each point p in the polygon there exist a node of G in distance at most r from p.

Note that any search strategy $\mathcal{S}'$ for a polygon P can be used to obtain a search strategy $\mathcal{S}$ for underlying partial grid network G as follows. For each searcher s used in $\mathcal{S}'$ introduce four searchers $s_1, \ldots, s_4$ that will 'mimic' its movements by going along edges of G. More precisely, the searchers $s_1, \ldots, s_4$ will ensure that at any point, if s is located at a point (x, y), then $s_1, \ldots, s_4$ will reside on nodes with coordinates $(\lfloor x/r \rfloor, \lfloor y/r \rfloor)$, $(\lfloor x/r \rfloor, \lceil y/r \rceil)$, $(\lceil x/r \rceil, \lfloor y/r \rfloor)$, $(\lceil x/r \rceil, \lceil y/r \rceil)$. In this way, area protected by s in $\mathcal{S}'$ is always protected by four searchers in $\mathcal{S}$. This gives us the following.

Observation 1. *Let P be a polygon and let G by an underlying partial grid network that covers P. Then, there exists a search strategy for G using k searchers such that its execution in G results in clearing P and $k = O(p)$, where p is the minimum number of searchers required for clearing P (in continuous way).*

6   Open Problems

In view of the lower bound shown in [20] that even in such simple networks as trees each distributed or on-line algorithm may be forced to use $\Omega(n/\log n)$ times more searchers than the connected search number of the underlying network, one possible line of research is to restrict attention to specific topologies that allow to obtain algorithms with good provable upper bounds. This work gives one such example. An interesting research direction is to find other non-trivial settings in which distributed or on-line search can be conducted efficiently. Also, we leave a logarithmic gap in our approximation ratio. Since there exist grids that require $\Omega(\sqrt{n})$ searchers the gap can be possibly closed by analyzing the grids that require few (e.g. $O(\log n)$) searchers.

The above questions related to network topologies can be stated more generally: what properties of the on-line model are crucial for such search for fast and invisible fugitive to be efficient? This work and also a recent one [6] suggest that a 'sense of direction' may be one such factor. Possibly interesting directions may be to analyze the influence of visibility on search scenarios.

We finally note that the only optimization criterion that was of interest in this work is the number of searchers. This coincides with the research done in off-line search problems where this was the most important criterion giving nice ties between graph searching theory and structural graph theory. However, one may consider adding different optimization criteria like time (defined as the maximum number of synchronized steps) or cost (the total number of moves performed by all searchers).

References

1. Altshuler, Y., Yanovski, V., Wagner, I., Bruckstein, A.: Multi-agent cooperative cleaning of expanding domains. Int. J. Robot. Res. **30**(8), 1037–1071 (2011)
2. Barrière, L., Flocchini, P., Fomin, F., Fraigniaud, P., Nisse, N., Santoro, N., Thilikos, D.: Connected graph searching. Inf. Comput. **219**, 1–16 (2012)
3. Best, M.J., Gupta, A., Thilikos, D.M., Zoros, D.: Contraction obstructions for connected graph searching. Discret. Appl. Math. (2015). https://doi.org/10.1016/j.dam.2015.07.036
4. Bhadauria, D., Klein, K., Isler, V., Suri, S.: Capturing an evader in polygonal environments with obstacles: the full visibility case. Int. J. Robot. Res. **31**(10), 1176–1189 (2012)
5. Blin, L., Fraigniaud, P., Nisse, N., Vial, S.: Distributed chasing of network intruders. Theor. Comput. Sci. **399**(1–2), 12–37 (2008)
6. Borowiecki, P., Dereniowski, D., Kuszner, L.: Distributed graph searching with a sense of direction. Distrib. Comput. **28**(3), 155–170 (2015)
7. Chung, T., Hollinger, G., Isler, V.: Search and pursuit-evasion in mobile robotics - a survey. Auton. Robots **31**(4), 299–316 (2011)
8. Daadaa, Y., Flocchini, P., Zaguia, N.: Network decontamination with temporal immunity by cellular automata. In: Bandini, S., Manzoni, S., Umeo, H., Vizzari, G. (eds.) ACRI 2010. LNCS, vol. 6350, pp. 287–299. Springer, Heidelberg (2010). https://doi.org/10.1007/978-3-642-15979-4_31

9. Dereniowski, D.: Connected searching of weighted trees. Theor. Comput. Sci. **412**, 5700–5713 (2011)
10. Dereniowski, D.: Approximate search strategies for weighted trees. Theor. Comput. Sci. **463**, 96–113 (2012)
11. Dereniowski, D.: From pathwidth to connected pathwidth. SIAM J. Discret. Math. **26**(4), 1709–1732 (2012)
12. Durham, J., Franchi, A., Bullo, F.: Distributed pursuit-evasion without mapping or global localization via local frontiers. Auton. Robot. **32**(1), 81–95 (2012)
13. Ellis, J., Warren, R.: Lower bounds on the pathwidth of some grid-like graphs. Discret. Appl. Math. **156**(5), 545–555 (2008)
14. Flocchini, P., Huang, M., Luccio, F.: Decontamination of hypercubes by mobile agents. Networks **52**(3), 167–178 (2008)
15. Flocchini, P., Luccio, F., Pagli, L., Santoro, N.: Optimal network decontamination with threshold immunity. In: Spirakis, P.G., Serna, M. (eds.) CIAC 2013. LNCS, vol. 7878, pp. 234–245. Springer, Heidelberg (2013). https://doi.org/10.1007/978-3-642-38233-8_20
16. Flocchini, P., Luccio, F., Pagli, L., Santoro, N.: Network decontamination under m-immunity. Discret. Appl. Math. **201**, 114–129 (2016)
17. Fomin, F., Thilikos, D.: An annotated bibliography on guaranteed graph searching. Theor. Comput. Sci. **399**(3), 236–245 (2008)
18. Gonçalves, V.C.F., Lima, P.M.V., Maculan, N., França, F.M.G.: A distributed dynamics for webgraph decontamination. In: Margaria, T., Steffen, B. (eds.) ISoLA 2010. LNCS, vol. 6415, pp. 462–472. Springer, Heidelberg (2010). https://doi.org/10.1007/978-3-642-16558-0_39
19. Hollinger, G., Singh, S., Djugash, J., Kehagias, A.: Efficient multi-robot search for a moving target. Int. J. Robot. Res. **28**(2), 201–219 (2009)
20. Ilcinkas, D., Nisse, N., Soguet, D.: The cost of monotonicity in distributed graph searching. Distrib. Comput. **22**(2), 117–127 (2009)
21. Karaivanov, B., Markov, M., Snoeyink, J., Vassilev, T.: Decontaminating planar regions by sweeping with barrier curves. In: Proceedings of the 26th Canadian Conference on Computational Geometry, CCCG 2014 (2014)
22. Kolling, A., Carpin, S.: Multi-robot pursuit-evasion without maps. In: Proceedings of IEEE International Conference on Robotics and Automation, ICRA 2010, pp. 3045–3051 (2010)
23. LaPaugh, A.: Recontamination does not help to search a graph. J. ACM **40**(2), 224–245 (1993)
24. Markov, M., Haralampiev, V., Georgiev, G.: Lower bounds on the directed sweep-width of planar shapes. Serdica J. Comput. **9**(2), 151–166 (2015)
25. Megiddo, N., Hakimi, S., Garey, M., Johnson, D., Papadimitriou, C.: The complexity of searching a graph. J. ACM **35**(1), 18–44 (1988)
26. Moraveji, R., Sarbazi-Azad, H., Zomaya, A.: Performance modeling of Cartesian product networks. J. Parallel Distrib. Comput. **71**(1), 105–113 (2011)
27. Parsons, T.D.: Pursuit-evasion in a graph. In: Alavi, Y., Lick, D.R. (eds.) Theory and Applications of Graphs. LNCS, vol. 642, pp. 426–441. Springer, Heidelberg (1978). https://doi.org/10.1007/BFb0070400
28. Petrov, N.: A problem of pursuit in the absence of information on the pursued. Differ. Uravn. **18**, 1345–1352 (1982)
29. Robin, C., Lacroix, S.: Multi-robot target detection and tracking: taxonomy and survey. Auton. Robot. **40**(4), 729–760 (2016)
30. Sachs, S., LaValle, S., Rajko, S.: Visibility-based pursuit-evasion in an unknown planar environment. Int. J. Robot. Res. **23**(1), 3–26 (2004)

31. Stiffler, N., O'Kane, J.: A complete algorithm for visibility-based pursuit-evasion with multiple pursuers. In: Proceedings of the IEEE International Conference on Robotics and Automation, ICRA 2014, pp. 1660–1667 (2014)
32. Yang, B., Dyer, D., Alspach, B.: Sweeping graphs with large clique number. Discret. Math. **309**(18), 5770–5780 (2009)

Online Unit Clustering in Higher Dimensions

Adrian Dumitrescu[1] and Csaba D. Tóth[2(⊠)]

[1] University of Wisconsin–Milwaukee, Milwaukee, WI, USA
dumitres@uwm.edu
[2] California State University, Northridge, Los Angeles, CA, USA
csaba.toth@csun.edu

Abstract. We revisit the online UNIT CLUSTERING problem in higher dimensions: Given a set of n points in $\mathbb{R}^d$, that arrive one by one, partition the points into clusters (subsets) of diameter at most one, so as to minimize the number of clusters used. In this paper, we work in $\mathbb{R}^d$ using the L_∞ norm. We show that the competitive ratio of any algorithm (deterministic or randomized) for this problem must depend on the dimension d. This resolves an open problem raised by Epstein and van Stee (WAOA 2008). We also give a randomized online algorithm with competitive ratio $O(d^2)$ for UNIT CLUSTERING of integer points (i.e., points in $\mathbb{Z}^d$, $d \in \mathbb{N}$, under L_∞ norm). We complement these results with some additional lower bounds for related problems in higher dimensions.

1 Introduction

Covering and clustering are ubiquitous problems in the theory of algorithms, computational geometry, optimization, and others. Such problems can be asked in any metric space, however this generality often restricts the quality of the results, particularly for online algorithms. Here we study lower bounds for several such problems in a high dimensional Euclidean space and mostly in the L_∞ norm. We first consider their *offline* versions.

Problem 1. k-CENTER. Given a set of n points in $\mathbb{R}^d$ and an integer k, cover the set by k congruent balls centered at the points so that the diameter of the balls is minimized.

The following two problems are dual to Problem 1.

Problem 2. UNIT COVERING. Given a set of n points in $\mathbb{R}^d$, cover the set by balls of unit diameter so that the number of balls is minimized.

Problem 3. UNIT CLUSTERING. Given a set of n points in $\mathbb{R}^d$, partition the set into clusters of diameter at most one so that number of clusters is minimized.

Research supported in part by the NSF awards CCF-1422311 and CCF-1423615.

Problems 1 and 2 are easily solved in polynomial time for points on the line, i.e., for $d = 1$; however, both problems become NP-hard already in the Euclidean plane [17, 22]. Factor 2 approximations are known for k-CENTER in any metric space (and so for any dimension) [16, 18]; see also [23, Chap. 5], [24, Chap. 2], while polynomial-time approximation schemes are known for UNIT COVERING for any fixed dimension [20].

Problems 2 and 3 look similar; indeed, one can go from clusters to balls in a straightforward way; and conversely one can assign multiply covered points to unique balls. As such, the two problems are identical in the offline setting.

We next consider their *online* versions. In this paper we focus on problems 2 and 3 in particular. It is worth emphasizing two common properties: (i) a point assigned to a cluster must remain in that cluster; and (ii) two distinct clusters cannot merge into one cluster, i.e., the clusters maintain their identities.

The performance of an online algorithm ALG is measured by comparing it to an optimal offline algorithm OPT using the standard notion of competitive ratio [5, Chap. 1]. The competitive ratio of ALG is defined as $\sup_\sigma \frac{\text{ALG}(\sigma)}{\text{OPT}(\sigma)}$, where σ is an input sequence of request points, $\text{OPT}(\sigma)$ is the cost of an optimal offline algorithm for σ and $\text{ALG}(\sigma)$ denotes the cost of the solution produced by ALG for this input. For randomized algorithms, $\text{ALG}(\sigma)$ is replaced by the expectation $E[\text{ALG}(\sigma)]$, and the competitive ratio of ALG is $\sup_\sigma \frac{E[\text{ALG}(\sigma)]}{\text{OPT}(\sigma)}$. Whenever there is no danger of confusion, we use ALG to refer to an algorithm or the cost of its solution, as needed.

Charikar et al. [8] have studied the online version of UNIT COVERING. The points arrive one by one and each point needs to be assigned to a new or to an existing unit ball upon arrival; the L_2 metric is used in $\mathbb{R}^d$, $d \in \mathbb{N}$. The location of each new ball is fixed as soon as it is opened. The authors provided a deterministic algorithm of competitive ratio $O(2^d d \log d)$ and gave a lower bound of $\Omega(\log d / \log \log \log d)$ on the competitive ratio of any deterministic online algorithm for this problem. Tight bounds of 2 and 4 are known for $d = 1$ and $d = 2$, respectively.

Chan and Zarrabi-Zadeh [7] introduced the online UNIT CLUSTERING problem at WAOA 2006. While the input and the objective of this problem are identical to those for UNIT COVERING, this latter problem is more flexible in that the algorithm is not required to produce unit balls at any time, but rather the smallest enclosing ball of each cluster should have diameter *at most* 1; moreover, the ball may change (grow or shift) in time. The L_∞ metric is used in $\mathbb{R}^d$, $d \in \mathbb{N}$. The authors showed that several standard approaches for UNIT CLUSTERING, namely the deterministic algorithms Centered, Grid, and Greedy, all have competitive ratio at most 2 for points on the line ($d = 1$). Moreover, the first two algorithms above are applicable for UNIT COVERING, with a competitive ratio at most 2 for $d = 1$, as well.

In fact, Chan and Zarrabi-Zadeh [7] showed that no online algorithm (deterministic or randomized) for UNIT COVERING can have a competitive ratio better than 2 in one dimension ($d = 1$). They also showed that it is possible to get better results for UNIT CLUSTERING than for UNIT COVERING. Specifically, they developed the first algorithm with competitive ratio below 2 for $d = 1$, namely a randomized algorithm with competitive ratio 15/8. Moreover, they developed a general method to achieve competitive ratio below 2^d in $\mathbb{R}^d$ under L_∞ metric for any $d \geq 2$, by "lifting" the one-dimensional algorithm to higher dimensions. In particular, the existence of an algorithm for UNIT CLUSTERING with competitive ratio ρ_1 for $d = 1$ yields an algorithm with competitive ratio $\rho_d = 2^{d-1}\rho_1$ for every $d \geq 2$ for this problem. The current best competitive ratio for UNIT CLUSTERING in $\mathbb{R}^d$ is obtained in exactly this way: the current best ratio 5/3, for $d = 1$, is due to Ehmsen and Larsen [13], and this gives a ratio of $2^{d-1}\frac{5}{3}$ for every $d \geq 2$.

As such, the 1-dimensional case becomes quite important since no other significantly faster method is currently available for dealing with the problem in higher dimensions. It is however worth noting that Algorithm Grid is easily extendable to $\mathbb{R}^d$. In higher dimensions, its competitive ratio is only marginally worse than that obtained by "lifting" the one-dimensional algorithm to higher dimensions.

> Algorithm Grid. Build a uniform grid in $\mathbb{R}^d$ where cells are unit cubes of the form $\prod [i_j, i_j + 1)$, where $i_j \in \mathbb{Z}$ for $j = 1, \ldots, d$. For each new point p, if the grid cell containing p is nonempty, put p in the corresponding cluster; otherwise open a new cluster for the grid cell and put p in it.

Since in $\mathbb{R}^d$ each cluster of OPT can be split to at most 2^d grid-cell clusters created by the algorithm, its competitive ratio is at most 2^d, and this analysis is tight.

The 15/8 ratio [7] has been subsequently reduced to 11/6 by the same authors [25]; that algorithm is still randomized. Epstein and van Stee [15] gave the first deterministic algorithm with ratio below 2, namely one with ratio 7/4, and further improving the earlier 11/6 ratio. In the latest development, Ehmsen and Larsen [13] provided a deterministic algorithm with competitive ratio 5/3, which holds the current record in both categories.

From the other direction, the lower bound for deterministic algorithms has evolved from 3/2 in [7] to 8/5 in [15], and then to 13/8 in [21]. Whence the size of the current gap for the competitive ratio of deterministic algorithms for the one-dimensional case of UNIT CLUSTERING is quite small, namely $\frac{5}{3} - \frac{13}{8} = \frac{1}{24}$, but remains nonzero. The lower bound for randomized algorithms has evolved from 4/3 in [7] to 3/2 in [15].

For points in the plane (i.e., $d = 2$), the lower bound for deterministic algorithms has evolved from 3/2 in [7] to 2 in [15], and then to 13/6 in [13]. The lower bound for randomized algorithms has evolved from 4/3 in [7] to 11/6 in [15].

As such, the best lower bounds on the competitive ratio for $d \geq 2$ prior to our work are 13/6 for deterministic algorithms [13] and 11/6 for randomized algorithms [15].

Notation and terminology. Throughout this paper the L_∞-norm is used. Then the UNIT CLUSTERING problem is to partition a set of points in $\mathbb{R}^d$ into clusters (subsets), each contained in a unit cube, i.e., a cube of the form $\mathbf{x} + [0,1]^d$ for some $\mathbf{x} \in [0,1]^d$, so as to minimize the number of clusters used. $\mathbb{E}[X]$ denotes the expected value of a random variable X.

Contributions. We obtain the following results:

(i) The competitive ratio of every online algorithm (deterministic or random-ized) for UNIT CLUSTERING in $\mathbb{R}^d$ under L_∞ norm is $\Omega(d)$ for every $d \geq 2$ (Theorem 1 in Sect. 2). We thereby give a positive answer to a question of Epstein and van Stee; specifically, they asked whether the competitive ratio grows with the dimension [15, Sect. 4]. The question was reposed in [13, Sect. 7].

(ii) The competitive ratio of every online algorithm (deterministic or random-ized) for UNIT COVERING in $\mathbb{R}^d$ under L_∞ norm is at least $d + 1$ for every $d \geq 1$ (Theorem 2 in Sect. 3). This generalizes a result of Chan and Zarrabi-Zadeh [7] from $d = 1$ to higher dimensions. We also give a randomized algorithm with competitive ratio $O(d^2)$ for UNIT COVERING in $\mathbb{Z}^d$, $d \in \mathbb{N}$, under L_∞ norm (Theorem 3 in Sect. 3). The algorithm applies to UNIT CLUSTERING in $\mathbb{Z}^d$, $d \in \mathbb{N}$, with the same competitive ratio.

(iii) The competitive ratio of `Algorithm Greedy` for UNIT CLUSTERING in $\mathbb{R}^d$ under L_∞ norm is unbounded for every $d \geq 2$ (Theorem 4 in Sect. 4). The competitive ratio of `Algorithm Greedy` for UNIT CLUSTERING in $\mathbb{Z}^d$ under L_∞ norm is at least 2^{d-1} and at most $2^{d-1} + \frac{1}{2}$ for every $d \geq 2$ (Theorem 5 in Sect. 4).

Related work. Several other variants of UNIT CLUSTERING have been studied in [14]. A survey of algorithms for UNIT CLUSTERING in the context of online algorithms appears in [9]. Clustering with variable sized clusters has been studied in [10,11]. Grid-based online algorithms for clustering problems have been developed by the same authors [12].

UNIT COVERING is a variant of SET COVER. Alon et al. [1] gave a deter-ministic online algorithm of competitive ratio $O(\log m \log n)$ for SET COVER, where n is the number of possible points (the size of the ground set) and m is the number of sets in the family. If every element appears in at most Δ sets, the competitive ratio of the algorithm can be improved to $O(\log \Delta \log n)$. Buch-binder and Naor [6] improved these competitive ratio to $O(\log m \log(n/\mathsf{OPT}))$ and $O(\log \Delta \log (n/\mathsf{OPT}))$, respectively, under the same assumptions. For sev-eral combinatorial optimization problems (e.g., covering and packing), the clas-sic technique that rounds a fractional linear programming solution to an integer solution has been adapted to the online setting [2–4,6,19].

In these results, the underlying set system for the covering and packing prob-lem must be finite: The online algorithms and their analyses rely on the size of the ground set. For UNIT CLUSTERING and UNIT CLUSTERING over infinite sets, such as $\mathbb{R}^d$ or $\mathbb{Z}^d$, these techniques could only be used after a suitable discretiza-tion and a covering of the domain with finite sets, and it is unclear whether they can beat the trivial competitive ratio of 2^d in a substantive way.

2   Lower Bounds for Online Unit Clustering

In this section, we prove the following theorem.

Theorem 1. *The competitive ratio of every (i) deterministic algorithm (with an adaptive deterministic adversary), and (ii) randomized algorithm (with a randomized oblivious adversary), for* UNIT CLUSTERING *in* $\mathbb{R}^d$ *under* L_∞ *norm is* $\Omega(d)$ *for every* $d \geq 1$.

Proof. Let ϱ be the competitive ratio of an online algorithm. We may assume $\varrho \leq d$, otherwise there is nothing to prove. We may also assume that $d \geq 4$ since this is the smallest value for which the argument gives a nontrivial lower bound. Let K be a sufficiently large even integer (that depends on d).

Deterministic Algorithm. We first prove a lower bound for a deterministic algorithm, assuming an adaptive deterministic adversary. We present a total of $\lfloor d/2 \rfloor K^d$ points to the algorithm, and show that it creates $\Omega(d \cdot \mathsf{OPT})$ clusters, where OPT is the offline minimum number of clusters for the final set of points. Specifically, we present the points to the algorithm in $\lfloor d/2 \rfloor$ rounds. Round $i = 1, \ldots, \lfloor d/2 \rfloor$ consists of the following three events:

(i) The adversary presents (inserts) a set S_i of K^d points; S_i is determined by a vector $\sigma(i) \in \{-1, 0, 1\}^d$ to be later defined.
(ii) The algorithm may create new clusters or expand existing clusters to cover S_i.
(iii) If $i < \lfloor \frac{d}{2} \rfloor$, the adversary computes $\sigma(i+1)$ from the clusters that cover S_i.

In the first round, the adversary presents points of the integer lattice; namely $S_1 = [K]^d$, where $[K] = \{x \in \mathbb{Z} : 1 \leq x \leq K\}$. In round $i = 2, \ldots, \lfloor d/2 \rfloor$, the point set S_i will depend on the clusters created by the algorithm in previous rounds. We say that a cluster *expires in round* i if it contains some points from S_i but no additional points can (or will) be added to it in any subsequent round. We show that over $\lfloor d/2 \rfloor$ rounds, $\Omega(d \cdot \mathsf{OPT})$ clusters expire, which readily implies $\varrho = \Omega(d)$.

Optimal solutions. For $i = 1, \ldots, \lfloor d/2 \rfloor$, denote by OPT_i the offline optimum for the set $\bigcup_{j=1}^{i} S_j$ of points presented up to (and including) round i. Since $S_1 = [K]^d$ and K is even, $\mathsf{OPT}_1 = K^d/2^d$. The optimum solution for S_1 is unique, and each cluster in the optimum is a Cartesian product $\prod_{i=1}^{d}\{a_i, a_i + 1\}$, where $a_i \in [K]$ is odd for $i = 1, \ldots, d$ (Fig. 1(a)).

Consider $2^d - 1$ additional near-optimal solutions for S_1 obtained by translating the optimal clusters by a d-dimensional 0–1 vector, and adding new clusters along the boundary of the cube $[K]^d$. We shall argue that the points inserted in round i, $i \geq 2$, can be added to some but not all of these solutions. To make this precise, we define these solutions a bit more carefully. First we define an infinite set of hypercubes

$$\mathcal{Q} = \left\{ \prod_{i=1}^{d}[a_i, a_i + 1] : a_i \in \mathbb{Z} \text{ is odd for } i = 1, \ldots, d \right\}.$$

For a point set $S \subset \mathbb{R}^d$ and a vector $\tau \in \{0,1\}^d$, let the clusters be the subsets of S that lie in translates $Q + \tau$ of hypercubes $Q \in \mathcal{Q}$, that is, let

$$C(S, \tau) = \{S \cap (Q + \tau) : Q \in \mathcal{Q}\}.$$

Since S_1 is an integer grid, the clusters $C(S_1, \tau)$ contain all points in S_1 for all $\tau \in \{0,1\}^d$. See Fig. 1(a–d) for examples. Due to the boundary effect, the number of clusters in $C(S_1, \tau)$ is

$$\frac{K^d + O(dK^{d-1})}{2^d} = \mathsf{OPT}_1 \cdot \left(1 + O\left(\frac{d}{K}\right)\right) = (1 + o(1))\, \mathsf{OPT}_1,$$

if K is sufficiently large with respect to d.

In round $i = 2, \ldots, \lfloor d/2 \rfloor$, the point set S_i is a perturbation of the integer grid S_1 (as described below). Further, we ensure that the final point set $S = \bigcup_{i=1}^{\lfloor d/2 \rfloor} S_i$ is covered by the clusters $C(S, \tau)$ for at least one vector $\tau \in \{0,1\}^d$. Consequently,

$$\mathsf{OPT}_i = \mathsf{OPT}_1(1 + o(1)) = (1 + o(1)) \frac{K^d}{2^d}, \text{ for all } i = 1, \ldots, \lfloor d/2 \rfloor.$$

At the end, we have $\mathsf{OPT} = \mathsf{OPT}_{\lfloor d/2 \rfloor} = (1 + o(1)) \frac{K^d}{2^d}$.

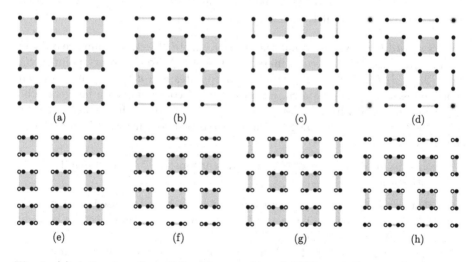

Fig. 1. (a) A 6×6 section of the integer grid and $\mathsf{OPT}_1 = 9$ clusters. (b–d) Near-optimal solutions $C(S_1, \tau)$ for $\tau = (0, 1)$, $(1, 0)$, and $(1, 1)$. (e–f) The perturbation with signature $\sigma = (-1, 0)$, and clusters $C(S, \tau)$ for $\tau = (0, 0)$ and $\tau = (0, 1)$, where S is the union of the perturbed points (full dots), and grid points (empty circles). (g–h) The perturbation with signature $\sigma = (1, 0)$ and clusters $C(S, \tau)$ for $\tau = (1, 0)$ and $\tau = (1, 1)$ and the same S.

Perturbation. A perturbation of the integer grid S_1 is encoded by a vector $\sigma \in \{-1,0,1\}^d$, that we call the *signature* of the perturbation. Let $\varepsilon \in (0, \frac{1}{2})$. For an integer point $p = (p_1, \ldots, p_d) \in S_1$ and a signature σ, the perturbed point p' is defined as follows; see Fig. 1(e–h) for examples in the plane: For $j = 1, \ldots, d$, let p'_j be

- p_j when $\sigma_j = 0$;
- $p_j + \varepsilon$ if p_j is odd, and $p_j - \varepsilon$ if p_j is even when $\sigma_j = -1$;
- $p_j - \varepsilon$ if p_j is odd, and $p_j + \varepsilon$ if p_j is even when $\sigma_j = 1$.

For $i = 2, \ldots, \lfloor d/2 \rfloor$, the point set S_i is a perturbation of S_1 with signature $\sigma(i)$, for some $\sigma(i) \in \{-1,0,1\}^d$. The signature of S_1 is $\sigma(1) = (0, \ldots 0)$ (and so S_1 can be viewed as a null perturbation of itself). At the end of round $i = 1, \ldots, \lfloor d/2 \rfloor - 1$, we compute $\sigma(i+1)$ from $\sigma(i)$ and from the clusters that cover S_i. The signature $\sigma(i)$ determines the set S_i, for every $i = 2, \ldots, \lfloor d/2 \rfloor$. Note the following relation between the signatures $\sigma(i)$ and the clusters $C(S_i, \tau)$.

Observation 1. *Consider a point set S_i with signature $\sigma(i) \in \{-1,0,1\}^d$. The clusters $C(S_i, \tau)$ cover S_i if and only if for all $j = 1, \ldots d$,*

- $\sigma_j(i) = 0$, *or*
- $\sigma_j(i) = -1$ *and* $\tau_j = 0$, *or*
- $\sigma_j(i) = 1$ *and* $\tau_j = 1$.

It follows from Observation 1 that the final point set $S = \bigcup_{i=1}^{\lfloor d/2 \rfloor} S_i$ is covered by the clusters $C(S, \tau)$ for at least one vector $\tau \in \{0,1\}^d$.

Adversary strategy. At the end of round $i = 1, \ldots, \lfloor d/2 \rfloor - 1$, we compute $\sigma(i+1)$ from $\sigma(i)$ by changing a 0-coordinate to -1 or $+1$. Note that every point in S_i, $i = 1, 2, \ldots, \lfloor d/2 \rfloor$, has $i - 1$ perturbed coordinates and $d - i + 1$ unperturbed coordinates. For all points in S_i, all unperturbed coordinates are integers. The algorithm covers S_i with at most $\varrho \cdot \mathsf{OPT}_i$ clusters. Project these clusters to the subspace $\mathbb{Z}^{d+1-i}$ corresponding to the unperturbed coordinates. We say that a cluster is

- *small* if its projection to $\mathbb{Z}^{d+1-i}$ contains at most $2^{d-i}/\varrho$ points, and
- *big* otherwise.

Note that we distinguish small and big clusters in round i based on how they cover the set S_i (in particular, a small cluster in round i may become large in another round, or vice versa).

Since the L_∞-diameter of a cluster is at most 1, a small cluster contains at most $(2^{d-i}/\varrho) \cdot 2^{i-1} = 2^d/(2\varrho)$ points of S_i (by definition, it contains at most $2^{d-i}/\varrho$ points in the projection to $\mathbb{Z}^{d+1-i}$, each of these points is the projection of K^{i-1} points of S_i; since S_i is a perturbation of the integer grid, any cluster contains at most 2^{i-1} of these preimages). The total number of points in S_i that lie in small clusters is at most

$$(\varrho \cdot \mathsf{OPT}_i) \frac{2^d}{2\varrho} = \mathsf{OPT}_i \cdot 2^{d-1} = \left(\frac{1}{2} + o(1) \right) K^d.$$

Consequently, the remaining $\left(\frac{1}{2} - o(1)\right) K^d$ points in S_i are covered by big clusters. For a big cluster C, let $s(C)$ denote the number of unperturbed coordinates in which its extent is 1. Then the number of points in C satisfies

$$2^{d-i}/\varrho \le 2^{s(C)}$$
$$d - i - \log_2 \varrho \le s(C).$$

We say that a big cluster C *expires* if no point can (or will) be added to C in the future. Consider the following experiment: choose one of the zero coordinates of the signature $\sigma(i)$ uniformly at random (i.e., all $d + 1 - i$ choices are equally likely), and change it to -1 or $+1$ with equal probability $1/2$. If the j-th extent of a cluster C is 1, then it cannot be expanded in dimension j. Consequently, a big cluster C expires with probability at least

$$\frac{s(C)}{d+1-i} \cdot \frac{1}{2} = \frac{d - i - \log_2 \varrho}{2(d+1-i)} \ge \frac{d - \lfloor d/2 \rfloor - \log_2 d}{2d} = \Omega(1), \qquad (1)$$

as $i \le \lfloor d/2 \rfloor$ and we assume $\varrho \le d$. It follows that there exists an unperturbed coordinate j, and a perturbation of the j-th coordinate such that

$$\Omega(1) \cdot \left(\frac{1}{2} - o(1)\right) \frac{K^d}{2^d} = \Omega(\mathsf{OPT})$$

big clusters expire in (at the end of) round $i = 1, \ldots, \lfloor d/2 \rfloor - 1$. The adversary makes this choice and the corresponding perturbation. In round $i = \lfloor d/2 \rfloor$, all clusters that cover any point in $S_{\lfloor d/2 \rfloor}$ expire, because no point will be added to any of these clusters. Since $S_{\lfloor d/2 \rfloor}$ is a perturbation of S_1, at least $\mathsf{OPT}_1 = \Omega(\mathsf{OPT})$ clusters expire in the last round, as well.

If a cluster expires in round i, then it contains some points of S_i but does not contain any point of S_j for $j > i$. Consequently, each cluster expire in at most one round, and the total number of expired clusters over all $\lfloor d/2 \rfloor$ rounds is $\Omega(d \cdot \mathsf{OPT})$. Since each of these cluster was created by the algorithm in one of the rounds, we have $\varrho \cdot \mathsf{OPT} = \Omega(d \cdot \mathsf{OPT})$, which implies $\varrho = \Omega(d)$, as claimed.

Randomized Algorithm. We modify the above argument to establish a lower bound of $\Omega(d)$ for a randomized algorithm with an oblivious randomized adversary. The adversary starts with the integer grid $S_1 = [K]^d$, with signature $\sigma(1) = \mathbf{0}$ as before. At the end of round $i = 1, \ldots, \lfloor d/2 \rfloor - 1$, it chooses an unperturbed coordinate of $\sigma(i)$ uniformly at random, and switches it to -1 or $+1$ with equal probability (independently of the clusters created by the algorithm) to obtain $\sigma(i+1)$. By (1), the expected number of big clusters that expire in round i, $1 \le i < \lfloor d/2 \rfloor$, is $\Omega(\mathsf{OPT}_i) = \Omega(\mathsf{OPT})$; and all $(1 - o(1))\mathsf{OPT}_{\lfloor d/2 \rfloor} = \Omega(\mathsf{OPT})$ big clusters expire in round $\lfloor d/2 \rfloor$. Consequently, the expected number of clusters created by the algorithm is $\Omega(d \cdot \mathsf{OPT})$, which implies $\varrho = \Omega(d)$, as required. $\qquad \square$

3   Lower Bounds and Algorithms for Online Unit Covering

The following theorem extends a result from [7] from $d = 1$ to higher dimensions.

Theorem 2. *The competitive ratio of every deterministic online algorithm (with an adaptive deterministic adversary) for* UNIT COVERING *in* $\mathbb{R}^d$ *under* L_∞ *norm is at least* $d + 1$ *for every* $d \geq 1$.

Proof. We construct an input sequence $p_1, \ldots, p_{d+1} \in \mathbb{Z}^d$ for which OPT $= 1$ and ALG $= d + 1$ using an adaptive adversary (which knows the actions of the algorithm). We construct such a sequence inductively, so that

- each new point p_i requires a new cube, $Q_i \subset \mathbb{R}^d$, and
- all points presented can be covered by one integer unit cube incident to the origin.

Let $x_1, \ldots, x_d$ be the d coordinate axes in $\mathbb{R}^d$; and x_{d+1} be the new axis in $\mathbb{R}^{d+1}$. The induction basis is $d = 1$. We may assume for concreteness that $p_1 = 0$, and suppose that the algorithm opens a unit interval $[x, x + 1]$ to cover this point. If $x = -1$, let $p_2 = 1$, else let $p_2 = -1$. The algorithm now opens a new unit interval to cover p_2. It is easily seen that $p_1, p_2 \in \mathbb{Z}$ and $\{p_1, p_2\}$ define a unit interval.

For the induction step, assume the existence of a sequence $\sigma = p_1, \ldots, p_{d+1} \in \mathbb{Z}^d$ that forces the algorithm to open a new unit cube, $Q_i \subset \mathbb{R}^d$, to cover each new point p_i, $i = 1, \ldots, d + 1$ (and so ALG $= d + 1$), while OPT $= 1$ with σ being covered by a single cube $U_d \subset \mathbb{Z}^d$. Present the following sequence of $d + 2$ points to the algorithm in $\mathbb{R}^{d+1}$: $(p_1, 0), \ldots, (p_{d+1}, 0)$. The algorithm must use $d + 1$ cubes, say, $Q_1, \ldots, Q_{d+1} \subset \mathbb{R}^{d+1}$ to cover these points. As such, the $d+1$ unit cubes $\pi(Q_1), \ldots, \pi(Q_{d+1}) \subset \mathbb{R}^d$, cover $p_1, \ldots, p_{d+1} \in \mathbb{Z}^d$, where $\pi(Q_i)$ is the projection onto the first d coordinates of Q_i; moreover, the unit cubes $\pi(Q_1), \ldots, \pi(Q_d)$ do not cover $(p_{d+1}, 0)$. Only $\pi(Q_d)$ contains p_{d+1}, but the cube Q_d cannot contain both $(p_{d+1}, -1)$ and $(p_{d+1}, 1)$. Consequently, $(p_{d+1}, -1)$ or $(p_{d+1}, 1)$ is not covered by $\bigcup_{i=1}^{d+1} Q_i$. The adversary presents such an uncovered point, which requires a new cube Q_{d+2}. (Note that the points $p_1, \ldots, p_{d+2}$ form a lattice path, where p_i and p_{i+1} differ in the $(i + 1)$-th coordinate.) This completes the inductive step, and thereby the proof of the theorem.   □

Remark. Since the proof of Theorem 2 uses integer points, the result also holds for UNIT COVERING in $\mathbb{Z}^d$ under L_∞ norm.

Online algorithm for UNIT COVERING *over* $\mathbb{Z}^d$. Note that the lower bound construction used sequences of integer points (i.e., points in $\mathbb{Z}^d$). We substantially improve on the trivial 2^d upper bound on the competitive ratio of UNIT COVERING over $\mathbb{Z}^d$ (or the $2^{d-1} + \frac{1}{2}$ upper bound of the greedy algorithm, see Sect. 4).

The online algorithm by Buchbinder and Naor [6] for SET COVER, for the unit covering problem over $\mathbb{Z}^d$, yields an algorithm with $O(d \log (n/\text{OPT}))$ competitive ratio under the assumption that a set of n possible integer points is given in advance. Recently, Gupta and Nagarajan [19] gave an online random-ized algorithm for a broad family of combinatorial optimization problems that can be expressed as sparse integer programs. For unit covering over the integers in $[n]^d$, their results yield a competitive ratio of $O(d^2)$, where $[n] = \{1, 2, \ldots, n\}$. The competitive ratio does not depend on n, but the algorithm must know n in advance.

We now remove the dependence on n so as to get a truly online algorithm for UNIT COVERING over $\mathbb{Z}^d$. Consider the following randomized algorithm.

Algorithm Iterative Reweighing. Let $P \subset \mathbb{Z}^d$ be the set of points presented to the algorithm and $\mathcal{C}$ the set of cubes chosen by the algorithm; initially $P = \mathcal{C} = \emptyset$. The algorithm chooses cubes for two different reasons, and it keeps them in sets $\mathcal{C}_1$ and $\mathcal{C}_2$, where $\mathcal{C} = \mathcal{C}_1 \cup \mathcal{C}_2$. It also maintains a third set of cubes, $\mathcal{B}$, for bookkeeping purposes; initially $\mathcal{B} = \emptyset$. In addition, the algorithm maintains a weight function on all integer unit cubes. Initially $w(Q) = 2^{-(d+1)}$ for all integer unit cubes (this is the default value for all cubes that are disjoint from P).

We describe one iteration of the algorithm. Let $p \in \mathbb{Z}^d$ be a new point; put $P \leftarrow P \cup \{p\}$. Let $\mathcal{Q}(p)$ be the set of 2^d integer unit cubes that contain p.
1. If $p \in \bigcup \mathcal{C}$, then do nothing.
2. Else if $p \in \bigcup \mathcal{B}$, then let $Q \in \mathcal{B} \cap \mathcal{Q}(p)$ be an arbitrary cube and put $\mathcal{C}_1 \leftarrow \mathcal{C}_1 \cup \{Q\}$.
3. Else if $\sum_{Q \in \mathcal{Q}(p)} w(Q) \geq 1$, then let Q be an arbitrary cube in $\mathcal{Q}(p)$ and put $\mathcal{C}_2 \leftarrow \mathcal{C}_2 \cup \{Q\}$.
4. Else, the weights give a probability distribution on $\mathcal{Q}(p)$. Successively choose cubes from $\mathcal{Q}(p)$ at random with this distribution in $2d$ inde-pendent trials and add them to $\mathcal{B}$. Let $Q \in \mathcal{B} \cap \mathcal{Q}(p)$ be an arbitrary cube and put $\mathcal{C}_1 \leftarrow \mathcal{C}_1 \cup \{Q\}$. Double the weight of every cube in $\mathcal{Q}(p)$.

Theorem 3. *The competitive ratio of* `Algorithm Iterative Reweighing` *for* UNIT COVERING *in* $\mathbb{Z}^d$ *under* L_∞ *norm is* $O(d^2)$ *for every* $d \in \mathbb{N}$.

Proof. Suppose that a set P of n points is presented to the algorithm sequentially, and the algorithm created unit cubes in $\mathcal{C} = \mathcal{C}_1 \cup \mathcal{C}_2$. Note that $\mathcal{C}_1 \subseteq \mathcal{B}$. We show that $\mathbb{E}[|\mathcal{B}|] = O(d^2 \cdot \mathsf{OPT})$ and $\mathbb{E}[|\mathcal{C}_2|] = O(\mathsf{OPT})$. This immediately implies that $\mathbb{E}[|\mathcal{C}|] \leq \mathbb{E}[|\mathcal{C}_1|] + \mathbb{E}[|\mathcal{C}_2|] \leq \mathbb{E}[|\mathcal{B}|] + \mathbb{E}[|\mathcal{C}_2|] = O(d^2 \cdot \mathsf{OPT})$.

First consider $\mathbb{E}[|\mathcal{B}|]$. New cubes are added to $\mathcal{B}$ in step 4. In this case, the algorithm places at most $2d$ cubes into $\mathcal{B}$, and doubles the weight of all 2^d cubes in $\mathcal{Q}(p)$ that contain p. Let $\mathcal{C}_{\mathsf{OPT}}$ be an offline optimum set of unit cubes. Each point $p \in P$ lies in some cube $Q_p \in \mathcal{C}_{\mathsf{OPT}}$. The weight of Q_p is initially $2^{-(d+1)}$, and it never exceeds 2; indeed, since $Q_p \in \mathcal{Q}(p)$, its weight before the last doubling must have been at most 1 in step 4 of the algorithm; thus its weight is doubled in at most $d+2$ iterations. Consequently, the algorithm invokes step 4 in at most $(d+2)\mathsf{OPT}$ iterations. In each such iteration, it adds at most $2d$ cubes to $\mathcal{B}$. Overall, we have $|\mathcal{B}| \leq (d+2) \cdot 2d \cdot \mathsf{OPT} = O(d^2 \cdot \mathsf{OPT})$, as required.

Next consider $\mathbb{E}[|\mathcal{C}_2|]$. A new cube is added to $\mathcal{C}_2$ in step 3. In this case, none of the cubes in $\mathcal{Q}(p)$ is in $\mathcal{B}$ and $\sum_{Q \in \mathcal{Q}(p)} w(Q) \geq 1$ when point p is presented, and the algorithm increments $|\mathcal{C}_2|$ by one. At the beginning of the algorithm, we have $\sum_{Q \in \mathcal{Q}(p)} w(Q) = \sum_{Q \in \mathcal{Q}(p)} 2^{-(d+1)} = 2^d \cdot 2^{-(d+1)} = 1/2$. Assume that the weights of the cubes in $\mathcal{Q}(p)$ were increased in t iterations, starting from the beginning of the algorithm, and the sum of weights of the cubes in $\mathcal{Q}(p)$ increases by $\delta_1, \ldots, \delta_t > 0$ (the weights of several cubes may have been doubled in an iteration). Since $\sum_{Q \in \mathcal{Q}(p)} w(Q) = 1/2 + \sum_{i=1}^{t} \delta_i$, then $\sum_{Q \in \mathcal{Q}(p)} w(Q) \geq 1$ implies $\sum_{i=1}^{t} \delta_i \geq 1/2$. For every $i = 1, \ldots, t$, the sum of weights of some cubes in $\mathcal{Q}(p)$, say, $\mathcal{Q}_i \subset \mathcal{Q}(p)$, increased by δ_i in step 4 of a previous iteration. Since the weights doubled, the sum of the weights of these cubes was δ_i at the beginning of that iteration, and the algorithm added one of them into $\mathcal{B}$ with probability at least δ_i in one random draw, which was repeated $2d$ times independently. Consequently, the probability that the algorithm did *not* add any cube from $\mathcal{Q}_i$ to $\mathcal{B}$ in that iteration is at most $(1 - \delta_i)^{2d}$. The probability that none of the cubes in $\mathcal{Q}(p)$ has been added to $\mathcal{B}$ before point p arrives is (by independence) at most

$$\prod_{i=1}^{t} (1 - \delta_i)^{2d} \leq e^{-2d \sum_{i=1}^{t} \delta_i} \leq e^{-d}.$$

The total number of points p for which step 3 applies is at most $|P|$. Since each unit cube contains at most 2^d points, we have $|P| \leq 2^d \cdot \mathsf{OPT}$. Therefore $\mathbb{E}[|\mathcal{C}_2|] \leq |P| e^{-d} \leq (2/e)^d \mathsf{OPT} \leq \mathsf{OPT}$, as claimed. $\square$

The above algorithm applies to UNIT CLUSTERING of integer points in $\mathbb{Z}^d$ with the same competitive ratio:

Corollary 1. *The competitive ratio of* `Algorithm Iterative Reweighing` *for* UNIT CLUSTERING *in* $\mathbb{Z}^d$ *under* L_∞ *norm is* $O(d^2)$ *for every* $d \in \mathbb{N}$.

4   Lower Bound for `Algorithm Greedy` for Unit Clustering

Chan and Zarrabi-Zadeh [7] showed that the greedy algorithm for UNIT CLUSTERING on the line ($d = 1$) has competitive ratio of 2 (this includes both an upper bound on the ratio and a tight example). Here we show that the competitive ratio of the greedy algorithm is unbounded. We first recall the algorithm:

Algorithm Greedy. For each new point p, if p fits in some existing cluster, put p in such a cluster (break ties arbitrarily); otherwise open a new cluster for p.

Theorem 4. *The competitive ratio of* `Algorithm Greedy` *for* UNIT CLUSTERING *in* $\mathbb{R}^d$ *under* L_∞ *norm is unbounded for every* $d \geq 2$.

Proof. It suffices to consider $d = 2$; the construction extends to arbitrary dimensions $d \geq 2$. The adversary presents $2n$ points in pairs $\{(1 + i/n, i/n), (i/n, 1 + i/n)\}$ for $i = 0, 1, \ldots, n - 1$. Each pair of points spans a unit square that does not contain any subsequent point. Consequently, the greedy algorithm will create

n clusters, one for each point pair. However, $\mathsf{OPT} = 2$ since the clusters $C_1 = \{(1 + i/n, i/n) : i = 0, 1, \ldots, n-1\}$ and $C_2 = \{(i/n, 1 + i/n) : i = 0, 1, \ldots, n-1\}$ are contained in the unit squares $[1, 2] \times [0, 1]$ and $[0, 1] \times [1, 2]$, respectively. $\square$

When we restrict Algorithm Greedy to integer points, its competitive ratio is exponential in d.

Theorem 5. *The competitive ratio of Algorithm Greedy for* UNIT CLUSTERING *in* $\mathbb{Z}^d$ *under* L_∞ *norm is at least* 2^{d-1} *and at most* $2^{d-1} + \frac{1}{2}$ *for every* $d \geq 1$.

Proof. We first prove the lower bound. Consider an integer input sequence implementing a barycentric subdivision of the space, as illustrated in Fig. 2. Let K be a sufficiently large positive multiple of 4 (that depends on d). We present a point set S, where $|S| = (2 + o(1))(K/2)^d$ points to the algorithm, and show that it creates $(1 + o(1)) 2^{d-1}\mathsf{OPT}$ clusters.

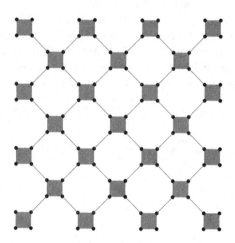

Fig. 2. A planar instance for the greedy algorithm with $K = 12$; the edges in E are drawn in red. (Color figure online)

Let $S = B \cup D$, where

$$A = \{(x_1, \ldots, x_d) \mid x_i \equiv 0 \pmod 4, \ 0 \leq x_i \leq K, \ i = 1, \ldots, d\},$$
$$B = A + \{0, 1\}^d,$$
$$C = \{(x_1, \ldots, x_d) \mid x_i \equiv 2 \pmod 4, \ 0 \leq x_i \leq K, \ i = 1, \ldots, d\},$$
$$D = C + \{0, 1\}^d,$$
$$E = \{\{u, v\} : u \in B, v \in D, \|u - v\|_\infty \leq 1\}.$$

Note that each element of C is the barycenter (center of mass) of 2^d elements of A, namely the vertices of a cell of $(4\mathbb{Z})^d$ containing the element. Here E is a set of pairs of lattice points (edges) that can be put in one-to-one correspondence with the points in D. As such, we have

$$|A| = \left(\frac{K}{4} + 1\right)^d, \quad |B| = 2^d|A| = (1 + o(1))\frac{K^d}{2^d},$$

$$|C| = \left(\frac{K}{4}\right)^d, \quad |D| = 2^d|C| = (1 + o(1))\frac{K^d}{2^d},$$

$$|E| = |D| = (1 + o(1))\frac{K^d}{2^d},$$

$$\mathsf{OPT} = |A \cup C| = |A| + |C| = (2 + o(1))\left(\frac{K}{4}\right)^d.$$

It follows that $|E| = (1 + o(1))\, 2^{d-1}\mathsf{OPT}$. The input sequence presents the points in pairs, namely those in E. The greedy algorithm makes one new non-extendable cluster for each such "diagonal" pair (each cluster is a unit cube), so its competitive ratio is at least 2^{d-1} for every $d \geq 2$.

An upper bound of 2^d follows from the fact that each cluster in OPT contains at most 2^d integer points; we further reduce this bound. Let $\Gamma_1, \ldots, \Gamma_k$ be the clusters of an optimal partition ($k = \mathsf{OPT}$). Assume that the algorithm produces m clusters of size at least 2 and s singleton clusters. Since each cluster of OPT contains at most one singleton cluster created by the algorithm, we have

$$\mathsf{ALG} = m + s \leq \frac{(k - s)2^d + s(2^d - 1)}{2} + s = \frac{k\,2^d - s}{2} + s$$

$$= k\,2^{d-1} + \frac{s}{2} \leq k\,2^{d-1} + \frac{k}{2} = k\left(2^{d-1} + \frac{1}{2}\right),$$

as required.   □

5   Conclusion

Our results suggest several directions for future study. For instance, the gaps between the linear lower bounds and the exponential upper bounds in the competitive ratios for UNIT COVERING and UNIT CLUSTERING are intriguing. We conclude by listing a few specific questions of interest.

Question 1. Is there a lower bound on the competitive ratio for UNIT COVERING or UNIT CLUSTERING that is exponential in d? Is there a superlinear lower bound?

Question 2. Do our lower bounds for UNIT CLUSTERING in $\mathbb{R}^d$ and $\mathbb{Z}^d$ under the L_∞ norm carry over to the L_2 norm (or the L_p norm for $1 \leq p < \infty$)?

Question 3. Is there an online algorithm for UNIT CLUSTERING whose competitive ratio is sub-exponential in d?

Question 4. Are there online algorithms for UNIT CLUSTERING that do not fit into the paradigm of "lifting" the one-dimensional algorithm?

References

1. Alon, N., Awerbuch, B., Azar, Y., Buchbinder, N., Naor, J.: The online set cover problem. SIAM J. Comput. **39**(2), 361–370 (2009)
2. Azar, Y., Buchbinder, N., Hubert Chan, T.-H., Chen, S., Cohen, I.R., Gupta, A., Huang, Z., Kang, N., Nagarajan, V., Naor, J., Panigrahi, D.: Online algorithms for covering and packing problems with convex objectives. In: Proceedings of the 57th IEEE Symposium on Foundations of Computer Science (FOCS), pp. 148–157. IEEE (2016)
3. Azar, Y., Bhaskar, U., Fleischer, L., Panigrahi, D.: Online mixed packing and covering. In: Proceedings of the 24th ACM-SIAM Symposium on Discrete Algorithms (SODA), pp. 85–100. SIAM (2013)
4. Azar, Y., Cohen, I.R., Roytman, A.: Online lower bounds via duality. In: Proceedings of the 28th ACM-SIAM Symposium on Discrete Algorithms (SODA), pp. 1038–1050. SIAM (2017)
5. Borodin, A., El-Yaniv, R.: Online Computation and Competitive Analysis. Cambridge University Press, Cambridge (1998)
6. Buchbinder, N., Naor, J.: Online primal-dual algorithms for covering and packing. Math. Oper. Res. **34**(2), 270–286 (2009)
7. Chan, T.M., Zarrabi-Zadeh, H.: A randomized algorithm for online unit clustering. Theory Comput. Syst. **45**(3), 486–496 (2009)
8. Charikar, M., Chekuri, C., Feder, T., Motwani, R.: Incremental clustering and dynamic information retrieval. SIAM J. Comput. **33**(6), 1417–1440 (2004)
9. Chrobak, M.: SIGACT news online algorithms column 13. SIGACT News Bull. **39**(3), 96–121 (2008)
10. Csirik, J., Epstein, L., Imreh, C., Levin, A.: Online clustering with variable sized clusters. Algorithmica **65**(2), 251–274 (2013)
11. Divéki, G., Imreh, C.: An online 2-dimensional clustering problem with variable sized clusters. Optim. Eng. **14**(4), 575–593 (2013)
12. Divéki, G., Imreh, C.: Grid based online algorithms for clustering problems. In. Proceedings of the 15th IEEE International Symposium on Computational Intelligence and Informatics (CINTI), p. 159. IEEE (2014)
13. Ehmsen, M.R., Larsen, K.S.: Better bounds on online unit clustering. Theor. Comput. Sci. **500**, 1–24 (2013)
14. Epstein, L., Levin, A., van Stee, R.: Online unit clustering: variations on a theme. Theor. Comput. Sci. **407**(1–3), 85–96 (2008)
15. Epstein, L., van Stee, R.: On the online unit clustering problem. ACM Trans. Algorithms **7**(1), 1–18 (2010)
16. Feder, T., Greene, D.H.: Optimal algorithms for approximate clustering. In: Proceedings of the 20th Annual ACM Symposium on Theory of Computing (STOC), pp. 434–444 (1988)
17. Fowler, R.J., Paterson, M., Tanimoto, S.L.: Optimal packing and covering in the plane are NP-complete. Inf. Process. Lett. **12**(3), 133–137 (1981)
18. Gonzalez, T.F.: Clustering to minimize the maximum intercluster distance. Theor. Comput. Sci. **38**, 293–306 (1985)
19. Gupta, A., Nagarajan, V.: Approximating sparse covering integer programs online. Math. Oper. Res. **39**(4), 998–1011 (2014)
20. Hochbaum, D.S., Maass, W.: Approximation schemes for covering and packing problems in image processing and VLSI. J. ACM **32**(1), 130–136 (1985)

21. Kawahara, J., Kobayashi, K.M.: An improved lower bound for one-dimensional online unit clustering. Theor. Comput. Sci. **600**, 171–173 (2015)
22. Megiddo, N., Supowit, K.J.: On the complexity of some common geometric location problems. SIAM J. Comput. **13**(1), 182–196 (1984)
23. Vazirani, V.: Approximation Algorithms. Springer, New York (2001). https://doi.org/10.1007/978-3-662-04565-7
24. Williamson, D.P., Shmoys, D.B.: The Design of Approximation Algorithms. Cambridge University Press, Cambridge (2011)
25. Zarrabi-Zadeh, H., Chan, T.M.: An improved algorithm for online unit clustering. Algorithmica **54**(4), 490–500 (2009)

On Conceptually Simple Algorithms
for Variants of Online Bipartite Matching

Allan Borodin[1], Denis Pankratov[1]([⊠]), and Amirali Salehi-Abari[2]

[1] University of Toronto, Toronto, ON, Canada
{bor,denisp}@cs.toronto.edu
[2] Faculty of Business and IT, UOIT, Oshawa, ON, Canada
abari@uoit.ca

Abstract. We present a series of results regarding conceptually simple algorithms for bipartite matching in various online and related models. We first consider a deterministic adversarial model. The best approximation ratio possible for a single-pass deterministic online algorithm is $1/2$, which is achieved by any greedy algorithm. Dürr et al. [15] recently presented a 2-pass algorithm called CATEGORY-ADVICE that achieves approximation ratio $3/5$. We extend their algorithm to multiple passes. We prove the exact approximation ratio for the k-PASS CATEGORY-ADVICE algorithm for all $k \geq 1$, and show that the approximation ratio converges to the inverse of the golden ratio $2/(1 + \sqrt{5}) \approx 0.618$ as k goes to infinity. The convergence is extremely fast—the 5-PASS CATEGORY-ADVICE algorithm is already within 0.01% of the inverse of the golden ratio. We then consider two natural greedy algorithms—MINDEGREE and MINRANKING. We analyze MINDEGREE in the online IID model and show that its approximation ratio is exactly $1 - 1/e$. We analyze MINRANKING in the priority model and show that this natural algorithm cannot obtain the approximation of the RANKING algorithm in the ROM model.

Keywords: Online bipartite matching · Adversarial model
Online IID model · Priority model · CATEGORY-ADVICE algorithm
MINGREEDY algorithm

1 Introduction

Maximum bipartite matching (MBM) is a classical graph problem. Let $G = (U, V, E)$ be a bipartite graph, where U and V are the vertices on the two sides, and $E \subseteq U \times V$ is a set of m edges. The celebrated $O(m\sqrt{n})$ Hopcroft-Karp algorithm [21] was discovered in 1973 where n is the number of vertices. The first improvement in the regime of relatively sparse graphs came forty years

A. Borodin and D. Pankratov—Research is supported by NSERC.

A. Salehi-Abari—Research was done while the author was a postdoctoral fellow at the University of Toronto. Research was also supported by NSERC.

R. Solis-Oba and R. Fleischer (Eds.): WAOA 2017, LNCS 10787, pp. 253–268, 2018.
https://doi.org/10.1007/978-3-319-89441-6_19

later when Madry [26] developed a $\widetilde{O}(m^{10/7})$ algorithm based on electrical flows. For dense graphs, i.e., when $m \approx n^2$, Mucha and Sankowski [31] describe an algorithm running in time $O(n^\omega)$, where $\omega \leq 2.373$ is the matrix multiplication constant. We refer the interested reader to [14,27] and references therein for more information on MBM in the offline setting. While current algorithms for solving MBM optimally in the offline setting are reasonably efficient, they still fall short of linear time algorithms. For large graphs, linear or near linear time algorithms might be required. In that regard, a $(1 - \epsilon)$-approximation can be computed in $O(m/\epsilon)$ time by a version of the Hopcroft-Karp algorithm in the offline setting [14]. Arguably, such algorithms are not conceptually simple and require a reasonable understanding of the problem.

Our focus in this paper[1] is on "conceptually simple algorithms" with regard to variants of the online MBM problem. We will not define "conceptual simplicity" but claim that certain types of algorithms (e.g., greedy and local search) usually fall within this informal "you know it when you see it" concept. The online model is sometimes necessitated by applications, and can be studied with respect to a completely adversarial model, the random order model (ROM), or a distributional input model (e.g., known and unknown IID input models). In these models, the algorithm has no control over the ordering of the inputs and must make irrevocable decisions for each input item as it arrives. As such, online algorithms are a prime example of a conceptually simple algorithmic paradigm that can be extended in various ways leading to simple offline algorithms. These online extensions can provide improved performance both in terms of worst-case approximation ratios and in terms of performance on real data. See, for example, the experimental analysis of MaxSat provided by Poloczek and Williamson [35]. This still begs the question as to why we should restrict ourselves to conceptually simple algorithms when better offline algorithms are known.

While conceptually simple algorithms may not match the best approximations realized by more complex methods, they are usually very efficient (i.e., linear or near linear time with small constant factors) and often work well on realistic data exceeding worst-case approximation bounds. Conceptually simple algorithms can also be used as a preprocessing step for initializing a local search algorithm as in Chandra and Halldórsson [11]. Moreover, conceptually simple algorithms are easy to implement and modify with relatively little knowledge about the problem domain. Conceptual simplicity is arguably the main reason for the use of simple mechanisms in auctions (see, for example, Lucier and Syrgkanis [25]) and the success of MapReduce [13] in distributed parallel applications.

We will consider two departures from the adversarial and distributional one-pass online models. In the first departure, we consider a multi-pass online algorithm generalizing the two-pass algorithm in Dürr et al. [15]. In this regard we are also motivated by the Poloczek et al. [34] two-pass algorithm for MaxSat. These multi-pass algorithms can be viewed as de-randomizations of known online algorithms and the Dürr et al. algorithm (and our extension) can also be viewed

---

[1] The full version of the paper can be found on arXiv [9].

as an $O(n)$ space semi-streaming algorithm. The second departure is that of priority algorithms [8], a model for greedy and more generally myopic algorithms that extend online algorithms by allowing the algorithm to determine (in some well-defined way) the order of input arrivals. Other conceptually simple generalizations of the traditional online model are clearly possible, such as the ability to modify previous decisions and parallel executions of online algorithms.

Adversarial Online Model. In 1990, Karp et al. [24] studied MBM in the adversarial online setting. Any greedy algorithm (yielding a maximal matching) achieves a $1/2$ approximation and no deterministic algorithm can do asymptotically better in the adversarial online model. Karp et al. gave a randomized online algorithm called RANKING and showed that it achieves a $1 - 1/e \approx 0.632$ expected approximation ratio. Moreover, they proved that no randomized algorithm can beat $1 - 1/e$ in the adversarial online model. After 17 years, a mistake in the analysis of RANKING was found independently by Krohn and Varadarajan and by Goel and Mehta. A correct proof was first given by Goel and Mehta [19], followed by many alternative proofs.

Online Stochastic Models. Feldman et al. [17] introduced the known IID model for MBM. Feldman et al. model statistics about the upcoming queries by the notion of a type graph $G = (U, V, E)$ and a probability distribution $p : U \to [0, 1]$. The online nodes are sampled from p independently one at a time. An algorithm knows G and p beforehand. As before, an algorithm is required to make an irrevocable decision on how to match each arriving online node. In this setting, the adversary can only choose the type graph and the distribution p but doesn't have further control over the online sequence of nodes. Feldman et al. [17] provide an algorithm beating the $1 - 1/e$ barrier achieving approximation ratio ≈ 0.67. This work was followed by a long line of work including [3,10,20,22,29]. So far, the best approximation ratio for arbitrary arrival rates is ≈ 0.706 due to Jaillet and Lu [22].

Semi-streaming Model. Streaming algorithms are motivated by the necessity to process extremely large data streams. Much of the streaming literature concerns various forms of counting and statistics gathering. Semi-streaming algorithms are streaming algorithms designed for (say) graph search and optimization problems where the output itself requires $\Omega(n)$ space and hence a realistic goal is to maintain $\tilde{O}(n)$ space rather than space $O(m)$. Eggert et al. [16] provide a FPTAS multi-pass semi-streaming algorithm for MBM using space $\tilde{O}(n)$ in the edge input model. In the vertex input semi-streaming model, Goel et al. [18] give a *deterministic* $1 - 1/e$ approximation and Kapralov [23] proves that no semi-streaming algorithm can improve upon this ratio. (See also the recent survey by McGregor [30].) Streaming algorithms do not have to make online decisions but must maintain small space throughout the computation while online algorithms must make irrevocable decisions for each input item but have no space requirement. In some cases, streaming algorithms are designed so as to make results available at any time during the computation and hence some streaming algorithms can also be viewed as online algorithms as well. Conversely, any online algorithm that restricts itself to $\tilde{O}(n)$ space can be considered as a streaming algorithm.

Priority Model. In the priority model [8], an input to the algorithm is represented as a set of items coming from some universe. The universe of items in the MBM problem is the set of all pairs (online node, its neighborhood). The algorithm orders the entire universe by defining a priority function mapping each possible input item to a real number. Then, the adversary picks an instance G and reveals the online nodes in increasing order as specified by the priority function value. In a fixed-order algorithm, there is one initial ordering of the items whereas in an adaptive-order algorithm, the ordering can be redefined in each iteration. The algorithm makes an irrevocable decision about each input item. This captures many offline greedy algorithms that make a single pass over input items. Many problems have been studied in the priority model [1,5–7,12,33,37]. In a *fully randomized priority algorithm* [1] both the ordering of the input items and the decisions for each item are randomized. When only the decisions are randomized (and the ordering is deterministic) we will simply say *randomized priority algorithm*. With regards to maximum matching, Aronson et al. [2] proved that an algorithm that picks a random vertex and matches it to a random available neighbor (if it exists) achieves approximation ratio $1/2 + \epsilon$ for some $\epsilon > 0$ in general graphs. Besser and Poloczek [5] show that the algorithm that picks a random vertex of minimum degree and matches it to a randomly selected neighbor cannot improve upon the $1/2$ approximation ratio (with high probability) even for bipartite graphs. Pena and Borodin [32] show that no deterministic priority algorithm can achieve approximation ratio better than $1/2$. See [4,32] with respect to the difficulty of proving inapproximation results for randomized priority algorithms.

Advice Model. Dürr et al. [15] studied the online MBM problem in the adversarial (tape) *advice model*. In the most unrestricted advice setting, the advice bits are set by an all-powerful oracle. Dürr et al. show that $\Theta_\epsilon(n)$ advice bits are necessary and sufficient to guarantee approximation ratio $1 - \epsilon$ for MBM. They also show that $O(\log n)$ advice bits are sufficient for a deterministic advice algorithm to guarantee a $1 - 1/e$ approximation ratio. Construction of the $O(\log n)$ advice bits is based on examining the behavior of the RANKING algorithm on all $n!$ possible random strings for a given input of length n, which requires exponential time. It is not known if there is an efficient way to construct $O(\log n)$ advice bits. One may put computational or information-theoretic restrictions on the advice string, and ask what approximation ratios are achievable by online algorithms with restricted advice. Beyond their theoretical value, advice algorithms also give rise to classes of conceptually simple offline algorithms if the advice string is restricted to be efficiently computable. The Dürr et al. [15] deterministic advice algorithm CATEGORY-ADVICE achieves approximation ratio $3/5$ using an n-bit advice string, where the advice string itself is computable by an online algorithm. This algorithm can be viewed as a 2-pass online algorithm.

Our Online Multipass Results. We generalize the CATEGORY-ADVICE algorithm to a k-PASS CATEGORY-ADVICE algorithm for $k \geq 1$. For each $k \geq 1$, we prove that the exact approximation ratio of k-PASS CATEGORY-ADVICE algorithm is F_{2k}/F_{2k+1}, where F_n is the nth Fibonacci number. Our bounds show that the analysis of Dürr et al. for the 2-PASS CATEGORY-ADVICE algorithm was tight. Our result immediately implies that the approximation ratio of k-PASS CATEGORY-ADVICE converges to the inverse of the golden ratio $2/(1 + \sqrt{5}) \approx 0.618$ as k goes to infinity.

Our Results for the Known IID Model. A greedy algorithm always matches an online vertex if it has at least one available neighbor. In the known IID model, we can consider greedy algorithms without loss of generality (see Remark 1). Moreover, greedy algorithms satisfying natural consistency conditions achieve approximation ratio at least $1 - 1/e$. Ties may occur in a greedy algorithm when an online node has more than one available neighbor. A good tie-breaking rule can improve the approximation ratio. Algorithms beating the $1 - 1/e$ ratio are known in this model [3,10,20,22,29].[2] They are usually stated as non-greedy algorithms, but using a general conversion they can be turned into greedy algorithms, albeit with somewhat unnatural tie-breaking rules. These algorithms have polynomial time preprocessing step and, thus, are feasible from a theoretical point of view, but might not be feasible from a practical point of view on large inputs. Moreover, we argue that these algorithms are not that "conceptually simple." We study a deterministic greedy online algorithm MINDEGREE using a natural, conceptually simple, and efficient tie-breaking rule. This algorithm is motivated by the offline matching algorithm MINGREEDY [36]. We show that MINDEGREE does not beat the $1 - 1/e$ approximation achieved by any consistent greedy algorithm.

Our Results for the Priority Model. The RANKING algorithm in the ROM model is an instance of a fully randomized priority algorithm. The ordering of the online vertices is simply a uniform random permutation of the set of adversarially chosen input items. Is there a more "informed" way to deterministically or randomly choose the ordering within the priority framework? A natural idea is to give priority to the online nodes having the smallest degree since intuitively they seem to be the hardest to match if not seen early. Our intuition turns out to be misleading. We show that giving priority to nodes having the smallest degree using some deterministic (or the uniform random tie-breaking) rule cannot match the approximation achieved by a uniform ordering of the online nodes. In contrast to the 2-PASS CATEGORY-ADVICE 3/5 approximation, our analysis can also be used to show that a deterministic two-pass algorithm that computes the degree of the offline vertices in the first pass and then reorders the offline vertices according to non-decreasing degree (breaking ties by the initial ordering) will not achieve an asymptotic approximation ratio better than 1/2.

---

[2] In fact, it is easy to see that there is an optimal tie-breaking rule that can be computed by dynamic programming. Unfortunately, the size of the dynamic programming table is exponentially large, and moreover, currently it is not known how to analyze such optimal tie-breaking rules.

2 Preliminaries

$G = (U, V, E)$ denotes a bipartite graph where $U, V \subset \mathbb{N}$ form a partition of the vertices, and edges are $E \subseteq U \times V$. We consider the MBM in various online models. U represents the online vertices that are revealed one at a time, and V represents the offline vertices. A vertex u from U is revealed together with all edges incident on it. The online models differ in how vertices in U and their arrivals are chosen—adversarially, sampled from a known distribution, or via a limited adversary.

Let $M \subseteq E$ be some matching in G. For $u \in U$, we write $u \in M$ when there exists $v \in V$ such that $(u, v) \in M$. Similarly, we write $v \in M$ when there exists $u \in U$ such that $(u, v) \in M$. For $u \in M$, we write $M(u)$ to indicate the neighbor of u in M. We write $\text{OPT}(G)$ to stand for an offline optimum matching in G. Given an algorithm ALG, we let $\text{ALG}(G)$ stand for the matching returned by the algorithm ALG on input G. Abusing notation, we will also use $\text{ALG}(G)$ (resp. $\text{OPT}(G)$) to stand for the size of the matching returned by the algorithm on G (resp. by $\text{OPT}(G)$).

Definition 1. *We define* asymptotic approximation ratio *of an algorithm ALG as* $AR(ALG) = \lim_{\mathbb{E}[OPT(G)] \to \infty} \inf_G \frac{\mathbb{E}[ALG(G)]}{\mathbb{E}[OPT(G)]}$, *where the expectation is over the input distribution and randomness of the algorithm. Approximation ratios in other models (e.g., adversarial input, deterministic algorithms, etc.) are defined analogously.*

An online (or priority) MBM algorithm is greedy if whenever a newly arriving online node has at least one available neighbor the algorithm matches the arrived node to one of its neighbors.

Remark 1. Any online (or priority) algorithm for MBM achieving approximation ratio ρ (in an adversarial or stochastic input setting) can be turned into a *greedy* algorithm achieving approximation ratio $\geq \rho$. Informally, the idea is a simulation of the non-greedy algorithm in which we replace any non-greedy decision by forcing a match while remembering the configuration of the non-greedy algorithm.

In the graphs $G = (U, V, E)$ that we consider, we shall often refer to the so-called "parallel edges." Let $U' = \{u'_1, \ldots, u'_k\} \subseteq U$ and $V' = \{v'_1, \ldots, v'_k\} \subseteq V$ be two distinguished subsets such that for all $i \in [k]$ we have $(u'_i, v'_i) \in E$ (there might be other edges incident on U' and V'). The parallel edges between U' and V' are precisely the edges (u'_i, v'_i).

2.1 The Ranking Algorithm

In the adversarial model, graph G as well as an order π of its online vertices U is chosen by an adversary. Karp et al. [24] presented RANKING (see Algorithm 1). They showed that $AR(\text{RANKING}) = 1 - 1/e$ and that no randomized algorithm can do better in the adversarial model. RANKING samples a permutation σ of the offline nodes uniformly at random and runs a greedy algorithm on G breaking ties using σ. That is, when u arrives, if u has many unmatched neighbors v then u is matched with v that minimizes $\sigma(v)$. We write RANKING(π, σ) to denote the matching returned by RANKING given π and σ. When π is clear from the context we will omit it and simply write RANKING(σ).

2.2 Known IID Model with Uniform Probability Distribution

In the known IID model, $G = (U, V, E)$ is a *type graph* and nodes U are referred to as "types." The type graph specifies the distribution from which the actual *instance graph* $\widehat{G} = (\widehat{U}, \widehat{V}, \widehat{E})$ is generated. An instance graph is generated by setting $\widehat{V} = V$, sampling each $\widehat{u} \in \widehat{U}$ IID uniformly from U, and letting $(\widehat{u}, \widehat{v}) \in \widehat{E}$ if and only if the

---
Algorithm 1. The RANKING algorithm.

---
procedure RANKING$(G = (U, V, E))$
  Sample permutation $\sigma : V \to V$
    uniformly at random
  for all $u \in U$ **do**
    When u arrives, set
    $N(u) = \{$unmatched v s.th. $(u, v) \in E\}$
    if $N(u) \neq \emptyset$ **then**
      $v^* = \arg\min_v\{\sigma(v) \mid v \in N(u)\}$
      match u with v^*

---

corresponding $(u, v) \in E$. Each $\widehat{u}$ is drawn IID from U with replacement, thus it is possible that the same $u \in U$ appears multiple times in $\widehat{U}$. The type graph is chosen adversarially and is revealed to the algorithm in advance. Nodes $\widehat{u}$ are generated and presented to the algorithm one at a time. The algorithm makes an irrevocable decision on how to match $\widehat{u}$ (if at all). More general versions of this model have been defined and studied, but we won't need them in this paper.

A basic guarantee on the performance of a greedy algorithm in the known IID model holds as long as its tie-breaking rules satisfy natural consistency conditions. These conditions are somewhat technical, so we defer them to the full version of the paper [9]. Goel and Mehta [19] proved that such greedy algorithms achieve approximation at least $1 - 1/e$ in the known IID model.[3]

3 Deterministic Multipass Online Algorithms

To break through the 1/2 barrier of the adversarial online model, Dürr et al. [15] modify the model to allow for a second pass over the input. They give a deterministic 2-pass algorithm, called CATEGORY-ADVICE, that achieves a 3/5 approximation ratio. CATEGORY-ADVICE belongs to the class of RANKING-based algorithms called category algorithms that were introduced in the work of Dürr et al. A category algorithm considers a permutation σ of the offline nodes. Instead of running RANKING directly with σ, a category function $c : V \to \mathbb{Z} \cup \{\pm\infty\}$ is

---

[3] The Goel and Mehta [19] result is even stronger as it holds for the ROM model.

computed first. The updated permutation σ_c is the unique permutation satisfying the following defining property: for all $v_1, v_2 \in V$, we have $\sigma_c(v_1) < \sigma_c(v_2)$ if and only if $c(v_1) < c(v_2)$ or $(c(v_1) = c(v_2)$ and $\sigma(v_1) < \sigma(v_2))$. Then RANKING($\sigma_c$) is performed with σ_c as the permutation of the offline nodes.

The CATEGORY-ADVICE algorithm starts with an arbitrary permutation σ, e.g., σ could be induced by the names of nodes V. In the first pass, the algorithm runs RANKING with σ. Let M be the matching obtained in the first pass. The category function $c : V \rightarrow [2]$ is then defined as follows: $c(v) = 1$ if $v \notin M$ and $c(v) = 2$ otherwise. In the second pass, the CATEGORY-ADVICE algorithm runs RANKING(σ_c). The output of the second run of RANKING is declared as the output of the CATEGORY-ADVICE algorithm. In other words, in the second pass the algorithm gives preference to those vertices that were **not** matched in the first pass. (We observe that the Besser-Poloczek graphs, see the full version of the paper [9], show that the category function $c(v) =$ degree of v will not yield an asymptotic approximation better than $1/2$.)

We present a natural extension of the CATEGORY-ADVICE algorithm to multiple passes, called k-PASS CATEGORY-ADVICE (see Algorithm 2). Let F_n denote the nth Fibonacci number, i.e., $F_1 = F_2 = 1$, and $F_n = F_{n-1} + F_{n-2}$ for $n \geq 3$. For each $k \geq 1$, we prove that the k-PASS CATEGORY-ADVICE algorithm[4] achieves the approximation ratio F_{2k}/F_{2k+1}. Moreover, we show that this is tight, i.e., there exists a family of bipartite graphs, one for each $k \geq 1$, such that the k-PASS CATEGORY-ADVICE algorithm computes a matching that is F_{2k}/F_{2k+1} times the size of the maximum matching. In particular, we show that the analysis of the $3/5$ approximation ratio in [15] is tight. It immediately follows that as k goes to infinity, the approximation guarantee of the k-PASS CATEGORY-ADVICE algorithm converges (quickly) to the inverse of the golden ratio $2/(1 + \sqrt{5}) \approx 0.618$. The main theorem of this section is

Theorem 1. *The exact approximation ratio of the k-PASS CATEGORY-ADVICE algorithm is F_{2k}/F_{2k+1}, where F_n is the nth Fibonacci number. Thus, the approximation ratio of the k-PASS CATEGORY-ADVICE algorithms tends to the inverse of the golden ration $2/(1 + \sqrt{5}) \approx 0.618$ as k goes to infinity. This holds even when k depends on n arbitrarily. The algorithm can be realized as a k-pass semi-streaming algorithm using $O(n \log k)$ space, which is $O(n)$ for constant k.*

3.1   Positive Result

The pseudocode for k-PASS CATEGORY-ADVICE appears in Algorithm 2. The algorithm is defined iteratively with each iteration corresponding to a new pass. The algorithm initializes σ of the offline nodes to the identity permutation. The algorithm maintains a category function $c : V \rightarrow \mathbb{Z} \cup \{\pm\infty\}$. Initially, c is set to

---

[4] A notable feature of this multi-pass algorithm is that after pass i, the algorithm can deterministically commit to matching a subset of size $\frac{F_{2i}}{F_{2i+1}}|M|$ where M is a maximum matching. This follows from a certain monotonicity property. See the full version of the paper for details [9].

$-\infty$ everywhere. In the ith pass, the algorithm runs RANKING on σ_c. Let M_i be the resulting matching. For each $v \in V$, if $c(v) = -\infty$ and $v \in M_i$ then $c(v)$ is updated to $-i$. The algorithm uses the updated c in the next pass. In words, c records for each vertex v the (negative of the) first pass in which v was matched. In the subsequent pass, the algorithm gives preference to the nodes that were unmatched, followed by nodes that were matched for the first time in the latest round, etc.

Lemma 1. *The k-PASS CATEGORY-ADVICE algorithm achieves approximation ratio F_{2k}/F_{2k+1}.*

Proof (by induction on k).
Base case: $k = 1$. The 1-PASS CATEGORY-ADVICE algorithm is the simple deterministic greedy algorithm, which achieves a $1/2 = F_2/F_3$ approximation ratio.

Inductive step. Consider the $(k + 1)$-PASS CATEGORY-ADVICE algorithm running on $G = (U, V, E)$. WLOG, we may assume that $|U| = |V|$

Algorithm 2. k-PASS CATEGORY-ADVICE.

procedure k-PASS CATEGORY ADVICE
  $\sigma \leftarrow$ the identity permutation of V
  initialize array c of size V with $-\infty$
  for i from 1 to k **do**
    Run pass i: $M_i \leftarrow$ RANKING(σ_c)
    for $v \in V$ **do**
      if $c(v) = -\infty$ and $v \in M_i$ **then**
        $c(v) \leftarrow -i$
  return M_k

---

and G has a perfect matching (see the full version of the paper [9]). Let $U_1 \subseteq U$ be the set of nodes matched in the first pass of the algorithm. Let $V_1 \subseteq V$ be the nodes that U_1 nodes are matched to. Define $U_2 = U \setminus U_1$ and $V_2 = V \setminus V_1$. Note that there are no edges between U_2 and V_2; otherwise some node of U_2 would have been matched to some node of V_2 in the first round. Let M_i be the matching found by the $(k + 1)$-PASS CATEGORY-ADVICE algorithm in round i, where $i \in [k + 1]$. Also define $M_{11} = M_{k+1} \cap U_1 \times V_1$, $M_{12} = M_{k+1} \cap U_1 \times V_2$, and $M_{21} = M_{k+1} \cap U_2 \times V_1$. We are interested in bounding $|M_{k+1}| = |M_{11}| + |M_{12}| + |M_{21}|$. See Fig. 1.

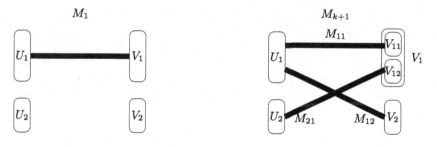

Fig. 1. G is the input graph. On the left we show the matching constructed in the first pass. On the right we show the matching constructed in the $k + 1$st pass.

After the first round, nodes in U_1 prefer nodes from V_2 to those from V_1. Moreover, nodes in V_2 are only connected to nodes in U_1 and there is a perfect matching between V_2 and a subset of U_1. Thus, the matching constructed between U_1 and V_2 in the next k passes is the same as if we ran k-PASS CATEGORY-ADVICE algorithm on the subgraph of G induced by $U_1 \cup V_2$. This implies that $|M_{12}| \geq (F_{2k}/F_{2k+1})|V_2| = (F_{2k}/F_{2k+1})|U_2|$.

By the monotonicity property (see the full version of the paper [9]), in the $k + 1$st pass, all nodes from U_1 that were not matched with V_2 will be matched with some nodes in V_1, i.e., $|U_1| = |M_{12}| + |M_{11}|$. Let V_{11} be such a set, and let $V_{12} = V_1 \setminus V_{11}$. To lower bound $|M_{21}|$, we first lower bound the size of a maximum matching between U_2 and V_{12}. Since U_2 is only connected to V_1 and since there is a perfect matching, a maximum matching between U_2 and V_1 is of size $|U_2|$. Thus, the size of a maximum matching between U_2 and V_{12} is at least $|U_2| - |V_{11}|$. Also, observe that $|V_{11}| = |V_1| - |V_{12}|$ and $|V_{12}| = |M_{12}|$. Therefore, the size of a maximum matching between U_2 and V_{12} is at least $|U_2| - (|V_1| - |M_{12}|) = |U_2| - |U_1| + |M_{12}|$ (note that $|U_1| = |V_1|$). Finally in the last round, the algorithm constructs a maximal matching between U_2 and V_{12} guaranteeing that $|M_{21}| \geq (1/2)(|U_2| - |U_1| + |M_{12}|)$. Putting it all together, we obtain $|M_{k+1}| = |M_{11}| + |M_{12}| + |M_{21}| = |U_1| - |M_{12}| + |M_{12}| + |M_{21}| \geq |U_1| + \frac{1}{2}(|U_2| - |U_1| + |M_{12}|) \geq \frac{1}{2}\left(|U_2| + |U_1| + \frac{F_{2k}}{F_{2k+1}}|U_2|\right) = \frac{1}{2}\left(n + \frac{F_{2k}}{F_{2k+1}}(n - |M_1|)\right) = \left(\frac{1}{2} + \frac{F_{2k}}{2F_{2k+1}}\right)n - \frac{F_{2k}}{2F_{2k+1}}|M_1|$.

By the monotonicity property we also have that $|M_{k+1}| \geq |M_1|$. Thus, we derive $|M_{k+1}| \geq \max\left\{|M_1|, \left(\frac{1}{2} + \frac{F_{2k}}{2F_{2k+1}}\right)n - \frac{F_{2k}}{2F_{2k+1}}|M_1|\right\}$. This implies that $|M_{k+1}| \geq \frac{1/2 + F_{2k}/2F_{2k+1}}{1 + F_{2k}/(2F_{2k+1})}n = \frac{F_{2k+1} + F_{2k}}{2F_{2k+1} + F_{2k}}n = \frac{F_{2(k+1)}}{F_{2(k+1)+1}}n$.

3.2   Negative Result

In this subsection, we construct a family of bipartite graphs $(G_k)_{k=1}^{\infty}$ with the following properties[5]: (1) G_k has exactly F_{2k+1} online nodes and F_{2k+1} offline nodes, (2) G_k has a perfect matching, (3) the k-PASS CATEGORY-ADVICE algorithm finds a matching of size F_{2k}, (4) for all $k' > k$, the k'-PASS CATEGORY-ADVICE algorithm finds a matching of size $F_{2k} + 1$.

This family of graphs shows that our analysis of k-PASS CATEGORY-ADVICE is tight. The construction of the G_k is recursive. The base case is given by G_1, which is depicted in Fig. 2. The online nodes are shown on the left whereas the offline nodes are on the right. The online nodes always arrive in the order shown on the diagram from top to bottom, and the initial permutation σ of the offline nodes is also given by the top to bottom ordering of the offline nodes on the diagram.

---

[5] The last property allows us to conclude that the approximation ratio of k-PASS CATEGORY-ADVICE converges to the inverse of the golden ratio even when k is allowed to depend on n arbitrarily.

In the recursive step, we construct G_{k+1} from G_k. The online nodes U of G_{k+1} are partitioned into three disjoint sets $U = U_1 \cup U_2 \cup U_3$ such that $|U_1| = |U_3| = F_{2k+1}$ and $|U_2| = F_{2k}$. Similarly, the offline nodes V of G_{k+1} are partitioned into three disjoint sets $V = V_1 \cup V_2 \cup V_3$ such that $|V_1| = |V_3| = F_{2k+1}$ and $|V_2| = F_{2k}$. There is a copy of G_k between U_1

Fig. 2. G_1 is used for the basis of the induction.

and V_3. U_2 and V_2 are connected by parallel edges. There is a complete bipartite graph between U_1 and V_1, as well as between U_2 and V_1. Finally, U_3 is connected to V_1 by parallel edges. The construction is depicted in Fig. 3.

Lemma 2. *Properties (1–4) mentioned at the beginning of this subsection hold for the G_k.*

Proof (by induction on k). Base case: $k = 1$. G_1 has 2 online and 2 offline nodes and $F_3 = 2$. Clearly, G_1 has a perfect matching. The 1-PASS CATEGORY-ADVICE algorithm is the regular greedy algorithm, which returns a matching of size 1 on G_1, and $F_2 = 1$. Lastly, 2-PASS CATEGORY-ADVICE finds a matching of size $F_2 + 1 = 2$ and adding more passes does not change anything.

Inductive step. Assume that properties (1–4) hold for G_k. The number of online vertices of G_{k+1} is equal to the number of offline vertices and is equal to $F_{2k+1} + F_{2k} + F_{2k+1} = F_{2k+2} + F_{2k+1} = F_{2k+3} = F_{2(k+1)+1}$. By inductive assumption, G_k has a perfect matching. Therefore, U_1 vertices can be matched with V_3 via the perfect matching given by G_k. In addition, U_2 can be matched with V_2 by parallel edges, and U_3 can be

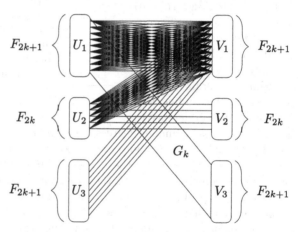

Fig. 3. G_{k+1} shows the inductive construction.

matched with V_1 by parallel edges as well. Thus, G_{k+1} has a perfect matching. Thus, we proved properties 1 and 2. To prove the 3rd property, observe that in the first pass, the $(k + 1)$-PASS CATEGORY-ADVICE algorithm matches U_1 and V_1 by parallel edges, U_2 with V_2 by parallel edges, and leaves U_3 and V_3 unmatched. Since V_3 is only connected to the nodes U_1, in the next k passes the behavior of the algorithm between U_1 and V_3 nodes is exactly that of the k-PASS CATEGORY-ADVICE algorithm. Therefore, by the inductive assumption, the algorithm is going to match exactly F_{2k} nodes from U_1 with the nodes in V_3. The remaining $F_{2k+1} - F_{2k}$ nodes from U_1 will be matched to the nodes in V_1 (since those are the only neighbors of U_1 nodes besides the nodes from V_3).

The nodes from U_2 in all passes behave the same way – they prefer V_1 nodes to V_2 nodes. Thus, since V_1 will have F_{2k} nodes unmatched after processing all nodes of U_1 in the last round, all of U_2 nodes will be matched to V_1 in the last round. This implies that after processing U_1 and U_2 in the last round, all of V_1 nodes are matched. Therefore, none of U_3 nodes can be matched. Thus, the matching found by the $(k+1)$-pass CATEGORY-ADVICE algorithm on G_{k+1} is of size $|U_1| + |U_2| = F_{2k+1} + F_{2k} = F_{2k+2} = F_{2(k+1)}$. The last property is proved similarly.

4   MinDegree Algorithm in the Known IID Model

Our algorithm is MINDEGREE (see Algorithm 3). The motivation for studying this algorithm is as follows. It is easy to see that in the adversarial setting we can take any greedy algorithm, modify the tie-breaking rule to always give more preference to the offline nodes of degree 1, and this will not decrease the approximation ratio of the algorithm. Generalizing this, we conclude that online vertices should give preference to

---
Algorithm 3. The MINDEGREE algorithm.

---
procedure MINDEGREE($G = (U, V, E)$)
    Let $S = V$ be the set of active offline nodes
  repeat
      Let $\hat{u}$ denote the arriving online node
      Let $N(\hat{u}) = \{v \in S \mid (u, v) \in E\}$
      if $N(\hat{u}) \neq \emptyset$ **then**
         Let $v = \arg\min_{v \in N(\hat{u})} \deg(v)$
         Match $\hat{u}$ with v
         Remove v from S
  until all online nodes have been processed

---

the offline vertices of smaller degrees. The problem is that in the adversarial setting, degrees of the offline nodes are not known a priori. However, in the known IID setting we can estimate the degrees of the offline vertices from the type graph. This is precisely what MINDEGREE formalizes. The algorithm is given a type graph $G = (U, V, E)$ as input. It keeps track of a set S of currently "active", i.e., not yet matched, offline nodes. When a new node $\tilde{u}$ arrives, it is matched to its active neighbor of minimum degree in the type graph.

Remark 2. Our algorithm does not fully break ties, i.e., MINDEGREE takes *some* neighbor of currently minimum degree. In practice, it means that ties are broken in some arbitrary way, e.g., by names of vertices. In our theoretical analysis, this means that the adversary is more powerful, as it can decide how the algorithm breaks these ties.

MINDEGREE is a conceptually simple and promising algorithm in the known IID setting. Indeed, a version of MINDEGREE called MINGREEDY has been extensively studied in the offline setting (see Besser and Poloczek [5] and references therein). Unfortunately, in spite of having excellent empirical performance as well as excellent performance under various random input models, MINGREEDY has a bad worst-case approximation ratio of $1/2$ in the offline adversarial setting [5]. As far as we know, this algorithm has not been analyzed in the known IID model. We obtain a result that, in spirit, is similar to the offline result, but the proof is different. Namely, we show that MINDEGREE cannot achieve an approximation ratio better than $1 - 1/e$, which is guaranteed by any consistent greedy algorithm in the known IID model. See the full version of the paper [9] for the proof of the following result.

Theorem 2

$$AR(\text{MINDEGREE}) = 1 - \frac{1}{e}$$

The negative result holds no matter which rule is used to break (remaining) ties in MINDEGREE.

5   MinRanking – A Hybrid Algorithm in the Priority Model

We propose a conceptually simple greedy algorithm for MBM. Our algorithm is a natural hybrid between two well-known algorithms—RANKING and MINGREEDY. We have already encountered MINGREEDY in this paper (see Sects. 1 and 4). MINGREEDY is an offline algorithm that selects a random vertex of minimum degree in the input graph and matches it with a random neighbor, removes the matched vertices, and proceeds. In a natural adaptation of MINGREEDY to bipartite graphs $G = (U, V, E)$, the algorithm picks a random node of minimum degree from U and matches it to a random neighbor from V. Observe that this algorithm can be realized as a fully randomized priority algorithm, where the ordering of the input items is by increasing degree with ties broken randomly. See the full version of the paper [9] for the pseudocode.

Karp et al. [24] exhibited a family of graphs, on which RANKING gets its worst approximation ratio $1 - 1/e$. MINGREEDY finds a perfect matching on these graphs. Thus, it is natural to consider the performance of an algorithm that combines RANKING and MINGREEDY. This is our proposed MINRANKING algorithm (see Algorithm 4). In MINRANKING, a random permutation π of vertices V is initially sampled. Then, nodes in U are processed in the increasing order of their current degrees (cur deg) with ties broken randomly. When a node u is processed, it is matched with its neighbor appearing earliest in the ordering π.

MINRANKING modifies MINGREEDY in the same way that the online RANKING algorithm modifies the seemingly more natural online randomized algorithm that simply matches an online vertex to an available neighbor uniformly at random which surprisingly was shown to be (asymptotically) a $1/2$ approximation. So it is hopeful that MINRANKING can improve upon MINGREEDY. Like MINGREEDY, our algorithm can be implemented and analyzed in the fully randomized adaptive priority model [8]. Since our algorithm is a generalization of RANKING, its asymptotic approximation ratio is at least

$1 - 1/e \approx 0.632$. We show that the asymptotic approximation ratio of this algorithm is at most $1/2 + 1/(2e) \approx 0.684$, as witnessed by the construction of Besser and Poloczek [5].

Theorem 3

$$1 - \frac{1}{e} \leq AR(\text{MINRANKING}) \leq \frac{1}{2} + \frac{1}{2e}.$$

This negative result shows that MINRANKING falls short of the known bound for RANKING in the ROM model, where it achieves approximation ratio 0.696 [28]. From our result it follows that a deterministic ordering of the online nodes by non-decreasing degree (breaking ties by the given initial labeling of those nodes) will also fall short. That is (similar to the result in [32] for deterministic decisions), a naive randomized ordering can be better than a seemingly informed deterministic ordering. See the full version of the paper [9] for more details.

Algorithm 4. The MINRANKING algorithm.

---

procedure MINRANKING($G = (U, V, E)$)
    Pick a permutation $\pi : V \to V$ at random
    repeat
        Let $d = \min\{\text{cur deg}(i) \mid i \in U\}$
        Let $S = \{i \in U \mid \text{cur deg}(i) = d\}$
        Pick $i \in S$ uniformly at random
        Let $N(i)$ be the set of neighbors of i
        if $N(i) = \emptyset$ **then**
            i remains unmatched
            Delete i from G
        else
            $j = \arg\min_k\{\pi(k) \mid k \in N(i)\}$
            Match i with j
            Delete i and j from G
            Update cur deg
    until $U = \emptyset$

---

6 Conclusion and Open Problems

We have considered a number of "online-based" algorithms for the MBM problem. We believe that the algorithms considered in this paper all pass the intuitive "you know it when you see it" standard for conceptually simple algorithms. In particular, these algorithms take linear time in the number of edges and are very easy to implement. Even given the restricted nature of these algorithms, it is a challenge to understand their performance.

Our results for the MBM, in conjunction with the results in Poloczek et al. [34] for MaxSat, show both the promise and limitations of conceptually simple algorithms. Many open problems are suggested by this work. Clearly, any problem studied in the competitive online literature can be considered within the expanded framework of conceptually simple algorithms. For what problems is there a general method for de-randomizing online algorithms? Is there a precise algorithmic model that lends itself to analysis and captures multi-pass algorithms? And in addition to worst-case and stochastic analysis, how would any of the conceptually simple MBM algorithms perform "in practice?"

References

1. Angelopoulos, S., Borodin, A.: On the power of priority algorithms for facility location and set cover. In: Jansen, K., Leonardi, S., Vazirani, V. (eds.) APPROX 2002. LNCS, vol. 2462, pp. 26–39. Springer, Heidelberg (2002). https://doi.org/10.1007/3-540-45753-4_5

2. Aronson, J., Dyer, M., Frieze, A., Suen, S.: Randomized greedy matching. II. Random Struct. Algorithms 6(1), 55–73 (1995)

3. Bahmani, B., Kapralov, M.: Improved bounds for online stochastic matching. In: de Berg, M., Meyer, U. (eds.) ESA 2010. LNCS, vol. 6346, pp. 170–181. Springer, Heidelberg (2010). https://doi.org/10.1007/978-3-642-15775-2_15

4. Besser, B., Poloczek, M.: Erratum to: greedy matching: guarantees and limitations. Algorithmica 80, 1–4 (2017)

5. Besser, B., Poloczek, M.: Greedy matching: guarantees and limitations. Algorithmica 77(1), 201–234 (2017)

6. Borodin, A., Boyar, J., Larsen, K.S.: Priority algorithms for graph optimization problems. In: Persiano, G., Solis-Oba, R. (eds.) WAOA 2004. LNCS, vol. 3351, pp. 126–139. Springer, Heidelberg (2005). https://doi.org/10.1007/978-3-540-31833-0_12

7. Borodin, A., Ivan, I., Ye, Y., Zimny, B.: On sum coloring and sum multi-coloring for restricted families of graphs. Theor. Comput. Sci. 418, 1–13 (2012)

8. Borodin, A., Nielsen, M.N., Rackoff, C.: (Incremental) priority algorithms. Algorithmica 37(4), 295–326 (2003)

9. Borodin, A., Pankratov, D., Salehi-Abari, A.: On conceptually simple algorithms for variants of online bipartite matching. CoRR, abs/1706.09966 (2017)

10. Brubach, B., Sankararaman, K.A., Srinivasan, A., Xu, P.: New algorithms, better bounds, and a novel model for online stochastic matching. In: Proceedings of ESA, pp. 24:1–24:16 (2016)

11. Chandra, B., Halldórsson, M.M.: Greedy local improvement and weighted set packing approximation. J. Algorithms 39(2), 223–240 (2001)

12. Davis, S., Impagliazzo, R.: Models of greedy algorithms for graph problems. Algorithmica 54(3), 269–317 (2009)

13. Dean, J., Ghemawat, S.: MapReduce: simplified data processing on large clusters. Commun. ACM 51(1), 107–113 (2008)

14. Duan, R., Pettie, S.: Linear-time approximation for maximum weight matching. J. ACM 61(1), 1:1–1:23 (2014)

15. Dürr, C., Konrad, C., Renault, M.: On the power of advice and randomization for online bipartite matching. In: Proceedings of ESA, pp. 37:1–37:16 (2016)

16. Eggert, S., Kliemann, L., Munstermann, P., Srivastav, A.: Bipartite matching in the semi-streaming model. Algorithmica 63(1–2), 490–508 (2012)

17. Feldman, J., Mehta, A., Mirrokni, V., Muthukrishnan, S.: Online stochastic matching: beating 1-1/e. In: Proceedings of FOCS, pp. 117–126 (2009)

18. Goel, A., Kapralov, M., Khanna, S.: On the communication and streaming complexity of maximum bipartite matching. In: Proceedings of the Twenty-Third Annual ACM-SIAM Symposium on Discrete Algorithms, SODA 2012, Kyoto, Japan, 17–19 January 2012, pp. 468–485 (2012)

19. Goel, G., Mehta, A.: Online budgeted matching in random input models with applications to adwords. In: Proceeding of SODA, pp. 982–991 (2008)

20. Haeupler, B., Mirrokni, V.S., Zadimoghaddam, M.: Online stochastic weighted matching: improved approximation algorithms. In: Chen, N., Elkind, E., Koutsoupias, E. (eds.) WINE 2011. LNCS, vol. 7090, pp. 170–181. Springer, Heidelberg (2011). https://doi.org/10.1007/978-3-642-25510-6_15

21. Hopcroft, J.E., Karp, R.M.: An $n^{5/2}$ algorithm for maximum matchings in bipartite graphs. SIAM J. Comput. **2**(4), 225–231 (1973)

22. Jaillet, P., Lu, X.: Online stochastic matching: new algorithms with better bounds. Math. Oper. Res. **39**(3), 624–646 (2014)

23. Kapralov, M.: Better bounds for matchings in the streaming model. In: Proceedings of SODA, pp. 1679–1697 (2013)

24. Karp, R.M., Vazirani, U.V., Vazirani, V.V.: An optimal algorithm for on-line bipartite matching. In: Proceedings of STOC, pp. 352–358 (1990)

25. Lucier, B., Syrgkanis, V.: Greedy algorithms make efficient mechanisms. In: Proceedings of Conference on Economics and Computation, EC, pp. 221–238 (2015)

26. Madry, A.: Navigating central path with electrical flows: from flows to matchings, and back. In: Proceedings of FOCS, pp. 253–262 (2013)

27. Madry, A.: Computing maximum flow with augmenting electrical flows. In: Proceedings of FOCS, pp. 593–602 (2016)

28. Mahdian, M., Yan, Q.: Online bipartite matching with random arrivals: an approach based on strongly factor-revealing LPs. In: Proceedings of STOC, pp. 597–606 (2011)

29. Manshadi, V.H., Gharan, S.O., Saberi, A.: Online stochastic matching: online actions based on offline statistics. In: Proceedings of SODA, pp. 1285–1294 (2011)

30. McGregor, A.: Graph sketching and streaming: new approaches for analyzing massive graphs. In: Weil, P. (ed.) CSR 2017. LNCS, vol. 10304, pp. 20–24. Springer, Cham (2017). https://doi.org/10.1007/978-3-319-58747-9_4

31. Mucha, M., Sankowski, P.: Maximum matchings via Gaussian elimination. In: Proceedings of FOCS, pp. 248–255 (2004)

32. Pena, N., Borodin, A.: On the limitations of deterministic de-randomizations for online bipartite matching and max-sat. CoRR, abs/1608.03182 (2016)

33. Poloczek, M.: Bounds on greedy algorithms for MAX SAT. In: Demetrescu, C., Halldórsson, M.M. (eds.) ESA 2011. LNCS, vol. 6942, pp. 37–48. Springer, Heidelberg (2011). https://doi.org/10.1007/978-3-642-23719-5_4

34. Poloczek, M., Schnitger, G., Williamson, D.P., Van Zuylen, A.: Greedy algorithms for the maximum satisfiability problem: simple algorithms and inapproximability bounds. SICOMP, accepted for publication

35. Poloczek, M., Williamson, D.P.: An experimental evaluation of fast approximation algorithms for the maximum satisfiability problem. In: Proceedings of International Symposium on Experimental Algorithms, SEA, pp. 246–261 (2016)

36. Tinhofer, G.: A probabilistic analysis of some greedy cardinality matching algorithms. Ann. Oper. Res. **1**(3), 239–254 (1984)

37. Ye, Y., Borodin, A.: Priority algorithms for the subset-sum problem. J. Comb. Optim. **16**(3), 198–228 (2008)

Efficient Dynamic Approximate Distance Oracles for Vertex-Labeled Planar Graphs

Itay Laish and Shay Mozes[(✉)]

Efi Arazi School of Computer Science,
The Interdisciplinary Center Herzliya, Herzliya, Israel
itaylaish@post.idc.ac.il, smozes@idc.ac.il

Abstract. Let G be a graph where each vertex is associated with a label. A *Vertex-Labeled Approximate Distance Oracle* is a data structure that, given a vertex v and a label λ, returns a $(1+\varepsilon)$-approximation of the distance from v to the closest vertex with label λ in G. Such an oracle is *dynamic* if it also supports label changes. In this paper we present three different dynamic approximate vertex-labeled distance oracles for planar graphs, all with polylogarithmic query and update times, and nearly linear space requirements. No such oracles were previously known.

Keywords: Planar graphs · Approximate distance oracles
Vertex labels · Portals · ε-Cover

1 Introduction

Consider the following scenario. A 911 dispatcher receives a call about a fire and needs to dispatch the closest fire truck. There are two difficulties with locating the appropriate vehicle to dispatch. First, the vehicles are on a constant move. Second, there are different types of emergency vehicles, whereas the dispatcher specifically needs a fire truck. Locating the closest unit of certain type under these assumptions is the *dynamic vertex-labeled distance query problem* on the road network graph. Each vertex in this graph can be annotated with a label that represents the type of the emergency vehicle currently located at that vertex. An alternative scenario where this problem is relevant is when one wishes to find a service provider (e.g., gas station, coffee shop), but different locations are open at different times of the day.

A data structure that answers distance queries between a vertex and a label, and supports label updates is called a *dynamic vertex-labeled distance oracle*. We model the road map as a planar graph, and extend previous results for the static case (where labels are fixed). We present oracles with polylogarithmic update and query times (in the number of vertices) that require nearly linear space.

For a full version of this paper, see https://arxiv.org/abs/1707.02414. This research was supported by the ISRAEL SCIENCE FOUNDATION (grant No. 794/13).

We focus on approximate vertex-labeled distance oracles for fixed parameter $\varepsilon \geq 0$. When queried, such an oracle returns at least the true distance, but not more than $(1+\varepsilon)$ times the true distance. These are also known as stretch-$(1+\varepsilon)$ distance oracles. Note that, in our context, the graph is fixed, and only the vertex labels change.

1.1   Related Work

A seminal result on approximate vertex-to-vertex distance oracles for planar graphs is that of Thorup [11]. He presented a stretch-$(1 + \varepsilon)$ distance oracle for directed planar graphs. For any $0 < \varepsilon < 1$, his oracle requires $O(\varepsilon^{-1} n \log n \log(nN))$ space and answers queries in $O(\log \log (nN) + \varepsilon^{-1})$ time. Here N denotes the ratio of the largest to smallest arc length. For undirected planar graphs Thorup presented a $O(\varepsilon^{-1} n \log n)$ space oracle that answers queries in $O(\varepsilon^{-1})$ time. Klein [5,6] independently described a stretch-$(1 + \varepsilon)$ distance oracle for undirected graphs with the same bounds, but faster preprocessing time.

The first result for the static vertex-labeled problem for undirected planar graph is due to Li et al. [7]. They described a stretch-$(1 + \varepsilon)$ distance oracle that is based on Klein's results [5]. Their oracle requires $O(\varepsilon^{-1} n \log n)$ space, and answers queries in $O(\varepsilon^{-1} \log n \log \Delta)$ time. Here Δ is the hop-diameter of the graph, which can be $\Theta(n)$. Mozes and Skop [9], building on Thorup's oracle, described a stretch-$(1+\varepsilon)$ distance oracle for directed planar graphs that can be stored using $O(\varepsilon^{-1} n \log n \log(nN))$ space, and has $O(\log \log n \log \log nN + \varepsilon^{-1})$ query time.

Li et al. [7] considered the dynamic case, but their update time is $\Theta(n \log n)$ in the worst case. Łącki et al. [1] presented a different dynamic vertex-to-label oracle for undirected planar graphs, in the context of computing Steiner trees. Their oracle requires $O(\sqrt{n} \log^2 n \log D\varepsilon^{-1})$ amortized time per update or query (in expectation), where D is the stretch of the metric of the graph (could be nN). Their oracle however does not support changing the label of a specific vertex. It supports merging two labels, and splitting of labels in a restricted way. To the best of our knowledge, ours are the first approximate dynamic vertex-labeled distance oracles with polylogarithmic query and update times, and the first that support directed planar graphs.

1.2   Our Results and Techniques

We present three approximate vertex-labeled distance oracles with polylogarithmic query and update times and nearly linear space and preprocessing times. Our update and construction times are expected amortized due to the use of dynamic hashing.[1] Our solutions differ in the tradeoff between query and update times. One solution works for directed planar graphs, whereas the other two only work for undirected planar graphs.

---

[1] We assume that a single comparison or addition of two numbers takes constant time.

We obtain our results by building on and combining existing techniques for the static case. All of our oracles rely on recursively decomposing the graph using shortest paths separators. Our first oracle for undirected graphs (Sect. 3) uses uniformly spaced connections, and efficiently handles them using fast predecessor data structures. The upshot of this approach is that there are relatively few connections. The caveat is that this approach only works when working with bounded distances, so a scaling technique [11] is required.

Our second oracle for undirected graphs (Sect. 5) uses the approach taken by Li et al. [7] in the static case. Each vertex has a different set of connections, which are handled efficiently using a dynamic prefix minimum query data structure. Such a data structure can be obtained using a data structure for reporting points in a rectangular region of the plane [12].

Our oracle for directed planar graphs (Sect. 4) is based on the static vertex-labeled distance oracle of [9], which uses connections for sets of vertices (i.e., a label) rather than connections for individual vertices. We show how to efficiently maintain the connections for a dynamically changing set of vertices using a bottom-up approach along the decomposition of the graph.

Our data structures support both queries and updates in polylogarithmic time. No previously known data structure supported both queries and updates in sublinear time. The following table summarizes the comparison between our oracles and the relevant previously known ones (Table 1).

Table 1. Vertex-to-label distance oracles time bound comparison

	D/U	Query time	Update time
Li et al. [7]	U	$O(\varepsilon^{-1} \log n \log \Delta)$	$O(n \log n)$
Łącki et al. [1]	U	$O(\varepsilon^{-1} \sqrt{n} \log^2 n \log D)$	$O(\varepsilon^{-1} \sqrt{n} \log^2 n \log D)$
Section 3 (faster query)	U	$O(\varepsilon^{-1} \log n \log \log nN)$	$O(\varepsilon^{-1} \log n \log \log \varepsilon^{-1} \log nN)$
Section 5 (faster update)	U	$O(\varepsilon^{-1} \frac{\log^2 (\varepsilon^{-1} n)}{\log \log (\varepsilon^{-1} n)})$	$O(\varepsilon^{-1} \log^{1.51}(\varepsilon^{-1} n))$
Mozes and Skop [9]	D	$O(\varepsilon^{-1} + \log \log n \log \log nN)$	N/A
Section 4	D	$O(\varepsilon^{-1} \log n \log \log nN)$	$O(\varepsilon^{-1} \log^3 n \log nN)$

In the table above, D/U stands for Directed and Undirected graphs.

Due to space constraints we omit all formal proofs from this extended abstract. The proofs are available in the full version of the paper.

2   Preliminaries

We shall use the term edges and arcs when referring to undirected and directed graphs, respectively. Given an undirected graph G with a spanning tree T rooted at r and an edge uv not in T, the *fundamental cycle* of uv (with respect to T) is the cycle composed of the r-to-u and r-to-v paths in T, and the edge uv. By a spanning tree of a *directed graph* G we mean a spanning tree of the underlying undirected graph of G.

Let $\ell : E(G) \to \mathbb{R}^+$ be a non-negative length function. Let N be the ratio of the maximum and minimum values of $\ell(\cdot)$. We assume, only for ease of presentation, that shortest paths are unique. Let $\delta_G(u,v)$ denote the u-to-v distance in G (w.r.t. $\ell(\cdot)$). For a simple path Q and a vertex set $U \subseteq V(Q)$ with $|U| \geq 2$, we define Q_U, the *reduction* of Q to U as a path whose vertices are U. Consider the vertices of U in the order in which they appear in Q. For every two consecutive vertices u_1, u_2 of U in this order, there is an arc $u_1 u_2$ in Q_U whose length is the length of the u_1-to-u_2 sub-path of Q.

Let $\mathcal{L}$ be a set of labels. We say that a graph G is *vertex-labeled* if every vertex is assigned a single label from $\mathcal{L}$. For a label $\lambda \in \mathcal{L}$, let S_G^λ denote the set of vertices in G with label λ. We define the distance from a vertex $u \in V(G)$ to the label λ by $\delta_G(u, \lambda) = \min_{v \in S_G^\lambda} \delta_G(u, v)$. If G does not contain the label λ, or λ is unreachable from u, we say that $\delta_G(u, \lambda) = \infty$.

Definition 1. *For a fixed parameter $\varepsilon \geq 0$, a stretch-$(1 + \varepsilon)$ vertex-labeled distance oracle is a data structure that, given a vertex $u \in V(G)$ and a label $\lambda \in \mathcal{L}$, returns a distance d satisfying $\delta_G(u, \lambda) \leq d \leq (1 + \varepsilon)\delta_G(u, \lambda)$.*

Definition 2. *For fixed parameters $\alpha, \varepsilon \geq 0$, a scale-$(\alpha, \epsilon)$ vertex-labeled distance oracle is a data structure that, given a vertex $u \in V(G)$ and a label $\lambda \in \mathcal{L}$, such that $\delta_G(u, \lambda) \leq \alpha$, returns a distance d satisfying $\delta_G(u, \lambda) \leq d \leq \delta_G(u, \lambda) + \varepsilon\alpha$. If $\delta_G(u, \lambda) > \alpha$, the oracle returns ∞.*

The only properties of planar graphs that we use in this paper are the existence of shortest path separators (see below), and the fact that single source shortest paths can be computed in $O(n)$ time in a planar graph with n vertices [4].

Definition 3. *Let G be a directed graph. Let G' be the undirected graph induced by G. Let P be a path in G'. Let S be a set of vertex disjoint directed shortest paths in G. We say that P is composed of S if (the undirected path corresponding to) each shortest path in S is a subpath of P and each vertex of P is in some shortest path in S.*

Definition 4. *Let G be a directed embedded planar graph. An undirected cycle C is a balanced cycle separator of G if each of the strict interior and the strict exterior of C contains at most $2|V(G)|/3$ vertices. If, additionally, C is composed of a constant number of directed shortest paths, then C is called a shortest path separator.*

Let G be a planar graph. We assume that G is triangulated since we can triangulate G with infinite length edges, so that distances are not affected. It is well known [8,11] that for any spanning tree of G, there exists a fundamental cycle C that is a balanced cycle separator. The cycle C can be found in linear time. Note that, if T is chosen to be a shortest path tree, or if any root-to-leaf path of T is composed of a constant number of shortest paths, then the fundamental cycle C is a shortest path separator.

2.1 Existing Techniques for Approximate Distance Oracles for Planar Graphs

Thorup shows that to obtain a stretch-$(1 + \varepsilon)$ distance oracle, it suffices to show scale-(α, ε) oracles for so-called α-*layered* graphs. An α-layered graph is one equipped with a spanning tree T such that each root-to-leaf path in T is composed of $O(1)$ shortest paths, each of length at most α. This is summarized in the following lemma:

Lemma 1 [11, Lemma 3.9]. *For any planar graph G and fixed parameter ε, a stretch-$(1 + \varepsilon)$ distance oracle can be constructed using $O(\log nN)$ scale-(α, ε') distance oracles for α-layered graphs, where $\alpha = 2^i$, $i = 0, \ldots \lceil \log nN \rceil$ and $\varepsilon' \in \{1/2, \varepsilon/4\}$. If the scale-$(\alpha, \varepsilon')$ has query time $t(\epsilon')$ independent of α, the stretch-$(1 + \varepsilon)$ distance oracle can answer queries in $O(t(1/2)\varepsilon^{-1} + t(\varepsilon/4) \log \log (nN))$.*

All of our distance oracles are based on a recursive decomposition of G using shortest path separators. If G is undirected (but not necessarily α-layered), we can use any shortest path tree to find a shortest path separator in linear time. Similarly, if G is α-layered, we can use the spanning tree G is equipped with to find a shortest path separator in linear time.

We recursively decompose G into subgraphs using shortest path separators until each subgraph has a constant number of vertices. We represent this decomposition by a binary tree $\mathcal{T}_G$. To distinguish the vertices of $\mathcal{T}_G$ from the vertices of G we refer the former as *nodes*.

Each node r of $\mathcal{T}_G$ is associated with a subgraph G_r. The root of $\mathcal{T}_G$ is associated with the entire graph G. We sometimes abuse notation and equate nodes of $\mathcal{T}_G$ with their associated subgraphs. For each non-leaf node $r \in \mathcal{T}_G$, let C_r be the shortest path separator of G_r. Let Sep_r be the set of shortest paths C_r is composed of. The subgraphs G_{r_1} and G_{r_2} associated with the two children of r in $\mathcal{T}_G$ are the interior and exterior of C_r (w.r.t. G_r), respectively. Note that C_r belongs to both G_{r_1} and G_{r_2}. For a vertex $v \in V(G)$, we denote by r_v the leaf node of $\mathcal{T}_G$ that contains v.

See Fig. 1 for an illustration.

We now describe the basic building block used in our (and in many previous) distance oracle. Let u, v be vertices in G. Let Q be a path on the root-most separator (i.e., the separator in the node of $\mathcal{T}_G$ closest to its root) that is intersected by the shortest u-to-v path P. Let t be a vertex in $Q \cap P$. Note that $\delta_G(u, v) = \delta_G(u, t) + \delta_G(t, v)$. Therefore, if we stored for u the distance to every vertex on Q, and for v the distance from every vertex on Q, we would be able to find $\delta_G(u, v)$ by iterating over the vertices of Q, and finding the one minimizing the distance above. This, however, is not efficient since the number of vertices

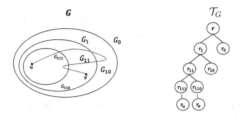

Fig. 1. An illustration of (part of) the recursive decomposition of a graph G using cycle separators, and the corresponding decomposition tree $\mathcal{T}_G$. The graph G is decomposed using a cycle separator into G_0, and G_1. Similarly, G_1 is decomposed into G_{10} and G_{11}, and G_{11} is decomposed into G_{110} and G_{111}. The node r is the root of $\mathcal{T}_G$ and is associated with $G_r = G$. Similarly, r_1 is associated with G_1, etc. The nodes r_u and r_v are the leaf nodes that contain u and v, respectively. The node r_1 is the root-most node whose separator is intersected by the shortest u-to-v path in G (indicated in blue). Hence, this path is fully contained in $G_{r_1} = G_1$. (Color figure online)

on Q might be $\theta(|V(G)|)$. Instead, we store the distances for a subset of Q. This set is called an (α, ε)-*covering connections set*.

Definition 5 ((α, ε)-covering connections set) [11, Sect. 3.2.1]. *Let $\varepsilon, \alpha \geq 0$ be fixed constants. Let G be a directed graph. Let Q be a shortest path in G of length at most α. For $u \in V(G)$ we say that $C_G(u, Q) \subseteq V(Q)$ is an (α, ε)-covering connections set from u to Q if and only if for every vertex t on Q s.t. $\delta_G(u, t) \leq \alpha$, there exists a vertex $q \in C_G(u, Q)$ such that $\delta_G(u, q) + \delta_G(q, t) \leq \delta_G(u, t) + \varepsilon\alpha$.*

One defines (α, ε)-covering connections sets $C_G(Q, u)$ from Q to u symmetrically. Thorup proves that there always exists an (α, ε)-covering connections set of size $O(\varepsilon^{-1})$:

Lemma 2 [11, Lemma 3.4]. *Let $G, Q, \varepsilon, \alpha$ and u be as in Definition 5. There exists an (α, ε)-covering connections set $C_G(u, Q)$ of size at most $\lceil 2\varepsilon^{-1}\rceil$. This set can be found in $O(|Q|)$ if the distances from u to every vertex on Q are given.*

We will use the term ε-covering connections set whenever α is obvious from the context. Thorup shows that (α, ε)-covering connections sets can be computed efficiently.

Lemma 3 [11, Lemma 3.15]. *Let H be an α-layered graph. In $O(\varepsilon^{-2} n \log^3 n)$ time and $O(\varepsilon^{-1} n \log n)$ space one can compute and store a decomposition $\mathcal{T}_H$ of H using shortest path separators, along with (α, ε)-covering connections sets $C_H(u, Q)$ and $C_H(Q, u)$ for every vertex $u \in V(H)$, every ancestor node r of r_u in $\mathcal{T}_H$, and every $Q \in Sep_r$.*

3   An Oracle for Undirected Graphs with Faster Query

Let H be an undirected α-layered graph,[2] and let T be the associated spanning tree of H. For any fixed parameter ε' we set $\varepsilon = \frac{\varepsilon'}{3}$. We decompose H using shortest path separators w.r.t. T. Let $\mathcal{T}_H$ be the resulting decomposition tree. For every node $r \in \mathcal{T}_H$ and every shortest path $Q \in Sep_r$, we select a set $C_Q \subseteq V(Q)$ of ε^{-1} connections evenly spread intervals along Q.[3] Thus, for every vertex $t \in V(Q)$ there is a vertex $q \in C_Q$ such that $\delta_H(t, q) \leq \varepsilon\alpha$.

For each $r \in \mathcal{T}_H$, for each shortest path $Q \in Sep_r$, for each $q \in C_Q$, we compute in $O(|H_r|)$ time a shortest path tree in H_r rooted at q using [4]. This computes the connection lengths $\delta_{H_r}(u, q)$, for all $u \in V(H_r)$.

Lemma 4. *Let $u \in V(H)$. For every ancestor node $r \in \mathcal{T}_H$ of r_u, and every $Q \in Sep_r$, C_Q is a 2ε-covering connections set from u to Q.*

3.1   Warm Up: The Static Case

We start by describing our data structure for the static case with a single fixed label λ (i.e., each vertex either has label λ or no label at all). For every node $r \in \mathcal{T}_H$, let S_r^λ be the set of λ-labeled vertices in H_r. For every separator $Q \in Sep_r$, every vertex $q \in C_Q$, and every vertex $v \in S_r^\lambda$ let $\hat{\delta}_{H_r}(q, v) = k\varepsilon\alpha$ where k is the smallest value such that $\delta_{H_r}(q, v) \leq k\varepsilon\alpha$. Thus, $\delta_{H_r}(q, v) \leq \hat{\delta}_{H_r}(q, v) \leq \delta_{H_r}(q, v) + \varepsilon\alpha$. Let $L_r(q, \lambda)$ be the list of the distances $\hat{\delta}_{H_r}(q, v)$ for all $v \in S_r^\lambda$. We sort each list in ascending order. Thus, the first element of $L_r(q, \lambda)$ denoted by $first(L_r(q, \lambda))$ is at most $\varepsilon\alpha$ more than the distance from q to the closest λ-labeled vertex in H_r. We note that each vertex $u \in V(H)$ may contribute its distance to $O(\varepsilon^{-1} \log n)$ lists. Hence, we have $O(\varepsilon^{-1} n \log n)$ elements in total. Since H is an α-layered graph, the length of each Q is bounded by α. Hence, the universe of these lists can be regarded as non-negative integers bounded by $\frac{\alpha}{\varepsilon\alpha} = \varepsilon^{-1}$. Thus, these lists can be sorted in total $O(\varepsilon^{-1} n \log n)$ time.

Query(u, λ). Given $u \in H$. We wish to find the closest λ-labeled vertex v to u in H. For each ancestor r of r_u, for each $Q \in Sep_r$, we perform the following search. We inspect for every $q \in C_Q$, the distance $\delta_{H_r}(u, q) + first(L_r(q, \lambda))$. We also inspect the λ-labeled vertices in H_{r_u} explicitly. We return the minimum distance inspected. See Fig. 2 for an illustration.

Lemma 5. *The query algorithm runs in $O(\varepsilon^{-1} \log n)$ time, and returns a distance d such that $\delta_H(u, \lambda) \leq d \leq \delta_H(u, \lambda) + 3\varepsilon\alpha$.*

---

[2] The discussion of α-layered graphs in Sect. 2 refers to directed graphs, and hence also applies to undirected graphs.

[3] We assume that the endpoints of the intervals are vertices on Q, since otherwise once can add artificial vertices on Q without asymptotically changing the size of the graph.

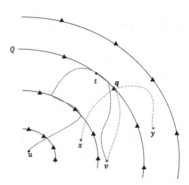

Fig. 2. Illustration of the query algorithm. The solid quarter-circles are shortest paths of separators in G. The vertices x, y and v have label λ, and v is the closest λ-labeled vertex to u. The path Q belongs to the root-most node r whose separator is intersected by the shortest u-to-λ path (solid blue). (Color figure online)

We now generalize to multiple labels. Let $\mathcal{L}$ be the set of labels in H. For $r \in \mathcal{T}_H$, let $\mathcal{L}_r$ be the restriction of $\mathcal{L}$ to labels that appear in H_r. For every label $\lambda \in \mathcal{L}_r$, every $Q \in Sep_r$ and every $q \in C_Q$, we store the list $L_r(q, \lambda)$. This does not affect the total size of our structure, since each vertex has one label, so it still contributes its distances to $O(\varepsilon^{-1} \log n)$ lists. Naively, we could store for every node r, every vertex q, and every label $\lambda \in \mathcal{L}$ the list $L_r(q, \lambda)$ in a fixed array of size $|\mathcal{L}|$. This allows $O(1)$-time access to each list, but increases the space by a factor of $|\mathcal{L}|$ w.r.t. the single label case. Instead, we use hashing. Each vertex q holds a hash table of the labels that contributed distances to q. For the static case, one can use perfect hashing [2] with expected construction time and constant query time. In the dynamic case, we will use a dynamic hashing scheme, e.g., [10], which provides query and deletions in $O(1)$ worst case, and insertions in $O(1)$ expected amortized time.

3.2 The Dynamic Case

We now turn our attention to the dynamic case. We wish to use the following method for updating our structure. When a node v changes its label from λ_1 to λ_2, we would like to iterate over all ancestors r of r_v in $\mathcal{T}_H$. For every $Q \in Sep_r$ and every $q \in C_Q$, we wish to remove the value contributed by v from $L_r(q, \lambda_1)$, and insert it to $L_r(q, \lambda_2)$. We must maintain the lists sorted, but do not wish to pay $O(\log n)$ time per insertion to do so. We will be able to pay $O(\log \log \varepsilon^{-1})$ per insertion/deletion by using a successor/predecessor data structure as follows.

For every $r \in \mathcal{T}_H$, $Q \in Sep_r$, and $q \in C_Q$, let $L_r(q)$ be the list containing *all distances* from all vertices in $V(H_r)$ to q sorted in ascending order. We note that since the distance for each specific vertex to q does not depend on its label, the list $L_r(q, \lambda)$ is a restriction of $L_r(q)$ to the λ-labeled vertices in H_r.

During the construction of our structure we build $L_r(q)$, and, for every vertex v in H_r, we store for v its corresponding index in $L_r(q)$. We denote this index as $ID_q(v)$. We also store for q a single lookup table from the IDs to the corresponding distances. We note that v has $O(\varepsilon^{-1}\log n)$ such identifiers, and in total we need $O(\varepsilon^{-1}n\log n)$ space to store them.

Now, instead of using linked list as before, we implement $L_r(q,\lambda)$ using a successor/predecessor structure over the universe $[1,\ldots,|V(H_r)|]$ of the IDs. For example, we can use y-fast tries [13] that support operations in $O(\log\log\varepsilon^{-1})$ expected amortized time and minimum query in $O(1)$ worst case.

Query(u, λ). The query algorithm remains the same as in the static case. For every ancestor r of r_u in T_H, every $Q \in Sep_r$, and every connection $q \in C_Q$, we retrieve the minimal ID from $L_r(q,\lambda)$, and use the lookup table to get the actual distance between q and the vertex with that ID.

Update. Assume that the vertex v changes its label from λ_1 to λ_2. For every ancestor r of r_v in T_H, every $Q \in Sep_r$, and every $q \in C_Q$, we remove $ID_q(v)$ from $L_r(q,\lambda_1)$ and insert it to $L_r(q,\lambda_2)$.

Lemma 6. *The update time is $O(\varepsilon^{-1}\log n \cdot \log\log\varepsilon^{-1})$ expected amortized.*

Lemma 7. *The data structure can be constructed in $O(\varepsilon^{-1}n\log n \cdot \log\log\varepsilon^{-1})$ expected amortized time, and stored using $O(\varepsilon^{-1}n\log n)$ space.*

We plug in this structure to Lemma 1 and obtain the following theorem:[4]

Theorem 1. *Let G be an undirected planar graph. There exists a stretch-$(1+\varepsilon)$ Approximate Dynamic Vertex-Labeled Distance Oracle that supports query in $O(\varepsilon^{-1}\log n\log\log nN)$ worst case and updates in $O(\varepsilon^{-1}\log n\cdot\log\log\varepsilon^{-1}\log nN)$ expected amortized. The construction time of that oracle is $O(\varepsilon^{-1}n\log n \cdot \log\log\varepsilon^{-1}\log nN)$ and it can be stored in $O(\varepsilon^{-1}n\log n\log nN)$ space.*

4   Oracle for Directed Graphs

For simplicity we only describe an oracle that supports queries from a given label to a vertex. Vertex to label queries can be handled symmetrically. To describe our data structure for directed graphs, we first need to introduce the concept of ε-covering set from a *set of vertices* to a directed shortest path.

Definition 6. *Let S be a set of vertices in a directed graph H. Let Q be a shortest path in H of length at most α. $C_H(S,Q) \subseteq V(Q) \times \mathbb{R}^+$ is an ε-covering set from S to Q in H if for every $t \in Q$ s.t. $\delta_H(S,t) \leq \alpha$, there exists $(q,\ell) \in C_H(S,Q)$ s.t. $\ell + \delta(q,t) \leq \delta_H(S,t) + \varepsilon\alpha$, and $\ell \geq \delta_H(S,q)$.*

In the definition above we use ℓ instead of $\delta(S,q)$ (compare to Definition 5) because we cannot afford to recompute exact distances as S changes. Instead, we store and use approximate distances ℓ.

---

[4] Formally, one needs to show that Lemma 1 holds for vertex-labeled oracles as well. See our full paper at https://arxiv.org/abs/1707.02414.

Lemma 8. *Let H be a directed planar graph. Let Q be a shortest path in H of length at most α. For every set of vertices $S \subseteq V(H)$ there is an ε-covering set $C_H(S, Q)$ of size $O(\varepsilon^{-1})$.*

Our construction relies on the following lemma.

Lemma 9 (Thinning Lemma). *Let H, S and Q be as in Lemma 8. Let $\{S_i\}_{i=1}^k$ be sets such that $S = \bigcup_{i=1}^k S_i$. For $1 \le i \le k$, let $D_H(S_i, Q)$ be an ε'-covering set from S_i to Q, ordered by the order of the vertices on Q. Then for every $\varepsilon > 0$, an $(\varepsilon + \varepsilon')$-covering set $C_H(S, Q)$ from S to Q of size $\lceil 2\varepsilon^{-1} \rceil$ can be found in $O(\varepsilon^{-1} + |\bigcup_{i=1}^k D_H(S_i, Q)|)$ time.*

Let H be a directed planar α-layered graph, equipped with a spanning tree T. For every fixed parameter ε, let $\hat{\epsilon} = \frac{\varepsilon}{8 \log n}$, and $\varepsilon^* = \frac{\varepsilon}{2}$. We apply Lemma 3 with $\hat{\epsilon}$ to H and obtain a decomposition tree $\mathcal{T}_H$, and $\hat{\epsilon}$-covering sets $C_{H_r}(v, Q)$ and $C_{H_r}(Q, v)$ for every $v \in V(H)$, every ancestor r of r_v in $\mathcal{T}_H$ and every $Q \in Sep_r$. For every $1 \le i \le \log n$, let $\varepsilon_i = \frac{\varepsilon \log n - i + 1}{4 \log n}$.

For every $r \in \mathcal{T}_H$, for every $\lambda \in \mathcal{L}_r$, and for every $Q \in Sep_r$, we apply Lemma 9 to the $\hat{\epsilon}$-covering connections sets $C_{H_r}(v, Q)$ for all $v \in S_r^\lambda$, with $\varepsilon' = \hat{\epsilon}$ and ε set to $\frac{\varepsilon}{4}$. Thus, we obtain an ε^*-covering set $C_{H_r}^*(S_r^\lambda, Q)$. Let i be the level of r in $\mathcal{T}_H$, we also store for r a set of ε_i-covering sets as follows. For every ancestor node t of r in $\mathcal{T}_H$ and every $Q \in Sep_t$, we store $C_{H_t}(S_r^\lambda, Q)$. We assume for the moment that these sets are given. We defer the description of their construction (see the update procedure and the proof of Lemma 12). We will use the ε^*-covering sets for efficient queries, and the more accurate ε_i-covering sets to be able to perform efficient updates. See Fig. 3.

4.1 *Query*(λ, u)

The query algorithm is straightforward. For every ancestor r of r_u we find $(q, \ell) \in C_{H_r}^*(S_r^\lambda, Q)$ and $t \in C_{H_r}(Q, u)$ that minimize the distance $\ell + \delta_{H_r}(q, t) + \delta_{H_r}(t, u)$. We also inspect the distance to the λ-labeled vertices in r_u explicitly. We return the minimum distance inspected. To see that the query time is $O(\varepsilon^{-1} \log n)$, we note that for every one of the $O(\log n)$ ancestors of r_u we inspect $O(\varepsilon^{-1})$ distances on constant number of separators. Inspecting the distances in r_u itself takes constant time.

Lemma 10. $Query(\lambda, u) \le \delta_H(\lambda, u) + \varepsilon\alpha.$

4.2 Update

Assume that some vertex u changes its label from λ_1 to λ_2. For every ancestor r of r_u and every $Q \in Sep_r$, we would like to remove $C_{H_r}(u, Q)$ from $C_{H_r}(S_r^{\lambda_1}, Q)$, and combine $C_{H_r}(u, Q)$ into $C_{H_r}(S_r^{\lambda_2}, Q)$. While the latter is straightforward using Lemma 9, removing $C_{H_r}(u, Q)$ from $C_{H_r}(S_r^{\lambda_1}, Q)$ is more difficult. For example, if u was the closest λ_1 labeled vertex to every vertex on Q, it is possible

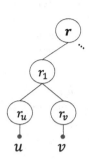

	Vertex connections $\hat{e}$-covering sets	Query connections ε^*-covering sets	Update connections ε_i-covering sets
r	N/A	From S_G^λ to $Q \in Sep_r$	From S_G^λ to $Q \in Sep_r$
r_1	N/A	From $S_{r_1}^\lambda = S_{r_u} \cup S_{r_v}$ to $Q \in Sep_{r_1}$	From $S_{r_1}^\lambda = S_{r_u} \cup S_{r_v}$ to $Q \in \{Sep_{r_1}, Sep_r\}$
r_u	N/A	From $S_{r_u}^\lambda = u$ to $Q \in Sep_{r_u}$	From $S_{r_u}^\lambda = u$ to $Q \in \{Sep_{r_u}, Sep_{r_1}, Sep_r\}$
r_v	N/A	From $S_{r_v}^\lambda = v$ to $Q \in Sep_{r_v}$	From $S_{r_v}^\lambda = v$ to $Q \in \{Sep_{r_v}, Sep_{r_1}, Sep_r\}$
u	From u to every $Q \in \{Sep_{r_u}, Sep_{r_1}, Sep_r\}$, and from every such Q to u.	N/A	N/A
v	From v to every $Q \in \{Sep_{r_v}, Sep_{r_1}, Sep_r\}$, and from every such Q to v.	N/A	N/A

Fig. 3. A summary of the connections-sets stored by the directed oracle. On the left, part of a decomposition tree of a graph is shown. The vertices u and v are the only λ labeled vertices. On the right, a table listing all the covering sets that are stored for the label λ.

that $C_{H_r}(u, Q) = C_{H_r}(S_r^{\lambda_1}, Q)$. In that case, we will have to rebuild $C_{H_r}(S_r^{\lambda_1}, Q)$ from the other $O(|V(H_r)|)$ vertices of $S_r^{\lambda_1}$. Instead of removing the connections of u, we will rebuild $C_{H_r}(S_r^{\lambda_1}, Q)$ bottom-up starting from the leaf node r_u.

We therefore start by describing how to update r_u. There is a constant number of vertices in r_u, and hence $|S_{r_u}^{\lambda_1}| = O(1)$. Let $v_1, v_2, \ldots v_k$ be the vertices in $S_{r_u}^{\lambda_1}$. We stress that for every $1 \le j \le k$, v_j has an $\hat{e}$-covering set $C_{H_t}(v_j, Q)$ of size $O(\hat{e}^{-1})$ from v_j to Q, for *every ancestor t of r_u in $\mathcal{T}_H$*, and for every $Q \in Sep_t$.

We apply the Thinning Lemma 9 to each such t and Q on $\{C_{H_t}(v_j, Q)\}_{j=1}^k$ with $\varepsilon' = \hat{e}$ and ε set to $\hat{e}$. Lemma 9 yields a $2\hat{e}$-covering set $C_{H_t}(S_{r_u}^\lambda, Q)$.

We next handle the ancestors r of r_u in $\mathcal{T}_H$ in bottom up order. Let x and y be the children of $r \in \mathcal{T}_H$. We first note that $H_r = H_x \cup H_y$ and hence, $S_r^{\lambda_1} = S_x^{\lambda_1} \cup S_y^{\lambda_1}$. Therefore, by Lemma 9, for every ancestor t of r, and every $Q \in Sep_t$, $C_{H_t}(S_r^{\lambda_1}, Q)$ can be obtained from $C_{H_t}(S_x^{\lambda_1}, Q) \cup C_{H_t}(S_y^{\lambda_1}, Q)$. Let i by the level of r in $\mathcal{T}_H$, and hence the level of x and y is $i+1$. Since t is an ancestor of r, it is also an ancestor of x and y. Hence, x (y) stores an ε_{i+1}-covering set $C_{H_t}(S_x^{\lambda_1}, Q)$ ($C_{H_t}(S_y^{\lambda_1}, Q)$). We apply Lemma 9 on $C_{H_t}(S_x^{\lambda_1}, Q)$ and $C_{H_t}(S_y^{\lambda_1}, Q)$ with $\varepsilon' = \varepsilon_{i+1}$ and $\varepsilon = 2\hat{e}$ to get an $(\varepsilon_{i+1} + 2\hat{e})$-covering set $C_{H_t}(S_r^{\lambda_1}, Q)$. The following lemma shows that $C_{H_t}(S_r^{\lambda_1}, Q)$ is an ε_i-covering set.

Lemma 11. *Let r be a node in level i in $\mathcal{T}_H$. For every ancestor t of r, and every $Q \in Sep_t$, $C_{H_t}(S_r^{\lambda_1}, Q)$ is an ε_i-covering set from $S_r^{\lambda_1}$ to Q.*

To finish the update process, we need to update the ε^*-covering sets that we use for queries. Let r be an ancestor node of r_u in level i on $\mathcal{T}_H$. By

Lemma 11, for every $Q \in Sep_r$, we have an ε_i-covering set $C_{H_r}(S_r^{\lambda_1}, Q)$. Since $\varepsilon_i < \varepsilon^*$, $C_{H_r}(S_r^{\lambda_1}, Q)$ is also an ε^*-covering set. However, it is too large. We apply Lemma 9 on $C_{H_r}(S_r^{\lambda_1}, Q)$ with $\varepsilon' = \varepsilon_i$, and ε set to $\frac{\varepsilon}{4}$ to get $(\varepsilon_i + \frac{\varepsilon}{4})$-covering set. We note that since $\varepsilon_i \leq \frac{\varepsilon}{4}$ for every $1 \leq i \leq \log n$, we get that $\varepsilon_i + \frac{\varepsilon}{4} \leq 2\frac{\varepsilon}{4} \leq \frac{\varepsilon}{2} = \varepsilon^*$. Hence the output of Lemma 9 is the desired ε^*-covering set $C_{H_r}^*(S_r^{\lambda_1}, Q)$. We repeat the entire process for λ_2.

Lemma 12. *There exists a scale-(α, ε) distance oracle for directed α-layered planar graph, with query time $O(\varepsilon^{-1} \log n)$ worst case, and update time of $O(\varepsilon^{-1} \log^3 n)$ expected amortized. The oracle can be constructed in $O(\varepsilon^{-2} n \log^5 n)$ time and stored using $O(\varepsilon^{-1} n \log^3 n)$ space.*

We can now apply Lemma 1 to get the following theorem:

Theorem 2. *For any directed planar graph and fixed parameter ε, there exists a $(1 + \varepsilon)$ approximate vertex-labeled distance oracle that support queries in $O(\varepsilon^{-1} \log n \log \log nN)$ worst case and updates in $O(\varepsilon^{-1} \log^3 n \log nN)$ expected amortized time. This oracle can be constructed in $O(\varepsilon^{-2} n \log^5 n \log nN)$ expected amortized time, and stored using $O(\varepsilon^{-1} n \log^3 n \log nN)$ space.*

5   Oracle for Undirected Graphs with Faster Update

Both Thorup [11, Lemma 3.19] and Klein [5] independently presented efficient vertex-vertex distance oracles for undirected planar graph that use connections sets. Klein later improved the construction time [6]. They show that, in undirected planar graph, one can avoid the scaling approach that uses α-layered graphs. Instead, there exist connections sets that approximate distance with $(1 + \varepsilon)$ multiplicative factor rather than $\varepsilon\alpha$ additive factor. We use the term *portals* [6] to distinguish this type of connections from the previous one.

Definition 7. *Let G be an undirected planar graph, and let Q be a shortest path in G. For every vertex $v \in V(G)$ we say that a set $C_G(v, Q)$ is an ε-covering set of portals if and only if, for every vertex t on Q there exist a vertex q on Q such that: $\delta_G(v, q) + \delta_G(q, t) \leq (1 + \varepsilon)\delta_G(v, t)$.*

We use a recursive decomposition $\mathcal{T}_G$ with shortest path separators, and use Klein's algorithm [6] to select all the portal sets $C_{G_r}(u, Q)$ efficiently. We cannot use the lists of Sect. 3 because there may be too many portals, and we cannot use the thinning lemma (Lemma 9) of Sect. 4 because its proof uses a directed construction, and hence, cannot be applied in undirected graphs. Instead, we take the approach used by Li et al. for the static vertex-labeled case [7]. We work with all portals of vertices with the appropriate label, and find the closest one using dynamic Prefix/Suffix Minimum Queries.

Definition 8 (Dynamic Prefix Minimum Data Structure). *A Dynamic Prefix Minimum Data Structure is a data structure that maintains a set A of n pairs in $[1, n] \times \mathbb{R}$, under insertions, deletions, and Prefix Minimum Queries (PMQ) of the following form: given $l \in [1, n]$ return a pair $(x, y) \in A$ s.t. $x \in [1, l]$, and for every other pair (x', y') with $x' \in [1, l]$, $y \leq y'$.*

Suffix minimum queries (SMQ) are defined analogously. Let $PMQ(A, l)$ and $SMQ(A, l)$ denote the result of the corresponding queries on the set A and l.

We assume that for every $u, v \in V(G_r)$, $C_{G_r}(u, Q) \cap C_{G_r}(v, Q) = \emptyset$. This is without loss of generality, since if x is a portal of a set of vertices $v_0, \ldots, v_k$, we can split x to k copies. This does not increase $|G|$ by more than a factor of ε^{-1}. See Fig. 4.

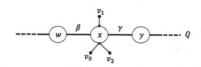

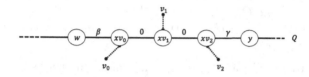

Fig. 4. Illustration of the reduction to unique portals. Above, the path Q with the portal x that is used by v_0, v_1, and v_2. Below, x was replaced by xv_0, xv_1, and xv_2, inner connected with zero length edges. Here, xv_0, xv_1, xv_2 are the portals of v_0, v_1, and v_2 respectively. Note that this reduction does not introduce new paths in the graph, nor changes the distance along Q.

To describe our data structure, we first need the following definitions. Let $Q \in Sep_r$ for some $r \in T_G$. Let $q_0, \ldots, q_k$ be the vertices on Q by their order along Q. G is undirected, hence the direction of Q is chosen arbitrarily. For every $0 \leq j \leq k$, let $h(q_j)$ denote the distance from q_0 to q_j on Q. We note that since Q is a shortest path in G, $h(q_i) = \delta_G(q_0, q_j)$. For every $\lambda \in \mathcal{L}_r$ we maintain a dynamic prefix minimum data structure $Pre_{Q,\lambda}$ over $\{(j, -h(q_j) + \delta_{G_r}(q_j, \lambda))\}_{j=0}^{k}$. We similarly maintain a a dynamic suffix minimum data structure $Suf_{Q,\lambda}$ over $\{(j, h(q_j) + \delta_{G_r}(q_j, \lambda))\}_{j=0}^{k}$.

Query(u, λ). For every ancestor r of r_u in T_G, every $Q \in Sep_r$, and every $q_j \in C_{G_r}(u, Q)$ we wish to find the index i that minimizes $\delta_{G_r}(u, q_j) + \delta_{G_r}(q_j, q_i) + \delta_{G_r}(q_i, \lambda)$. Observe that for $i \leq j$, $\delta_{G_r}(q_j, q_i) = h(j) - h(i)$, while for $i \geq j$, $\delta_{G_r}(q_j, q_i) = h(i) - h(j)$. We therefore find the optimal $i \leq j$ and $i \geq j$ separately. Note that $min_{i \leq j}(\delta_{G_r}(u, q_j) + \delta_{G_r}(q_j, q_i) + \delta_{G_r}(q_i, \lambda)) = \delta_{G_r}(u, q_j) + h(j) + PMQ(Pre_{Q,\lambda}, j)$. Similarly, we handle the case where $i \geq j$ using $SMQ(Suf_{Q,\lambda}, j)$. Thus, we have two queries for each portal of u. We also compute the distance from u to λ in r_u explicitly. We return the minimum distance computed.

Lemma 13. *The query algorithm returns a distance d such that $\delta_G(u, \lambda) \leq d \leq (1 + \varepsilon)\delta_G(u, \lambda)$.*

Update. Assume that the label of u changes from λ_1 to λ_2. For every ancestor r of $r_u \in T_G$, and $Q \in Sep_r$, and for $q_i \in C_{G_r}(u, q)$, we remove from Pre_{Q,λ_1} and Suf_{Q,λ_1} the element (x, y) with $x = i$, and insert the element $(i, -h(i) + \delta_{G_r}(u, q_i))$ into Pre_{Q,λ_2}, and $(i, h(i) + \delta_{G_r}(u, q_i))$ into Suf_{Q,λ_2}. We note that since we assume that every vertex q_i is a portal of at most one vertex, the removals are well defined, and the insertions are safe.

The time and space bounds for the oracle described above are given in the following lemma.

Lemma 14. *Assume there exists a dynamic prefix/suffix minimum data structure in the word RAM model, that for a set of size m, supports PMQ/SMQ in $O(T_Q(m))$ time, and updates in $O(T_U(m))$ time, can be constructed in $O(T_C(m))$ time, where $T_C(m) \geq m$, and can be stored in $O(S(m))$ space. Then there exist a dynamic vertex-labeled stretch-$(1 + \varepsilon)$ distance oracle for planar graphs with worst case query time $O(\varepsilon^{-1} \log(n) T_Q(\varepsilon^{-1} n))$, and expected amortized update time $O(\varepsilon^{-1} \log(n) T_U(\varepsilon^{-1} n))$. The oracle can be constructed using $O(\varepsilon^{-1} n \log^2 n + \log(n) T_c(\varepsilon^{-1} n))$ expected amortized time, and stored in $O(\log(n) S(\varepsilon^{-1} n))$ space.*

It remains to describe a fast prefix/suffix minimum query structure. We use a result due to Wilkinson [12] for solving the 2-sided reporting problem in $\mathbb{R}^2$ in the word RAM model. In this problem, we maintain a set A of n points in $\mathbb{R}^2$ under an online sequence of insertions, deletions and queries of the following form. Given a rectangle $B = [l_1, h_1] \times [l_2, h_2]$ such that exactly one of l_1, l_2 and one of h_1, h_2 is ∞ or $-\infty$, we report $A \cap B$. Here, $[l_1, h_1] \times [l_2, h_2]$ represents the rectangle $\{(x, y) : l_1 \leq x \leq l_2, h_1 \leq y \leq h_2\}$. Since Wilkinson assumes the word RAM model, it is assumed that the coordinates of the points in A are integers that fit in a single word. Wilkinson's data structure is captured by the following lemma.

Lemma 15 [12, Theorem 5]. *For any $f \in [2, \log n / \log \log n]$, there exists a data structure for 2-sided reporting with update time $O((f \log n \log \log n)^{1/2})$, query time $O((f \log n \log \log n)^{1/2} + \log_f(n) + k)$ where k is the number of points reported. The structure requires linear space.*

In fact, Wilkinson's structure first finds the point with the minimum y-coordinate in the query region, and then reports the other points. Using this fact, and setting $f = \log^\gamma n$ for some arbitrary small constant γ. We get the following lemma, in which we also state Wilkinson's construction time explicitly.

Lemma 16. *There exists a linear space data structure for 2-sided reporting on n points, with update time $O(\log^{1/2+\gamma} n)$ and query time $O(\frac{\log n}{\log \log n})$. This data structure can be constructed in $O(n \log^{1/2+\gamma} n)$ time. Moreover, upon query the data structure returns the minimum y-coordinate of a point in the query region.*

The prefix/suffix queries required by Lemma 17 correspond to one-sided range reporting in the plane, which can be solved using 2-sided queries, by setting the upper limit of the query rectangle to nN.

Lemma 17. *For any constant $\gamma > 0$, there exists a linear space dynamic prefix/suffix minimum data structure over n elements with update time $O(\log^{1/2+\gamma} n)$, and query time $O(\frac{\log n}{\log \log n})$. This data structure can be constructed in $O(n \log^{1/2+\gamma} n)$ time.*

We therefore obtain the following theorem.

Theorem 3. *For any undirected planar graph and fixed parameters ε, γ, there exists a stretch-$(1 + \varepsilon)$ vertex-labeled distance oracle that approximates distances in $O(\varepsilon^{-1} \frac{\log n \log(\varepsilon^{-1} n)}{\log \log (\varepsilon^{-1} n)})$ time worst case, and supports updates in $O(\varepsilon^{-1} \log n \log^{\frac{1}{2}+\gamma}(\varepsilon^{-1} n))$ expected amortized time. This data structure can be constructed using $O(n \log^2 n + \varepsilon^{-1} n \log n \log^{\frac{1}{2}+\gamma}(\varepsilon^{-1} n))$ expected amortized time and stored using $O(\varepsilon^{-1} n \log n)$ space.*

6 Conclusion

In this paper we presented approximate vertex-labeled distance oracles for directed and undirected planar graphs with polylogarithmic query and update times and nearly linear space. All of our oracles have $\Omega(\log n)$ query and updates since we handle root-to-leaf paths in the decomposition tree. It would be interesting to study whether this can be avoided, as done in the vertex-to-vertex case, where approximate distance oracles with faster query times exist (see e.g., [3,11,14] and references therein). Another interesting question that arises is that of faster dynamic prefix minimum data structures. In Sect. 5 we used Wilkinson's 2-sided reporting [12] as a dynamic prefix/suffix minimum data structure. Can other approaches to this problem be used to obtain a faster solution?

Acknowledgements. We thank Paweł Gawrychowski and Oren Weimann for fruitful discussions.

References

1. Łącki, J., Ocwieja, J., Pilipczuk, M., Sankowski, P., Zych, A.: The power of dynamic distance oracles: efficient dynamic algorithms for the steiner tree. In: STOC, pp. 11–20 (2015)
2. Fredman, M.L., Komlós, J., Szemerédi, E.: Storing a sparse table with $O(1)$ worst case access time. J. ACM **31**(3), 538–544 (1984). https://doi.org/10.1145/828.1884
3. Gu, Q.-P., Xu, G.: Constant query time $(1+\epsilon)$-approximate distance oracle for planar graphs. In: Elbassioni, K., Makino, K. (eds.) ISAAC 2015. LNCS, vol. 9472, pp. 625–636. Springer, Heidelberg (2015). https://doi.org/10.1007/978-3-662-48971-0_53
4. Henzinger, M.R., Klein, P.N., Rao, S., Subramanian, S.: Faster shortest-path algorithms for planar graphs. J. Comput. Syst. Sci. **55**(1), 3–23 (1997). https://doi.org/10.1006/jcss.1997.1493
5. Klein, P.N.: Preprocessing an undirected planar network to enable fast approximate distance queries. In: SODA, pp. 820–827 (2002)

6. Klein, P.N.: Multiple-source shortest paths in planar graphs. In: SODA, pp. 146–155 (2005)
7. Li, M., Ma, C.C.C., Ning, L.: $(1 + \epsilon)$-distance oracles for vertex-labeled planar graphs. In: Chan, T.-H.H., Lau, L.C., Trevisan, L. (eds.) TAMC 2013. LNCS, vol. 7876, pp. 42–51. Springer, Heidelberg (2013). https://doi.org/10.1007/978-3-642-38236-9_5
8. Lipton, R.J., Tarjan, R.E.: A separator theorem for planar graphs. SIAM J. Appl. Math. **36**(2), 177–189 (1979)
9. Mozes, S., Skop, E.E.: Efficient vertex-label distance oracles for planar graphs. In: Sanità, L., Skutella, M. (eds.) WAOA 2015. LNCS, vol. 9499, pp. 97–109. Springer, Cham (2015). https://doi.org/10.1007/978-3-319-28684-6_9
10. Pagh, R., Rodler, F.F.: Cuckoo hashing. In: auf der Heide, F.M. (ed.) ESA 2001. LNCS, vol. 2161, pp. 121–133. Springer, Heidelberg (2001). https://doi.org/10.1007/3-540-44676-1_10
11. Thorup, M.: Compact oracles for reachability and approximate distances in planar digraphs. J. ACM **51**(6), 993–1024 (2004). https://doi.org/10.1145/1039488.1039493
12. Wilkinson, B.T.: Amortized bounds for dynamic orthogonal range reporting. In: Schulz, A.S., Wagner, D. (eds.) ESA 2014. LNCS, vol. 8737, pp. 842–856. Springer, Heidelberg (2014). https://doi.org/10.1007/978-3-662-44777-2_69
13. Willard, D.E.: Log-logarithmic worst-case range queries are possible in space $\Theta(N)$. Inf. Process. Lett. **17**(2), 81–84 (1983). https://doi.org/10.1016/0020-0190(83)90075-3
14. Wulff-Nilsen, C.: Approximate distance oracles for planar graphs with improved query time-space tradeoff. In: SODA, pp. 351–362 (2016)

A Communication-Efficient Distributed Data Structure for Top-k and k-Select Queries

Felix Biermeier, Björn Feldkord, Manuel Malatyali$^{(\boxtimes)}$,
and Friedhelm Meyer auf der Heide

Computer Science Department, Heinz Nixdorf Institute, Paderborn University,
Paderborn, Germany
{felixbm,bjoernf,malatya,fmadh}@mail.upb.de

Abstract. We consider the scenario of n sensor nodes observing streams of data. The nodes are connected to a central server whose task it is to compute some function over all data items observed by the nodes. In our case, there exists a total order on the data items observed by the nodes. Our goal is to compute the k currently lowest observed values or a value with rank in $[(1-\varepsilon)k, (1+\varepsilon)k]$ with probability $(1-\delta)$. We propose solutions for these problems in an extension of the distributed monitoring model where the server can send broadcast messages to all nodes for unit cost. We want to minimize communication over multiple time steps where there are m updates to a node's value in between queries. The result is composed of two main parts, which each may be of independent interest:

1. Protocols which answer Top-k and k-Select queries. These protocols are memoryless in the sense that they gather all information at the time of the request.
2. A dynamic data structure which tracks for every k an element close to k.

We describe how to combine the two parts to receive a protocol answering the stated queries over multiple time steps. Overall, for Top-k queries we use $\mathcal{O}(k + \log m + \log \log n)$ and for k-Select queries $\mathcal{O}(\frac{1}{\varepsilon^2} \log \frac{1}{\delta} + \log m + \log^2 \log n)$ messages in expectation. These results are shown to be asymptotically tight if m is not too small.

1 Introduction

Consider a distributed sensor network which is a system consisting of a huge amount of nodes. Each node continuously observes its environment and measures information (e.g. temperature, pollution or similar parameters). We are interested in aggregations describing the current observations at a central server. To keep the server's information up to date, the server and the nodes can communicate with each other. In sensor networks, however, the amount of such

This work was partially supported by the German Research Foundation (DFG) within the Priority Program "Algorithms for Big Data" (SPP 1736).

© Springer International Publishing AG, part of Springer Nature 2018
R. Solis-Oba and R. Fleischer (Eds.): WAOA 2017, LNCS 10787, pp. 285–300, 2018.
https://doi.org/10.1007/978-3-319-89441-6_21

communication is particularly crucial, as communication has the largest impact to energy consumption, which is limited due to battery capacities [12]. Therefore, algorithms aim at minimizing the (total) communication required for computing the respective aggregation function at the server.

We consider several ideas to potentially lower the communication used. Each single computation of an aggregate should use as little communication as possible. Computations of the same aggregate should *reuse* parts of previous computations. Only compute aggregates, if necessary. Recall that the continuous monitoring model creates a new output as often as possible.

1.1   Model

We consider the distributed monitoring model, introduced by Cormode et al. in [9], in which there are n distributed nodes, each uniquely identified by an identifier (ID) from the set $\{1, \ldots, n\}$, connected to a single server. Each node observes a stream of data items over time, i.e. at any discrete time step t node i observes a data item d_i^t. We assume that the data items have a total order and denote by *rank(d)* the position of data item d in the sorted ordering. Furthermore, we assume that the sorted ordering is unique, i.e. for each data items d_i and d_j either $d_i \leq d_j$ or $d_i \geq d_j$ holds.

The server is asked to, given a query at time t, compute an output $f(t)$ which depends on the data items d_i^t with $i = 1, \ldots, n$ observed across all distributed streams. The exact definition of $f(\cdot)$ depends on the concrete problems under consideration, which are defined in the section below. For the solution of these problems, we are interested in both, exact and approximation algorithms. An exact algorithm computes the (unique) output $f(t)$ with probability 1. An ε-approximation of $f(t)$ is an output $\tilde{f}(t)$ of the server such that $(1 - \varepsilon)f(t) \leq \tilde{f}(t) \leq (1 + \varepsilon)f(t)$ holds. We call an algorithm that provides an ε-approximation with probability at least $1 - \delta$, an (ε, δ)-approximation algorithm. We say an algorithm is correct *with high probability*, if for a given constant $c > 1$ it is correct with probability at least $1 - n^{-c}$.

Communication Network. To be able to compute the output, the nodes and the server can communicate with each other by exchanging single cast messages or by broadcast messages sent by the server and received by all nodes. Both types of communication are instantaneous and have unit cost per message. That is, sending a single message to one specific node incurs cost of one and so does one broadcast message. Each message has a size of $\mathcal{O}\left(\mathscr{B} + \log(n) + \log(\log(\frac{1}{\delta}))\right)$ bits, where $\mathscr{B}$ denotes the number of bits needed to encode a data item. A message will usually, besides a constant number of control bits, consist of a data item, a node ID and an identifier to distinguish between messages of different instances of an algorithm applied in parallel (as done when using standard probability amplification techniques). A broadcast channel is an extension to [9], which was originally proposed in [8] and afterwards applied in [3,10,11]. Between any two time steps we allow a communication protocol to take place, which may use a polylogarithmic number of rounds. The optimization goal is the minimization of the communication complexity, given by the total number of exchanged messages, required to answer the posed requests or rebuild the data structure.

Problem Description. In this work, we consider two basic problems that are related to the rank of the data items: (1) Compute exactly (all of) the k smallest data items observed by the nodes at the current time step t and (2) output an (ε, δ)-approximation of the data item with rank k.

Formally, let π_t denote the permutation of the node IDs $\{1, \ldots, n\}$ such that $\pi_t(i)$ gives the index of the data item with rank i at time t, i.e. $i = rank(d^t_{\pi_t(i)})$. First, we denote the Top-k-Problem as the output $\{d^t_{\pi_t(1)}, \ldots, d^t_{\pi_t(k)}\}$ for a given $1 \le k \le n$ and we consider exact algorithms for this problem. Second, we consider the approx. k-Select Problem which is to output one data item $d \in \{d^t_{\pi_t((1-\varepsilon)k)}, \ldots, d^t_{\pi_t((1+\varepsilon)k)}\}$. We consider (ε, δ)-approximation algorithms, i.e. an algorithm outputs such a data item correctly with probability at least $1 - \delta$.

Distributed Data Structure. We develop a data structure which supports the following operations:

UPDATE(i, d): Node i receives a new data item d.

INITIALIZE(): Set up the data structure (the nodes may already have observed data items).

REFRESH(): Is called by the server if a request is made and the data structure is already initialized.

ROUGH-RANK(k): Returns a data item d where $rank(d) \in [k, k \cdot \log^c(n)]$ holds with probability at least $1 - \log^{c'}(n)$ for some suitable constants c, c'.

ROUGH-RANK queries are used to receive an element as a basis for further computations in our protocols. We consider a distributed data stream in our model as a sensor node observing its environment by explicitly calling an UPDATE operation to overwrite its previous observation by a new one.

1.2 Our Contribution

In this paper we propose exact algorithms for the k currently lowest observed values or approximation algorithms for a value with rank in $[(1 - \varepsilon)k, (1 + \varepsilon)k]$ with probability $(1 - \delta)$. Our data structure is based on single-shot computations which are of independent interest (Table 1):

With notion of the data structure this relates to the results as presented in Table 2.

In Sect. 4.3 we describe how to combine the one-shot computations with the given data item from the ROUGH-RANK operation supported by our data structure. This leads to the overall bound of $\Theta(k + \log(m) + \log(\log(n)))$ messages in expectation for Top-k queries. The bound on the number of messages is asymptotically tight if $m \ge \log(n)$ holds where m denotes the number of UPDATEs since the last query. For k-Select queries and applying the same combination as above, this protocol uses $\Theta\left(\frac{1}{\varepsilon^2} \log(\frac{1}{\delta}) + \log(m) + \log^2(\log(n))\right)$ messages in

Table 1. Summary of results for single-shot computations.

Problem	Type	Bound	(Total) Comm.	Comm. rounds	Reference
Top-k	Exact	Upper	$\mathcal{O}(k \cdot \log(n))$	$\mathcal{O}(k \cdot \log(n))$	[10]
Top-k	Exact	Upper	$k + \log(n) + 1$	$\mathcal{O}(k + \log(n))$	Section 2
Top-k	Exact	Upper	$\mathcal{O}(k + \log(n))$	$\mathcal{O}(\log(\frac{n}{k}))$	Section 3.2
Top-k	Exact	Lower	$k + \Omega(\log(n))$	/	Section 5
k-select	Aprx.	Upper	$\mathcal{O}(\frac{1}{\varepsilon^2}\log(\frac{1}{\delta}) + \log(n))$	$\mathcal{O}(\log(\frac{n}{k}))$	Section 3.2
k-select	Aprx.	Lower	$\Omega(\log(n))$	/	Section 5

Table 2. Summary of results using our new data structure.

Operation	(Total) Comm.	Comm. rounds	Reference
INITIALIZE	$\mathcal{O}(\log(n))$, $\Omega(\log(n))$	$\mathcal{O}(\log(n))$	Section 4.1
REFRESH	$\mathcal{O}(\log(m))$, $\Omega(\log(m))$	$\mathcal{O}(\log(n))$	Section 4.2
UPDATE	Amortized	1	Section 4
ROUGH-RANK	0	0	Section 4

expectation, where the bound on the number of messages is asymptotically tight if $m \geq \log^{\log(\log(n))}(n)$ holds.

Furthermore, we parameterize our algorithms such that it is possible to choose a trade-off between the number of messages used and the number of communication rounds, i.e. time, used to process the queries.

1.3 Related Work

Cormode et al. introduce the Continuous Monitoring Model [9] with an emphasis on systems consisting of n nodes generating or observing *distributed* data streams and a designated coordinator. In this model the coordinator is asked to continuously compute a function, i.e. to compute a new output with respect to all observations made up to that point. The objective is to aim at minimising the total communication between the nodes and the coordinator. We enhance the continuous monitoring model (as proposed by Cormode et al. in [8]) by a broadcast channel. Note, that we are not strictly continuous in the sense that we introduce a dynamic data structure which only computes a function, if there is a query for it. However, there is still a continuous aspect: In every time step, our data structure maintains elements close to all possible ranks in order to quickly answer queries if needed.

An interesting area of problems within this model are threshold functions: The coordinator has to decide whether the function value (based on all observations) has reached a given threshold τ. For well structured functions (e.g. count-distinct or the sum-problem) asymptotically optimal bounds are known [8,9]. Functions which do not provide such structures (e.g. the entropy [1]), turn out to require much more communication volume.

A related problem is considered in [2]. In their work, Babcock and Olston consider a variant of the distributed top-k monitoring problem: There is a set of objects $\{O_1, \ldots, O_n\}$ given, in which each object has a numeric value. The stream of data items updates these numeric values (of the given objects). In case each object is associated with exactly one node, their problem is to monitor the k largest values. Babcock and Olston have shown by an empirical evaluation, that the amount of communication is by an order of magnitude lower than that of a naive approach.

A model related to our (sub-)problem of finding the k-th largest values, and exploiting a broadcast channel is investigated by the shout-echo model [13,15]: A communication round is defined as a broadcast by a single node, which is replied by all remaining nodes. The objective is to minimise the number of communication rounds, which differs from ours.

2   One Shot Computation: Top-K

In this section we present an algorithm which identifies all data items with rank at most k currently observed by the sensor nodes. Note that when we later apply this protocol for multiple time steps, not all sensor nodes might participate. In this section, we denote by N the number of participating nodes.

Our protocol builds a simple search-tree-like structure based on a height the nodes draw from a geometric distribution. Afterwards, a simple strategy comparable to an in-order tree walk is applied. In order to identify the smallest data item this idea is implemented as follows: The protocol draws a uniform sample of (expected) size $\frac{1}{\phi}$ and broadcasts the smallest data item. Successively, the protocol chooses a uniform sample until the smallest item is identified. In this description, each drawing of a sample corresponds to consider all children of the current root of the search tree and then continue with the left-most child as the new root.

The protocol is given a maximal height h_{max} for the search tree which corresponds to the number of repetitions of the protocol described above. We define a specific value for h_{max} in Theorem 1. Furthermore, the algorithm is given a parameter ϕ which defines the failure probability of the geometric distribution.

The algorithm starts by drawing a random variable h_i from a geometric distribution, i.e. $\Pr[h_i = h] = \phi^{h-1}(1 - \phi)$. We discuss the choice of ϕ at the end of this section. Note that ϕ enables a trade off between the number of messages sent in expectation and communication rounds used.

The protocol can be implemented in our distributed setting by having the server broadcast the values ℓ, u, and h such that each node with the corresponding height values and data items responds. Note, the variables $r_1, \ldots, r_j$ used in Steps 8, 9, 11, 12, and 13 refer to responses of the current call of Top-k-Rec.

Analysis. In the following we show that the expected number of messages used by the Top-k Protocol is upper bounded by $k + \frac{1-\phi}{\phi} \cdot \log_{1/\phi}(N) + 1$ in Theorem 1. Afterwards, an upper bound of $\mathcal{O}(\phi \cdot k + h_{max})$ on the number of communication rounds is presented in Lemma 1. Defining $\phi := 1/2$ the bound

Algorithm 1. Top-k Protocol (ϕ, h_{max})

Initialization()	Top-k-Rec(ℓ, u, h)
1. Each node i draws a random variable h_i, i.i.d. from a geometric distribution with $p = 1 - \phi$	1. **If** $h = 0$ **then**
	2. **if** $|S| = k$ **then** return S,
	3. **Else** end recursion
2. Server defines $\ell := -\infty$, $u := \infty$, $h := h_{max}$ and $S \leftarrow \emptyset$	4. Server probes sensor nodes i with $\ell < d_i < u$ and $h_i \geq h$
3. **Call** Top-k-Rec(ℓ, u, h)	Let $r_1 < \ldots < r_j$ be the responses
4. Raise an **error**, if $|S| < k$	5. **If** there was no response **then**
	6. **Call** Top-k-Rec$(\ell, u, h - 1)$
	7. **Else**
	8. **Call** Top-k-Rec$(\ell, r_1, h - 1)$
	9. $S \leftarrow S \cup r_1$
	10. **For** $i = 1$ to $j - 1$ do
	11. **Call** Top-k-Rec$(r_i, r_{i+1}, h - 1)$
	12. $S \leftarrow S \cup r_{i+1}$
	13. **Call** Top-k-Rec$(r_j, u, h - 1)$

on the communication translates to a tight bound of $k + \log(n) + 1$ in Corollary 1 complemented by a simple lower bound of $k + \Omega(\log(n))$ in Sect. 5.

We show an upper bound on the communication used by the Top-k Protocol analyzing the expected value of a mixed distribution. The analysis works as follows: We sort the nodes by their rank and determine the number of nodes with height $\leq h$ before the first node with a height $> h$ in this ordering by a geometric-sequence in Definition 1. For each height h, this number can then be used to determine the number of nodes which send a message on height h, which we model by a geocoin-experiment in Definition 2. Note that this analysis turns out to be very simple since independence can be exploited in a restricted way and leads to a proper analysis with respect to exact constants.

Definition 1. *We call a sequence $G = (G_1, \ldots, G_m)$ of m random experiments a geometric-sequence, if each G_h is chosen from a geometric distribution with $p_h^{geo} := \phi^h$. We denote its size$(G) := \sum_h G_h$ and say it covers all nodes, if size$(G) \geq N$.*

For the analysis, we choose a fixed length of $m := \log_{1/\phi}(N)$ and modify G to $G' = (G_1, \ldots, G_{m-1}, N)$ such that G' covers all nodes with probability 1.

Based on a given geometric-sequence, we define a sequence describing the number of messages send by the nodes on a given height. We take the number of nodes G_i as a basis for a Bernoulli experiment where the success probability is the probability a node sends a message on height i. This is $\Pr[h = h_i \mid h \leq h_i] = \frac{\phi^{h-1}(1-\phi)}{1-\phi^h}$.

Definition 2. *We denote a geocoin-experiment by $C = (C_1, \ldots, C_m)$ of random variables C_h which are drawn from $Binom(n = G_h, p_h^{bin} = \frac{\phi^{h-1}(1-\phi)}{1-\phi^h})$, i.e. C_h out of G_h successful coin tosses where each coin toss is successful with probability p_h^{bin}.*

Theorem 1. *Let $N > k$ and $h_{max} \geq \log_{1/\phi}(N)$ hold. The Top-k Protocol uses at most $k + \frac{1-\phi}{\phi} \log_{1/\phi}(N) + 1$ messages in expectation.*

Proof. The probability to send a message of a node v within the Top-k is 1. It remains to show that the overhead is bounded by $\frac{1-\phi}{\phi} \log_{1/\phi}(N) + 1$.

The number of messages sent by Algorithm 1 (excluding the k nodes observing the k smallest data items) is upper bounded by a geocoin-experiment C. Let $\mathcal{H} := \log_{1/\phi}(N)$. For $h < \mathcal{H}$ we use that the geometric distribution is memory-less and hence $\mathbb{E}[C_h] = (1 - p_h^{geo}) \cdot (p_h^{bin} + \mathbb{E}[C_h]) = (1 - \phi^h) \cdot \left(\frac{\phi^{h-1}(1-\phi)}{1-\phi^h} + \mathbb{E}[C_i] \right)$. This can simply be rewritten as $\mathbb{E}[C_i] = (1 - \phi)/\phi$.

For $i \geq \mathcal{H}$ we bound the number of messages by the total number of nodes with height at least $\mathcal{H}$. These can be described as the expectation of a Bernoulli experiment with N nodes and success probability $\phi^{\mathcal{H}-1}$ and hence $\mathbb{E}[C_{\geq \mathcal{H}}] \leq \phi^{\mathcal{H}-1} \cdot N = 1/\phi$.

In total, we get $\sum_h \mathbb{E}[C_h] = \left(\sum_{i=h}^{\mathcal{H}-1} \mathbb{E}[C_i] \right) + \mathbb{E}[C_{\geq \mathcal{H}}] \leq \frac{1-\phi}{\phi} \log_{1/\phi}(N) + 1$. $\square$

Lemma 1. *The Top-k Protocol needs $\mathcal{O}(\phi \cdot k + h_{max})$ communication rounds in expectation.*

Note that our bounds describe a trade off between the number of messages and communication rounds, where the number of messages decreases with a small success probability $1 - \phi$. Intuitively speaking, this stems from more larger resulting height values such that the search structure has a smaller breadth.

Corollary 1. *For $N = n$, $\phi := \frac{1}{2}$, and $h_{max} := \log(n)$, the Top-k Protocol uses an amount of $k + \log(n) + 1$ number of messages in expectation and $\mathcal{O}(k + \log(n))$ communication rounds.*

3 One Shot Computation: Approximate k-Select

In this section we present an algorithm which gives an (ε, δ)-approximation for the k-Select Problem, i.e. a data item d is identified with a rank between $(1 - \varepsilon)k$ and $(1 + \varepsilon)k$ with probability at least $1 - \delta$.

In Sect. 3.1, we introduce an algorithm which identifies a data item with rank $\Theta(k)$. This is done to reduce the number of messages for the algorithm proposed in Sect. 3.2 which uses a standard sampling technique to achieve the desired approximation.

3.1 Constant Factor Approximation

The following algorithm employs a similar strategy as Algorithm 1. However, the protocol terminates and outputs a data item as soon as the targeted height of h_{min} is reached. This data item is one of the responses on height h_{min}, dependent on the value of ϕ. Note that it may not be sufficient to output the smallest value, since the number of responses may be very large if ϕ is small.

Algorithm 2. $\mathrm{C}_{\mathrm{O}}\mathrm{FASEL}(h_{max}, \phi, k)$ (ConstantFactorSelect)

---

1. Each node i defines a random variable h_i, i.i.d. drawn from
   a geometric distribution with $p = (1 - \phi)$, and redefines $h_i := \min\{h_i, h_{max}\}$.
2. Server defines $d_{min} := \infty$. $\triangleright \forall$ data items d: $\infty > d$
3. Server defines $0 < \alpha < 1$, s.th. $\lfloor \log_{1/\phi}(7k) \rfloor = \log_{1/\phi}(7k) - \alpha \in \mathbb{N}$ holds.
4. **for** $h := h_{max}$ **to** $h_{min} = \log_{1/\phi}(7\,k) - \alpha + 1$ **do**
5.   Server probes all nodes i with $d_i < d_{min}$ and $h_i = h$.
6.   Let $r_1 < r_2 < \ldots < r_j$ be the responses, ordered by their values.
7.   **If** $h > h_{min}$ **then** Server redefines $d_{min} := r_1$ **else** $d_{min} := r_{(1/\phi)^\alpha}$.
8. Output d_{min}

We show that Algorithm 2 outputs a data item with a rank larger than k and smaller than $42\,k$ with constant success probability in Lemma 2. Furthermore, we state that Algorithm 2 determines a data item which is at most by a polylogarithmic factor larger than the expectation with high probability in Lemma 3. We need this result later when reusing the protocol in Sect. 4. An upper bound on the number of used messages is presented in Lemma 4. We shortly state how to amplify the success probability to $1 - \delta'$ for a given $\delta' > 0$ in Theorem 2.

Lemma 2. *The* $\mathrm{C}_{\mathrm{O}}\mathrm{FASEL}$ *Protocol outputs a data item* d, $rank(d) \in [k, 42k]$ *with probability at least* 0.6.

Proof. The algorithm outputs the $(1/\phi)^\alpha$ smallest data item d the server gets as a response on height $h = h_{min}$. To analyze its rank simply consider the random number X of nodes i that observed smaller data items $d_i < d$. The claim follows by simple calculations: (i) $\Pr[X < k] \le \frac{1}{5}$ and (ii) $\Pr[42k > X] \le \frac{1}{5}$.

The event that X is (strictly) smaller than k holds, if there are $(1/\phi)^\alpha$ out of k nodes with a random height at least h_{min}. Let X_1 be drawn by a binomial distribution $Bin(n = k, p = \phi^{h_{min}-1})$. It holds $\mathbb{E}[X_1] = k \cdot \phi^{h_{min}-1} = \frac{1}{7} \cdot (\frac{1}{\phi})^\alpha$. Then, $\Pr[X < k] \le \Pr[X_1 \ge (\frac{1}{\phi})^\alpha] = \Pr[X_1 \ge (1+6)\frac{1}{7\phi^\alpha}] \le \exp(-\frac{1}{3}\frac{1}{7\phi^\alpha}6^2) \le \frac{1}{5}$.

On the other hand, the event that X is (strictly) larger than $42k$ holds, if there are less than $(1/\phi)^\alpha$ out of $42k$ nodes with a random height of at least h_{min}. Let X_2 be drawn by a binomial distribution $Bin(n = 42k, p = \phi^{h_{min}-1})$. It holds $\mathbb{E}[X_2] = (42k)\phi^{h_{min}-1} = (42k)(7k)^{-1}\phi^{-\alpha} = \frac{6}{\phi^\alpha}$. Then, $\Pr[X > 42k] \le \Pr[X_2 < \frac{1}{\phi^\alpha}] = \Pr[X_2 < (1-(1-\frac{1}{6}))\frac{6}{\phi^\alpha}] \le \exp(-\frac{1}{2}(\frac{6}{\phi^\alpha}(1-\frac{1}{6})^2) \le \exp(-\frac{25}{12}) \le \frac{1}{5}$. $\square$

Lemma 3. *For a given constant* $c > 8$ *there exist constants* $c_1, c_2 > 1$, *such that the* $\mathrm{C}_{\mathrm{O}}\mathrm{FASEL}$ *Protocol as given in Algorithm 2 outputs a data item* d *with a rank in* $[\log^{c_1}(n) \cdot 7k, \log^{c_2}(n) \cdot 7k]$ *with probability at least* $1 - n^{-c}$.

Lemma 4. *Let* $N > k$ *and* $h_{max} \ge \log_{1/\phi}(N)$ *hold. The* $\mathrm{C}_{\mathrm{O}}\mathrm{FASEL}$ *Protocol presented in Algorithm 2 uses an amount of at most* $\mathcal{O}(\frac{1}{\phi}(\log_{1/\phi}(\frac{N}{k}) + 1))$ *messages in expectation.*

We apply a standard boosting technique, i.e. we use $\mathcal{O}(\log(\frac{1}{\delta'}))$ independent instances of Algorithm 2, and consider the median of the outputs of all instances to be the overall output. We denote this amplified version of CoFaSel by CoFaSelAmp. Thus, an output in the interval $[k, 42\,k]$ with probability at least $1 - \delta'$ is determined.

Since we run the $\mathcal{O}(\log(\frac{1}{\delta'}))$ instances in parallel, and the server is able to process all incoming messages within the same communication round, the number of communication rounds does not increase by this extension of the protocol. These simple observations lead to the following theorem summarizing a first result for the k-select problem:

Theorem 2. *Let $N > k$ and $h_{max} \geq \log_{1/\phi}(N)$ hold. Let δ' be a given constant. The algorithm CoFaSelAmp determines a data item d with rank at least k and at most $42k$ with probability at least $1 - \delta'$ using $\mathcal{O}(\frac{1}{\phi} \log_{1/\phi}(\frac{N}{k}) \log(\frac{1}{\delta'}))$ messages in expectation and $h_{max} - h_{min} + 2$ communication rounds.*

3.2 Approximate k-Select

In this section we propose an algorithm which is based on the algorithm from the previous section. Here, we aim for an (ε, δ)-approximation of the k-Selection problem for a single time step. Using the approximation given by CoFaSelAmp, which gives a data item d with a rank between k and $42k$ with probability at least $1 - \delta'$, a simple standard sampling strategy is applied afterwards. Note that only those nodes take place in this strategy which observed a data item d_i smaller than d.

Algorithm 3. Approx. k-Select Protocol ApproKSel($k, \phi, \varepsilon, \delta', \delta, h_{max}$)

---
1. Call CoFaSelAmp($h_{max}, \phi, k, \delta'$) and obtain data item d.
2. Each node i with $d_i < d$:
3.   Toss a coin with $p := \min\left(1, \frac{c}{k}\mathcal{S}_{\varepsilon,\delta}\right)$.
4.   On success send d_i to the server.
5. The server sorts these values and outputs d_K, the $p \cdot k$-th smallest item.

In the following, we show that Algorithm 3 is an (ε, δ)-approximation of the k-Selection protocol in Theorem 3. We discuss a possible choice of parameters in Corollary 2.

Theorem 3. *Let $N > k$ and $h_{max} \geq \log_{1/\phi}(N)$ hold. The Approx. k-Select Protocol selects data item d_K with rank in $[(1 - \varepsilon)\,k, (1 + \varepsilon)\,k]$ with probability at least $1 - \delta$ using $\mathcal{O}((1 + \frac{\log^c(n)}{k}\delta')\mathcal{S}_{\varepsilon,\delta} + \frac{1}{\phi} \log_{1/\phi}(\frac{N}{k}) \log(\frac{1}{\delta'}))$ msg. in exp. and $h_{max} - h_{min} + 3$ comm. rounds.*

For the sake of self containment we propose a bound which considers all nodes to take part in the protocol ($N = n$). Note, that the CoFaSelAmp protocol outputs a value with rank smaller than $7k \cdot$ polylog (n) w.h.p. (c.f. Lemma 3).

Corollary 2. *Let c be a sufficiently large constant. Furthermore, let $N = n$, $\phi := \frac{1}{2}$, $h_{max} := \log n$, and $\delta' := \frac{1}{\log^c(n)}$. The protocol uses an amount of at most $\mathcal{O}(\mathscr{S}_{\varepsilon,\delta} + \log(n)\log(\log(n)))$ messages in expectation and $\log(\frac{n}{k})$ rounds of communication.*

This represents the case that a small number of messages and a large number of communication rounds are used. This observation is complemented by a lower bound of $\Omega(\log(n))$ in Sect. 5. Note, that this bound can be reduced to $\mathcal{O}(\mathscr{S}_{\varepsilon,\delta} + \log(n))$ by running one instance of CoFaSel until $h'_{min} := \lceil \log(7k) \rceil + c$ and denote the output (i.e. the smallest data item) as d. The Approx. k-Select Protocol is then applied only on nodes that observed data items smaller than d.

Corollary 3. *Let $N = n$, $\phi := \frac{1}{2}$, $h_{max} := \log n$, $k' := 2k$, $\epsilon := \frac{1}{2}$ and $\delta' := \frac{1}{\log^c(n)}$. The protocol uses $\mathcal{O}(k + \log(n))$ messages in expectation to solve the Top-k-Problem.*

4   Multiple Time Step Computation: A Fully Dynamic Distributed Data Structure

In this section we consider computations of Top-k or approx. k-Select for multiple time steps. We use a dynamic data structure to keep rough information such that a required computation can be executed more efficiently. The main part of this section focuses on computing an element with rank close to k utilizing our data structure. The final results for answering the queries on the basis of this element are described in Sect. 4.3.

Our basic idea is to maintain a structure similar to the trees (to be more precise, only the left-most path) used to identify the (approximately) k'th smallest items in the previous chapters. The data structure maintains the rough rank sketch which is defined as follows:

Definition 3 (Rough Rank Sketch (RRS)). *A data structure for the approximate k-select problem fulfills the RRS property if a request for the data item of rank k will be answered with an item of rank in $[k, \log^c(n) \cdot k]$ with probability at least $1 - \log^{-c}(n)$.*

We divide the ranks $1, \ldots, n$ into classes. The goal is that a data item of each class (representative) is contained in our data structure. The height of a class represents the expected maximum height found within this class, such that our representative will have a height value within the noted bounds.

Let $\mathscr{H} := \log_{1/\phi}(\log(n))$. The idea of classes is captured in the following definition:

Definition 4. *Let κ be sufficiently large. A Class C_ℓ^t consists of all data items d_j^t with $rank(d_j^t) \in [\log^{\ell 8\kappa}(n), \log^{(\ell+1)8\kappa}(n))$. We denote by $h(C_\ell^t) = (\ell 8\kappa\mathscr{H}, (\ell+1)8\kappa\mathscr{H}]$ the height of the class C_ℓ^t.*

By abuse of notation we introduce $d_i^t \in C_\ell^t$ which shortens $rank(d_i^t) \in C_\ell^t$.
We divide each class into sub-classes as follows:

Definition 5. *Let κ be set as before. We denote by a sub class $C_{\ell,\tau}^t$, with $\tau \in \{0, \ldots, 3\}$, the set of data items d_i^t with a rank between $\log^{\ell 8\kappa + 2\tau\kappa}(n)$ and $\log^{\ell 8\kappa + (2\tau + 2)\kappa}(n)$. The height of $C_{\ell,\tau}^t$ is $h(C_{\ell,\tau}^t) = ((\ell 8\kappa + (2\tau + 1)\kappa)\mathscr{H}, (\ell 8\kappa + (2\tau + 3)\kappa)\mathscr{H}]$.*

We omit the time step t in our notation whenever it is clear from the context.

Definition 6. *The data items in a class C_ℓ are well-shaped, if for each data item d_i with $rank(d_i) \in [\log^{\ell 8\kappa + 2\tau\kappa}(n), \log^{\ell 8\kappa + (2\tau + 2)\kappa}(n)]$ it holds $h_i \le (\ell 8\kappa + (2\tau + 3)\kappa)\mathscr{H}$.*

Algorithm 4. SeleMon(ϕ)   [Select and Monitor]

INITIALIZE() [Repeat until all classes are filled, i.e. $\forall \ell \; \exists a_\ell \in S_{\ell,1}, r_\ell \in S_{\ell,2}$]

1. **Call** CoFaSel($\phi, h_{max} = \log_{1/\phi} n, k = 1$) (Algorithm 2), and keep a data structure DS with all (h_i, d_i) pairs, where for each h_i the smallest response d_i is kept.
2. Assign data item d_i with height h_i to its sub class $S_{\ell,\tau'}$, if $h_i \in h(C_{\ell,\tau'})$ holds.
3. Choose $a_\ell \in S_{\ell,1}$ and $r_\ell \in S_{\ell,2}$ uniformly at random.

UPDATE(i, d) [Executed by node i]

1. Update d_i^t by $d_i^{t+1} = d$, delete (d_i^t, h_i^t) from DS (if it was in DS).
2. **If** $d_i = a_\ell$ or $d_i = r_\ell$ or ($h_i \in [h(a_\ell), h(r_\ell)]$ and $d_i < a_\ell$)
   then delete all (h_j, d_j) pairs from DS, where $d_j \in S_\ell$ holds.
3. Draw a new value from the geometric distribution with $p = (1 - \phi)$ and redefine $h_i := \min\{h_i, h_{max}\}$.

REFRESH() [Repeat until all classes are filled, i.e. $\forall \ell \; \exists a_\ell \in S_{\ell,1}, r_\ell \in S_{\ell,2}$]

1. define $t := t + 1$
2. Determine level ℓ such that all classes $C_{\ell'}$, $\ell' > \ell$ are filled, more formally $\forall \ell' > \ell, S_{\ell,1} \ne \emptyset \wedge S_{\ell,2} \ne \emptyset$.
3. Determine maximal height h of all nodes i that observed an UPDATE since the last INITIALIZE or REFRESH operation. Let ℓ'' be the level with $h \in h(C_{\ell''})$. Define $\ell := max(\ell, \ell'')$.
4. **Call** INITIALIZE() (only on sensor nodes i with $d_i < r_\ell$)

ROUGH-RANK(k)   ▷ k denotes a rank

1. Determine ℓ such that $k \in C_{\ell - 1}$ holds.
2. **Output** representative $r_\ell \in S_{\ell,2}$.

4.1 Correctness of INITIALIZE

We start by analyzing the outcome of the INITIALIZE operation. In this, we show that a class is well-shaped with sufficiently large probability in Lemma 5 and argue that the data structure yields a RR-Sketch in Theorem 4, afterwards.

Lemma 5. *Let $\ell \in \mathbb{N}$. After an execution of INITIALIZE, the class C_ℓ is well-shaped with probability at least $1 - \log^{-c}(n)$, for some constant c.*

Lemma 6. *Consider a sub class $C_{\ell,\tau'}$, with $\tau' \in \{1, 2\}$. There is a data item $d_i \in C_{\ell,\tau'}$ with $h_i > (\ell 8\kappa + (2\tau' + 1)\kappa)\mathcal{H}$ with high probability.*

Theorem 4. *After execution of INITIALIZE for each rank k exists a data item in the data structure with rank between k and $k \cdot \log^c(n)$ with probability at least $1 - \log^{-c'}(n)$ for constants c, c'.*

Proof. First consider a fixed class C_ℓ for a fixed $\ell \in \mathbb{N}$. Based on Lemma 5 we can show that the distribution of the random heights is well-shaped with probability at least $1 - \log^{-c^*}(n)$ for a constant c^*. Now, with high probability there is a data item with such a height for sufficiently large κ and n due to Lemma 6. We may in fact choose c^* such that the probabilities for both to occur is at least $1 - \log^{-c^*}(n)$. These observations together show that there is a data item d'_τ identified and stored in DS and thus, for each request $k \in C_{\ell-1}$ the algorithm has identified a representative in C_ℓ as a response with a rank only by a polylogarithmic factor larger than k.

Furthermore, note that there are at most $\log(n)$ number of classes. The argument stated above applied to each class leads to the desired result, where (applying union bound) also shows the desired success probability of $1 - \log^{-(c^* - 1)}(n)$. □

Now we have shown that the Rough Rank Sketch is calculated by executing INITIALIZE with certain probability. To analyze the number of messages in expectation and the number communication rounds we refer to Lemma 4 and Theorem 2, respectively. Since INITIALIZE is strongly based on the CoFaSel protocol, similar arguments hold for this section, again. However, note that the repetitions of the algorithm to obtain representatives a_ℓ and r_ℓ for each level ℓ and thus for each class C_ℓ is not straight forward. A complete recomputation from scratch until all representatives are obtained introduces a factor of $\log^*(n)$ to the communication costs and rounds. For this simply observe that there are at most $\log(n)$ different classes C_ℓ, for which However, here only for those levels the CoFaSel protocol is called, where a_ℓ and r_ℓ is not known leading to additional constant factor overhead (in expectation).

4.2   Correctness of REFRESH

In the previous subsection we have shown that the algorithm INITIALIZE computes a Rough Rank Sketch for a fixed time step. In this section we show that the REFRESH method preserves and/or rebuild parts of the data structure such that a Rough Rank Sketch is achieved after m UPDATES took place. We analyze two different scenarios and analyze the probability of the scenario to occur: The representative of the class itself is UPDATED and thus, the class gets deleted and the case that the representative does not get an UPDATE, but the rank does not reflect the situation correctly at the next time step $t + 1$.

UPDATE to a Representative. We analyze the probability that the alarm a_ℓ or the representative r_ℓ of a class C_ℓ is updated and thus, the class gets deleted from the data structure. This is okay, if there are sufficiently many UPDATEs, i.e. m is sufficiently large. However, if m is small compared to the number of data items sub classes $C_{\ell,1}$ and $C_{\ell,2}$ consists of, the probability to choose exactly a_ℓ or r_ℓ for an UPDATE and thus delete from the data structure is small, as Lemma 7 states. The proof can be found in the full version [4].

Lemma 7. *Let m be the number of UPDATE operations since INITIALIZE or REFRESH is called. Let C_ℓ be a class with $m < \log^{\ell 8\kappa}(n)$. The representative of C_i did not get an update with probability at least $1 - \log^{-c}(n)$, for a constant c.*

Push Representative Out of Class. We want to estimate the probability that some representative d_i from our data structure was in C_ℓ^t but not in C_ℓ^{t+1}.

We start our analysis with a result on the rank of a data item, given the randomly drawn height in Lemma 8. Afterwards, we show in Lemma 9 that the representative of C_ℓ^t is still in C_ℓ^{t+1}, if m is not too large. However, if the number of updates is large, we analyze that the representative is deleted from DS in Lemma 10 with sufficiently large probability. We conclude that the desired properties of the data structure are restored after a REFRESH operation (Theorem 5).

Lemma 8. *Fix a time t. Consider a data item $d_i \in DS$ and let $h_i \in h(C_{\ell,\tau'})$ be the height of data item d_i. It holds $d_i \in C_{\ell,\tau'}$ with high probability.*

Lemma 9. *Fix an ℓ with $m < \log^{\ell 8\kappa}(n)$. Let t_0 be the time step INITIALIZE was called and $t - t_0 \le \log(n)$ hold. If at every time $t' \in [t_0, t]$ it holds $m_{t'} < \log^{\ell 8\kappa}(n)$, then the representative of $C_\ell^{t_0}$ is a valid representative of C_ℓ^t (w.h.p.).*

Lemma 10. *Fix an ℓ with $m \ge \log^{\ell 8\kappa}(n)$. Let t_0 be the time step INITIALIZE was called and $t - t_0 \le \log(n)$ hold. If $r_\ell \in C_{\ell,2}^t$ is no longer in C_ℓ^{t+1} the protocol deletes r_ℓ with high probability.*

Theorem 5. *After each REFRESH operation the Rough Rank Sketch is restored, i.e. for each rank k there exists a data item d in the data structure DS as defined in Algorithm 4 with rank between k and $k \cdot \log^c(n)$ with probability at least $1 - \log^{-c'}(n)$.*

Finally, we show the number of messages the data structure uses in order to build or rebuild the Rough Rank Sketch.

Theorem 6. *The operations INITIALIZE and REFRESH use $\mathcal{O}(\frac{1}{\phi} \log_{1/\phi}(n))$ and $\mathcal{O}(\frac{1}{\phi} \log_{1/\phi}(m))$ messages in expectation, respectively.*

4.3   Implications for Top-k and Approximate k-Select

Here we shortly describe how the data structure can be used to efficiently answer a Top-k or k-Select request. Note that the k does not need to be known beforehand, each request can be posed with a different parameter.

Both computations start with a REFRESH operation if there were UPDATEs since the last REFRESH. We then obtain an item with rank close to k with a ROUGH-RANK(k) operation.

For determining the k smallest items, the response from the data structure, denoted by d, is broadcasted such that all sensor nodes with a larger data item do not take place in the call of Top-k Protocol with parameter k. If this call was not successful, a second call of Top-k Protocol is executed on all sensor nodes. Considering the expected costs conditioned on whether the first or second call was successful and applying the law of total expectation, the simple bound on the communication follows:

Corollary 4. *One computation of the Top-k needs an amount of $k + \mathcal{O}(\log(m) + \log(\log(n)))$ messages in expectation assuming m UPDATEs are processed since the last Top-k query.*

Exactly the same approach is used to solve the k-Select problem. We define the (internal) failure probability $\delta' := \log^{-1}(n)$ and obtain the following simple bound by applying the same arguments as for the Top-k Protocol.

Corollary 5. *One computation of approx. k-Select Problem uses $\mathcal{O}(\mathscr{S}_{\varepsilon,\delta} + \log(m) + \log^2(\log(n)))$ msg. in expectation assuming m UPDATES are processed since the last query.*

5   Lower Bounds

In this section we consider lower bounds for the problems considered in the past sections. We show that our main results in the previous sections are asymptotically tight up to additive costs of $\mathcal{O}(\log(\log(n)))$ per time step for a constant choice of ϕ, and $\mathcal{O}(\log^2(\log(n)))$, respectively. For scenarios in which the adversary changes a polylogarithmic number of values in each time step, the proposed bounds are asymptotically tight.

Lemma 11 ([10]). *Every algorithm needs $\Omega(\log(n))$ messages in expectation to output the maximum in our setting.*

We extend this lemma to multiple time steps and to monte carlo algorithms which solve the problem at each time step with a fixed probability:

Lemma 12. *Every algorithm that outputs for a given $c > 1$, a data item d where $rank(d) \in \mathcal{O}(\log^c(n))$ holds and success probability at least $1 - \frac{1}{\log(n)}$ uses $\Omega(\log(n))$ messages in expectation in our setting.*

Proof. Assume there is an algorithm ALG which achieves the above requirements with $o(\log(n))$ messages in expectation. We construct an algorithm $\mathcal{A}$ which first applies ALG and then the Top-k Protocol for $k := 1$, and $h_{max} := \log(n)$.

Observe that $\mathcal{A}$ uses $\mathcal{O}(\log(\log(n)))$ messages in expectation if the event $m \leq \log^c(n)$ occurs and $\log(n)$ messages, else. The probability that the latter event occurs is upper bounded by $\frac{1}{\log(n)}$ and thus, $\mathcal{A}$ uses $o(\log(n))$ messages in expectation which contradicts Lemma 11. $\qquad\square$

We further extend the lower bounds to multiple time steps in which an adversary is allowed to change values of at most m nodes between two consecutive time steps. It is easy to see if the instance always changes the smallest m nodes between each time steps and chooses random permutations, the following holds:

Theorem 7. *Every algorithm $\mathcal{A}$ that tracks the minimum over T time steps uses an amount of $\Omega(\log(n) + T \log(m))$ messages in expectation, if m UPDATES per step are processed in our setting.*

Proof. For the proof we assume, that T is at most $\log(n)$, split T in $T_1, T_2, \ldots$, each of size $\log(n)$ (except for the last one).

Now, construct an instance as follows: Initially define sets of data items $\mathcal{S}_0, \mathcal{S}_1, \ldots, \mathcal{S}_{\log(n)}$. In each time step the data items for an UPDATE are chosen from these sets. For each consecutive set $\mathcal{S}_t, \mathcal{S}_{t+1}$ holds that the largest data item in $\mathcal{S}_{t+1}$ is smaller than the smallest data item in $\mathcal{S}_t$. Furthermore, the size of each $\mathcal{S}_t$ is m, for $t \geq 1$ and n for $t = 0$. The adversary chooses a random permutation π_0 which defines which node initially observes a data item from $\mathcal{S}_0$. Denote the set of nodes with the m smallest data items given by the random permutation by $\mathcal{N}$. For each consecutive time step $t > 0$ the adversary chooses a random permutation π_t and the m nodes to process these UPDATES.

Observe that in the first step ($t = 0$) based on Lemma 11 we argue that at least $\Omega(\log(n))$ messages are used to identify the minimum. For each consecutive time step ($t > 0$) the adversary chooses the same set of nodes $\mathcal{N}$ to process UPDATES. Based on the construction of $\mathcal{S}_t$ and $\mathcal{S}_{t-1}$ no information of $\mathcal{S}_{t-1}$ can be exploited by any algorithm to proceed the permutation of $\mathcal{S}_t$ using less than $\Omega(\log(m))$ messages in expectation, concluding the proof. $\qquad\square$

References

1. Arackaparambil, C., Brody, J., Chakrabarti, A.: Functional monitoring without monotonicity. In: Albers, S., Marchetti-Spaccamela, A., Matias, Y., Nikoletseas, S., Thomas, W. (eds.) ICALP 2009. LNCS, vol. 5555, pp. 95–106. Springer, Heidelberg (2009). https://doi.org/10.1007/978-3-642-02927-1_10
2. Babcock, B., Olston, C.: Distributed Top-K monitoring. In: International Conference on Management of Data, pp. 28–39. ACM (2003)
3. Bemmann, P., Biermeier, F., Bürmann, J., Kemper, A., Knollmann, T., Knorr, S., Kothe, N., Mäcker, A., Malatyali, M., Meyer auf der Heide, F., Riechers, S., Schaefer, J., Sundermeier, J.: Monitoring of domain-related problems in distributed data streams. In: 24th International Colloquium on Structural Information and Communication Complexity (to appear)

4. Biermeier, F., Feldkord, B., Malatyali, M., Meyer auf der Heide, F.: A communication-efficient distributed data structure for top-k and k-select queries. arXiv preprint arXiv:1709.07259 (2017)
5. Canetti, R., Even, G., Goldreich, O.: Lower bounds for sampling algorithms for estimating the average. Inf. Process. Lett. **53**(1), 17–25 (1995)
6. Cormode, G., Korn, F., Muthukrishnan, S., Srivastava, D.: Space-and time-efficient deterministic algorithms for biased quantiles over data streams. In: Proceedings of the Twenty-Fifth ACM SIGMOD-SIGACT-SIGART Symposium on Principles of Database Systems, pp. 263–272. ACM
7. Cormode, G., Korn, F., Muthukrishnan, S., Srivastava, D.: Effective computation of biased quantiles over data streams. In: 21st International Conference on Data Engineering, ICDE 2005 Proceedings, pp. 20–31. IEEE (2005)
8. Cormode, G., Muthukrishnan, S., Yi, K.: Algorithms for distributed functional monitoring. ACM Trans. Algorithms **7**, 21 (2011)
9. Cormode, G., Muthukrishnan, S., Yi, K.: Algorithms for distributed functional monitoring. In: Proceedings of the Nineteenth Annual ACM-SIAM Symposium on Discrete algorithms, SODA 2008. Society for Industrial and Applied Mathematics, Philadelphia, PA, USA (2008)
10. Mäcker, A., Malatyali, M., Meyer auf der Heide, F.: Online Top-k-position monitoring of distributed data streams. In: 29th International Parallel and Distributed Processing Symposium. IEEE (2015)
11. Mäcker, A., Malatyali, M., Meyer auf der Heide, F.: On competitive algorithms for approximations of Top-k-position monitoring of distributed streams. In: 30th International Parallel and Distributed Processing Symposium. IEEE (2016)
12. Madden, S., Franklin, M., Hellerstein, J., Hong, W.: The design of an acquisitional query processor for sensor networks. In: Proceedings of the 2003 ACM SIGMOD International Conference on Management of data, pp. 491–502 (2003)
13. Marberg, J.: An optimal shout-echo algorithm for selection in distributed sets. UCLA (1985)
14. Muthukrishnan, S.: Data Streams: Algorithms and Applications. Now Publishers Inc., Breda (2005)
15. Rotem, D., Santoro, N., Sidney, J.: Shout echo selection in distributed files. Networks **16**, 77–86 (1986)
16. Yi, K., Zhang, Q.: Optimal tracking of distributed heavy hitters and quantiles. Algorithmica **65**(1), 206–223 (2013)
17. Zengfeng, H., Yi, K., Zhang, Q.: Randomized algorithms for tracking distributed count, frequencies, and ranks. In: Proceedings of the 31st Symposium on Principles of Database Systems (2012)
18. Zhang, Z., Cheng, R., Papadias, D., Tung, A.K.H.: Minimizing the communication cost for continuous skyline maintenance. In: Proceedings of the ACM SIGMOD International Conference on Management of Data, pp. 495–508. ACM, New York (2009)
19. Zhang, Q.: Communication-efficient computation on distributed noisy datasets. In: Proceedings of the 27th ACM Symposium on Parallelism in Algorithms and Architectures, pp. 313–322. ACM (2015)

Strategyproof Mechanisms for Additively Separable Hedonic Games and Fractional Hedonic Games

Michele Flammini[1], Gianpiero Monaco[2(✉)], and Qiang Zhang[3]

[1] GSSI Institute, University of L'Aquila, L'Aquila, Italy
`michele.flammini@univaq.it`
[2] University of L'Aquila, L'Aquila, Italy
`gianpiero.monaco@univaq.it`
[3] Sapienza University of Rome, Rome, Italy
`csqzhang@gmail.com`

Abstract. Additively separable hedonic games and fractional hedonic games have received considerable attention. These are coalition forming games of selfish agents based on their mutual preferences. Most of the work in the literature characterizes the existence and structure of stable outcomes (i.e., partitions in coalitions), assuming that preferences are given. However, there is little discussion on this assumption. In fact, agents receive different utilities if they belong to different partitions, and thus it is natural for them to declare their preferences strategically in order to maximize their benefit. In this paper we consider strategyproof mechanisms for additively separable hedonic games and fractional hedonic games, that is, partitioning methods without payments such that utility maximizing agents have no incentive to lie about their true preferences. We focus on social welfare maximization and provide several lower and upper bounds on the performance achievable by strategyproof mechanisms for general and specific additive functions. In most of the cases we provide tight or asymptotically tight results. All our mechanisms are simple and can be computed in polynomial time. Moreover, all the lower bounds are unconditional, that is, they do not rely on any computational or complexity assumptions.

Keywords: Noncooperative games: computation
Strategyproof mechanisms · Coalitions formation
Additively separable hedonic games · Fractional hedonic games

1 Introduction

Teamwork, clustering and group formations, have been important and widely investigated issues in computer science research. In many economic, social and political situations, individuals carry out activities in groups rather than by themselves. In these scenarios, it is of crucial importance to consider the satisfaction of the members of the groups. For example, the utility of an individual

© Springer International Publishing AG, part of Springer Nature 2018
R. Solis-Oba and R. Fleischer (Eds.): WAOA 2017, LNCS 10787, pp. 301–316, 2018.
https://doi.org/10.1007/978-3-319-89441-6_22

in a group sharing a resource, depends both, on the consumption level of the resource, and on the identity of the members in the group; similarly, the utility for a party belonging to a political coalition depends both, on the party trait, and on the identity of its members.

Hedonic games, introduced in [17], model the formation of coalitions (groups) of players (or agents). These are games in which agents have preferences over the set of all possible agent coalitions, and the utility of an agent depends on the composition of the cluster she belongs to.

In this paper we consider *additively separable hedonic games* (ASHGs), which constitute a natural and succinctly representable class of hedonic games. Each player in an ASHG has a value for any other player, and the utility of a coalition to a particular player is simply the sum of the values she assigns to the members of her coalition. Additive separability satisfies a number of desirable axiomatic properties [3] and ASHGs are the non-transferable utility generalization of graph games studied by Deng and Papadimitriou [16]. We further consider *fractional hedonic games* (FHGs), introduced in [2], which are similar to ASHGs, with the difference that the utility of each agent is divided by the size of her cluster. This allows to model behavioral dynamics in social environments that are not captured by ASHGs: one usually prefers having a couple of good friends in a cluster composed by few other people rather than being part of a crowded cluster populated by uninteresting agents.

Coalition formation in ASHGs and FHGs, has received growing attention, but mainly from the perspective of coalition stability, i.e., core, Nash equilibria, etc, or from a classical offline optimization point of view, i.e., where solutions are not necessarily stable (see Related Work), with little emphasis on mechanism design. We consider such games where agents have private preferences. A major challenge is to design algorithms that work well even when the input is reported by selfish agents aiming only at maximizing their personal utility. An interesting approach is to use strategyproof mechanisms [18,27], that is designing algorithms (not using payments) where selfish utility maximizing agents have no incentive to lie about their true preferences.

Our Contribution. We present strategyproof mechanisms for ASHGs and FHGs, both for general and specific additive valuation functions. In particular, we consider: (i) *general valuations* where additive valuations among agents can get any values; (ii) *non-negative valuations* where additive valuations among agents can only get positive values; (iii) *duplex valuations* where additive valuations among agents can only get values in $\{-1, 0, 1\}$ (we can think about setting where each agent i can express for any other agent j if she is an enemy, neutral or a friend); (iv) *simple valuations* where additive valuations among agents can only get values in $\{0, 1\}$ (we can think about setting where each agent i can express for any other agent j if she is neutral or a friend). The latter setting has been also considered in other papers since it models a basic economic scenario referred to in the literature as Bakers and Millers [2,10]. See Sect. 2 for more details about the considered valuations.

We focus on the classical utilitarian social welfare, that is the sum of individual utilities of the players in a coalition, and provide several lower and upper bounds on the performance achievable by strategyproof mechanisms.

In general, showing upper and lower bounds of truthful mechanisms is important, since by the revelation principle, every non-truthful mechanism can be transformed in a truthful one with the same performance. We believe that, applying this framework in our setting is particularly relevant, given the considerable interest on the hedonic games variants considered here. We are mainly interested in deterministic mechanisms, however we also provide some randomized lower bounds (notice that randomized lower bounds are stronger than deterministic ones). Our results are summarized in Table 1. In most of the cases (except the case of duplex valuations) we provide tight or asymptotically tight results.

We point out that, on the one hand, all our mechanisms are simple and can be computed in polynomial time. On the other hand, all the lower bounds (some of them randomized) are unconditional, that is, they do not rely on any computational or complexity assumptions.

Related Work. In the literature, a significant stream of research considered hedonic games (see [5]), and in particular ASHGs, from a strategic cooperative point of view [7,12,20], with the purpose of characterizing the existence and the properties of coalitions structures such as the core, and from a non-cooperative point of view [11,21] with special focus on pure Nash equilibria. Computational complexity issues related to the problem of computing stable outcomes have been considered in [3,22,25,26,30]. Finally, hedonic games have also been considered in [6,8,14–16] from a classical optimization point of view, i.e., where solutions are not necessarily stable. Concerning FHGs, Aziz et al. [2], give some properties guaranteeing the (non-)emptiness of the core. Moreover, Brandl et al. [13], study the computational complexity of understanding the existence of core and individual stable outcomes. From a non-cooperative point of view, the papers [9,10], study the existence, efficiency and computational complexity of Nash equilibria. Further results on the price of stability for specific FHGs have been presented in [23]. Other stability notions have been also investigated, like in [1,19], where the authors focused on Pareto stability. Finally, Aziz et al. [4], consider the computational complexity of computing welfare maximizing partitions (not necessarily stable).

The design of *truthful* mechanisms, that is of algorithms that use payments to convince the selfish agents to reveal the truth and that then compute the outcome on the basis of their reported values, has been studied in innumerable scenarios. However, there are settings where monetary transfers are not feasible, because of either ethical or legal issues [24], or practical matters in enforcing and collecting payments [27]. A growing stream of research focuses on the design of the more applicable *strategyproof* mechanisms, that lead agents to report their true preferences, without using payments.

Wright and Vorobeychik [31] focus on strategyproof mechanisms for ASHGs. They only consider positive preferences. Under this assumption, a trivial optimal strategyproof mechanism just puts all the agents in the same grand coalition.

Therefore, they consider coalition size constraints and (approximate) envy-freeness. Their main contribution is a mechanism that, despite not having theoretical guarantees, achieves good experimental performance. We point out that in this paper we focus on theoretical results concerning ASHGs and FHGs, for which to the best of our knowledge no strategyproof mechanism has been proposed in the scientific literature, yet.

Vallée et al. [29] consider classical hedonic games with general preference relationships, and characterize the conditions of the game structure that allow rational false-name manipulations. However, they do not provide mechanisms. Aziz et al. [1] show that the serial dictatorship mechanism is Pareto optimal, and strategyproof for general hedonic games when appropriate restrictions are imposed on agents. Finally, Rodríguez-Álvarez [28], studies strategyproof core stable solutions properties for hedonic games.

Paper organization. The paper is organized as follows. In Sect. 2, we formally describe the problems and introduce some useful definitions. The studies on the performance of strategyproof mechanisms are then presented in Sects. 3, 4, 5, and 6, which address, respectively, general, non-negative, duplex and simple valuations. In Sect. 7 we discuss some extension of our results. Finally, in Sect. 8, we resume our results and list some interesting open problems.

Due to space constraints, some proofs have been removed. All the details will appear in the full version of the paper.

Table 1. Our results for the different cases. *Stands for randomized mechanisms. L. B. stands for lower bounds. U. B. stands for upper bounds.

		$[-1,1]$	$[0,1]$	$\{-1,0,1\}$	$\{0,1\}$
ASHGs	L. B	*Unbounded**	OPT	$\Omega(n), 2 - \epsilon^*$	OPT
	U. B			$O(n^2)$	
FHGs	L. B	*Unbounded**	$\frac{n}{2}$	$2 - \epsilon$	$\frac{6}{5}$
	U. B		$\frac{n}{2}$	$O(n)$	2

2 Preliminaries

In additive separable hedonic games (ASHGs) and fractional hedonic games (FHGs), we are given a set $N = \{1, \ldots, n\}$ of selfish agents. The objective or outcome of the game is a partition of the agents into disjoint coalitions $\mathcal{C} = \{C_1, C_2, \ldots\}$, where each coalition C_j is a subset of agents and each agent is in exactly one coalition. Let $\mathscr{C}$ be the collection of all the possible outcomes. Given a partition $\mathcal{C} \in \mathscr{C}$, we denote by $|\mathcal{C}|$ the number of its coalitions and by $\mathcal{C}^i$ the coalition of $\mathcal{C}$ containing agent i. Similarly, given a coalition C, we let $|C|$ be the size or number of agents in C. The *grand coalition* is the outcome in which all the agents are in the same coalition, i.e., $|\mathcal{C}| = 1$. We assume that each agent has a privately known valuation $v_i : N \to \mathbb{R}$, mapping every agent to a real (possibly negative) value. In ASHGs, for any $\mathcal{C} \in \mathscr{C}$, the preference or utility of agent i is $u_i(\mathcal{C}) = \sum_{j \in \mathcal{C}^i} v_i(j)$, that is, it is additively induced by her valuation function. Similarly, in FHGs, for any $\mathcal{C} \in \mathscr{C}$, the utility of agent i is $u_i(\mathcal{C}) = \frac{\sum_{j \in \mathcal{C}^i} v_i(j)}{|\mathcal{C}^i|}$.

We are interested in four basic classes of valuation functions. Namely, for any pair of agents $i, j \in N$, we consider:

General valuations: $v_i(j) \in [-1, 1]$;
Non-negative valuations: $v_i(j) \in [0, 1]$;
Duplex valuations: $v_i(j) \in \{-1, 0, 1\}$;
Simple valuations: $v_i(j) \in \{0, 1\}$.

In every case, we assume that $v_i(i) = 0$, for every $i \in N$. Notice that any valuation function can be represented by using values in the range $[-1, 1]$.

Agents are self-interested entities. Thus, they may strategically misreport their valuation functions in order to maximize their utilities. Let $\mathbf{d}$ denote the preferences (valuation functions) declared by all the agents.

A deterministic mechanism $\mathcal{M}$ maps every set (or list) of preferences $\mathbf{d}$ to a set of disjoint coalitions $\mathcal{M}(\mathbf{d}) \in \mathscr{C}$. We denote by $\mathcal{M}^i(\mathbf{d})$ the coalition assigned to agent i by $\mathcal{M}$. The utility of agent i is given by $u_i(\mathcal{M}(\mathbf{d}))$. Let $\mathbf{d}_{-i}$ be the valuation functions declared by all agents except agent i and d_i be a possible declaration of valuation function by i. A deterministic mechanism $\mathcal{M}$ is *strategyproof* if for any $i \in N$, any list of preferences $\mathbf{d}_{-i}$, any v_i and any d_i, it holds that $u_i(\mathcal{M}(\mathbf{d}_{-i}, v_i)) \geq u_i(\mathcal{M}(\mathbf{d}_{-i}, d_i))$. In other words, a strategyproof mechanism prevents any agent i from benefiting by declaring a valuation different from v_i, whatever the other declared valuations are.

A randomized mechanism $\mathcal{M}$ maps every set of agents' preferences $\mathbf{d}$ to a distribution Δ over the set of all the possible outcomes $\mathscr{C}$. The expected utility of agent i is given by $\mathbb{E}[u_i(\mathcal{M}(d))] = \mathbb{E}_{\mathcal{C} \sim \Delta}[u_i(\mathcal{C})]$. A randomized mechanism $\mathcal{M}$ is strategyproof (in expectation) if for any $i \in N$, any preferences $\mathbf{d}_{-i}$, any v_i and any d_i, $\mathbb{E}[u_i(\mathcal{M}(\mathbf{d}_{-i}, v_i))] \geq \mathbb{E}[u_i(\mathcal{M}(\mathbf{d}_{-i}, d_i))]$.

In this paper, we are interested in strategyproof mechanisms that perform well with respect to the goal of maximizing the classical utilitarian social welfare, that is, the sum of the utilities achieved by all the agents. Namely, the social welfare of a given outcome $\mathcal{C}$ is $SW(\mathcal{C}) = \sum_{i \in N} u_i(\mathcal{C})$. We denote by $SW(C) = \sum_{i \in C} u_i(\mathcal{C})$ the overall social welfare achieved by the agents belonging to a given coalition C. We measure the performance of a mechanism by comparing the social welfare it achieves with the optimal one. More precisely, the approximation ratio of a deterministic mechanism $\mathcal{M}$ is defined as $r^{\mathcal{M}} = \sup_{\mathbf{d}} \frac{\text{OPT}(\mathbf{d})}{SW(\mathcal{M}(\mathbf{d}))}$, where $\text{OPT}(\mathbf{d})$ is the social welfare achieved by an optimal set of coalitions in the instance induced by $\mathbf{d}$. For randomized mechanisms, the approximation ratio is computed with respect to the expected social welfare, that is $r^{\mathcal{M}} = \sup_{\mathbf{d}} \frac{\text{OPT}(\mathbf{d})}{\mathbb{E}[SW(\mathcal{M}(\mathbf{d}))]}$.

We say that a deterministic mechanism $\mathcal{M}$ is *acceptable* if it always guarantees a non negative social welfare, i.e., $SW(\mathcal{M}(\mathbf{d})) \geq 0$ for any possible list of preferences $\mathbf{d}$. Similarly, a randomized mechanism $\mathcal{M}$ is *acceptable* if

$\mathbb{E}[SW(\mathcal{M}(\mathbf{d}))] \geq 0$ holds for every $\mathbf{d}$. In the following, we will always implicitly restrict to acceptable mechanisms. In fact, a simple acceptable strategyproof mechanism for all the considered classes of valuations can be trivially obtained by putting every agent in a separate singleton coalition, regardless of all the declared valuations.

Graph representation. ASHGs and FHGs have a very intuitive graph representation. In fact, any instance of the games can be expressed by a weighted directed graph $G = (V, E)$, where nodes in V represent the agents, and arcs or directed edges are associated to non null valuations. Namely, if $v_i(j) \neq 0$, an arc (i, j) is contained in E of weight $w(i, j) = v_i(j)$. As an example, in case of simple valuations, if $(i, j) \notin E$ then $v_i(j) = 0$, while if $(i, j) \in E$ then $w(i, j) = v_i(j) = 1$.

Throughout the paper we will sometimes describe an instance of the considered game by its graph representation. In the following sections, we provide our results for all of the four considered classes of valuation functions.

3   General Valuations

In this section, we consider the setting where agents have general valuations. We are able to prove that there is no randomized strategyproof mechanism with bounded approximation ratio both for ASHGs and FHGs. Clearly, the theorem applies also to deterministic mechanisms, since they are special cases of randomized ones.

Theorem 1. *For general valuation functions, there is no randomized strategyproof acceptable mechanism with bounded approximation ratio both for ASHGs and FHGs.*

Proof. We prove the theorem only for ASHGs. However, the same arguments directly apply also to FHGs.

(a) Instance I_1 (b) Instance I_2

Fig. 1. The lower bound instance for general valuations.

Let $\mathcal{M}$ be a given randomized strategyproof mechanism. Provided that $\mathcal{M}$ is strategyproof, we implicitly assume that the agents' declared preferences $\mathbf{d}$ correspond to the true valuation functions. Let us then consider the instance I_1 depicted in Fig. 1a, and let p be the probability that $\mathcal{M}$ returns an outcome for I_1 where agents 2 and 3 are together in the same coalition. Then, the expected social welfare is $E[SW(\mathcal{M}(\mathbf{d}))] \leq p(\epsilon-0.1)+(1-p)\epsilon = \epsilon-0.1p$, while the optimal solution has social welfare ϵ. Therefore, the randomized mechanism has bounded approximation ratio only if $\epsilon - 0.1p > 0$, that implies $p < 10\epsilon$. Let us now consider the instance I_2 depicted in Fig. 1b, and let q be the probability that mechanism $\mathcal{M}$ returns an outcome where agents 2 and 3 are together in the same coalition. Then the expected social welfare is $E[SW(\mathcal{M}(\mathbf{d}))] \leq 0.9q + (1 - q)\epsilon$. We notice that $\mathcal{M}$ can be strategyproof only if $p \geq q$, otherwise agent 2 could improve her utility by declaring value -1 for agent 3, since in such a case she would get utility $-p\epsilon > -q\epsilon$. The optimal solution of instance I_2 has value 0.9. Thus, the approximation ratio of $\mathcal{M}$ is $\frac{\text{OPT}(\mathbf{d})}{E[SW(\mathcal{M}(\mathbf{d}))]} \geq \frac{0.9}{0.9q + (1-q)\epsilon} \geq \frac{0.9}{0.9q + \epsilon} > \frac{0.9}{10\epsilon}$. As ϵ can be arbitrarily small, we can then conclude that $\mathcal{M}$ has an unbounded approximation ratio. The claim then follows by the arbitrariness of $\mathcal{M}$. $\qquad\square$

4   Non-negative Valuations

In this section, we consider the setting where agents have non-negative valuations. Let us first present a simple optimal mechanism for non-negative valuations in ASHGs.

Mechanism $\mathcal{M}_1$. *Given as input a list of agents' valuations* $\mathbf{d} = \langle d_1, \ldots, d_n \rangle$, *the mechanism outputs the grand coalition, i.e.* $\mathcal{M}(\mathbf{d}) = \{\{1, \ldots, n\}\}$.

It is trivial to see that, in ASHGs with non-negative valuations, the above mechanism $\mathcal{M}_1$ is acceptable, strategyproof, and achieves the optimal social welfare. Therefore, we now focus on FHGs. We are able to show that any deterministic strategyproof mechanism cannot have an approximation better than $\frac{n}{2}$.

Theorem 2. *For FHGs with non-negative valuations, no deterministic strategyproof acceptable mechanism can achieve approximation ratio r, with $r < \frac{n}{2}$.*

Given the above result, it is easy to show that, returning the grand coalition is the best we can do.

Proposition 1. *For FHGs with non-negative valuations, Mechanism $\mathcal{M}_1$ is a deterministic strategyproof acceptable mechanism with approximation ratio $\frac{n}{2}$.*

Proof. As valuations are non-negative and Mechanism $\mathcal{M}_1$ always outputs the grand coalition, the mechanism is clearly acceptable and strategyproof. Let us now focus on its approximation ratio for the social welfare. Notice that, given any $\mathbf{d}$, then $\text{OPT}(\mathbf{d}) \leq \frac{\sum_{i \in N} \sum_{j \in N} v_i(j)}{2}$. This is because any coalition in the optimal coalitions with positive social welfare consists of at least two agents. Otherwise, the coalition has zero social welfare since $v_i(i) = 0$ for any $i \in N$. On the other hand the grand coalition has social welfare equal to $\frac{\sum_{i \in N} \sum_{j \in N:} v_i(j)}{n}$. The approximation ratio follows. $\qquad\square$

5   Duplex Valuations

In this section, we consider the setting where agents have duplex valuations. We first present deterministic lower bounds for ASHGs and FHGs.

Theorem 3. *For ASHGs with duplex valuations, no deterministic strategyproof acceptable mechanism has approximation ratio less than* $n - 2$.

Proof. Let us consider the instance I_1 depicted in Fig. 2a, where the valuations of the n agents are as follows:

- for $i = 1, \ldots, n - 2$, $v_i(j) = 1$ if $j = n - 1$ and $v_i(j) = 0$ otherwise;
- $v_{n-1}(j) = 1$ if $j = n$ and $v_{n-1}(j) = -1$ otherwise;
- $v_n(j) = -1$ for $j = 1, \ldots, n - 2$ and $v_n(n - 1) = 0$.

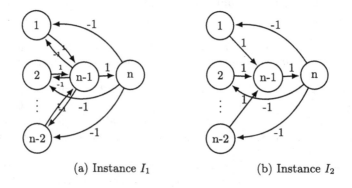

(a) Instance I_1 (b) Instance I_2

Fig. 2. The lower bound instance for duplex valuations.

In the optimal outcome agents $n - 1$ and n are in the same coalition and all other agents are in different coalitions. The resulting social welfare is 1, and in particular it is due to agent $n - 1$ having utility 1. It is easy to see that any mechanism having bounded approximation has to return the optimal outcome, as any other solution would have social welfare at most zero. Let us now consider the other instance I_2 depicted in Fig. 2b, where agent $n - 1$ is the only one with a different valuation function with respect to I_1, that is $v_{n-1}(n) = 1$ and $v_{n-1}(j) = 0$ for $j \neq n$. Any strategyproof mechanism with bounded approximation ratio for I_2 has to put agents $n - 1$ and n in the same coalition, otherwise $n - 1$ would have null utility and could increase her utility by declaring her valuation function as it is in instance I_1. Moreover, any outcome in which $n - 1$ and n are together, independently from the other coalitions, has social welfare 1. However, the optimal outcome, by putting $1, 2, \ldots, n - 1$ all together in a same coalition and agent n alone, achieves social welfare $n - 2$. This proves the $n - 2$ lower bound for any deterministic strategyproof mechanism.   □

Theorem 4. *For FHGs with duplex valuations, no deterministic strategyproof acceptable mechanism can achieve approximation $2 - \epsilon$, for any $\epsilon > 0$.*

Proof. The proof is very similar to Theorem 3, but here the optimal solution has value $\frac{n-2}{n-1}$ and the best strategyproof acceptable mechanism returns an outcome of social welfare $\frac{1}{2}$. It follows that for big value of n, the ratio tends to 2, and thus proving the theorem.   □

We are also able to prove the following randomized lower bound.

Theorem 5. *For ASHGs with duplex valuations, no randomized strategyproof acceptable mechanism can achieve approximation $2 - \epsilon$, for any $\epsilon > 0$.*

We now present a deterministic strategyproof acceptable mechanism $\mathcal{M}_2$ with approximation $O(n^2)$ for ASHGs and $O(n)$ for FHGs. We doubt the existence of deterministic strategyproof acceptable mechanisms with approximation ratio $O(n)$ for ASHGs and $O(1)$ for FHGs. We provide some discussion supporting it, at the end of the section. Closing the gap for duplex valuations, is one of the main open problems.

The following definition is crucial for Mechanism $\mathcal{M}_2$.

Definition 1. *Given $\mathbf{d} = \langle d_1, \ldots, d_n \rangle$ declared by the set of agents N, we say that an agent $i \in N$ is a sink if there is no agent $j \in N$ such that $d_i(j) = 1$ and $d_j(i) \neq -1$.*

The idea of the mechanism is as follows. It considers the agents in an arbitrary ordering. If the considered agent i has value 1 for some other agent j, such that j also has value 1 for i, or j is a sink, or j is before i in the ordering, then it returns agents i and j together in a coalition, and each other agent in a coalition alone. If, after considering all the agents, the mechanism does not create the coalition with two agents, then returns each agent in a coalition alone. It follows the formal description of the mechanism $\mathcal{M}_2$.

Mechanism $\mathcal{M}_2$. *Given any declared valuation $\mathbf{d} = \langle d_1, \ldots, d_n \rangle$, the mechanism performs as follows:*

1. *Consider any ordering of the agents and, for the sake of simplicity, let i be the i-th agent in such ordering.*
2. *For $i = 1$ to n:*
   a. *If there exists $j \in N$ such that $d_i(j) = 1 \wedge d_j(i) = 1$: put agents i and j together into a coalition and any other agent alone, and terminate.*
   b. *If there exists $j \in N$ such that $d_i(j) = 1 \wedge d_j(i) = 0 \wedge j$ is a sink: put agents i and j together into a coalition and any other agent alone, and terminate.*
   c. *If there exists $j \in N$ such that $d_i(j) = 1 \wedge d_j(i) = 0 \wedge j < i$: put agents i and j together into a coalition and any other agent alone, and terminate.*
3. *If no coalition of two agents has been created during the step 2: return each agent in a coalition on its own.*

Theorem 6. *For ASHGs and FHGs with duplex valuations, Mechanism $\mathcal{M}_2$ is a deterministic strategyproof acceptable mechanism. The approximation ratio is $O(n^2)$ for ASHGs with duplex valuations, and $O(n)$ for FHGs with duplex valuations.*

Proof. The mechanism $\mathcal{M}_2$ returns at most one coalition composed by two agents and all the other coalitions are composed by one agent alone. Moreover, no agent i is put together with another agent j in the same coalition if there is a value of -1 between them, that is if $d_i(j) = -1$ or $d_j(i) = -1$. This implies that no agent gets negative utility in the outcome returned by $\mathcal{M}_2$, i.e., $\mathcal{M}_2$ is acceptable. More specifically, if a coalition of two agents is created, then such a coalition has positive (i.e., strictly greater than zero) social welfare. In particular, in ASHGs every agent gets utility 1 or zero, while in FHGs $\frac{1}{2}$ or zero. Furthermore notice that, given the valuations declared by agents, if all the agents are sinks, then the optimal solution has social welfare zero and also $\mathcal{M}_2$ returns the outcome where each agent is in a coalition alone. On the other hand, if there is at least one agent that is not a sink, then it is not difficult to see that the optimal solution has positive social welfare. We now prove that, in such a case $\mathcal{M}_2$ would return a coalition with two agents together with positive social welfare.

Lemma 1. *Given the valuations $\mathbf{d} = \langle d_1, \ldots, d_n \rangle$ declared by agents, if there exists an agent i that is not a sink, then Mechanism $\mathcal{M}_2$ returns an outcome where two agents are put together in the same coalition, thus yielding positive social welfare.*

We are now ready to show that Mechanism $\mathcal{M}_2$ is strategyproof. The following argument is valid for both ASHGs and FHGs. The proof relies on the analysis of different cases.

Assume an agent i gets positive (i.e., greater than zero) utility when she declares her valuations truthfully. Then, agent i cannot improve her utility by declaring valuations $d_i \neq v_i$. In fact getting positive utility, that is utility 1 or $\frac{1}{2}$ depending on whether we consider ASHGs or FHGs respectively, is the best she can obtain.

Assume now that an agent i gets utility zero when she declares her valuations truthfully. We show that agent i cannot improve her utility by declaring valuations $d_i \neq v_i$. If the agent i is a sink then she has no incentive to lie. In fact, in this case i would get positive utility only if she is put together with an agent j such that $d_i(j) = 1$ and $d_j(i) = -1$. However the outcome returned by Mechanism $\mathcal{M}_2$ is such that no agent gets negative utility. Moreover i has no incentive to declare a value of 1 for some agent (in order to become not a sink anymore) if the real value is indeed different than 1. It remains to consider the case where the agent i is not a sink. By Lemma 1 we know that in this case our mechanism always returns a coalition of two agents. Let us first suppose that such coalition, that we call $C_{j,z}$, is formed by agents j and z together. If i has not been considered by the mechanism, that is, for instance the coalition $C_{j,z}$ has been created while considering agent j that appears before i in the ordering, then there is nothing that agent i can do in order to get positive utility. Indeed the only thing that i could do is (mis)-declaring $d_i(j) = 1$ (if we suppose that $v_i(j) \neq 1$). In such a case, if also $d_j(i) = 1$, the mechanism could return the coalition with i and j together. However agent i would still not get positive utility. If i has been considered by the mechanism but has not been put in a coalition together with another agent, then it means that while $\mathcal{M}_2$ was considering agent i, for any j such that $d_i(j) = 1$, j is not a sink and j was not considered by the mechanism yet. We notice that j has no incentive to declare a value of 1 for some agents z if the real value is not 1 (i.e., $v_j(z) \neq 1$). Still there is nothing that i can do.

Let us finally suppose that the coalition of two agents returned by $\mathcal{M}_2$ contains agent i (but still i gets utility zero). This is only possible if, while mechanism $\mathcal{M}_2$ was considering agent i, it was not able to put i together with another agent and (for the same reasons as in the previous case), there is nothing that agent i can do to change it. In fact, agent i could be put together with another agent j, that appears after i in the ordering, when the mechanism considers j. In this case the mechanism could put i together with j only if $d_j(i) = 1$. However it must be that $d_i(j) \neq 1$, otherwise the mechanism would have put i and j together while considering i, and therefore agent i still does not get positive utility.

We now show the approximation ratio of the mechanism. If the optimal solution has social welfare zero, then also our mechanism returns an outcome (i.e., all the agents alone) with social welfare zero. If the optimal solution has positive social welfare (and thus there exists an agent that is not a sink), then by Lemma 1, we know that our mechanism returns an outcome with social welfare at least 1 for ASHGs, and at least $\frac{1}{2}$ for FHGs. The theorem follows by noticing that, any agent can get utility at most $n-1$ for ASHGs and at most 1 for FHGs.
□

We point out that, if we consider ASHGs, there exists an instance and an ordering of the agents for that instance, such that the optimal solution has value order of n^2, while $\mathcal{M}_2$ puts two agents in a coalition in the last iteration of the loop *For*. Thus, even if $\mathcal{M}_2$ does not terminate after putting two agents in a coalition, the analysis cannot be improved. Clearly, $\mathcal{M}_2$ could perform more loops *For* in order to match more than one pair of agents. However, in such a case we can show that the mechanism is not strategyproof anymore. In fact, consider a cycle of 4 nodes with arcs $\{(1,2), (2,3), (3,4), (4,1)\}$, and all the weights 1. The ordering is $1, 2, 3, 4$. If the mechanism iterates the loop *For*, it would return in the first iteration agents $\{4,1\}$ in a coalition, and then, in a second iteration of the *For*, agents $\{2,3\}$ together. Notice that agent 1 has utility zero. However, agent 1 can improve her utility by declaring a further arc of weight -1 to agent 4. In fact, in this case, in the first iteration the mechanism would put agents $\{3,4\}$ together, and then, in the second one, agents $\{1,2\}$.

6 Simple Valuations

Exactly as in the case of non-negative valuations, for ASHGs with simple valuations, Mechanism $\mathcal{M}_1$ is acceptable and strategyproof and it also achieves the optimal social welfare. Therefore, we focus on FHGs. We first prove that any deterministic strategyproof mechanism cannot approximate better than $\frac{6}{5}$ the social welfare.

Theorem 7. *For FHGs with simple valuations, no deterministic strategyproof acceptable mechanism has approximation ratio less than $\frac{6}{5}$.*

Proof. Let us consider the instance I_1 depicted in Fig. 3a. The reader can easily check (by considering all the possible coalitions) that an optimal solution has social welfare $\frac{5}{3}$. It is composed by the three coalitions where, two of them contain two consecutive agents, and the remaining one contains three consecutive agents. For instance, an optimal solution could be $C_1 = \{1, 2\}, C_2 = \{3, 4\}, C_3 = \{5, 6, 7\}$. Notice that the grand coalition has social welfare 1. Therefore, a mechanism achieving an approximation better than $\frac{5}{3}$, has to return more than one coalition. In such a solution there always exists at least one agent, say agent k, having utility zero. Let us now consider the instance I_2 depicted in Fig. 3b, where without loss of generality we suppose that $k = 2$. Again the reader can easily check (by considering all the possible coalitions) that an optimal solution has social welfare 2. Such optimal solution is $C_1 = \{2, 3, 4\}, C_2 = \{5, 6\}, C_3 = \{1, 7\}$. Once again the reader can check that any solution where agents 2 and 3 are not in the same coalition, (i.e., any solution where agent 2 has utility equal to 0 in the instance I_1) can achieve a social welfare of at most $\frac{5}{3}$, and therefore an approximation not better than $\frac{6}{5}$. We conclude that any mechanism achieving an approximation ratio strictly better than $\frac{6}{5}$, in both instances I_1 and I_2, is not strategyproof. □

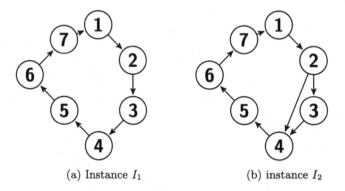

(a) Instance I_1 (b) instance I_2

Fig. 3. The lower bound instance for simple valuations.

We now show a deterministic strategyproof acceptable mechanism with nearly optimal social welfare. Given the preferences declared by the agents $\mathbf{d} = d_1, \ldots, d_n$, and the associated directed graph representation $G = (V, E)$ (notice that since we are considering simple valuations, d_i represents (indeed is) the set of arcs outgoing from node i in G), we construct an undirected weighted graph $\bar{G} = (\bar{V}, \bar{E})$, where $\bar{V} = V$. There is an (undirected) edge $\{i, j\} \in \bar{E}$, if

$(i,j) \in E$ or $(j,i) \in E$. Finally, for each $\{i,j\} \in \bar{E}$, we have that the weight $w(i,j) = 1$ if either $(i,j) \in E$ or $(j,i) \in E$, and $w(i,j) = 2$ if both $(i,j) \in E$ and $(j,i) \in E$ (otherwise $w(i,j) = 0$, i.e., $\{i,j\} \notin \bar{E}$). A matching m of $\bar{G}$ naturally induces an outcome for fractional hedonic games, that is, any edge $\{i,j\} \in m$ induces the coalition $C_{i,j} = \{i,j\}$, and for any node i not matched in m we have the coalition $C_i = \{i\}$. Notice that the coalitions induced by the matching are such that each agent can have utility either $\frac{1}{2}$ or 0. It is possible to show that, finding the maximum weighted matching of $\bar{G} = (\bar{N}, \bar{E})$, using a consistent tie-breaking rule, gives a strategyproof mechanism.

Lemma 2. *Fix a total order $\prec$ on matchings in the complete graph induced by all the agents. Let $\mathcal{M}$ be the mechanism that, given the input $\mathbf{d} = \langle d_1, \ldots, d_n \rangle$, finds the $\prec$-minimal matching m on $\bar{G} = (\bar{V}, \bar{E})$, such that $\sum_{\{i,j\} \in m} w(i,j)$ is maximized. Then $\mathcal{M}$ is strategyproof.*

The proof uses a similar idea to the one proposed in [18], which also shows that $\prec$-minimal matching can be found in poly-time.

Now we prove the approximation ratio of the mechanism. Given an undirected graph $G = (V, E)$, where w is the edges weight function, we denote by $w(E)$ the sum of the weights of the edges belonging to E, i.e., $w(E) = \sum_{\{i,j\} \in E} w(i,j)$.

Theorem 8. *The deterministic mechanism outputting the maximum matching as described in Lemma 2 is strategyproof and acceptable with approximation ratio of 2.*

Proof. Let m be the matching computed by the mechanism and $\mathcal{C}^m$ be the coalitions induced by m. Let $\mathcal{C}^* = \{C_1^*, \ldots, C_p^*\}$ be optimal coalitions (we do not consider optimal coalitions having social welfare equal to zero, indeed we can ignore them). Let $m' = m_1' \cup \ldots \cup m_p'$ where m_h', $1 \le h \le p$, is a maximum matching in the graph induced by the vertices of C_h^*. Let $\mathcal{C}^{m'}$ be the coalitions induced by m'. Let A_h be the vertices matched in m_h' and $B_h = C_h^* \setminus A_h$. Notice that B_h is a stable set and that $|A_h|$ is an even number.

Proposition 2. *When $|B_h| > 0$, then for any $h = 1, \ldots, p$, and any edge $\{i,j\} \in m_h'$, we have that $\sum_{b \in B_h} w(i,b) + w(j,b) \le w(i,j)(|B_h| + 1)$.*

Let $\hat{E}_h$ be the set of edges of the graph induced by the vertices of A_h minus the edges belonging to the matching m_h'. Moreover, let $w(\hat{E}_h) = \sum_{\{i,j\} \in \hat{E}_h} w(i,j)$.

Proposition 3. *For any $h = 1, \ldots, p$, then $w(\hat{E}_h) \le w(m_h')(|A_h| - 2)$.*

Then, when $|B_h| > 0$, by using Propositions 2 and 3 we can bound the social welfare of C_h^*, for any $h = 1, \ldots, p$, as follows:

$$SW(C_h^*) = \frac{1}{|C_h^*|} \Big[\sum_{\{i,j\} \in m_h'} (w(i,j) + \sum_{b \in B_h} w(i,b) + w(j,b)) + w(\hat{E}_h) \Big]$$

$$\le \frac{1}{|C_h^*|} [w(m_h') + w(m_h')(|B_h| + 1) + w(m_h')(|A_h| - 2)]$$

$$= w(m_h').$$

Similarly, when $|B_h| = 0$ we can get that $SW(C_h^*) \leq w(m_h')$. Therefore, overall we have that $SW(C^*) \leq w(m')$. Since it is easy to see that $w(m) \geq w(m')$, then we have that the social welfare of $\mathcal{C}^m$ is

$$SW(\mathcal{C}^m) = \frac{w(m)}{2} \geq \frac{w(m')}{2} \geq \frac{SW(C^*)}{2}.$$

□

We point out that, when dealing with FHGs, it is natural to resort on matchings. Many papers (for instance [2,4,9,10]) used them. The challenge is how to exploit their properties, and in this sense we make some steps forward. Indeed, we better exploit properties of maximum weighted matchings. This is proved by the fact that, our analysis can be used to improve from the 4-approximation (Theorem 7 of the paper [4]) of maximum weighted matching for symmetric valuations, i.e., undirected graph, to a 2-approximation (see Sect. 7). Another remark is that, our results are not only working for the approximation of asymmetric FHGs, i.e., directed graphs, but also include the strategyproofness, which was not considered before for FHGs.

We finally notice that, 2 is the best approximation achievable by using matchings, when dealing with the problem of computing the maximum social welfare in symmetric fractional hedonic games. In fact, consider a complete graph of n nodes. In the grand coalition, each node has utility $\frac{n-1}{n}$ (consider big n), while in a matching, each node has utility at most $\frac{1}{2}$.

7 Extension

In this section, we do not consider strategyproof mechanisms. Here, we want to emphasize that, by using techniques very similar to Theorem 8, it is possible to prove that, given a weighted and undirected graph inducing an instance of the fractional hedonic game with additive symmetric valuation function (in the symmetric game, for any couple of agents i and j, it holds that $v_i(j) = v_j(i)$), the outcome induced by a maximum weighted matching is a 2-approximation of the maximum social welfare. This improves the result in [4], where they prove a bound of 4.

Theorem 9. *For the symmetric fractional hedonic games, the coalitions induced by the maximum weighted matching (that can be computed in polynomial time) is a 2-approximation of the maximum social welfare.*

We finally notice that, 2 is the best approximation achievable by using matchings, when dealing with the problem of computing the maximum social welfare in symmetric fractional hedonic games. In fact, consider a complete graph of n nodes. In the grand coalition, each node has utility $\frac{n-1}{n}$ (consider big n), while in a matching, each node has utility at most $\frac{1}{2}$.

8   Conclusion and Future Work

In this paper, we studied strategyproof mechanisms for ASHGs and FHGs, under general and specific additive valuation functions. Despite the theoretical interest for specific valuations, for which we were able to show better bounds with respect to generic valuations, specific valuations also model realistic scenarios.

Our paper leaves some appealing open problems. First of all, it would be nice to close the gaps of Table 1, and in particular the gap of deterministic strategyproof mechanisms for duplex valuations. Moreover, it is worth to understand whether randomized strategyproof mechanisms can achieve significantly better performance than deterministic ones. It would be also important to understand what happens when valuations are drawn at random from some distribution (in order to avoid the bad instances), or when there are size constraints to the coalitions. Finally, another research direction, is that of considering more general valuation functions than additive ones.

References

1. Aziz, H., Brandt, F., Harrenstein, P.: Pareto optimality in coalition formation. Games Econ. Behav. **82**, 562–581 (2013)
2. Aziz, H., Brandt, F., Harrenstein, P.: Fractional hedonic games. In: Proceedings of the 13th International Conference on Autonomous Agents and Multiagent Systems (AAMAS), pp. 5–12 (2014)
3. Aziz, H., Brandt, F., Seedig, H.G.: Stable partitions in additively separable hedonic games. In: Proceedings of the Tenth International Conference on Autonomous Agents and Multiagent Systems (AAMAS), pp. 183–190 (2011)
4. Aziz, H., Gaspers, S., Gudmundsson, J., Mestre, J., Täubig, H.: Welfare maximization in fractional hedonic games. In: Proceedings of the Twenty-Fourth International Joint Conference on Artificial Intelligence (IJCAI), pp. 461–467 (2015)
5. Aziz, H., Savani, R.: Hedonic games (Chap. 15). In: Computational Social Choice. Cambridge University Press, Cambridge (2016)
6. Bachrach, Y., Kohli, P., Kolmogorov, V., Zadimoghaddam, M.: Optimal coalition structure generation in cooperative graph games. In: Proceedings of the Twenty-Seventh AAAI Conference on Artificial Intelligence (AAAI), pp. 81–87 (2013)
7. Banerjee, S., Konishi, H., Sönmez, T.: Core in a simple coalition formation game. Soc. Choice Welf. **18**, 135–153 (2001)
8. Bansal, N., Blum, A., Chawla, S.: Correlation clustering. Mach. Learn. **56**(1–3), 89–113 (2004)
9. Bilò, V., Fanelli, A., Flammini, M., Monaco, G., Moscardelli, L.: Nash stability in fractional hedonic games. In: Liu, T.-Y., Qi, Q., Ye, Y. (eds.) WINE 2014. LNCS, vol. 8877, pp. 486–491. Springer, Cham (2014). https://doi.org/10.1007/978-3-319-13129-0_44
10. Bilò, V., Fanelli, A., Flammini, M., Monaco, G., Moscardelli, L.: On the price of stability of fractional hedonic games. In: Proceedings of the 14th International Conference on Autonomous Agents and Multi-Agent Systems (AAMAS), pp. 1239–1247 (2015)
11. Bloch, F., Diamantoudi, E.: Noncooperative formation of coalitions in hedonic games. Int. J. Game Theory **40**, 262–280 (2010)
12. Bogomolnaia, A., Jackson, M.O.: The stability of hedonic coalition structures. Games Econ. Behav. **38**, 201–230 (2002)

13. Brandl, F., Brandt, F., Strobel, M.: Fractional hedonic games: individual and group stability. In: Proceedings of the 14th International Conference on Autonomous Agents and Multiagent Systems, (AAMAS), pp. 1219–1227 (2015)
14. Charikar, M., Guruswami, V., Wirth, A.: Clustering with qualitative information. J. Comput. Syst. Sci. **71**, 360–383 (2005)
15. Demaine, E.D., Emanuel, D., Fiat, A., Immorlica, N.: Correlation clustering in general weighted graphs. Theoret. Comput. Sci. **361**, 172–187 (2006)
16. Deng, X., Papadimitriou, C.H.: On the complexity of cooperative solution concepts. Math. Oper. Res. **19**(2), 257–266 (1994)
17. Dréze, J.H., Greenberg, J.: Hedonic coalitions: optimality and stability. Econometrica **48**, 987–1003 (1980)
18. Dughmi, S., Ghosh, A.: Truthful assignment without money. In: Proceedings of the 11th ACM Conference on Electronic Commerce (EC), pp. 325–334 (2010)
19. Elkind, E., Fanelli, A., Flammini, M.: Price of Pareto optimality in hedonic games. In: Proceedings of The Thirtieth AAAI Conference on Artificial Intelligence (AAAI), pp. 475–481 (2016)
20. Elkind, E., Wooldridge, M.: Hedonic coalition nets. In: Proceedings of the Eighth International Conference on Autonomous Agents and Multiagent Systems (AAMAS), pp. 417–424 (2009)
21. Feldman, M., Lewin-Eytan, L., Naor, J.S.: Hedonic clustering games. ACM Trans. Parallel Comput. **44**, 1384–1402 (2015)
22. Gairing, M., Savani, R.: Computing stable outcomes in hedonic games. In: Kontogiannis, S., Koutsoupias, E., Spirakis, P.G. (eds.) SAGT 2010. LNCS, vol. 6386, pp. 174–185. Springer, Heidelberg (2010). https://doi.org/10.1007/978-3-642-16170-4_16
23. Kaklamanis, C., Kanellopoulos, P., Papaioannou, K.: The price of stability of simple symmetric fractional hedonic games. In: Gairing, M., Savani, R. (eds.) SAGT 2016. LNCS, vol. 9928, pp. 220–232. Springer, Heidelberg (2016). https://doi.org/10.1007/978-3-662-53354-3_18
24. Nisan, N., Roughgarden, T., Tardos, E., Vazirani, V.: Algorithmic Game Theory. Cambridge University Press, Cambridge (2007)
25. Peters, D.: Graphical hedonic games of bounded treewidth. In: Proceedings of the Thirtieth Conference on Artificial Intelligence (AAAI), pp. 586–593 (2016)
26. Peters, D., Elkind, E.: Simple causes of complexity in hedonic games. In: Proceedings of the Twenty-Fourth International Joint Conference on Artificial Intelligence (IJCAI), pp. 25–31 (2015)
27. Procaccia, A.D., Tennenholtz, M.: Approximate mechanism design without money. In: Proceedings of the 10th ACM Conference on Electronic Commerce (EC), pp. 177–186 (2009)
28. Rodríguez-Álvarez, C.: Strategy-proof coalition formation. Int. J. Game Theory **38**(3), 431–452 (2009)
29. Vallée, T., Bonnet, G., Zanuttini, B., Bourdon, F.: A study of sybil manipulations in hedonic games. In: Proceedings of the Thirteenth International Conference on Autonomous Agents and Multi-Agent Systems (AAMAS), pp. 21–28 (2014)
30. Woeginger, G.J.: Core stability in hedonic coalition formation. In: van Emde Boas, P., Groen, F.C.A., Italiano, G.F., Nawrocki, J., Sack, H. (eds.) SOFSEM 2013. LNCS, vol. 7741, pp. 33–50. Springer, Heidelberg (2013). https://doi.org/10.1007/978-3-642-35843-2_4
31. Wright, M., Vorobeychik, Y.: Mechanism design for team formation. In: Proceedings of the Twenty-Ninth Conference on Artificial Intelligence (AAAI), pp. 1050–1056 (2015)

The Asymptotic Price of Anarchy
for *k*-uniform Congestion Games

Jasper de Jong, Walter Kern, Berend Steenhuisen, and Marc Uetz$^{(\boxtimes)}$

Faculty of Electrical Engineering, Mathematics and Computer Science,
University of Twente, P.O. Box 217, 7500, AE Enschede, The Netherlands
{j.dejong-3,w.kern,m.uetz}@utwente.nl,
b.a.steenhuisen@student.utwente.nl

Abstract. We consider the atomic version of congestion games with affine cost functions, and analyze the quality of worst case Nash equilibria when the strategy spaces of the players are the set of bases of a *k*-uniform matroid. In this setting, for some parameter *k*, each player is to choose *k* out of a finite set of resources, and the cost of a player for choosing a resource depends affine linearly on the number of players choosing the same resource. Earlier work shows that the price of anarchy for this class of games is larger than 1.34 but at most 2.15. We determine a tight bound on the asymptotic price of anarchy equal to ≈1.35188. Here, asymptotic refers to the fact that the bound holds for all instances with sufficiently many players. In particular, the asymptotic price of anarchy is bounded away from 4/3. Our analysis also yields an upper bound on the price of anarchy <1.4131, for all instances.

Keywords: Congestion games · Uniform matroid
Asymptotic price of anarchy

1 Introduction

The effect of selfish behaviour on the overall system performance is a fundamental problem in the analysis of traffic networks, and more generally congestion games since the early works of Pigou [13], Wardrop [17] and Braess [4]. The problem also lies at the foundation of algorithmic game theory via Roughgarden and Tardos's analysis of equilibria in network routing games [16].

In network routing games, players have routing demands in a network and interact with each other by sharing network links with load-dependent costs. A Wardrop (or Nash) equilibrium is a routing of the demands such that no player can decrease her cost by unilaterally deviating. It is well known that equilibrium solutions may cause higher total costs than the system optimum, defined as the solution that minimizes the total cost of all players. If the cost functions are affine, Pigou's simple example shows that equilibrium solutions can exceed the cost of the system optimum by a factor as large as 4/3 [13]. Pigou's example is surprisingly simple, as it consists of only two players, each of which can choose

© Springer International Publishing AG, part of Springer Nature 2018
R. Solis-Oba and R. Fleischer (Eds.): WAOA 2017, LNCS 10787, pp. 317–328, 2018.
https://doi.org/10.1007/978-3-319-89441-6_23

one of two parallel links, with constant and linear cost functions, respectively. Roughgarden and Tardos then showed that the relative gap can be at most 4/3 in the worst case, for any network routing game with affine costs. The ratio between the total cost of a worst equilibrium solution and the minimum total cost has been named the *price of anarchy* (PoA) by Koutsoupias and Papadimitriou [11]. As to the PoA in network routing games, Roughgarden [15] also showed that the PoA is independent of the network topology for network routing games, as it is always attained on simple, Pigou-style networks.

In Wardrop's model, as well as in [15,16], the demand of any player may be distributed arbitrarily on minimum cost paths. A discrete version of Wardrop's model is that of a network routing game with *unsplittable* demands, also referred to as *atomic* network routing games. It has been introduced in its most general form by Rosenthal [14], who proved the existence of pure strategy Nash equilibria via potential functions, now often referred to as Rosenthal potential functions.

This type of *atomic congestion games with affine cost functions* are addressed in this paper. There is a finite set of players, a finite set of resources with affine cost functions, and a strategy of each player is to choose a set of resources from a given collection of subsets of the resources feasible for that player. In network routing games, this strategy space of a player is the set of all paths from a players' origin to destination, but in general it could be any set system specific to that player. When the strategy spaces of all players are identical, this is referred to as *symmetric* congestion games. Without any further restrictions on the strategy spaces of the players, the price of anarchy for affine cost functions is larger than 4/3, namely 5/2, as shown by Christodoulou and Koutsoupias [6] and Awerbuch et al. [1]. This price of anarchy is already attained for a network routing game with only three players, yet asymmetric strategy spaces. Correa et al. [7] eventually gave a class of instances of a *symmetric*, affine network routing game with price of anarchy asymptotically equal to 5/2 (when the number of players tends to infinity).

In this context, an obvious question to ask is if restrictions on the strategy spaces of the players allows a substantial improvement of the price of anarchy bound of 5/2. That this is generally possible is well known. For example, for instances where players choose only a singleton resource, and with symmetric strategy spaces, Lücking et al. [12] and Fotakis [10] showed that the price of anarchy is only 4/3. However, if players choose a singleton resource but the strategy spaces are no longer symmetric, there is again an asymptotic lower bound 5/2 (when the number of players tends to infinity) [5].

Our Contribution. The results discussed so far imply in particular that for atomic congestion games with affine cost functions, the price of anarchy *does* depend on the combinatorial structure of players' strategy spaces. Moreover, in light of the lower bound of [5] for asymmetric, singleton congestion games, there is room for improvement only for symmetric strategy spaces. Therefore, in [8] the case of k-uniform matroid congestion games with affine cost functions was studied. Here, each player is allowed to choose any of the k-element subsets of resources. They were able to show that the price of anarchy lies strictly in between 1.34 and 2.15. The present paper continues this line of research, and asks what the price of anarchy is if the number of players grows large. We answer this question, and prove that the asymptotic price of anarchy for k-uniform, affine congestion games equals ≈ 1.35188. The interpretation of our asymptotic PoA bound is that it holds for any k and any instance with a sufficient number of players. Tightness of our analysis follows by a matching lower bound, obtained through a parametric set of instances for which the price of anarchy exactly matches this bound. Our analysis also allows us to obtain an upper bound on the price of anarchy for *any* instance with any finite number of players. The bound we obtain here is slightly larger, yet <1.4131. That means that a finite number of instances may exist with a price of anarchy larger than 1.35189, but never as large as 1.4131.

Note that the idea to study the asymptotic price of anarchy is not new. For instance, recent work on nonatomic network routing games considers the limit that the total demand grows large, referred to as a "high congestion regime" [3]. They show for example in the parallel link setting and under some regularity conditions on the cost functions, that the price of anarchy converges to 1. See also [2] for further generalizations of these results. Another viewpoint on the asymptotics of the price of anarchy was taken by Feldman et al. [9]. Their work implies that for affine network routing games, the price of anarchy of atomic games converges to the $4/3$ bound of nonatomic games when the number of players grows large. Our result shows that k-uniform strategy spaces do not share this $4/3$ limit.

The main technical ingredient to obtain our results on the asymptotic price of anarchy is to interpret the solutions to k-uniform congestion games as k-matchings, and by observing that the symmetric difference between equilibrium and optimum solutions is the union of disjoint alternating paths. The bound is then obtained by carefully analysing the cost differences that these paths represent. We are convinced that this new way of analysing the problem might prove to be valuable also for other, more general problems.

2 Preliminaries and Notation

An instance of the *affine k-uniform congestion game* is given by a parameter $k \in \mathbb{N}$, a complete bipartite graph $G = (N \cup R, E)$ and non-negative affine cost functions (*i.e.*, functions of type $ax + b$ with $a, b \in \mathbb{R}_+$) c_r for each $r \in R$. We interpret N as a set of players and R as a set of resources. Each player $i \in N$ has to pick a prescribed number k of resources in R. We refer to $|N| + |R|$ as the *size* of the instance.

It is natural to model this situation in the context of matchings. Each player is to choose a set of k resources $r \in R$ or, equivalently, a set of k incident edges. Slightly misusing the standard notation, let us define a *k-matching* to be a subset $M \subseteq E$ whose degree vector $d = d^M$ satisfies $d_i = k$ for all $i \in N$. Each such k-matching induces corresponding costs $c_r^M = c_r(d_r)$, $r \in R$. Player $i \in N$ experiences corresponding cost $C_i = C_i^M := \sum_{ir \in M} c_r^M$ and we define the *total* cost to be $c^M := \sum_i C_i^M = \sum_R d_r c_r(d_r)$. Let $OPT \subseteq E$ denote a k-matching that minimizes the total cost. Sometimes it is also convenient to work with *edge costs* defined by $c_{ir} = c_{ir}^M := c_r^M$. Thus C_i then equals the total cost of the edges in M that are incident to i.

Given a k-matching $M \subseteq E$ and a player $i \in N$, we let $M(i) \subseteq R$ denote the set of resources assigned to i. Similarly, for $r \in R$ we let $M(r) \subseteq N$ denote the set of players that use resource r. A k-matching $M \subseteq E$ is called a *Nash equilibrium* if each player $i \in N$ is satisfied with her set $M(i) \subseteq R$ in the sense that no other choice would strictly decrease her cost C_i (assuming that the remaining sets $M(j)$ stay unchanged). It is straightforward to verify that this condition is satisfied iff C_i cannot be decreased by exchanging a single resource, i.e., switching from $M(i)$ to $M'(i) = M(i) - r + s$ for some $r \in M(i), s \notin M(i)$ (thereby increasing the degree of s to $d_s^M + 1$). In other words, Nash equilibria are characterized by the *Nash condition*: For all $i \in N$,

$$ir \in M, is \notin M \implies c_s(d_s^M + 1) \geq c_r^M(d_r).$$

The costs $c_s(d_s^M + 1)$ that a player experiences when he replaces one of his current resources r by s is also called the *opportunity cost* of s.

In what follows we assume that $NE \subseteq E$ is a Nash equilibrium. To simplify the notation, we let d^* and $\bar{d}$ denote the degree vectors of OPT and NE, resp. To simplify even more, let us denote the *Nash costs* by $\bar{c}_r := c_r(\bar{d}_r)$ and the corresponding opportunity costs by $\bar{c}_r^+ := c_r(\bar{d}_r + 1)$. So the Nash condition simply reads as

$$ir \in NE, is \notin NE \implies \bar{c}_s^+ \geq \bar{c}_r. \tag{1}$$

We seek to bound the PoA $= c^{NE}/c^{OPT}$. Our main result is

Theorem 1. *Any sequence of instances* $(k_t, G_t = (N_t, R_t))$ *with an increasing number of players* $|N_t| \to \infty$ *has* $\limsup c^{NE}/c^{OPT} < 1.35189$.

Section 3 provides some simplifications and "without loss of generality" assumptions. In Sect. 4 we upper bound the *gap* $c^{NE} - c^{OPT}$ and in Sect. 5 we lower bound the *Nash cost* c^{NE}. Together, these two results prove $\limsup c^{NE}/c^{OPT} \leq 1.35188\ldots$. Corresponding tight instances (with $|N|, k \to \infty$) are presented in Sect. 6. We conclude with some remarks and open problems in Sect. 7.

3   Simplifications of Worst-Case Instances

As in the previous section, we consider an instance of the affine k-uniform congestion game of size $m = |N| + |R|$ with cost minimal k-matching $OPT \subseteq E$ and Nash equilibrium $NE \subseteq E$. As before, we denote by d^* and $\bar{d}$ the degree vectors of OPT and NE, resp. Let $\rho = c^{NE}/c^{OPT}$ be the PoA of the instance. We call the instance k-*critical* (or *critical* for short) if no instance of size $< m$ has PoA $\geq \rho$ and no instance of size $= m$ has PoA $> \rho$. Clearly, in order to upper bound the PoA, we may restrict our attention to critical instances. In the following, we will derive some helpful properties of critical instances. Pigou's classic example is obviously a critical instance of size 4. Hence, a critical instance of size ≥ 5 must have PoA $> 4/3$.

OPT and NE induce a partition of R into three sets:

$$O := \{o \in R | \bar{d}_o > d_o^*\} \quad \text{(``overloaded resources'')}$$
$$U := \{u \in R | \bar{d}_u < d_u^*\} \quad \text{(``underloaded resources'')}$$
$$B := \{b \in R | \bar{d}_b = d_b^*\} \quad \text{(``balanced resources'')}$$

The symmetric difference $OPT \oplus NE$ consists of *exclusive Nash edges* $e \in NE \backslash OPT$ and *exclusive OPT edges* $e \in OPT \backslash NE$. The set $OPT \oplus NE$ can be partitioned into a set $\mathcal{P}$ of pairwise edge disjoint alternating paths joining O to U and a set $\mathcal{C}$ of alternating (even) cycles. (Here, "alternating" means that consecutive edges alternate between exclusive OPT and exclusive Nash edges.) We may assume w.l.o.g. that $\mathcal{C} = \emptyset$: If $C \in \mathcal{C}$, then $OPT \oplus C$ is obviously also cost minimal (as the d_r^* remain unchanged). We may continue this process of "switching along alternating cycles" until no one is left.

Another w.l.o.g. assumption is the following

Lemma 1. *We may assume that all cost functions c_r for $r \in O \cup B$ are linear, i.e., $c_r(x) = c_r \cdot x$ for some constant $c_r \geq 0$.*

Proof. Assume to the contrary that $c_r(x) = ax + b$ with $b > 0$. Define $\tilde{c}_r(x) := x \cdot \bar{c}_r / \bar{d}_r$ and observe that $\tilde{c}_r(\bar{d}_r) = \bar{c}_r$ and $\tilde{c}(\bar{d}_r + 1) > \bar{c}_r^+$. This implies that any Nash equilibrium remains Nash w.r.t. the modified cost functions and c^{NE} is unchanged. Furthermore, $\tilde{c}(d_r^*) \leq c_r(d_r^*)$, since $d_r^* \leq \bar{d}_r$. Thus c^{OPT} can only decrease. $\square$

The gap $c^{NE} - c^{OPT}$ can be best understood by observing how the cost increases while we move from OPT to NE by switching along alternating paths $P \in \mathcal{P}$. Let $\mathcal{P}_o \subseteq \mathcal{P}$ denote the set of paths $P \in \mathcal{P}$ that start in $o \in O$. So $|\mathcal{P}_o| = \bar{d}_o - d_o^*$. Switching OPT and NE along all paths in P_o simultaneously has the following effect on the edge costs of $P = (o, i, \ldots, j, u) \in \mathcal{P}_o$: Edge io with cost $\bar{c}_o$ enters and edge ju with cost $c_u \geq \bar{c}_u^+$ leaves the current k-matching. Apart from that, switching OPT and NE along P does not affect any costs of intermediate resources $r \neq o, u$ that P may visit. Thus, when passing from OPT to NE, the edges of P contribute a total increase of $\bar{c}_o - c_u \leq \bar{c}_o - \bar{c}_u^+$. We refer to the latter as the *internal* cost bound of path P. In addition to these $\bar{d}_o - d_o^*$ internal path costs, there are *external* costs experienced by d_o^* edges whose edge costs are raised from $c_o \cdot d_o^*$ to $c_o \cdot \bar{d}_o$, due to the increase in degree of o.

Summarizing, the gap can be bounded by

$$c^{NE} - c^{OPT} \leq \sum_{P=(o,\ldots,u)\in\mathcal{P}} \bar{c}_0 - \bar{c}_u^+ + \sum_O d_o^*(\bar{d}_o - d_o^*)c_o \qquad (2)$$

The external part can be upper bounded by $\frac{1}{4} \sum_O \bar{d}_o^2 c_o \leq \frac{1}{4} c^{NE}$ (the maximum being attained for $d_o^* = \bar{d}_o/2$). Thus, if the internal part is close to zero, we would get a PoA $\approx 4/3$. The difficult part is to estimate the internal cost bounds $\bar{c}_o - \bar{c}_u^+$. To provide some intuition, we observe the following simple fact:

Lemma 2. *Each $P = (o, i, \ldots, j, u) \in \mathcal{P}$ with positive internal cost bound $\bar{c}_o - \bar{c}_u^+ > 0$ has length at least 4.*

Proof. Being an alternating path that starts in a Nash edge io and ends in the *OPT* edge ju, P must have even length. If P had length 2, i.e., $P = (o, i, u)$, with $io \in NE$ and $iu \notin NE$, then $\bar{c}_o - \bar{c}_u^+ > 0$ would contradict the Nash condition (1). □

Thus each path that contributes a positive amount to the gap also "increases" the total Nash cost c^{NE}. (There must be at least a second Nash edge $e \in NE$ on P, apart from the first edge io.) Considerations of this kind in Sects. 4 and 5 will allow us to bound the gap in terms of c^{NE}, i.e., to bound the relative gap $(c^{NE} - c^{OPT})/c^{NE}$, which then yields a corresponding bound for c^{NE}/c^{OPT}. As a little exercise, consider the following

Lemma 3. *The Pigou example of size 4 is (up to scaling) the only critical instance for $k = 1$. Thus $PoA \leq 4/3$ for $k = 1$.*

Proof. Consider any path $P = (o, i, \ldots, u) \in \mathcal{P}$. Then $io \in NE$, implying that $iu \notin NE$ (since $k = 1$). Hence $\bar{c}_o - \bar{c}_u^+ \leq 0$ follows from the Nash property. Then (2) yields $c^{NE} - c^{OPT} \leq \sum_o d_o^*(\bar{d}_o - d_o^*)c_o \leq \frac{1}{4}c^{NE}$, i.e., $c^{NE}/c^{OPT} \leq 4/3$, as in Pigou's example. Criticality then implies that the instance has size 4, and it is then straightforward to show that (up to scaling) only Pigou's example achieves this bound. □

Thus we are left to deal with the case $k \geq 2$. Obviously, for $k \geq 2$, a k-critical instance with PoA $> 4/3$ must have at least size 5 (three resources) by definition of criticality. We end this section with some more properties of such instances, which were already shown in [8].

Lemma 4 ([8]). *For $k \geq 2$, no k-critical instance has $\bar{d}_r = |N|$ for some $r \in R$.*

The following result helps to bound the internal part in Eq. (2). Let $c_O^{max} := \max\{\bar{c}_o \mid o \in O\}$.

Lemma 5 ([8]). *In a critical instance, $\bar{c}_r^+ \geq c_O^{max}/2$ for all $r \in R$. Hence $\bar{c}_o - \bar{c}_u^+ \leq \bar{c}_o/2$ for all $o \in O, u \in U$.*

Note that, as the cost functions are affine, Lemma 5 also implies $\bar{c}_r \geq \frac{1}{2}\bar{c}_r^+ \geq \frac{1}{4}c_O^{max}$ for all $r \in R$.

4   An Upper Bound on the Relative Gap

We order the paths $P = (o, \ldots, u) \in \mathcal{P}$ according to both non-decreasing values of $\bar{c}_o$ as well as non-increasing values of $\bar{c}_u^+$. Let $\Delta := |\mathcal{P}|$ and assume $P_1, \ldots, P_\Delta$ where $P_t = (o_t, \ldots, u_t)$ is an ordering with $\bar{c}_{o_1} \leq \ldots \leq \bar{c}_{o_\Delta}$ and let $P_1', \ldots, P_\Delta'$ with $P_t' = (o_t', \ldots, u_t')$ satisfy $\bar{c}_{u_1'}^+ \geq \ldots \geq \bar{c}_{u_\Delta'}^+$.

In case $\bar{c}_{o_\Delta} \leq \bar{c}^+_{u'_\Delta}$, all paths have internal cost bound $\bar{c}_o - \bar{c}^+_u \leq \bar{c}_{o_\Delta} - \bar{c}^+_{u'_\Delta} \leq 0$, so the PoA is no more than $4/3$ (*cf.* Eq. (2)). So the interesting (critical) case is when $\bar{c}_{o_\Delta} > \bar{c}^+_{u'_\Delta}$. Let $t \geq 0$ be the first index for which $\bar{c}_{o_{t+1}} > \bar{c}^+_{u'_{t+1}}$. Let

$$\mathcal{P}^{-+} := \{P_1, \ldots, P_t\} \cap \{P'_1, \ldots, P'_t\}.$$

Loosely speaking, one might say that $\mathcal{P}^{-+}$ consists of those paths in $\mathcal{P}$ that join some $o \in \{o_1, \ldots, o_t\}$ to some $u \in \{u'_1, \ldots, u'_t\}$. (Here, we slightly misuse the notation, as, *e.g.*, $\{o_1, \ldots, o_t\}$ is actually a multiset whose elements may also appear among $o_{t+1}, \ldots, o_\Delta$.) Thus each $P \in \mathcal{P}^{-+}$ has internal cost bound $\bar{c}_o - \bar{c}^+_u \leq 0$. Similarly, each path in

$$\mathcal{P}^{+-} := \{P_{t+1}, \ldots, P_\Delta\} \cap \{P'_{t+1}, \ldots, P'_\Delta\}$$

has internal cost bound $\bar{c}_o - \bar{c}^+_u > 0$. The remaining paths in

$$\mathcal{P}^{++} := \{P_{t+1}, \ldots, P_\Delta\} \cap \{P'_1, \ldots, P'_t\} \text{ and}$$
$$\mathcal{P}^{--} := \{P_1, \ldots, P_t\} \cap \{P'_{t+1}, \ldots, P'_\Delta\}$$

may have positive or negative internal cost bounds. Figure 1 illustrates the idea behind these definitions.

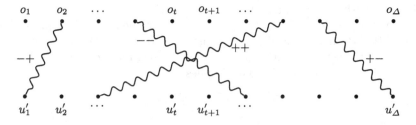

Fig. 1. Possible augmenting paths in $\mathcal{P}^{-+}, \mathcal{P}^{--}, \mathcal{P}^{++}$, and $\mathcal{P}^{+-}$, respectively.

In any case, however, $P = (o, \ldots, u) \in \mathcal{P}^{++}$ has internal cost bound $\bar{c}_o - \bar{c}^+_u \leq \bar{c}_o - \bar{c}^+_{u_t}$ and any $P' = (o', \ldots, u') \in \mathcal{P}^{--}$ has internal cost $\bar{c}_o - \bar{c}^+_{u'} \leq \bar{c}_{o_t} - \bar{c}^+_{u'} \leq \bar{c}^+_{u_t} - \bar{c}^+_{u'}$. Thus the internal costs of such P and P' add to at most $\bar{c}_o - \bar{c}^+_{u'} \leq \bar{c}_o/2$ (*cf.* Lemma 5). Since $|\mathcal{P}^{++}| = |\mathcal{P}^{--}|(= \Delta - t - |\mathcal{P}^{+-}|)$, we may group the paths $\mathcal{P}^{++} \cup \mathcal{P}^{--}$ in pairs $P \in \mathcal{P}^{++}, P' \in \mathcal{P}^{--}$ with total internal cost at most $\bar{c}_o/2$, where o is the starting point of P. Similarly, each (single) path $P = (o, \ldots, u) \in \mathcal{P}^{+-}$ contributes $\bar{c}_o - \bar{c}^+_u \leq \bar{c}_o/2$ to the internal costs. Thus if we let $O^+ \subseteq O$ denote the set of resources that appear among $o_{t+1}, \ldots, o_\Delta$, let $\mathcal{P}^+ := \mathcal{P}^{+-} \cup \mathcal{P}^{++}$, let $\mathcal{P}^+_o$ denote the paths in $\mathcal{P}^+$ that start in o, and write $\Delta^+_o := |\mathcal{P}^+_o|$, the gap is bounded by

$$c^{NE} - c^{OPT} \leq \sum_{O^+} \Delta^+_o \bar{c}_o/2 + \sum_O d^*_o(\bar{d}_o - d^*_o)c_o \qquad (3)$$

For further use we also introduce the corresponding set $U^- \subseteq U$ of resources that appear among $\{u'_{t+1}, \ldots, u'_\Delta\}$ and, for each $u \in U^-$ the set $\mathcal{P}^- := \mathcal{P}^{--} \cup \mathcal{P}^{+-}$, $\mathcal{P}_u^-$ the paths in $\mathcal{P}^-$ that end in u along with the corresponding $\Delta_u^- := |\mathcal{P}_u^-|$. Note that $\sum_{U^-} \Delta_u^- = \sum_{O^+} \Delta_o^+ = \Delta - t$. Furthermore, $\bar{c}_o > \bar{c}_u^+$ holds for all $o \in O^+, u \in U^-$.

Remark. As we will see in Sect. 6, tight examples for the PoA $= 1.35188\ldots$ are obtained with $O = O^+$ (and $U = U^-, \mathcal{P} = \mathcal{P}^{+-}$).

5   Lower Bounding the Cost of a Nash Equilibrium c^{NE}

As mentioned earlier, we seek to bound the relative gap $(c^{NE} - c^{OPT})/c^{NE}$ - which then yields a corresponding upper bound for the PoA. To this end, we have to lower bound c^{NE}, which is the purpose of this section. We first lower bound the number of edges in NE and afterwards deal with their costs.

Let $I^+ := NE(O^+)$ denote the set of players $i \in N$ that are joined to O^+ by a Nash edge. As $\bar{c}_o > \bar{c}_u^+$ for all $o \in O^+, u \in U^-$, the Nash property implies $iu \in NE$ for all $i \in I^+, u \in U^-$. Thus we obtain

$$|I^+| \geq \sum_{O^+} \bar{d}_o/(k - |U^-|), \tag{4}$$

since each $i \in I^+$ can receive at most $k - |U^-|$ Nash edges from O^+ (in addition to its $|U^-|$ Nash edges from U^-).

Let $J := \{j \in N \mid \exists P = (o, \ldots, j, u) \in \mathcal{P}^-\}$ be the set of "last players" on paths in $\mathcal{P}^-$. By definition of $\mathcal{P}^-$, the last edge ju of a path in $\mathcal{P}^-$ joins j to U^-. Hence at most $|U^-|$ paths in $\mathcal{P}^-$ can go through a fixed $j \in J$. There are $\Delta - t = \sum_{O^+} \Delta_o^+$ paths in $\mathcal{P}^-$. This implies

$$|J| \geq \sum_{U^-} \Delta_u^-/|U^-| = \sum_{O^+} \Delta_o^+/|U^-| \tag{5}$$

Next, we observe that

$$I^+ \cap J = \emptyset. \tag{6}$$

Indeed, if $i \in I^+ \cap J$, then $io \in NE$ for some $o \in O^+$ and $iu \in OPT\backslash NE$ for some $u \in U^-$, contradicting the Nash property (as $\bar{c}_o > \bar{c}_u^+$). Hence we can estimate the number of edges (due to (6) without any double counting):

$$|NE| \geq \sum_{O^+} \bar{d}_o + |I^+||U^-| + |J|k. \tag{7}$$

To estimate the corresponding edge costs is a bit more involved: First, observe that if $\sum_{O^+} \bar{d}_o \leq \epsilon \sum_R \bar{d}_r$ for a sufficiently small $\epsilon > 0$ (say, $\epsilon = 0.001$ is certainly sufficient), then we deduce from Eq. (3) and Lemma 5 (implying $\bar{c}_r \geq \frac{1}{4} c_O^{max}$) that

$$c^{NE} - c^{OPT} \leq \epsilon \sum_R \bar{d}_r^2 c_O^{max} + \frac{1}{4} \sum_O \bar{d}_o^2 c_o \leq 4\epsilon \sum_R \bar{d}_r \bar{c}_r + \frac{1}{4} c^{NE}, \tag{8}$$

so the instance has a gap close to $\frac{1}{4}c^{NE}$, corresponding to, say, a PoA ≤ 1.34 - thus an uninteresting (uncritical) instance in view of Sect. 6 below.

Thus, in what follows, we restrict ourselves to instances where $\sum_{O^+} \bar{d}_o \geq \epsilon \sum_R \bar{d}_r = \epsilon k|N|$ for some small $\epsilon > 0$. Since we seek to analyze instances with $|N| \to \infty$, we may thus assume that $(\sum_{O^+} \bar{d}_o)/k \to \infty$. Then also $|I^+| \to \infty$ (cf. Eq. (4)). This further implies $\bar{d}_u \to \infty$ and, consequently (using Lemma 5), $\bar{c}_u \geq c_O^{max}/2$ in the limit for $u \in U^-$. Therefore, in our estimate of c^{NE} below, we let each $u \in U^-$ have $\bar{c}_u = c_O^{max}/2$.

Next we deal with the costs of the $k|J|$ Nash edges incident to J. We intend to show that each of these edges also has Nash cost $\geq c_O^{max}/2$, or, at least, that we can account such an amount for each of these edges. Let $jr \in NE$ be any such edge. Pick any $o \in O^+$ with $\bar{c}_o = c_O^{max}$ and any $i \in I^+$ with $io \in NE$. If $ir \notin NE$, then, by the Nash condition, we have $\bar{c}_r^+ \geq c_O^{max}$, implying $\bar{c}_r \geq c_O^{max}/2$, and we are done. Otherwise, suppose $ir \in NE$. Assume first that jr is the only Nash edge from J to r. Then $\bar{d}_r \geq 2$ (one edge from i, one from j), hence $\bar{c}_r \geq c_O^{max}/3$ (using $\bar{c}_r \geq \frac{2}{3}\bar{c}_r^+$ and Lemma 5). In this case we *discharge* the cost of ir to jr, resulting in an accounted cost of $\frac{2}{3}c_O^{max}$ on the edge jr. In general, if there are $d \geq 1$ edges joining J to r, each of these has already Nash cost $\bar{c}_r \geq \frac{d+1}{d+2}\bar{c}_r^+ \geq \frac{d+1}{d+2}c_O^{max}/2$. (Note that, together with the edge ir, there are at least $d+1$ Nash edges incident to r.) The edge $ir \in NE$ has the same Nash cost $\bar{c}_r \geq \frac{d+1}{d+2}c_O^{max}/2$. Thus if we discharge a fraction $\frac{1}{d}$ of its cost to each of the edges $jr \in NE$, then each of the latter gets an accounted total cost of $\geq (\frac{d+1}{d+2} + \frac{d+1}{d(d+2)})c_O^{max}/2 \geq c_O^{max}/2$.

Accounting $\frac{1}{2}c_O^{max}$ for all edges $jr \in J \times R$ and $iu \in I^+ \times U^-$, we may estimate the relative gap as follows (abbreviating $u^- := |U^-|$):

$$(c^{NE} - c^{OPT})/c^{NE} \leq \frac{\sum_{O^+} \Delta_o^+ \bar{c}_o/2 + \sum_O d_o^*(\bar{d}_o - d_o^*)c_o}{\sum_O \bar{d}_o^2 c_o + (\frac{u^-}{k-u^-}\sum_{O^+} \bar{d}_o + \frac{k}{u^-}\sum_{O^+}\Delta_o^+)c_O^{max}/2}, \quad (9)$$

The three sums in the denominator correspond to the cost of Nash edges in $N \times O$, $I^+ \times U^-$, and $J \times R$, resp. Since $d_o^*(\bar{d}_o - d_o^*) \leq \frac{1}{4}\bar{d}_o^2$, removing the terms corresponding to $o \in O \backslash O^+$ in both the numerator and the denominator in (9) can only increase the right hand side. After replacing O by O^+ in the two sums in (9), we may estimate the fraction by considering the corresponding fractions per $o \in O^+$:

$$(c^{NE} - c^{OPT})/c^{NE} \leq \max_{o \in O^+} \frac{\Delta_o^+ \bar{c}_o/2 + d_o^*(\bar{d}_o - d_o^*)c_o}{\bar{d}_o^2 c_o + (\frac{u^-}{k-u^-}\bar{d}_o + \frac{k}{u^-}\Delta_o^+)c_O^{max}/2}$$

$$(10)$$

$$\leq \max_{o \in O^+} \frac{\Delta_o^+ \bar{d}_o/2 + d_o^*(\bar{d}_o - d_o^*)}{\bar{d}_o^2 + (\frac{u^-}{k-u^-}\bar{d}_o + \frac{k}{u^-}\Delta_o^+)/2}.$$

The latter inequality is obtained by replacing c_O^{max} with the smaller or equal $\bar{d}_o c_o$ and dividing by c_o. Now, fix any $o \in O^+$ where the maximum is attained. Again, $d_o^*(\bar{d}_o - d_o^*) \leq \frac{1}{4}$ implies that the maximum is attained when Δ_o^+ is as large as possible (unless $k/u^- > 4$, in which case the whole fraction is less than $\frac{1}{4}$). Thus we may assume $\Delta_o^+ = \bar{d}_o - d_o^*$. Writing $\bar{d}_o = \beta d_o^*$, $k = \alpha u^-$ (with $\alpha, \beta \geq 1$), the maximum in (10) can be bounded by

$$\mu = \max_{\alpha, \beta} \frac{(\beta - 1)\beta/2 + \beta(\beta - 1)}{\beta^2 + (\beta^2/(\alpha - 1) + (\beta - 1)\beta\alpha)/2} = 0.260292\ldots \tag{11}$$

(according to WolframAlpha, for $\alpha \approx 2.3, \beta \approx 2.5$). The corresponding upper bound for c^{NE}/c^{OPT} is $\rho = 1/(1 - \mu) \leq 1.35188\ldots$. This finishes the proof of Theorem 1. □

6   Tight Lower Bound Construction

We construct examples with PoA $= 1.35188\ldots$ as suggested by the analysis in Sects. 4 and 5. A close look reveals that a tight example must have $O = O^+$ (and, correspondingly, $U = U^-$) and $\bar{d}_o - d_o^*$ many paths in $\mathcal{P}$ starting in $o \in O$. We thus define $N := I \cup J$ and $R = O \cup B \cup U$. The number $|I| = |U| \in \mathbb{N}$ is a free parameter fixing the problem instance. (We thus let $|I| \to \infty$, which also increases the number of players to infinity.) In addition, we have the two parameters $\alpha \approx 2.3, \beta \approx 2.5$ as in Sect. 5. These determine $k := \alpha|U|$ and $\bar{d}_o = \beta d_o^*$.

The cost functions are $c_o(x) = \frac{1}{|U|}x$ for $o \in O$. For $b \in B$ we have $c_b(x) = \frac{1}{2}x$ and for $u \in U$ we have constant costs $c_u(x) = 1/2$.

Each $i \in I$ is joined by Nash edges to all nodes in O and U (and nothing else). Hence $k = \alpha|U|$ implies $|O| = (\alpha - 1)|U|$. For each $i \in I$, all Nash edges to U are also in OPT. In addition, a $\frac{1}{\beta}$ fraction of its Nash edges to O is also in OPT. No other OPT edges are incident to O. Assuming that we distribute the $OPT \cap NE$ edges from I to O in a regular manner, we may thus assume that each $o \in O$ receives d_o^* OPT edges and $|U| = \bar{d}_o = \beta d_o^*$ Nash edges, as required. The Nash cost of $o \in O$ is thus $\bar{c}_o = |U|\frac{1}{|U|} = 1$ and its OPT cost is $c_o^* = \frac{|U|}{\beta}\frac{1}{|U|} = \frac{1}{\beta}$. Thus the total Nash cost of O is $\bar{c}_O = |O|\bar{d}_o\bar{c}_o = (\alpha - 1)|U||U|$, while the OPT cost equals $c_O^* = |O|d_o^*c_o^* = (\alpha - 1)\frac{|U|}{\beta}|U|\frac{1}{\beta} = (\alpha - 1)/\beta^2|U|^2$.

Each $i \in I$ now still misses an amount of $k - \frac{|O|}{\beta} - |U|$ edges in OPT. These are exclusive OPT edges leading to $|I|(k - \frac{|O|}{\beta} - |U|)$ resources $b \in B$, each of degree $\bar{d}_b = d_b^* = 1$. We denote this set of resources b by $B_{\mathcal{P}}$ to indicate that these are the balanced resources on alternating paths. Thus $|B_{\mathcal{P}}| = |I|(k - |O|/\beta - |U|) = |U|(\alpha|U| - (\alpha - 1)/\beta|U| - |U|) = (\alpha - 1)(1 - 1/\beta)|U|^2$.

Each $j \in J$ is joined to all of U by exclusive OPT edges. In addition, it is joined by $NE \cap OPT$ edges to $k - |U|$ nodes in $B \setminus B_{\mathcal{P}}$, each of which has $\bar{d}_b = d_b^* = 1$. Now each $j \in J$ is still missing $|U|$ exclusive Nash edges. So we partition $B_{\mathcal{P}}$ into sets of size $|U|$ each and join each of these sets to some $j \in J$. This fixes the size $|J| := |B_{\mathcal{P}}|/|U| = (\alpha - 1)(1 - 1/\beta)|U|$ and completes the description of our instance. Figure 2 illustrates this construction.

To finish, we need to calculate the remaining costs. For $u \in U$ we have constant costs $c_u = 1/2$, so the Nash costs are $\bar{c}_u = |I||U|\frac{1}{2} = |U|^2\frac{1}{2}$. The OPT costs are $c_U^* = (|J||U| + |I||U|)\frac{1}{2} = (|B_{\mathcal{P}}| + |U|^2)\frac{1}{2} = ((\alpha - 1)(1 - 1/\beta) + 1)|U|^2\frac{1}{2}$.

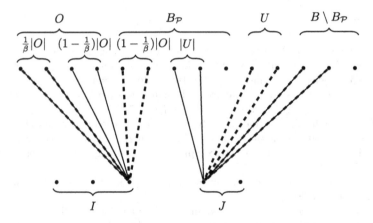

Fig. 2. Construction of a matching lower bound instance with PoA ≈ 1.35188. Edges in OPT are dashed lines, while edges in the NE are solid.

Finally, the costs of B are the same for both NE and OPT. They equal $k|J|\frac{1}{2} = \alpha(\alpha-1)(1-1/\beta)|U|^2\frac{1}{2}$.

Summarizing, according to WolframAlpha we obtain (dividing by $|U|^2$) with $\alpha \approx 2.3$ and $\beta \approx 2.5$

$$c^{NE}/c^{OPT} = \frac{(\alpha-1)+\frac{1}{2}+\alpha(\alpha-1)(1-1/\beta)\frac{1}{2}}{(\alpha-1)/\beta^2+(\alpha-1)(1-\frac{1}{\beta})\frac{1}{2}+\frac{1}{2}+\alpha(\alpha-1)(1-\frac{1}{\beta})\frac{1}{2}}$$

$$= \frac{2+1/(\alpha-1)+\alpha(1-1/\beta)}{2/\beta^2+1-1/\beta+1/(\alpha-1)+\alpha(1-1/\beta)} = 1.35188\ldots. \tag{12}$$

7 Remarks and Open Problems

Of course the most natural open problem is to find out whether the price of anarchy equals $\approx$1.35188 also for instances with only a few players. In our analysis in Sect. 5 we used the assumption that $|N| \to \infty$ only in order to conclude that $\bar{c}_u \geq \frac{1}{2}c_O^{max}$ in the limit. Without the assumption $|N| \to \infty$ we can still estimate $\bar{c}_u \geq \frac{1}{4}c_O^{max}$ with the help of Lemma 5. Performing the same analysis with this weaker estimate still leads to a reasonably small PoA < 1.4131 for all instances. Reconsidering the lower bound example in Sect. 5, we find that the PoA depends on the size of I (linearly related to $|N|$) determining the $\bar{c}_u, u \in U$ and how well α, β are approximated (*i.e.*, $|O|$ must be a multiple of β). If the parameters are well approximated for some small $|N|$ (implying small $|I|$ and hence $\bar{c}_u$ significantly less than $\frac{1}{2}$), it is conceivable that for these (finitely many) values of $|N|$ a PoA > 1.35189 might occur.

Other open questions relate to other subclasses of congestion games. Natural candidates are, *e.g.*, matroid congestion games, generalizing the special case of k-uniform matroids we study here. From the viewpoint of matchings, however, also asymmetric versions, where each player has a prescribed demand k_i of resources in R are fairly natural.

References

1. Awerbuch, B., Azar, Y., Epstein, A.: The price of routing unsplittable flow. In: Proceedings of 37th Annual ACM Symposium Theory Computing, pp. 57–66 (2005)
2. Colini-Baldeschi, R., Cominetti, R., Mertikopoulos, P., Scarsini, M.: On the asymptotic behavior of the price of anarchy: Is selfish routing bad in highly congested networks? https://arxiv.org/abs/1703.00927 (2017)
3. Colini-Baldeschi, R., Cominetti, R., Scarsini, M.: On the price of anarchy of highly congested nonatomic network games. In: Gairing, M., Savani, R. (eds.) SAGT 2016. LNCS, vol. 9928, pp. 117–128. Springer, Heidelberg (2016). https://doi.org/10.1007/978-3-662-53354-3_10
4. Braess, D.: Über ein Paradoxon aus der Verkehrsplanung. Unternehmensforschung **12**, 258–268 (1968). (German)
5. Caragiannis, I., Flammini, M., Kaklamanis, C., Kanellopoulos, P., Moscardelli, L.: Tight bounds for selfish and greedy load balancing. Algorithmica **61**(3), 606–637 (2010)
6. Christodoulou, G., Koutsoupias, E.: The price of anarchy of finite congestion games. In: Proceedings of 37th Annual ACM Symposium on Theory Computing, pp. 67–73 (2005)
7. Correa, J., de Jong, J., de Keijzer, B., Uetz, M.: The curse of sequentiality in routing games. In: Markakis, E., Schäfer, G. (eds.) WINE 2015. LNCS, vol. 9470, pp. 258–271. Springer, Heidelberg (2015). https://doi.org/10.1007/978-3-662-48995-6_19
8. de Jong, J., Klimm, M., Uetz, M.: Efficiency of equilibria in uniform matroid congestion games. In: Gairing, M., Savani, R. (eds.) SAGT 2016. LNCS, vol. 9928, pp. 105–116. Springer, Heidelberg (2016). https://doi.org/10.1007/978-3-662-53354-3_9
9. Feldman, M., Immorlica, N., Lucier, B., Roughgarden, T., Syrgkanis, V.: The price of anarchy in large games. In: Proceedings of 48th Annual ACM Symposium Theory Computing, pp. 963–976 (2016)
10. Fotakis, D.: Stackelberg strategies for atomic congestion games. ACM Trans. Comput. Syst. **47**, 218–249 (2010)
11. Koutsoupias, E., Papadimitriou, C.: Worst-case equilibria. In: Meinel, C., Tison, S. (eds.) STACS 1999. LNCS, vol. 1563, pp. 404–413. Springer, Heidelberg (1999). https://doi.org/10.1007/3-540-49116-3_38
12. Lücking, T., Mavronicolas, M., Monien, B., Rode, M.: A new model for selfish routing. Theor. Comput. Sci. **406**(3), 187–206 (2008)
13. Pigou, A.C.: The Economics of Welfare. Macmillan, London (1920)
14. Rosenthal, R.W.: A class of games possessing pure-strategy Nash equilibria. Internat. J. Game Theory **2**(1), 65–67 (1973)
15. Roughgarden, T.: The price of anarchy is independent of the network topology. J. Comput. Syst. Sci. **67**, 341–364 (2002)
16. Roughgarden, T., Tardos, É.: How bad is selfish routing? J. ACM **49**(2), 236–259 (2002)
17. Wardrop, J.G.: Some theoretical aspects of road traffic research. Proc. Inst. Civ. Eng. **1**(3), 325–362 (1952)

Author Index

Adiga, Abhijin 176
Alamdari, Soroush 66
Antoniadis, Antonios 164

Balogh, János 102
Becker, Amariah 1
Békési, József 102
Bienkowski, Marcin 132
Biermeier, Felix 285
Böhm, Martin 190
Borodin, Allan 253

Carmi, Paz 26

de Jong, Jasper 317
Dereniowski, Dariusz 223
Dósa, György 102
Dumitrescu, Adrian 238

Epstein, Leah 102

Farbstein, Boaz 52
Feldkord, Björn 285
Fernandes, Cristina G. 118
Flammini, Michele 301
Friedman, Alexander D. 176
Fujito, Toshihiro 17

Januszewski, Janusz 147
Jeż, Łukasz 190

Kamiyama, Naoyuki 90
Katz, Matthew J. 26
Kern, Walter 317
Kraska, Artur 132

Laish, Itay 269
Levin, Asaf 52, 102
Lintzmayer, Carla Negri 118

Mäcker, Alexander 207
Malatyali, Manuel 207, 285
Mehrabi, Saeed 76
Meyer auf der Heide, Friedhelm 207, 285
Monaco, Gianpiero 301
Mozes, Shay 269

Pankratov, Denis 253

Raghvendra, Sharath 176
Riechers, Sören 207

Saban, Rachel 26
Salehi-Abari, Amirali 253
San Felice, Mário César 118
Schewior, Kevin 164
Schmidt, Paweł 132
Sgall, Jiří 190
Shenmaier, Vladimir 41
Shmoys, David 66
Steenhuisen, Berend 317
Stein, Yael 26

Tóth, Csaba D. 238

Uetz, Marc 317
Urbańska, Dorota 223

Veselý, Pavel 190

Zhang, Qiang 301
Zielonka, Łukasz 147

Author Index

Printed in the United States
By Bookmasters